Atomic Transport and Defect Phenomena in Solids

University of Surrey, Guildford, UK
10-12 July 2006

FARADAY DISCUSSIONS
Volume 134, 2007

RSCPublishing

The Faraday Division of the Royal Society of Chemistry, previously the Faraday Society, founded in 1903 to promote the study of sciences lying between Chemistry, Physics and Biology.
http://www.rsc.org/science/physical.htm

EDITORIAL STAFF

Editor
Philip Earis

Assistant editor
Joanne Thomson

Publishing assistant
Rachel Dilworth

Team leader, serials production
Gisela Scott

Technical editor
Susan Batten

Publisher
Janet Dean

Faraday Discussions (Print ISSN 1359-6640, Electronic ISSN 1364-5498) is published 3 times a year by the Royal Society of Chemistry, Thomas Graham House, Science Park, Milton Road, Cambridge, UK CB4 0WF.
Volume 134 ISBN: 0 85404 953 3
ISBN-13: 978 0 85404 953 0

2007 annual subscription price: print+electronic £494, US $934; electronic only £445, US $840. Customers in Canada will be subject to a surcharge to cover GST. Customers in the EU subscribing to the electronic version only will be charged VAT. All orders, with cheques made payable to the Royal Society of Chemistry, should be sent to RSC Distribution Services, c/o Portland Customer Services, Commerce Way, Colchester, Essex, UK CO2 8HP.
Tel +44 (0) 1206 226050;
E-mail sales@rscdistribution.org

If you take an institutional subscription to any RSC journal you are entitled to free, site-wide web access to that journal. You can arrange access *via* Internet Protocol (IP) address at www.rsc.org/ip. Customers should make payments by cheque in sterling payable on a UK clearing bank or in US dollars payable on a US clearing bank. Periodicals postage is paid at Rahway, NJ and at additional mailing offices. Airfreight and mailing in the USA by Mercury Airfreight International Ltd., 365 Blair Road, Avenel, NJ 07001, USA.

US Postmaster: send address changes to *Faraday Discussions*, c/o Mercury Airfreight International Ltd., 365 Blair Road, Avenel, NJ 07001. All despatches outside the UK by Consolidated Airfreight.

PRINTED IN THE UK

Faraday Discussions documents a long-established series of *Faraday Discussion* meetings which provide a unique international forum for the exchange of views and newly acquired results in developing areas of physical chemistry, biophysical chemistry and chemical physics.

Atomic Transport and Defect Phenomena in Solids

Faraday Discussions
www.rsc.org/faraday_d

A General Discussion on Atomic Transport and Defect Phenomena in Solids was held at the University of Surrey, Guildford, UK on 10th, 11th and 12th July 2006.

RSC Publishing is a not-for-profit publisher and a division of the Royal Society of Chemistry. Any surplus made is used to support charitable activities aimed at advancing the chemical sciences. Full details are available from www.rsc.org

CONTENTS

ISSN 1359-6640; ISBN 0-85404-953-3
ISBN-13 978-085404-953-0

Atomic Transport and Defect Phenomena in Solids

RSCPublishing

Cover
See Sossina M. Haile, Calum R. I. Chisholm, Kenji Sasaki, Dane A. Boysen and Tetsuya Uda, *Faraday Discuss.*, 2007, **134**, 17–39. Rapid phosphate group reorientation facilitates proton transport in the superprotonic, disordered phase of CsH_2PO_4, a new electrolyte for fuel cell applications.

Image reproduced by permission of Professor Sossina M. Haile, from *Faraday Discuss.*, 2007, **134**, 17.

INTRODUCTORY LECTURE

PAPERS AND DISCUSSIONS

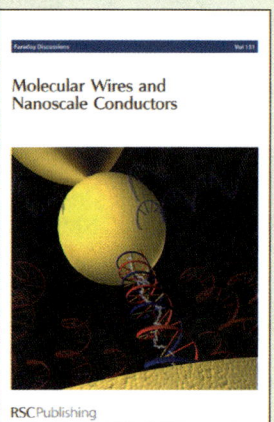

CONCLUDING REMARKS

ADDITIONAL INFORMATION

Optimisation of oxygen ion transport in materials for ceramic membrane devices

J. A. Kilner

Received 20th October 2006, Accepted 20th October 2006
First published as an Advance Article on the web 16th November 2006
DOI: 10.1039/b615306m

Oxygen transport in ceramic oxide materials has received much attention over the past few decades. Much of this interest has stemmed from the desire to construct high temperature electrochemical devices for energy conversion, an example being the solid oxide fuel cell. In order to achieve high performance for these devices, insights are needed in how to achieve optimum performance from the functional components such as the electrolytes and electrodes. This includes the optimisation of oxygen transport through the crystal lattice of electrode and electrolyte materials and across the homogeneous (grain boundary) and heterogeneous interfaces that exist in real devices. Strategies are discussed for the optimisation of these quantities and current problems in the characterisation of interfacial transport are explored.

1. Introduction

There are very few oxide materials can be classified as "fast" oxygen ion conductors; in fact they are a very select group of materials. Most conventional oxides that fulfil this criterion are limited to those with either the fluorite or perovskite crystal structures. These limitations however do not detract from their usefulness in practical devices. The applications of these materials include high temperature electrochemical devices such as solid oxide fuel cells (SOFCs), oxygen separation membranes, sensors and materials for syngas generators. All these devices must work at intermediate (5–600 °C) to high temperatures (700–1000 °C) and under exacting chemical and mechanical conditions. This means that the selection and subsequent optimization of the oxygen ion transport in these solids is difficult, however there is currently an urgent need for the development of new materials applicable to such novel electrochemical energy conversion devices to meet the twin needs of efficient and clean energy conversion to decrease the level of carbon emissions. One very important factor that is crucial to the device designer is an understanding of the limits of performance of these materials, *i.e.* are current materials close to the theoretical maximum of conductivity or are there new materials to be discovered that could offer significant enhancement in the oxygen transport properties?

2. Lattice oxygen ion conductivity

In this section we will examine how to optimise the bulk or lattice ionic conductivity using the fluorite material ceria as an example. This alone, of course, is not sufficient for a practical device because in most cases ceramic materials are used for device construction, and thus the grain boundary component must also be optimised as the devices operate in the dc mode.

Department of Materials, Imperial College, London, UK SW7 2AZ

Oxygen ion conductivity arises from mobile oxygen point defects. These can either be vacancies on the oxygen sub-lattice, *e.g.* ref. 1, or, in more recently discovered materials, arise from the movement of oxygen self interstitials.[2] In order to optimise the conductivity of an oxygen ion conducting material we need to be able to maximise both the number and the mobility of the ionic carriers. We will start be examining the fluorite structured materials based on the oxide CeO_2. These materials can be made into excellent oxygen ion conductors by the formation of solid solutions with many of the other trivalent rare earths, giving rise to the formation of oxygen vacancies as charge compensating defects as shown below:

$$Ln_2O_3 \Rightarrow 2Ln'_{Ce} + 3O_o^x + V_o^{\bullet\bullet}$$

with the neutrality condition given by;

$$2[V_o^{\bullet\bullet}] = [Ln'_{Ce}]$$

The question that subsequently arises is which is the best rare earth cation to use, and at what level of addition of the trivalent substitutional. Simplistic theory would imply that the conductivity would increase with the vacancy concentration and reach a maximum at 50% occupancy of the oxygen sub-lattice, and that all trivalent substitutionals would be equivalent. These questions need to be explored to understand how to optimise the bulk conductivity of the material. Let us first of all examine the question of the level of the additive. It has been well established for many years that the addition of the rare earths to ceria causes the oxygen ion conductivity, at any given temperature, to go through a distinct maximum[3] at much less than 50% occupancy of the oxygen sub-lattice. This conductivity maximum is associated with a minimum in the activation energy for conduction, as shown below in Fig. 1 for the ceria solid solutions, and has been ascribed by many authors to the effect of complex defect interactions.

It is clear that the minimum occurs at a few percent of the rare earth additive but that the position of the minimum and the minimum values of the activation energies are dependent upon the type of the rare earth, *i.e.* the assumption that all trivalent substitutionals are equivalent is invalid. This variation has been investigated over many years[5] and it has become clear that one of the major causes of these differences is the trapping of the oxygen vacancies by the substitutional cations to form a defect associate (dimer), *e.g.*

$$Ln'_{Ce} + V_o^{\bullet\bullet} \Leftrightarrow \{Ln'_{Ce}V_o^{\bullet\bullet}\}^{\bullet}$$

In such cases there are two contributions to the activation energy for conductivity, the migration energy ΔH_m and a contribution from the energy of association (or trapping)

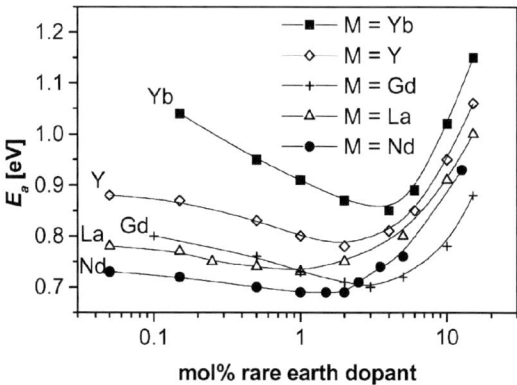

Fig. 1 Activation energies for conduction for the CeO_2–Ln_2O_3 solid solutions. Data taken from Faber *et al.*[4]

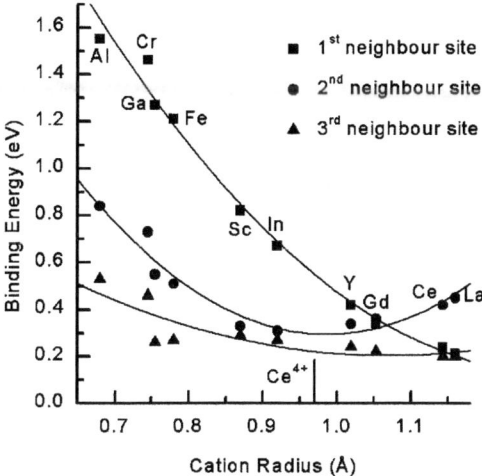

Fig. 2 Binding energy of dopant–vacancy complexes for ceria solid solutions. Reprinted from *Solid State Ionics*, **116**, L. Minervini, M. O. Zacate and R. W. Grimes, Defect cluster formation in M_2O_3-doped CeO_2, pp. 339–349, Copyright 1999, with permission from Elsevier.

energy ΔH_a. There are two further parts which contribute to the trapping energy ΔH_a: a Coulombic part which is dependent upon the effective charge of the two defects and is thus the same for all the rare earths and cannot account for the differences observed; and a component which arises from the elastic strain field surrounding a substitutional cation which is strongly dependent upon the size mismatch with the host. Recent calculations by Grimes *et al.* at Imperial College[6] have shown that the nature of this interaction is complex, however the main finding is that the trapping energy is minimised when the substitutional and the host have approximately the same ionic radius (Fig. 2). This seems to be intuitively correct because it is to be expected that the least disturbance to the parent lattice is found in such materials where there is a good size match, and in such cases it is often found that there is extensive solid solution giving rise to the high degrees of non stoichiometry needed for fast ionic conductors. For ceria the minimum in this dopant vacancy binding energy is found for Gd^{3+} in good accord with experimental findings that ceria–gadolinia solid solutions do give rise to some of the highest values of lattice conductivity found for ceria based materials.[5]

These observations imply that there are two minima in the activation energy to be found, the first with the concentration of the additive and the second with the ionic radius of the additive. These semi-empirical "rules" seem to work well when applied to other fluorite solid solutions They would suggest that zirconia–scandia solid solutions would be optimum for the zirconia based oxygen ion conductors and again this is in good accord with experimental findings.

An obvious next question is are these rules universal or do they only apply to the fluorite structured oxides? This is still a very open question, however some guidelines can be found by observing the behaviour of the perovskite structured oxides which, like the fluorites, can display fast oxygen ion conductivity. The perovskite oxides have been rather less comprehensively studied but there are some pointers towards similarities in their behaviour. Petric and Huang studied the conductivity of doped $NdGaO_3$ and noted that there was a minimum in the activation energy for conduction as Ga was substituted by Mg.[7] This minimum with concentration is similar to the trivalent substitution of the fluorite oxides, and suggests the formation of defect associates also occurs in the perovskites. If this is the case then there should also be an optimum size of substitutional atom for a given perovskite lattice. Hayashi *et al.*[8] has noted that the perovskite materials with the highest values of ionic conductivity are those where the substitutional ion is closely matched to the ion radius of the host cation that it is replacing. Thus for both the perovskite and the fluorite lattices it does appear that there is a strategy for optimising the lattice conductivity. It is a little too early to say that these "rules" for optimisation are universal but they do appear to give important guidelines to be followed when designing new materials.

One corollary to this analysis is that most of the materials that are commonly used in ceramic membrane devices, such as gadolinia doped ceria or scandia doped zirconia and lanthanum gallate are already optimised and thus it is difficult to see where the improvements in performance can be obtained, *i.e.* these materials are close to the limit of performance, certainly as regards to lattice transport of oxygen.

3. Interfacial oxygen ion transport

3.1 Homogeneous interfaces (grain boundaries)

In the early studies of oxygen ion conductivity it was difficult to unfold the various components that gave rise to the total ionic conductivity, mainly the grain boundary and grain interior or lattice component, by dc measurements alone. With the advent of modern automated impedance spectroscopy instrumentation the separation of the components of the resistance is relatively straightforward, there have now been many studies of the grain boundary conductivity and the effects of impurities[9] and even grain boundary structure[10] have been documented. In most cases what has been studied is the blocking effect of the grain boundaries transverse to the flux of oxygen ions. These ac impedance spectra are easily obtained for ionically conducting materials however for mixed conducting materials, for use in fuel cell cathodes and oxygen separation membranes, electrical methods cannot easily be used to investigate the grain boundaries. As a result the effects of the grain boundaries on the oxygen transport in these materials are not as well understood. Most data for these materials come from either indirect electrical studies or from oxygen isotopic exchange experiments. Indications that the transverse boundaries are blocking in these materials is suspected but not well documented. What has been established is that in certain materials the grain boundaries that lie parallel to the direction of the oxygen flux can have a short circuit effect and display much faster oxygen transport than in the bulk. In conventional (micron sized) ceramics this manifests mainly in the isotopic exchange experiments where "tails" appear in the isotopic penetration profiles. An example is given in Fig. 3 for a mixed conducting lanthanum manganate material, where the tail can be seen in the penetration profile and is clearly visible when the profile is re-plotted on a semi-logarithmic scale.[11]

Analysis of these data yields values of the grain boundary diffusivities several orders of magnitude faster than the bulk diffusion coefficient.[12] The presence of these "tails" is quite obvious for materials where the bulk diffusion is very low ($D \leq 10^{-12}$ $cm^2 \ s^{-1}$), however for the very fast diffusing materials the presence of a "tail" is not normally seen. This does not mean that the fast paths are not present; it is only that their effect is more limited. It is thus a logical next step to ask what would be the effect of having much smaller grain sizes where the proportion of grain boundaries is much higher. This can be done in two ways, by making nanosized ceramics or by making specially designed nanostructures. The evidence for enhanced ion transport from nanosized material is compelling but not unambiguous and the possibility exists that some of the enhancements seen, particularly for ceria based materials, are due to the presence of electronic conductivity at the grain boundaries. This is clearly an area where more detailed experimentation needs to be undertaken and in particular 3-D measurements of isotope distributions in exchanged nanomaterials would provide invaluable information as to the nature of the transport in these and other nanostructures. This type of analysis is not yet routine although a number of instruments are now being developed where this type of work should be possible.

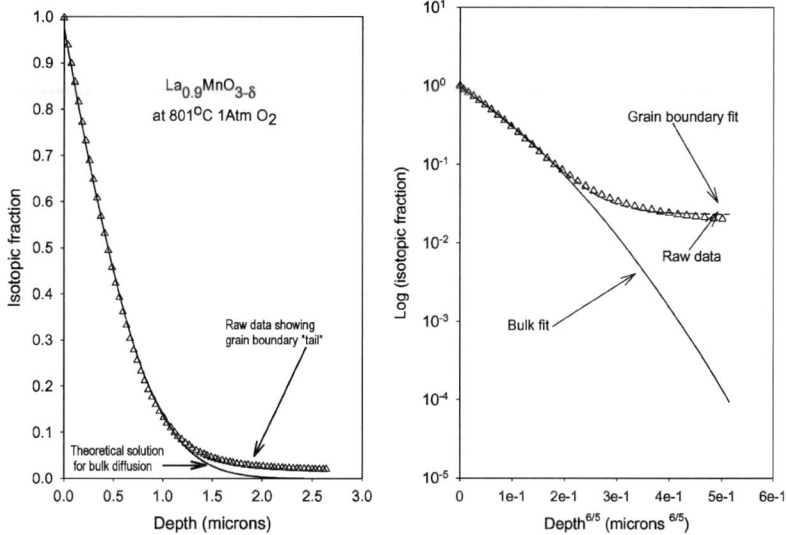

Fig. 3 Isotopic diffusion profile on linear and semi logarithmic scales for a LaMnO₃ sample showing the presence of a grain boundary tail. J. A. Kilner, in *Workshop on Grain Boundaries: Their Character Characterisation and Influence on Properties 1999*, ed. I. R. Harris and I. P. Jones, IOM Communications Ltd, 2001 – Reproduced by permission of IOM Communications Ltd.[12]

3.2 Heterogeneous interfaces (composites)

The final topic in this very brief review of oxygen conducting materials is the question of dissimilar materials and heterogeneous interfaces. These are seen as very important in devices such as fuel cell cathodes because the regions where these interfaces are exposed to the gas phase, the so-called three phase boundaries, are the sites where the oxygen reduction reaction is assumed to take place, particularly in materials such as $La_{1-x}Sr_xMnO_{3-\delta}$–yttria stabilised zirconia (LSM–YSZ) used as the cathode in the high temperature SOFC. Although many of the physical properties of these composites have been well established the oxygen exchange behaviour has not been systematically explored. This has mainly arisen because of the need for very high spatial resolution analyses of these materials, particularly by SIMS, to yield the microscopic distribution of the oxygen tracer in the components of the composite.

We have recently started such a study in order to understand the isotopic exchange behaviour and link this with the oxygen permeation obtained from dense samples of these composite materials.[13,14] The starting materials were LSM with $x = 0.2$ and 8 mol% YSZ, materials commonly used in high temperature SOFCs. These materials were sintered at high temperature to give dense ceramics. Two types of information can be obtained from these materials, the "effective" kinetic parameters for oxygen exchange and diffusion, and the microscopic data relating to the behaviour of each of the individual component phases.

Fig. 4 shows an isotopic penetration profile obtained as a result of a SIMS analysis of an isotopic exchange experiment on a YSZ–LSM composite.[13] The remarkable observation is that the profile is very smooth and actually fits well to the theoretical diffusion profile for a homogeneous medium. This implies that the SIMS analysis was performed over a large enough area (several hundred microns) to integrate over a large number of the individual grains of the composite to give these "effective" kinetic parameters. This approach seems to have some merit and we are pursuing the study of the composites to compare other "effective" quantities such as the oxygen permeation rates for dense ceramics[15] and the electrode resistance for porous

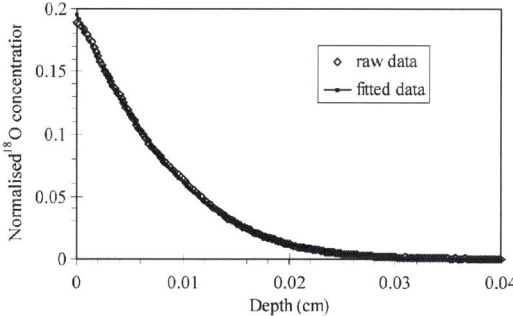

Fig. 4 Isotopic depth profile obtained for a YSZ–LSM composite material. Reprinted from *Solid State Ionics*, **176**, Y. Ji, J. A. Kilner and M. R. Carolan, Electrical properties and oxygen diffusion in yttria-stabilised zirconia (YSZ)-$La_{0.8}Sr_{0.2}MnO_{3\pm\delta}$ (LSM) composites, 937–943, Copyright 2005, with permission from Elsevier.

composite materials[16] to see if they correspond to the effective diffusion and exchange rates observed in the isotope experiments. However, it is also important to understand the microscopic behaviour of the materials to see if they are simply a sum of the behaviour of the components, or to see if there are any synergistic effects caused by the heterogeneous interfaces. At present we do not know enough to give a definite answer, but experiments are needed to try to answer these questions. Fig. 5 shows an attempt to gather such information for composites in the YSZ–LSM system. It shows a detailed surface map of an area, approximately 250 microns square, of an isotopically exchanged YSZ–LSM composite material imaged by a focused ion beam instrument. The map is able to resolve the isotopic concentration in each of the components of the composite, expressed here as a fraction normalised to the exchanging gas concentration. The next step is to obtain information about the isotopic distribution within the grains themselves and to ultimately build up a complete 3-D data set revealing the oxygen pathways through the composites. Experiments such as the one shown in Fig. 5 will allow the investigation of the behaviour of individual components within the composite and begin to answer some of the questions about the detailed transport of oxygen in such grossly heterogeneous materials.

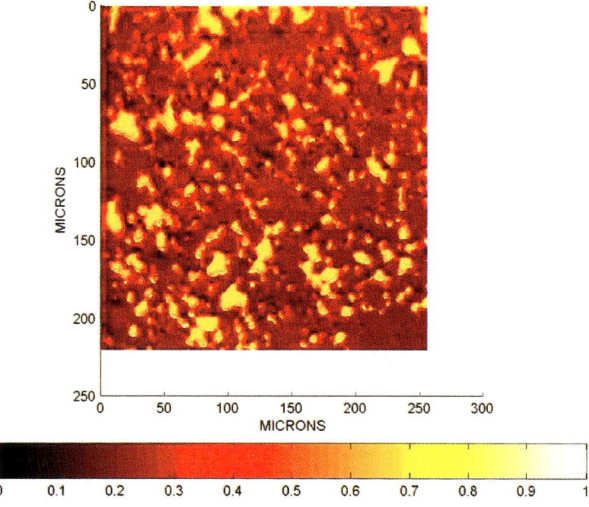

Fig. 5 False colour map of the normalised surface ^{18}O concentration of an isotopically exchanged LSM–YSZ composite.

4. Conclusions

The optimisation of the oxygen ion conductivity of oxide ceramic materials is a subject with many different and fascinating aspects that will continue to be the subject of investigations by pure and applied scientists alike. Clearly an understanding of the lattice conductivity requires that we have a clear formalism for the treatment of complex defect interactions and a good insight into what are the important factors that affect the interactions of the defects. This we are beginning to achieve, and we need to continue the undoubted successes that have arisen from the close interplay between theory and experimentation in this field. However, as mentioned above, the optimisation of the conductivity in a dense ceramic material requires an optimisation of both the lattice conductivity and that of any interfaces in the material or structure. The study of transport along and across these interfaces has rather been the Cinderella of this area, mainly because of the experimental difficulties in obtaining the sort of reliable data needed for systematic studies. However with the advent of high spatial resolution instrumentation this is becoming a reality and the prospect of fabricating fast oxygen transport structures through the nano-engineering of oxide structures is now an exciting possibility.

References

1 J. A. Kilner, Fast anion transport in solids, *Solid State Ionics*, 1983, **8**(3), 201–207.
2 V. V. Kharton, A. P. Viskup, A. V. Kovalevsky, E. N. Naumovich and F. M. B. Marques, Ionic transport in oxygen-hyperstoichiometric phases with K_2NiF_4-type structure, *Solid State Ionics*, 2001, **143**(3–4), 337–353.
3 J. A. Kilner and R. J. Brook, A study of oxygen ion conductivity in doped nonstoichiometric oxides, *Solid State Ionics*, 1982, **6**(3), 237–252.
4 J. Faber, C. Geoffroy, A. Roux, A. Sylvestre and P. Abelard, A systematic investigation of the dc-electrical conductivity of rare-earth doped ceria, *Appl. Phys. A: Mater. Sci. Process.*, 1989, **49**(3), 225–232.
5 J. A. Kilner, Fast oxygen transport in acceptor doped oxides, *Solid State Ionics*, 2000, **129**(1), 13–23.
6 L. Minervini, M. O. Zacate and R. W. Grimes, Defect cluster formation in M_2O_3-doped CeO_2, *Solid State Ionics*, 1999, **116**(3–4), 339–349.
7 A. Petric and P. N. Huang, Oxygen conductivity of $Nd(Sr/Ca)Ga(Mg)O_{3-\delta}$ perovskites, *Solid State Ionics*, 1996, **92**(1–2), 113–117.
8 H. Hayashi, H. Inaba, M. Matsuyama, N. G. Lan, M. Dokiya and H. Tagawa, Structural consideration on the ionic conductivity of perovskite-type oxides, *Solid State Ionics*, 1999, **122**(1–4), 1–15.
9 H. L. Tuller, Ionic conduction in nanocrystalline materials, *Solid State Ionics*, 2000, **131**(1–2), 143–157.
10 R. A. De Souza, J. Fleig, J. Maier, Z. L. Zhang, W. Sigle and M. Ruhle, Electrical resistance of low-angle tilt grain boundaries in acceptor-doped $SrTiO_3$ as a function of misorientation angle, *J. Appl. Phys.*, 2005, **97**(5), 053502.
11 A. V. Berenov, J. L. MacManus-Driscoll and J. A. Kilner, Oxygen tracer diffusion in undoped lanthanum manganites, *Solid State Ionics*, 1999, **122**(1), 41–49.
12 J. A. Kilner, Grain boundary phenomena in oxygen ion conductors, in *Workshop on Grain Boundaries: Their Character Characterisation and Influence on Properties 1999*, ed. I. R. Harris and I. P. Jones, IOM Communications Ltd, London, 2001.
13 Y. Ji, J. A. Kilner and M. F. Carolan, Electrical properties and oxygen diffusion in yttria-stabilised zirconia (YSZ)-$La_{0.8}Sr_{0.2}MnO_{3\pm\delta}$ (LSM) composites, *Solid State Ionics*, 2005, **176**(9–10), 937–943.
14 M. Dhallu and J. A. Kilner, Oxygen transport in YSZ/LSM composite materials, *J. Fuel Cell Sci. Technol.*, 2005, **2**(1), 29–33.
15 M. Dhallu, J. A. Kilner, V. V. Kharton and J. R. Frade, Oxygen permeation studies in YSZ-LSM composite materials for oxygen separators and syngas membranes, in *Solid State Ionics: the Science and Technology of Ions in Motion*, ed. B. V. R. Chowdari et al., World Scientific, Singapore, 2004, pp. 345–355.
16 A. Esquirol, J. Kilner and N. Brandon, Oxygen transport in $La_{0.6}Sr_{0.4}Co_{0.2}Fe_{0.8}O_{3-\delta}/Ce_{0.8}Ge_{0.2}O_{2-x}$ composite cathode for IT-SOFCs, *Solid State Ionics*, 2004, **175**(14), 63–67.

Solid acid proton conductors: from laboratory curiosities to fuel cell electrolytes

Sossina M. Haile,* Calum R. I. Chisholm,† Kenji Sasaki, Dane A. Boysen† and Tetsuya Uda‡

Received 24th March 2006, Accepted 4th May 2006
First published as an Advance Article on the web 7th August 2006
DOI: 10.1039/b604311a

The compound CsH_2PO_4 has emerged as a viable electrolyte for intermediate temperature (200–300 °C) fuel cells. In order to settle the question of the high temperature behavior of this material, conductivity measurements were performed by two-point AC impedance spectroscopy under humidified conditions ($p[H_2O] = 0.4$ atm). A transition to a stable, high conductivity phase was observed at 230 °C, with the conductivity rising to a value of 2.2×10^{-2} S cm^{-1} at 240 °C and the activation energy of proton transport dropping to 0.42 eV. In the absence of active humidification, dehydration of CsH_2PO_4 does indeed occur, but, in contradiction to some suggestions in the literature, the dehydration process is not responsible for the high conductivity at this temperature. Electrochemical characterization by galvanostatic current interrupt (GCI) methods and three-point AC impedance spectroscopy (under uniform, humidified gases) of CsH_2PO_4 based fuel cells, in which a composite mixture of the electrolyte, Pt supported on carbon, Pt black and carbon black served as the electrodes, showed that the overpotential for hydrogen electrooxidation was virtually immeasurable. The overpotential for oxygen electroreduction, however, was found to be on the order of 100 mV at 100 mA cm^{-2}. Thus, for fuel cells in which the supported electrolyte membrane was only 25 μm in thickness and in which a peak power density of 415 mW cm^{-2} was achieved, the majority of the overpotential was found to be due to the slow rate of oxygen electrocatalysis. While the much faster kinetics at the anode over those at the cathode are not surprising, the result indicates that enhancing power output beyond the present levels will require improving cathode properties rather than further lowering the electrolyte thickness. In addition to the characterization of the transport and electrochemical properties of CsH_2PO_4, a discussion of the entropy of the superprotonic transition and the implications for proton transport is presented.

Introduction

Solid acid proton conductors, based on tetrahedral oxyanion groups, have received attention as electrolytes in next generation fuel cells. Compounds within this class,

Materials Science, California Institute of Technology, Pasadena CA 91125, USA. E-mail: smhaile@caltech.edu

† Present Address: SuperProtonic Inc. Pasadena, CA 91101, USA.
‡ Present Address: Materials Science and Engineering, Kyoto University, Sakyo, Kyoto 606-8501, Japan.

such as $CsHSO_4$,[1] $Rb_3H(SeO_4)_2$,[2] and $(NH_4)_3H(SO_4)_2$,[3] exhibit anhydrous proton transport with conductivities of the order of 10^{-3} to 10^{-2} S cm^{-1} at moderate temperatures (120–300 °C). Unlike the polymers in more conventional proton exchange membrane fuel cells (PEMFCs), proton conduction in oxyanion solid acids does not rely on the migration of hydronium ions. Consequently, the requirement for humidification of the electrolyte is, in principle, eliminated as is the need for delicate water management.[4] The temperatures of operation accessible to fuel cells based on solid acids furthermore imply that catalysis rates will be enhanced relative to PEMFCs, opening up possibilities for reduction in precious metal loadings or even the elimination of precious metals entirely. These temperatures additionally imply a high tolerance of the catalysts to poisons, particularly CO, in the fuel stream. While these many features render solid acids very attractive as fuel cell electrolytes several challenges must be addressed in technologically relevant fuel cell systems. Prominent amongst these is the water solubility of all known solid acids with high conductivity, which requires the implementation of engineering designs to prevent condensed water from contacting the electrolyte, particularly during fuel cell shutdown.

In contrast to the application potential of solid acid proton conductors, which has been explored over only the past five years,[4] the fundamental physical and chemical characteristics of these materials have been studied for well over twenty years. This is a consequence of the fascinating sequence of phase transitions that occur in these compounds in response to heating, cooling or application of pressure. In general, these transitions involve changes to the network of hydrogen bonds which link the oxyanion groups to form dimers, chains, layers, or three-dimensional structures. In the particular case of proton transport, the dynamic disordering of the hydrogen bond network above the so-called superprotonic transition leads to a dramatic increase in proton conductivity by several orders of magnitude.[1–3,5] Subtle changes in the local hydrogen bond geometry similarly give rise to the well-known ferroelectric transition in compounds such as KH_2PO_4[6] and $Cs_3H(SeO_4)_2$.[7] This rich phase behavior has spawned the production of at least 500 papers that broadly address solid acids with stoichiometry $MHXO_4$, $M_3H(XO_4)_2$, $M_2H(X'O_4)$, or some variation thereof, where M = alkali metal or NH_4; X = S, Se; and X' = P, As.

In this work we review the scientific and technological status of selected solid acids, with emphasis on recent developments in the authors' laboratory in the study of CsH_2PO_4. After addressing the ongoing literature controversy regarding the high temperature properties of this material, in particular, the nature of the transformation occurring at approximately 230 °C, we present new data supporting the position that a true polymorphic transition occurs in this material. We then evaluate the behavior of CsH_2PO_4 as a fuel cell electrolyte, examining the relative rates of hydrogen electrooxidation and oxygen electroreduction. We then close with a speculative discussion of the configurational entropy of the high temperature phase of CsH_2PO_4, in which we propose that the disorder associated with the hydrogen bond network should be considered independently of the oxyanion group disorder. Thus, despite its rather innocent chemical formula, CsH_2PO_4 provides a rich variety of scientific challenges and technological opportunities.

Phase transition behavior

The literature debate

The controversy surrounding the high temperature properties of CsH_2PO_4 stem from the decomposition behavior of the material. Specifically, it has been argued by some that the dehydration of the compound

$$CsH_2PO_4(s) \rightarrow CsH_{2-2x}PO_{4-x}(s) + xH_2O(g), (0 \leq x \leq 1) \rightarrow CsPO_3 + H_2O\ (x = 1)$$

induces a transient rise in conductivity as water leaves the structure, but that there is no true polymorphic transition to a high conductivity phase. Others, however, have argued that, while decomposition can interfere with the observation of the polymorphic transition, it nevertheless occurs. A selection of the relevant papers documenting this controversy is provided in Table 1, along with a notation indicating whether the paper supports, refutes or remains neutral on the matter of a superprotonic phase transition.

The early thermal analyses of CsH_2PO_4 were very much in contradiction with one another. The first papers on the topic of the thermal behavior of CsH_2PO_4 (beyond simple weight loss measurements) appear to be two reports from Rashkovich et al., and even these are in disagreement. The earlier paper concludes that two transitions occur prior to decomposition,[8] whereas the latter concludes that the thermal events are entirely due to decomposition.[9] In two papers co-authored by Clark, two polymorphic transitions are reported for CsH_2PO_4.[10,11] The latter, occurring at 230 °C, was found to be fully reversible. However, it was associated with a slight weight loss ($\sim 1.5\%$) for powder samples examined under ambient conditions. This feature would become the point of significant controversy in later years. Wada subsequently confirmed the 230 °C transition by dilatometry measurements of single crystal samples, observing a sharp increase in lattice constants at this temperature.[12] Almost simultaneously, Gupta reported, again on the basis of calorimetry and thermal gravimetric analysis, a polymorphic transition in CsH_2PO_4 at 235 °C just prior to the maximum in the decomposition process.[13] Again, however, initiation of the weight loss coincided with the reported polymorphic transition. In contradiction to these results, Nirsha et al. published a study two years later concluding that thermal events at 233 °C and higher in CsH_2PO_4 are entirely due to decomposition.[14]

The matter of a polymorphic phase transition in CsH_2PO_4 may have remained an obscure point in the field of solid state chemistry were it not for the results of Baranov et al. showing a so-called superprotonic transition to occur at 230 °C,[5] precisely the temperature of the reversible, higher temperature transformation first reported by Clark.[10,11] The conductivity was shown to increase by five orders of magnitude at the transition, and apparently reliable data were obtained to temperatures of ~ 250 °C. In hindsight, it is clear that Baranov was able to observe the transition because single crystals, in which dehydration is slow compared to powdered materials, were utilized for the experiments. Shortly after Baranov's study, Bronowska and Pietraszko reported the structure of superprotonic CsH_2PO_4 and provided the first clear demonstration of the significance of water partial pressure in suppressing dehydration.[15] All of the peaks in the high temperature X-ray powder diffraction pattern, along with their relative intensities, could be explained on the basis of the proposed high temperature structure. The subsequent Raman study of Romain and Novak[16] supported Bronowska's conclusions regarding the structural features of the superprotonic state, while Vargas and Torijano in 1993 also agreed (initially) with the existence of a reversible, but hysteretic, phase transition at 227 °C on the basis of differential scanning calorimetry.[17]

The controversy surrounding the properties of CsH_2PO_4 began in earnest in 1996 with a publication by Lee suggesting that the observed conductivity effects were artifacts of thermal decomposition and partial polymerization at the surfaces of the CsH_2PO_4 particles.[19] The hypothesis was based on a review of literature data, without the benefit of new experimental results. A later paper from this same author repeated these conclusions, but in this case experimental support was provided in the form of a limited set of optical micrographs showing the degradation of single crystal surfaces.[24] Inspired by Lee's work, Ortiz, Vargas and Mellander published a series of papers also taking the view that only decomposition occurs at the supposed superprotonic transition.[21,22,33] In this case, the conclusions were based on thermal analysis and high temperature X-ray diffraction experiments performed on powdered samples and on conductivity measurements performed on single crystal samples. Thermal events were found to coincide with weight loss events, increases in conductivity at 230 °C were found to diminish in significance with repeated thermal cycling, and the high temperature diffraction data showed a rather messy

Table 1 Selected publications describing the high-temperature structural and/or transport properties of CsH_2PO_4. S = supporting the conclusion of a superprotonic transition; R = refuting the conclusion; and ? = uncommitted

Authors	Article Title	Year	Ref.	S/R/?*
L. N. Rashkovich, K. B. Meteva, Ya. É. Shevchik, V. G. Hoffman, and A. V. Mishchenko	Growing Single Crystals of Cesium Dihydrogen Phosphate and Some of Their Properties.	1977	8	S?
L. N. Rashkovich and K. B. Meteva	Properties of Cesium Dihydrophosphate.	1978	9	R
E. Rapoport, J. B. Clark and P. W. Richter	High-Pressure Phase Relations of RbH_2PO_4, CsH_2PO_4, and KD_2PO_4.	1978	10	S
B. Metcalfe and J. B. Clark	Differential Scanning Calorimetry of RbH_2PO_4 and CsH_2PO_4.	1978	11	S
M. Wada, A. Sawada and Y. Ishibashi	Some High-Temperature Properties and the Raman-Scattering Spectra of CsH_2PO_4.	1979	12	S
L. C. Gupta, U. R. K. Rao, K. S. Venkateswarlu and B. R. Wani	Thermal-Stability of CsH_2PO_4.	1980	13	S
B. M. Nirsha, E. N. Gudinitsa, A. A. Fakeev, V. A. Efremov, B. V. Zhadanov and V. A. Olikova	Thermal Dehydration Process of CsH_2PO_4.	1982	14	R
A. I. Baranov, V. P. Khiznichenko, V. A. Sandler and L. A. Shuvalov	Frequency Dielectric-Dispersion in the Ferroelectric and Superionic Phases of CsH_2PO_4.	1988	5	S
W. Bronowska and A. Pietraszko	X-Ray Study of the High-Temperature Phase-Transition of CsH_2PO_4 Crystals.	1990	15	S
F. Romain and A. Novak	Raman Study of the High-Temperature Phase-Transition in CsH_2PO_4.	1991	16	S
R. A. Vargas and E. Torijano	Phase-Behavior of RbH_2PO_4 and CsH_2PO_4 in the Fast-Ion Regime.	1993	17	S?
A. Preisinger, K. Mereiter, and W. Bronowska	The Phase Transition of CsH_2PO_4 (CDP) at 505 K.	1994	18	S
K. S. Lee	Hidden Nature of the High-Temperature Phase Transitions in Crystals of KH_2PO_4-Type: Is It a Physical Change?	1996	19	R
Y. Luspin, Y, Vaills, and G. Hauret	Discontinuities in the Elastic Properties of CsH_2PO_4 at the Superionic Transition.	1997	20	S
E. Ortiz, R. A. Vargas and B. E. Mellander	On the High-Temperature Phase Transitions of CsH_2PO_4: a Polymorphic Transition? A Transition to a Superprotonic Conducting Phase?	1999	21	R

Table 1 (*continued*)

Authors	Article Title	Year	Ref.	S/R/?*
E. Ortiz, R. A. Vargas and B. E. Mellander	On the High-Temperature Phase Transitions of Some KDP-Family Compounds: a Structural Phase Transition? A Transition to a Bulk-High Proton Conducting Phase?	1999	22	R
W. Bronowska	Does the Structural Superionic Phase Transition at 231 °C in CsH_2PO_4 Really Not Exist?	2001	23	S
K. S. Lee	Surface Transformation of Hydrogen-Bonded Crystals at High-Temperatures and Topochemical Nature.	2002	24	R
J. Otomo, N. Minagawa, C. J. Wen, K. Eguchi and H. Takahashi	Protonic Conduction of CsH_2PO_4 and Its Composite With Silica in Dry and Humid Atmospheres.	2003	25	S
D. A. Boysen, S. M. Haile, H. J. Liu and R. A. Secco	High-Temperature Behavior of CsH_2PO_4 Under Both Ambient and High Pressure Conditions.	2003	26	S
J. H. Park, C. S. Kim, B. C. Choi, B. K. Moon and H. J. Seo	Physical Properties of CsH_2PO_4 Crystal at High Temperatures.	2003	27	R
D. A. Boysen, T. Uda, C. R. I. Chisholm and S. M. Haile	High-Performance Solid Acid Fuel Cells Through Humidity Stabilization.	2004	28	S
J. H. Park	Possible Origin of the Proton Conduction Mechanism of CsH_2PO_4 Crystals at High Temperatures.	2004	29	R
K. Yamada, T. Sagara, Y. Yamane, H. Ohki and T. Okuda	Superprotonic Conductor CsH_2PO_4 Studied by H-1, P-31 NMR and X-Ray Diffraction.	2004	30	S
J. Otomo, T. Tamaki, S. Nishida, S. Q. Wang, M. Ogura, T. Kobayashi, C. J. Wen, H. Nagamoto and H. Takahashi	Effect of Water Vapor on Proton Conduction of Cesium Dihydrogen Phosphate and Application to Intermediate Temperature Fuel Cells.	2005	31	S
A. I. Baranov, V. V. Grebenev, A. N. Khodan, V. V. Dolbinina and E. P. Efremova	Optimization of Superprotonic Acid Salts for Fuel Cell Applications.	2005	32	S

evolution of peaks.[21] From the diffraction data the authors identified what they believed to correspond to the most intense peak of the first dehydration product, however, this peak (at $2\Theta \approx 25.3°$) coincides almost precisely with the most intense peak for superprotonic CsH_2PO_4, as reported previously by Bronowska *et al.*[18] Additional support for the view of dehydration rather than polymorphic phase transitions comes from the work of Park who has published two papers agreeing with the position that only decomposition occurs in CsH_2PO_4 upon heating.[27,29] Here, the conclusion is based primarily on the results of AC impedance measurements performed on single crystals which purportedly show no jump in conductivity at 230 °C. Thermal analysis results are also reported to support the conclusion.

The data analysis methodology employed by Park for interpreting the AC impedance response of CsH_2PO_4 requires some comment. In Park's study impedance data were collected over frequency ranges of 1 Hz to 100 kHz (2003)[27] and 1 Hz to 3 MHz (2004)[29] and the effective dc conductivity was extracted by assuming the measured response to be dominated by the properties of the bulk CsH_2PO_4 crystal. However, because the characteristic frequency ($\omega_o = 1/\varepsilon_o\varepsilon\rho$, where ε_o = permittivity of vacuum, ε = relative dielectric constant, and ρ = resistivity) of CsH_2PO_4 rises dramatically at the transition, even at frequencies of 3 MHz one measures, not the bulk properties of the material of interest, but the properties of the electrolyte–electrode interface. In the absence of higher frequency data, one can only estimate the bulk electrolyte properties by extrapolation (typically in the Nyquist, Z_{real} vs. $-Z_{imag}$ representation) to high frequencies. The situation is illustrated in Fig. 1. The schematic plot in part (a) reflects the kind of data expected when the bulk characteristic frequency of the material of interest is within the experimental measurement range whereas that in part (b) corresponds to that expected when the characteristic frequency is beyond the highest measurement frequency available. Park has essentially treated data of the form shown in Fig. 1(b) by procedures appropriate to data of the form in Fig. 1(a), and, accordingly, that author's conclusion that there is no substantial rise in the conductivity of CsH_2PO_4 at ~ 230 °C cannot be accepted.

An interesting feature of the argument of those in favor of the dehydration model is that all MH_2PO_4 (and indeed MH_2AsO_4) compounds are assumed to exhibit identical high temperature behavior. That is, it is assumed that if dehydration can be demonstrated in the case, for example, of KH_2PO_4, this implies dehydration occurs in CsH_2PO_4 also.[19,22] The danger of such an argument is immediately obvious. There are innumerable cases in which compounds with similar stoichiometries exhibit dramatically different phase transition behaviors. A notable example is the influence of deuteration on low temperature ferroelectric transitions in several $M_3H(XO_4)_2$ compounds: while $Rb_3H(SO_4)_2$, $Rb_3H(SeO_4)_2$, and $K_3H(SO_4)_2$ do not exhibit ferroelectric transitions to temperatures as low as liquid helium, the deuterated analogs do, despite the isomorphous structures of the protonated and deuterated compounds.[34]

In parallel to the studies dismissing a superprotonic transition in CsH_2PO_4, several recent papers have appearing supporting its existence. Luspin et al. showed a sharp change in the elastic constants of single crystal CsH_2PO_4 to occur at 233 °C, with reliable data being obtained to a temperature of 255 °C.[20] In 2003 Boysen et al. published a definitive set of studies showing the sensitivity of the thermal behavior of CsH_2PO_4 to particle size and heating rate when examined under dry nitrogen.[26] From the simultaneous measurement of thermal events and weight loss, combined with evolved gas analysis, a clear transition at 228 ± 2 °C, prior to any decomposition, was detected for all sets of experimental conditions. Furthermore, it was shown that, as would be expected, dehydration in large single crystal samples was significantly suppressed, such that a high temperature isotropic phase could be

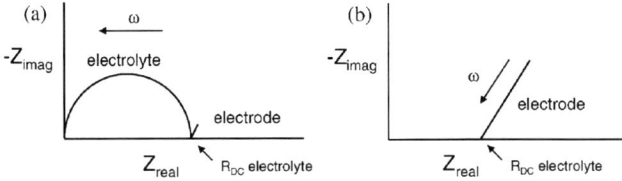

Fig. 1 Schematic impedance spectra resulting from electrode | electrolyte | electrode systems (a) at low temperatures and (b) at high temperatures.

easily observed optically under cross-polarizers. The decomposition could also be suppressed by the application of high hydrostatic pressure (1 GPa). Under these conditions a transition at 260 °C was observed and the high temperature phase found to be stable to temperatures as high as 375 °C.

Almost simultaneous with the publication of these results, Otomo et al. showed that water partial pressure could be used to suppress dehydration and by this method also provided convincing conductivity measurements of the superprotonic phase.[25] The raw impedance data collected both by Boysen et al.[26] and Otomo et al.[25] exhibit the features described in Fig. 1, supporting the hypothesis that the analysis procedure employed by Park[29] was not applicable to the high temperature phase. In a later study, Otomo et al. further showed that a water partial pressure of 0.3 atm was sufficient to suppress dehydration at 250 °C, and that the dehydration was, in fact, reversible, with slightly decomposed samples recovering their high conductivity upon exposure to sufficient humidity.[31] A preliminary evaluation of the complete CsH_2PO_4–H_2O–$CsPO_3$ phase diagram was reported in 2004 by Boysen et al., which supported the assertion that only slight levels of humidification are necessary to suppress dehydration.[28] Independently, Yamada et al. examined the structural properties of CsH_2PO_4 in 2004.[30] A high temperature diffraction pattern clearly corresponding to that reported earlier by Bronowska was obtained by sealing the powdered sample to prevent dehydration. In 2005 Baranov reproduced his earlier conductivity measurements of CsH_2PO_4 and also examined the influence of ammonium substitution on the transition behavior.[32] In addition to these fundamental studies of the properties of CsH_2PO_4, both Boysen et al.[28] and Otomo et al.[31] have shown that this material can be employed as the electrolyte in fuel cells with good long-term stability, behavior that would not be possible if the high conductivity were a transient artifact of dehydration.

In the cases where experimental procedures have been described in detail, it is apparent that weight loss and dehydration occur when CsH_2PO_4 is used in the form of loose powders, whereas single crystals typically show the strongest evidence of polymorphic phase transitions. On the basis of the thermodynamic measurements of the decomposition behavior of CsH_2PO_4, Fig. 2,[28] it is apparent that at the transition temperature of 230 °C, a water partial pressure of only ~0.026 atm, equivalent to 100% humidity at 22 °C, is sufficient to suppress dehydration. Because this value is close to ambient levels of humidity, without explicit control of water partial pressure, measured results will vary substantially depending on the laboratory climate. Furthermore, even gas flow rates, in addition to particle/crystallite size, can be expected to have a strong impact on dehydration kinetics. Thus, it is essential to explicitly control water partial pressure in any meaningful evaluation of the high temperature properties of CsH_2PO_4.

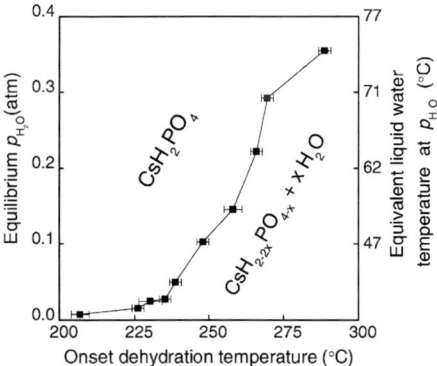

Fig. 2 Equilibrium phase diagram for the decomposition of CsH_2PO_4. Reproduced with permission from ref. 28.

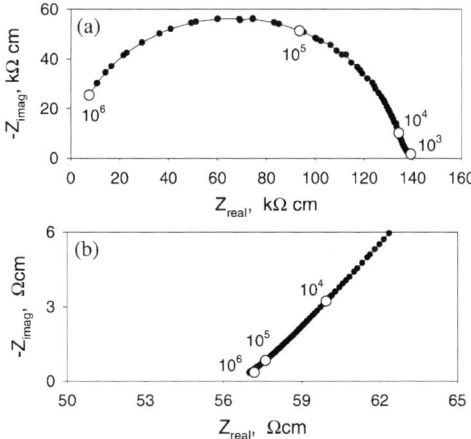

Fig. 3 Impedance spectra obtained from Ag | CsH₂PO₄ | Ag under humidified air (p[H₂O] = 0.4 atm) at (a) 220 °C and (b) 236 °C, showing the dramatic change in impedance properties upon passing through the superprotonic transition at 230 °C. Frequency values specified for selected data points.

Conductivity of CsH₂PO₄

While there is now an overwhelming body of evidence supporting the conclusion that CsH₂PO₄ undergoes a true, polymorphic transition to a cubic, superprotonic phase at ∼230 °C, we deemed it important here to provide additional conductivity data under ambient pressure conditions with controlled humidity levels. To date, such data have been reported only by Otomo et al.,[31] with all other experiments having been carried out without explicit control of humidity.

Powders of CsH₂PO₄ were prepared from aqueous solutions of the starting reagents Cs₂CO₃ (Alfa Aesar, 99.9%) and H₃PO₄ (ACS, 85% w/w aqueous solution) combined in a molar ratio of 1 : 2, to which methanol was added to induce rapid precipitation. After confirmation by X-ray powder diffraction that the desired material had been obtained, the powder was dried at ∼205 °C for several hours to remove surface adsorbed water so as to minimize its influence on transport properties. After drying, polycrystalline pellets of CsH₂PO₄, 5.1 mm in diameter and ∼2 mm in thickness, were prepared by cold uniaxial pressing at 630 kpsi. Final densities were 95 ± 2% of theoretical. Colloidal silver paste (Ted Pella 16032) was applied to either side of the samples to serve as electrodes for electrical characterization. AC impedance measurements were performed over the frequency range 20 Hz to 1 MHz, with an applied voltage amplitude of 1.0 V using an HP 4284A precision LCR (inductance–capacitance–resistance) meter. Data were collected over the temperature range of ∼170–255 °C, with the sample chamber maintained at p(H₂O) = 0.4 atm in air by passing the inlet gas (flow rate of ∼5 sccm) through water held at 75 ± 1 °C. This level of humidification ensures that decomposition is suppressed, Fig. 2. Impedance spectra were analyzed using the commercially available software package, ZView (Scribner & Assoc.).

Typical impedance spectra obtained from these measurements are presented in Fig. 3 in Nyquist form (Z_{real} vs. $-Z_{imag}$ as parametric functions of frequency, ω) for measurements both below (at 220 °C) and above (at 236 °C) the transition at 230 °C. Below 230 °C, the spectra exhibit a single arc that extends (almost) to the origin and the low frequency intercept of this arc with the real axis corresponds to the dc conductivity of the electrolyte. Above 230 °C, the behavior of the electrolyte can no longer be directly accessed. In this case, the (extrapolated) high frequency intercept with the real axis corresponds to the dc conductivity of the electrolyte. Quantitatively, the low temperature behavior is modeled using an equivalent circuit comprised of a resistor, R, with impedance $Z_R(\omega) = R$, and a constant phase element, Q, with impedance $Z_Q(\omega) = [Y(j\omega)^n]^{-1}$, where $j = \sqrt{-1}$, and Y and n are constants, placed in parallel with one another, whereas the high temperature behavior is represented by a resistor and constant phase element placed in series. That AC impedance data acquired from electrolyte materials require treatment in this manner is well-understood within the solid state ionics community, but is less widely appreciated amongst researchers in other fields.

The conductivity of CsH_2PO_4 so derived is presented in Fig. 4 in Arrhenius form for two heating and cooling cycles. This is a typical dataset, with many samples having been examined. Several features of the conductivity of CsH_2PO_4 are noteworthy. Most significant is the unambiguous existence of a high temperature phase, stable from the transition temperature of 230 °C to the highest measurement temperature of 254 °C. The conductivity rises from 8.5×10^{-6} at 223 °C to 1.8×10^{-2} Ω^{-1} cm^{-1} at 233 °C, in good agreement with the earlier work of Baranov et $al.$[5] Moreover, within the high conductivity regime, the activation energy for charge transport is 0.418 ± 0.002 eV, a value that is again consistent with the results of Baranov et $al.$ This low activation energy, combined with the overall high conductivity of CsH_2PO_4 above the transition warrant the identification of the high temperature phase as superprotonic. It is fair at this stage to conclude that the literature controversy regarding the high temperature behavior of CsH_2PO_4 has been put to rest.

Another important feature of the data in Fig. 4 is the rather large difference between the conductivity of CsH_2PO_4 measured in the first heating cycle and all other examinations of the low temperature phase. We attribute the high initial conductivity to residual surface water present in the grain boundary regions of the polycrystalline material. It has been shown elsewhere that even under dry conditions, removal of this water requires several days of exposure to temperatures of 205 °C and higher.[26] The conductivity results also reveal that there is tremendous hysteresis in the superprotonic transition behavior, with the reverse transformation occurring only after the material is undercooled by almost 30 °C relative to the transition temperature on heating. This behavior likely reflects the dramatic structural differences between the low and high temperature phases, as discussed in some detail below.

CsH_2PO_4 as a fuel cell electrolyte

The viability of CsH_2PO_4 based fuel cells was first demonstrated by the authors in 2004,[28] and later confirmed by Otomo et $al.$ in 2005.[35] The most recent work from

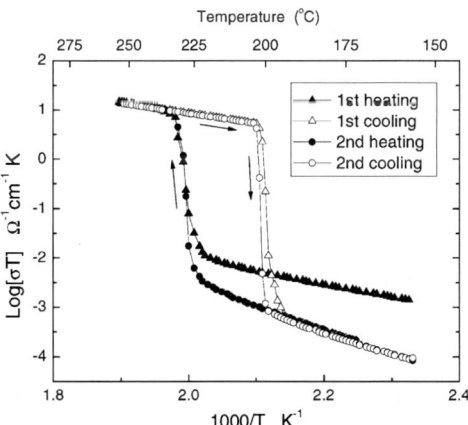

Fig. 4 Conductivity of polycrystalline CsH_2PO_4 under humidified air (p[H$_2$O] = 0.4 atm) for two heating and cooling cycles.

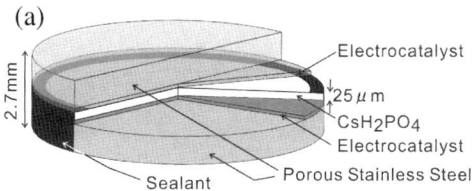

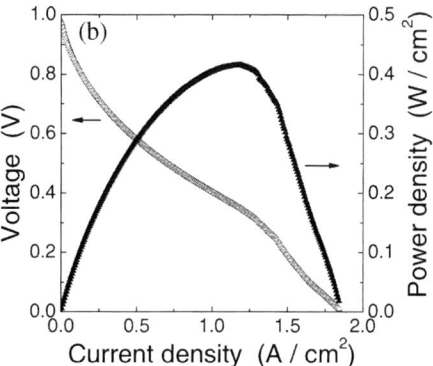

Fig. 5 (a) Schematic of thin-film fuel cell and (b) resulting polarization and power density curves for operation under humidified ($p[H_2O]$ = 0.3 atm) H_2/O_2 at 240 °C.[36] A peak power density of 415 mW cm^{-2} was obtained. Reproduced with permission from ref. 36.

the authors' laboratory indicates that supported electrolyte structures with membranes as thin as 25 μm can be fabricated and operated, Fig. 5, yielding single cell power densities of 415 mW cm^{-2} at peak power;[36] such performance is competitive with commercially available phosphoric acid fuel cells. These data were acquired from a cell operated at 240 °C with humidified hydrogen and humidified oxygen [$p(H_2O)$ = 0.3 atm] supplied to the anode and cathode, respectively. The electrodes were comprised of a composite mixture of CsH_2PO_4, Pt black and Pt supported on carbon, with a Pt loading of 7.7 mg cm^{-2} per electrode. The entire membrane–electrode-assembly was supported on a porous, stainless steel, gas-diffusion electrode, and from the absence of any influence of the gas flow rates on the polarization curves, it was demonstrated that mass transport through the support structure was not rate-limiting. The anomalous drop in voltage at high current densities (above ~1.33 A cm^{-2}) is due to irreversible mechanical degradation of the very thin electrolyte rather than any characteristics of the electrochemical reactions.

It is illustrative to consider the sources of overpotential in the fuel cell of Fig. 5. At 240 °C the conductivity (Fig. 4) is 2.2×10^{-2} Ω^{-1} cm^{-1} and thus for a membrane 25 μm in thickness the electrolyte area specific resistance is 0.11 Ω cm^2. Subtracting this contribution from the raw polarization curve yields the electrolyte corrected curve of Fig. 6. It is immediately evident that the membrane contributions to the polarization losses (only ~11 mV at a current density of 100 mA cm^{-2}) are much smaller than those of other sources (~130 mV at this same current density), which presumably represent slow electrocatalysis rates at the electrodes. The polarization measurements reported by Otomo *et al.*[35] suggest even greater electrode over-potentials, ~350 mV at 60 mA cm^{-2}. In PEMFCs electrocatalysis rates at the cathode are far slower than at the anode, and one might anticipate the same in the case of solid acid fuel cells, but such behavior is not an automatic characteristic of any fuel cell system. Accordingly, electrochemical experiments were performed to identify the source of electrocatalysis losses.

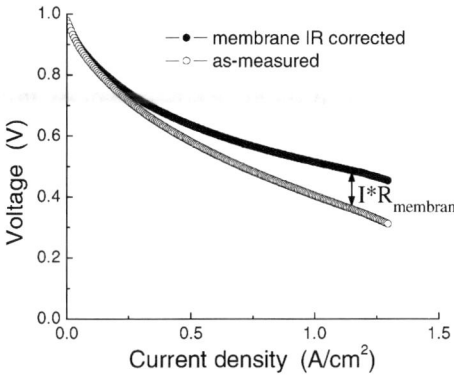

Fig. 6 Comparison between as-measured and membrane resistance (IR) corrected polarization curves for the fuel cell shown in Fig. 5.

For the examination of electrochemical characteristics of CsH_2PO_4 based fuel cells, a 700 μm thick cell configured with three electrodes was prepared, Fig. 7. As in the fuel cell measurements, the electrodes consisted of a composite of CsH_2PO_4, Pt black and Pt supported on carbon, added in a 3 : 3 : 1 weight ratio. Naphthalene, incorporated into the electrode composite as a fugitive pore-former, was removed under argon at a temperature of ~200 °C prior to data collection. Both AC impedance spectroscopy and galvanostatic current interrupt (GCI) methods were used to characterize electrochemical behavior. Measurements were performed with the cell exposed to a uniform atmosphere of either humidified hydrogen or humidified oxygen [$p(H_2O)$ = 0.57 atm, gas flow rate = 30 sscm] and held at a temperature of 238 °C. Impedance spectroscopy was carried out in the galvanostatic mode using a current amplitude of 2 mA, and a frequency range of 30 mHz to 5 kHz. With this frequency range (extending to lower values than in Fig. 3) it was possible to observe the low frequency intercept with the real axis in the Nyquist representation, and thereby determine the resistance due to the electrodes. Measurements were performed under an applied current bias of 0–28 mA cm^{-2}. The GCI data were collected using initial current values of 2–100 mA (corresponding to current densities of 1.1–113 mA cm^{-2}). The resulting voltage decay curves were fit to an appropriate exponential decay function to establish the proportion of the voltage due to capacitive (electrode) behavior for each value of initial current.

Examples of the impedance spectra and current decay curves obtained for cells exposed to humidified hydrogen and to humidified oxygen are presented in Fig. 8 and 9, respectively. Similar to the behavior of PEM fuel cells, hydrogen electro-oxidation proceeds far more rapidly than oxygen electroreduction. The GCI curve (Fig. 8b) suggests that at a current density of ~115 mA cm^{-2}, the hydrogen electrode is responsible for only ~0.3 mV out of the ~150 mV of overpotential associated with the electrodes (Fig. 6). Similarly, the impedance spectra indicate that the area specific anode polarization resistance is only ~0.06 Ω cm^2 (under zero bias).

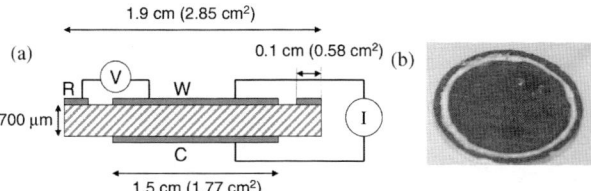

Fig. 7 Configuration for three-point electrical measurement (a) schematic and (b) photo of cell showing working and reference electrodes.

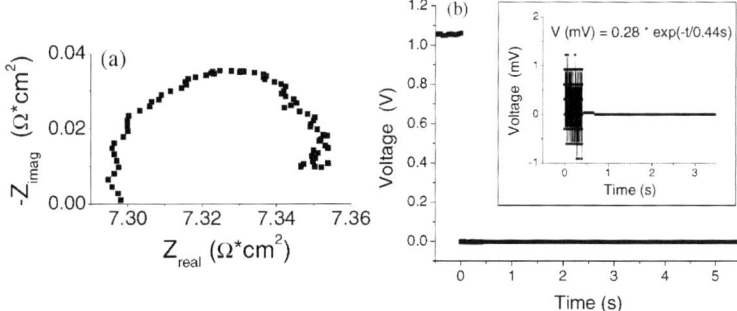

Fig. 8 Electrochemical characteristics of Pt–C | CsH$_2$PO$_4$ | Pt–C cells under humidified hydrogen ($p[$H$_2$O$]$ = 0.4 atm); (a) AC impedance spectrum, and (b) voltage decay curve after application of 113 mA cm^{-2} current. Inset shows short time data and fit to exponential decay.

In contrast, the GCI data indicate that the cathode overpotential is close to 80 mV at just 56.5 mA cm^{-2} (Fig. 9b), whereas under zero bias the area specific cathode polarization resistance, according to the impedance results, is ~9.5 Ω cm^2. In addition, both the voltage decay curves and impedance spectra suggest a two-step reaction process for oxygen electroreduction, with each step exhibiting a measurably different time constant.

From the GCI data collected under oxygen, the cathodic overpotential curve, η_C, was generated, Fig. 10. From a fit to a modified form of the Tafel equation, we obtain an exchange current density of 5.4 ± 0.4 mA cm^{-2}, which is about a factor of twenty lower than what has been reported for PEMFCs.[37] The η_C values determined in this manner correspond reasonably well to the total electrode overpotential curve obtained from the fuel cell measurements, while the slope of the overpotential curve at zero current density (6.2 Ω cm^2) is acceptably similar to the zero-bias area-specific polarization resistance measured by impedance spectroscopy (9.5 Ω cm^2), particularly given that different samples were used in each of the three experiments. In total, the results indicate that all sources of potential drop have been accounted for and further show that, although quantitative interpretation may not yet be possible, the cathode is clearly the rate-limiting step in state-of-the-art solid acid fuel cells. Increasing fuel cell power output will thus require enhancements in the oxygen electroreduction rates rather than further decreases in electrolyte thickness.

The 'warm' temperature of operation of CsH$_2$PO$_4$ based fuel cells opens up possibilities for the direct utilization of fuel cells other than hydrogen. In particular, at 250 °C one expects that methanol steam reforming will proceed rapidly and therefore that such fuel cells can be operated on methanol–water mixtures. In

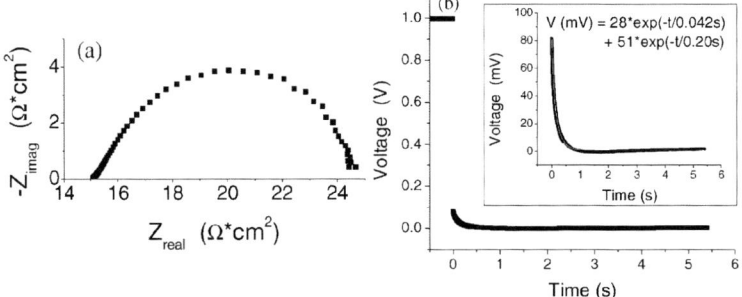

Fig. 9 Electrochemical characteristics of Pt–C | CsH$_2$PO$_4$ | Pt–C cells under humidified oxygen ($p[$H$_2$O$]$ = 0.4 atm); (a) AC impedance spectrum, and (b) voltage decay curve after application of 56.5 mA cm^{-2} current. Inset shows short time data and fit to exponential decay.

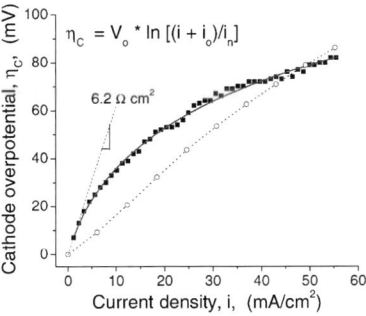

$$\eta_C = V_o * \ln[(i + i_o)/i_n]$$

Fig. 10 Comparison between the cathodic overpotential for oxygen electroreduction over Pt–C on CsH_2PO_4 determined by GCI methods (closed squares) with the total electrode overpotential determined from complete fuel cell measurements (open circles) at a temperature 240 °C and a water partial pressure of 0.4–0.6 atm. The solid line indicates the best fit to the GCI data according to the equation shown in the figure, where i_o is the exchange current density.

addition, the tolerance of the anode electrocatalyst to fuel stream poisons such as CO are expected to be high. It can further be expected that the solid nature of the electrolyte will ensure zero fuel crossover and thereby enable the use of high concentrations of alcohol in the fuel stream. With these considerations in mind, the authors recently demonstrated methanol and ethanol fuel cells with rather remarkable power densities, Fig. 11.[38] These cells, in which Pt–Ru served as the electrooxidation catalyst (rather than simple Pt) and the membrane was 47 μm in thickness, were constructed with a steam reforming catalyst (Cu–ZnO/Al$_2$O$_3$[39]) placed directly adjacent to the anode. In order to permit these direct comparisons of behavior under different fuels, the methanol and ethanol were supplied to the anode at rates equivalent to that at which hydrogen was supplied, assuming complete alcohol reformation. Quite notably, the open circuit potential under methanol (43 vol%) is as high as it is under hydrogen, while the peak power output under methanol is reduced only by 15% relative to that under hydrogen. The behavior is in stark contrast to polymer based fuel cells, in which direct operation on methanol produces power outputs that are only about 15% of what can be obtained from hydrogen. Furthermore, our analysis indicates that the major limitation in the CsH_2PO_4 based methanol fuel cell is the rate at which methanol reforming occurs over the Cu–ZnO/Al$_2$O$_3$ catalyst rather than poisoning of the anode electrocatalyst by CO.

Thus, the fuel cell experiments performed to date point out the potentially fruitful avenues for technological developments. As with all technological endeavors, basic

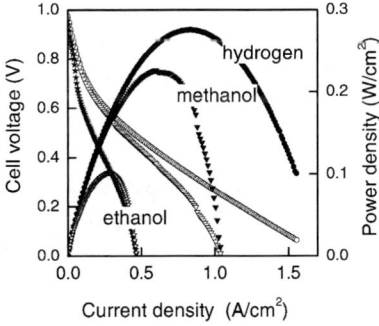

Fig. 11 Polarization and power density curves at 240 °C for fuel cells in which a steam reforming catalyst was placed directly adjacent to the fuel cell anode. Humidified oxygen was supplied to the anode and the fuel indicated (humidified) was supplied to the anode. Reproduced with permission from ref. 38

scientific studies of physical phenomena are expected to provide the insight required for further advances. With this philosophy in mind, we now turn to an analysis of the fundamental driving force for the superprotonic transition in CsH_2PO_4.

Configurational entropy of superprotonic CsH_2PO_4

In a recent study of $CsHSO_4$, we proposed a methodology for evaluating the configurational entropy of phases in which both global hydrogen bond disorder and local oxyanion group disorder occur and are independent of one another.[40] It was found that the measured transition entropy (determined from thermal analysis) could be explained by applying this methodology to the superprotonic structure of $CsHSO_4$ as proposed by Jirak *et al.*[41] The suitability of that approach for the evaluation of the configurational entropy of CsH_2PO_4 is now considered here. Doing so requires not only an analysis of the configurational entropy of superprotonic CsH_2PO_4 by this method, but also knowledge of the (experimental) transition entropy and the configurational entropy of the room temperature structure.

Experimental studies of the transition enthalpy and entropy

A review of the literature indicates that several values for the transition enthalpy have been reported, Table 2. The early reports of Metcalfe and Clark[11] and of Gupta *et al.*[13] suggested a transition enthalpy of ~ 8 kJ mol^{-1}, whereas later studies, which presumably utilized modern instrumentation, consistently indicate a transition enthalpy of 11.6 kJ mol^{-1}. Accordingly, this higher value is taken to be the physically correct value, implying a transition entropy ($\Delta S = \Delta H/T_C$) of 23 ± 1 J mol^{-1} K^{-1} for a transition temperature of 230 °C. For consistency, this single value of the transition temperature has been used for the conversion rather than the slightly differing transition temperatures reported in those studies.

Room temperature structure

At room temperature CsH_2PO_4 adopts a monoclinic structure, in which PO_4 groups are linked together by both asymmetric, single-minimum hydrogen bonds and symmetric, double-mimina bonds. These bonds, formed between oxygen atoms of neighboring phosphate groups, generate corrugated two-dimensional $[H_2PO_4^=]_\infty$ layers, in between which are located the Cs atoms, Fig. 12. The space group in the standard, primitive setting is $P2_1/m$ [$a = 7.912(2)$, $b = 6.383(1)$, $c = 4.8802(8)$ Å and $\beta = 107.73(2)°$].[43] Instead of this choice CsH_2PO_4 is often described in the non-primitive $B2_1/m$ setting because doing so reveals the pseudo–orthorhombic symmetry of the structure [$a = 4.8725(1)$, $b = 6.3689(1)$, $c = 15.0499(8)$ Å and $\beta = 90.22(1)°$].[44] The asymmetric unit contains one Cs atom, one P atom, three O atoms and two H atoms. Using the atom assignments of Nelmes and Choudhary[44] the asymmetric hydrogen bond has configuration O(1)–H(1)$\cdots$O(2), whereas the symmetric bond has configuration O(1)$\cdots$H(2)–H(2)$\cdots$(O1), with the H(2) atoms displaying an average site occupancy of 0.5. On cooling, CsH_2PO_4 undergoes a

Table 2 Reported values of the enthalpy and temperature of the superprotonic transition in CsH_2PO_4

ΔH/kJ mol^{-1}	T_c/ °C	Source
7.6	230	Metcalf (1978)[11]
8.4 ± 0.8	235	Gupta (1980)[13]
11.3 ± 0.5	230 ± 2	Chisholm (2002)[42]
11.3 ± 0.6	228 ± 2	Boysen (2003)[26]
11.9 ± 0.3	229	Yamada (2004)[30]

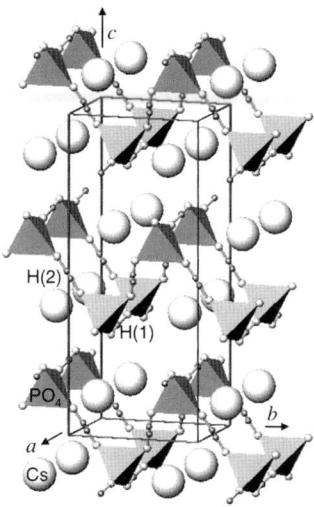

Fig. 12 Crystal structure of CsH_2PO_4 in its room temperature, paraelectric form, taken in space group setting $B2_1/m$.[44] The disordered hydrogen bonds formed by the H(2) atoms create a 1-dimensional chain of phosphate groups along [010], in turn, linked by the ordered hydrogen bonds formed by H(1) to generate a layered structure.

ferroelectric transition to a monoclinic structure of space group $P2_1$, at which the inversion symmetry about the O(1)–O(1) hydrogen bond is removed and the bond becomes asymmetric (with a single minimum).

The molar configurational entropy that one would expect to result from the disordered hydrogen bond in CsH_2PO_4 is $R\ln(2) = 5.76$ J mol^{-1} K^{-1}. On the basis of calorimetric measurements of CsH_2PO_4 cooled through the ferroelectric transition, however, Imai has suggested an entropy difference of only 3.2 ± 0.2 J mol^{-1} K^{-1} between the paraelectric and ferroelectric phases,[45] whereas Kanda *et al.* reported an even lower value of 1.05 ± 0.2 J mol^{-1} K^{-1}.[56] This apparent paradox may be the result of correlations between O(1)–O(1) hydrogen bonds at temperatures above the ferroelectric transition. Such correlations give rise to dynamic domains in which structural characteristics of the ferroelectric phase are retained at elevated temperatures, with the mean domain size decreasing with increasing temperature. As a consequence, the thermal signature of the ferroelectric phase transition may be spread out over a wide temperature regime and result in an artificially depressed value of the enthalpy (and hence entropy) of the transition. Although both sets of authors have attempted to explicitly account for these effects, the significant discrepancy between the reported values suggests that accurate treatment of the problem presents significant experimental and theoretical challenges. In particular, it appears that underestimation of the transition entropy occurs because too much of the thermal signal can be easily attributed to vibrational heat capacity and to experimental background.[45] In light of these challenges, we choose here to assign the full expected value of 5.76 J mol^{-1} K^{-1} to the configurational entropy of CsH_2PO_4 in the room temperature, paraelectric phase.

High temperature structure

The high temperature structure of CsH_2PO_4 is shown in Fig. 13. The compound adopts a CsCl-like structure with space group Pm-$3m$ and lattice constant 4.9549(4) Å at 515 K.[30] The Cs atoms reside at the corners of the primitive unit cell and the PO_4 group is orientationally disordered about the center, taking on one of six possible orientations. The oxygen positions are numbered so as to indicate the set of

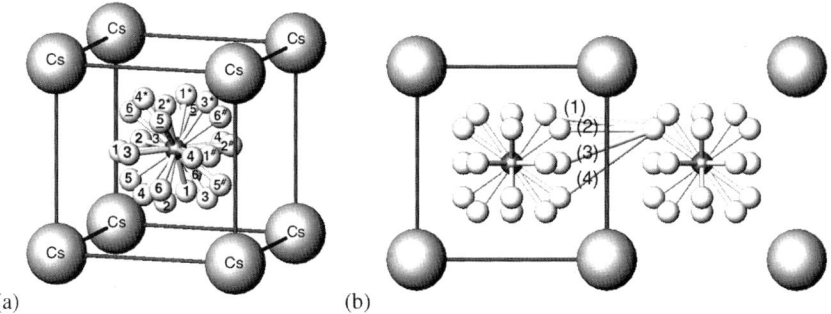

Fig. 13 High temperature structure of CsH_2PO_4; (a) single unit cell and (b) hydrogen bonds between neighboring cells. Hydrogen bonds (indicated by the numbers in parentheses) have respective lengths of 2.85, 2.46, 2.76 and 3.03 Å. The elevations of the oxygen atoms associated with these bonds are respectively, 0.5, 0.5, 0.634 and 0.5.

four sites associated with each of the six orientations. It is noteworthy that several related compounds also adopt this structure at slightly elevated temperature, including $Cs_2(HSO_4)(H_2PO_4)$, α-$Cs_3(HSO_4)_2(H_2PO_4)$, $Cs_5(HSO_4)_3(H_2PO_4)_2$ and even $CsH(PO_3H)$.[46] The orientational disorder of the oxyanion group can be understood to arise from the incompatibility of the tetrahedral unit with the octahedral symmetry at the cube center. In principle, it would be possible to orient the PO_4 group such that one of the P–O bonds were aligned with the $\langle 111 \rangle$ body diagonal and retain the overall cubic symmetry of the structure (removing the octahedral symmetry from the center and lowering the overall space group symmetry to $P432$). However, the distance between the phosphorous and caesium atoms along this direction is only 4.296 Å and indicates that the oxygen cannot lie directly on the line between them. Placement of the oxygen atoms at sites displaced from the body diagonal (to 1/2 1/4 0.366) in a manner that retains the cubic symmetry produces the structure shown in Fig. 13.

As noted by Yamada *et al.*[30] the shortest oxygen–oxygen distance between neighboring PO_4 groups is 2.46 Å, consistent with the length of a strong (single minimum) hydrogen bond. This bond, corresponding to bond (2) in Fig. 13b, links neighboring PO_4 groups *via* the face of the simple cubic unit cell. Each of the oxygen atoms marked with a number sign (#) in Fig. 13a has an oxygen neighbor in the phosphate group that resides directly to the right of the one shown with which a bond of this length could be formed. Similarly, the four oxygen atoms marked with asterisks (*) could form bonds of this length *via* the upper face of the cube. Several additional hydrogen bonds with somewhat longer oxygen–oxygen bond distances are also possible, which similarly provide linkages through the faces of the simple cubic unit cell. These bonds, if directed towards the phosphate group located above the one shown, involve the oxygen atoms identified with underlined labels in Fig. 13a.

Configurational entropy of systems with globally disordered hydrogen bonds

The approach we recently proposed[40] for the evaluation of the configurational entropy of $CsHSO_4$ follows the elegant model put forward by Pauling to describe the residual entropy of ice. Hexagonal ice (I_h) has a structure, shown schematically in a two-dimensional representation in Fig. 14, in which each oxygen atom is tetra-hedrally coordinated by four other oxygen atoms, and between each of these is a disordered hydrogen bond.[47] Pauling noted that the structure 'obeyed' several rules limiting the allowable hydrogen bond configurations, and then developed a means of calculating the number of configurations which conformed to those rules.[48] These so-called ice rules are: (1) each oxygen atom has two and only two protons (no

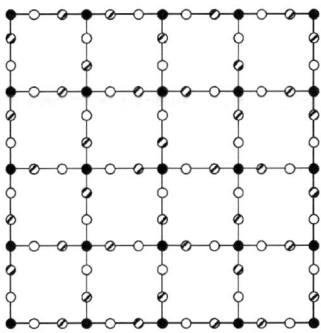

Fig. 14 Highly schematic rendition of the structure of hexagonal ice (I_h). Solid circles represent oxygen atoms, straight lines represent (double-minima) hydrogen bonds, hatched circles represent occupied proton sites and open circles represent unoccupied proton sites. Every oxygen atom has two protons as immediate neighbors so as to form water molecules, and every hydrogen bond is occupied by one and only one proton.

hydronium or hydroxyl ions are considered); (2) each hydrogen bond has one and only one proton (no L or D defects); (3) the hydrogen bonds are directed approximately towards the neighboring oxygen atoms; and (4) interaction between non-neighboring H_2O molecules does not influence the distribution of hydrogen bond configurations. If one considers a particular oxygen atom, then, there are four possible locations for its first proton, and three for its second proton. Of these configurations, only one-half are unique, bringing the number of possible configurations for an isolated H_2O group to $(4!)/[(2!)(2!)] = 6$. Within the structure of ice, one-half of all the possible hydrogen bonds are occupied, thus the probability that any particular bond is open or available for occupation by the protons of the H_2O group of interest is also one-half, reducing the number of H_2O group configurations by a factor of $\frac{1}{2}$ for each proton. With this final consideration the total number of possible configurations per H_2O molecule, Ω, is $(6)(\frac{1}{2})(\frac{1}{2}) = 6/4$. Formally, this result can be expressed as

$$\Omega_H = \left(\begin{array}{c} \#\text{of proton} \\ \text{configuration} \end{array}\right)\left(\begin{array}{c} \text{probability a proton} \\ \text{site is open} \end{array}\right)^{\left(\begin{array}{c} \#\text{of} \\ \text{protons} \end{array}\right)} \tag{1}$$

The residual entropy so determined, 3.37 J mol^{-1} K^{-1}, matches almost precisely the experimental value of 3.65 J mol^{-1} K^{-1},[49] and Pauling's insight resolved a major puzzle in statistical thermodynamics.

In assessing whether or not such an approach is appropriate for the evaluation of the configurational entropy of disordered oxyanion compounds, one must consider the following key question: is the orientation of the oxyanion related to the hydrogen bonds that the group forms? That is, are the XO_4 orientation and the location of hydrogen bonds correlated or independent? In the case of $CsHSO_4$, several models of the high temperature structure have been reported. In the Jirak model,[41] there is only one crystallographically distinct oxygen atom, a feature that argues towards the independence of the hydrogen bond locations and the tetrahedral group orientation. Other models, however, such as those of Belushkin[50] and Merinov[51,52] in particular, distinguish between donor and acceptor oxygen atoms, and thus the sulfate group orientation in these structural models is fixed by the location of the hydrogen bonds. Accordingly, these structures cannot legitimately be evaluated by an extension of the Pauling approach. In the case of CsH_2PO_4, there is only one crystallographically distinct oxygen atom,[30] and, like the Jirak structure for $CsHSO_4$, this atom can serve either as donor or acceptor in the hydrogen bond. It would appear that superprotonic CsH_2PO_4, for which there is no structural ambiguity, would then be a straightforward phase to evaluate in terms of the extended Pauling approach. But appearances are deceptive indeed!

Consider superprotonic, tetragonal $CsHSO_4$ according to the model proposed by Jirak.[41] The structure has characteristics of the zinc blende structure, in that each sulfate group has four neighboring sulfate groups, thus there are four directions in which hydrogen bonds can be formed. In addition, each sulfate group can reside in one of two orientations, and each orientation is compatible with any distribution of hydrogen bonds. Thus, the hydrogen bond distribution and sulfate group orientation can be evaluated separately, and they contribute in a multiplicative sense to the overall number of configurations available to $CsHSO_4$. Quantitatively, the number of hydrogen bond configurations can be treated exactly as given in eqn (1), but with the numerical values modified to reflect the specific structural characteristics of $CsHSO_4$, which also imply a modification to the ice rules.[40] Here we have (1) *one* and only *one* proton is associated with each *tetrahedron*; (2) one and only one proton occupies each hydrogen bond *that is formed*; (3) hydrogen bonds are directed towards oxygen atoms of neighboring *tetrahedra*; and (4) interaction between non-neighboring HSO_4 *tetrahedra* does not influence the distribution of hydrogen bond configurations. With these modifications, we find that the number of hydrogen bond configurations available for an isolated HSO_4 group is 4!/3! (the single proton may form a bond along any of the four bond directions), the probability that a site is available or open is $(\frac{3}{4})$, and this probability consideration must be applied to one proton. Thus, $\Omega_H = (4)(\frac{3}{4})^1$. Furthermore, as discussed above, $\Omega_{tetr} = 2$, giving the total number of configurations as $\Omega = \Omega_H \Omega_{tetr} = 6$. The implied transition entropy, 14.90 J mol^{-1} K^{-1}, is in excellent agreement with the measured value of 14.8 ± 0.2 J mol^{-1} K^{-1},[40] supporting the validity of this analysis under the assumption that all changes in entropy are configurational in nature.

Buried within the modified 'extended' ice rules are some rather profound implications for the proton transport behavior of $CsHSO_4$. The first rule is equivalent to stating that there is only one donor oxygen atom per sulfate group, a physically reasonable expectation, but one that, as in the case of ice, rules out the presence of defects such as $SO_4^=$ and $H_2SO_4^0$ in significant enough quantities to impact the total configurational entropy. The second rule, also in analogy to ice, rules out the presence of doubly occupied hydrogen bonds or D-defects, another physically reasonable expectation. In contrast to ice, however, L-defects have no meaning because on average, $\frac{1}{2}$ of the possible hydrogen bonds are unoccupied. Moreover, in further contrast to ice, the combination of the second and fourth rules implies that any number of the non-donor oxygen atoms of the HSO_4 group can serve as acceptor atoms in hydrogen bond formation. Thus, each sulfate group will form anywhere between one and four hydrogen bonds, with the sole restriction that one and only one will be formed as a result of a donor oxygen atom. A schematic illustrating the overall situation is presented in Fig. 15.

For this set of assumptions to be consistent with high conductivity in superprotonic $CsHSO_4$, it must be a sharp increase in the *mobility* of protonic defects, rather than a sharp increase in the *concentration* of such defects, that gives rise to the dramatic increase in conductivity at the transition. In fact, such an interpretation has been recently suggested on the basis of ^1H NMR studies. Yoshida *et al.*[53] observed the shape change in the ^1H NMR peak of single crystal $CsHSO_4$ as it was heated through the phase transition. A sharp Lorentzian peak, assigned to mobile protons, was superimposed on a broad Gaussian peak, assigned to immobile protons. The ratio of the integrated intensities of these two peaks were followed as a function of temperature, and the Gaussian contribution was found to gradually decrease with increasing temperature. This contribution finally disappeared at 117 °C, indicating that 100% of the protons are mobile a full 24 °C below the superprotonic transition.

Further insight on this point can be gained by comparison with the behavior of H_2SO_4. The electrical conductivity of anhydrous sulfuric acid at 25 °C is 1.04×10^{-2}

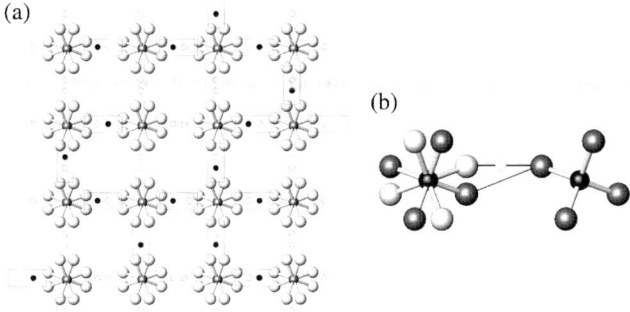

(a)

(b)

Fig. 15 Schematic rendition of the tetragonal structure of superprotonic CsHSO$_4$. Small circles represent proton positions, with closed circles being occupied and open circles being unoccupied. Rectangles indicate hydrogen bonds that are formed, which are fewer than the number of possible hydrogen bonds. Each SO$_4$ group resides in one of two possible orientations, and each has one and only one donor oxygen atom (*i.e.*, the HSO$_4$ is the basic unit). As shown in (b) either orientation of the sulfate group accommodates hydrogen bonds with neighboring sulfate groups.

Ω^{-1} cm^{-1},[54] comparable to that of superprotonic solid acids. Moreover, proton conduction is believed to occur *via* a mechanism similar to that operative in superprotonic CsHSO$_4$, which has rapid sulfate group reorientation facilitating long-range proton transport. Under ambient temperatures, the autopyrolysis of H$_2$SO$_4$ results in concentrations of H$_3$SO$_4{}^+$ and HSO$_4{}^-$ of only 0.15 and 0.11 mol%, respectively.[55] Thus, high concentrations of such defects are not required for attaining high proton conductivity. It is also noteworthy that the energetics of ionic defect formation must surely be dominated by Coulombic interactions, which, in turn, are governed by interatomic distances. Because the overall anion–cation distances in CsHSO$_4$ are relatively unchanged between the monoclinic (low conductivity) and tetragonal (superprotonic) phases, it is unlikely that the concentration of defects would change by an amount that could account for an increase in conductivity by four orders of magnitude upon passing through the superprotonic transition.

We now consider the situation of CsH$_2$PO$_4$. Unlike superprotonic CsHSO$_4$, the oxyanion groups of the phosphate each have *six* nearest oxyanion neighbors, as implied by the CsCl-like structure. Thus, there are six directions in which hydrogen bonds can be formed. To simplify the discussion, we replace, for the moment, the disordered H$_2$PO$_4$ group by a hypothetical H$_2$XO$_6$ oxyanion group in which the oxygen sites are fully occupied and oxygen atoms (as in the real structure) can equally well serve as donors or acceptors in hydrogen bond formation. By analogy to ice, the number of hydrogen bond configurations possible in this hypothetical compound results from the placement of two protons over six hydrogen positions, modified by the probability (4/6) that any proton site is open or available. Thus, $\Omega_{\mathrm{H}} = (6!/4!2!)(4/6)^2 = 20/3 = 6.67$ readily falls out of the analysis. The conceptual challenge arises when considering the configurational disorder of the tetrahedral groups to obtain the quantity Ω_{tetr}.

In CsHSO$_4$ a value of 2 is assigned to Ω_{tetr} because there are two sulfate group orientations compatible with any hydrogen bond configuration. This also coincides with the fact that there are two oxygen positions associated with the formation of any particular hydrogen bond, as indicated schematically in Fig. 15. The situation in CsH$_2$PO$_4$ differs in that there are several (depending on the distance criteria used) oxygen positions associated with any particular hydrogen bond, but this number does not coincide with the number of phosphate group orientations. Furthermore, the precise cut-off distance beyond which neighboring oxygen atoms should not be considered linked *via* a hydrogen bond is not entirely obvious. These points are clarified in the following examples.

Let the phosphate group, Fig. 13, have a hydrogen bond configuration such that one hydrogen bond extends out of the top of the cube in which the phosphate group resides and another one extends out of the bottom. Let us term this a 'straight-through' configuration of hydrogen bonds. Recall that there are no restrictions on acceptor oxygen atoms and, accordingly, only two bonds (which account for the two donor oxygen atoms of the phosphate group) need be specified. As indicated in the figure, there are four oxygen atom sites that are close to and equidistant from the top face of the cube (indicated by asterisks in the oxygen orientation labeling). Each of these four sites must therefore be viable positions for the donor oxygen atom in the upper hydrogen bond. Indeed, the O–O distances formed between these oxygen atoms and the equivalent four in the cube above that shown range from 2.46 to 3.03 Å (bonds 2–4 in Fig. 13b). Similarly, there are four oxygen atoms sites that are equidistant from the bottom face of the cube and that are viable positions for the donors in the downwards directed hydrogen bond. Furthermore, for each possible donor oxygen atom site for the upper hydrogen bond, the phosphate group geometry is such that one of the possible donor oxygen atom sides for the lower hydrogen bond would be part of the specified phosphate group orientation (orientations 1 through 4 in Fig. 13a). That is, the four upper oxygen atom positions are associated with four different phosphate group orientations, where each orientation incorporates one of the four lower oxygen atom positions. Thus, for hydrogen bond configurations in which the bonds extend out of opposing faces of the cube, there are four tetrahedral group orientations possible, if one only considers short hydrogen bonds ($\Omega_{tetr} = 4$).

Now let the phosphate group have a hydrogen bond configuration such that one hydrogen bond extends out of the top of the cube in which the phosphate group resides and another one extends to the right. Let us refer to this as an 'L' configuration of the hydrogen bonds. Once again, there are four clearly viable donor oxygen atom sites for the hydrogen bond extending upwards. There are also four such sites for the hydrogen bond extending to the right. In this case, however, there is not a one-to-one correspondence between the phosphate group orientations implied by these sites. That is, if the phosphate group is placed in any one of the four orientations implied by the four upper oxygen positions, only two of those orientations (orientations 1 and 2) incorporate oxygen atoms close to the right-side face of the cube (indicated by # in the oxygen atom label). Thus, if only oxygen atoms directly adjacent to a cube face can serve as donors to a bond extending out of that face, then the number of tetrahedral group orientations compatible with 'L' type hydrogen bond configurations is only two ($\Omega_{tetr} = 2$). Noting that there are three possible straight-through configurations of hydrogen bonds and twelve L-type configurations, the weighted average of Ω_{tetr} is 2.4. Again, this analysis is restricted to the consideration of short hydrogen bonds in the structure.

Another way in which to consider the tetrahedral group configurational entropy is to include O–O linkages that are formed by oxygen atoms not immediately neighboring the cube face of interest. In particular, the bond labeled as bond 1 in Fig. 13b represents the linkage of an oxygen atom primarily associated with the top face of the cube so as to form a hydrogen bond through the side face. The O–O distance of 2.85 Å falls with the range of bond lengths that can be considered for linkages only *via* the faces that the oxygen atoms are primarily associated with. In light of this additional possibility for bond formation, one can argue that any of the six orientations is possible with any hydrogen bond configuration and thus $\Omega_{tetr} = 6$. A complexity arises here from the possibility that, for a given hydrogen bond configuration and a given tetrahedral group orientation, there are, in fact, multiple ways in which the pair of donor oxygen atoms can be selected. This is illustrated as follows. Consider the tetrahedral group residing in orientation 6 and the possibility of a straight-through hydrogen bond configuration with the bonds extending through the top and bottom faces of the cube. Either of the upper two oxygen atoms associated with this orientation can serve as donors in the hydrogen bond that extends through the top face of the cube (atom labels underlined), if we consider the longer hydrogen bonds. Similarly, either of the lower two can serve as donors in the hydrogen bond than extends through the bottom face of the cube. Because the upper and lower donors can be selected independently, there are *four* possible ways of assigning the two donor oxygen atoms to conform with hydrogen bonds extending through the top and bottom faces, without changing the orientation of the phosphate group itself. This kind of consideration implies an even greater number of possibilities, specifically, *twelve*, for L-type hydrogen bond configurations. The weighted average then yields Ω_{tetr} = 8.8, which can be viewed as the maximum number of possible orientations for a liberal cut-off for hydrogen bond formation.

As an intermediate approach, one can retain the possibility of the formation of 'direct' hydrogen bonds of type (1), Fig. 13b, but hypothesize that they are unlikely to form as often as the 'indirect' bonds of type (2)–(4). In a sense, this treats the direct and indirect hydrogen bonds as energetically distinct, but without explicit assignment of energy values. Instead occupation values can be assigned. Doing so yields values of Ω_{tetr} intermediate between the 2.4 value obtained when one considers only the direct hydrogen bonds and the 8.8 value obtained when one considers the direct and indirect hydrogen bonds on an equal footing. In particular, a weighting scheme in which the indirect bonds are arbitrarily assigned an occupancy $\frac{1}{2}$ of that of the direct bonds yields a value of Ω_{tetr} of 5.2.

Overall, it is apparent there are several defensible ways of assigning a value to Ω_{tetr} in superprotonic CsH_2PO_4. If one considers the number of possible oxygen positions associated with the formation of a single hydrogen bond it is 4; if one considers the number of possible phosphate group orientations it is 6; if one considers the number of possible phosphate group orientations compatible with the hydrogen bond configurations it lies between 2.4 and 8.8, depending on the length of the allowable hydrogen bonds and the weighting scheme employed. Given the difficulty of defining Ω_{tetr}, it is not possible to use a comparison between the measured and computed values of the transition entropy to validate the proposed interpretation of superprotonic CsH_2PO_4. If, on the other hand, we take the approach to be valid, we can used the experimentally measured ΔS and the computed Ω_H (for which there is no ambiguity) to evaluate Ω_{tetr}. Doing so yields a value of 4.8, which certainly falls within the wide range anticipated. We tentatively interpret this in terms of the weighting of the short and long hydrogen bonds. Finally, we note that while this discussion does not present a definitive picture of the sources of configuration entropy in CsH_2PO_4, it is clear that the hydrogen bond disorder and phosphate group disorder must contribute in some independent fashion to the overall entropy. If these sources of disorder were to be considered entirely correlated such that the orientation of the phosphate group entirely fixed the location of hydrogen bonds, then the number of configurations per formula unit would be only 6, implying a transition entropy of only 9.13 J mol^{-1} K^{-1}, far less than the experimentally measured value of 23 J mol^{-1} K^{-1}.

Summary

This work provides an overview of the status of the science and technology of solid acid based fuel cells with particular emphasis on CsH_2PO_4. On the basis of an evaluation of the literature and our recent conductivity measurements, it is now beyond doubt that CsH_2PO_4 undergoes a superprotonic transition at 230 °C. Under typical laboratory climates, dehydration of CsH_2PO_4 initiates at temperatures close to the superprotonic transition and thus humidity control is essential to the evaluation of its true physical properties and to the use of this material in functioning devices. Fuel cell measurements to date demonstrate that it is possible to support thin-film CsH_2PO_4 electrolyte membranes on porous gas diffusion electrodes and obtain peak power densities as high as 415 mW cm^{-2}. At this stage, power output is limited by cathode activity rather than membrane conductivity. Indeed for a fuel cell operated at 240 °C with a 25 μm thin membrane, at a current density of 100 mA cm^{-2}, the voltage drop across the electrolyte is only 11 mV compared to well over 100 mV across the cathode. The voltage drop across the anode for fuel cells operated on hydrogen is immeasurable. The high temperature of operation of CsH_2PO_4 fuel cells has moreover enabled the demonstration of alcohol fueled cells with excellent power output. Further advances in this area will require improvements in the steam reforming catalysts placed within the anode chamber of such fuel cells.

With these results, it is clear that the technological importance of CsH_2PO_4 has been proven. While progress in developing commercially viable solid acid fuel cells continues, it is also clear that much of the fundamental physics of this material remain a mystery. Exactly what is the connection, if any, between hydrogen bond disorder and phosphate group disorder? How do these factors play into the remarkable proton conductivity, if at all? And can the answers be used to design materials with even more attractive properties?

Acknowledgements

This work has been supported by the US National Science Foundation, Division of Materials Research and the US Department of Energy through a subcontract *via* the DOE-funded Cornell Fuel Cell Institute, of Cornell University.

References

1   A. I. Baranov, L. A. Shuvalov and N. M. Shchagina, *JETP Lett. Engl. Transl.*, 1982, **36**(11), 459–462.
2   A. Pawlowski, Cz. Pawlaczyk and B. Hilzcer, *Solid State Ionics*, 1990, **44**, 17–19.
3   C. Ramasastry and A. S. Ramaiah, *J. Mater. Sci. Lett.*, 1981, **16**, 2011–2016.
4   S. M. Haile, D. A. Boysen, C. R. I. Chisholm and R. B. Merle, *Nature*, 2001, **410**, 910–913.
5   A. I. Baranov, V. P. Khiznichenko, V. A. Sandler and L. A. Shuvalov, *Ferroelectrics*, 1988, **81**, 1147–1150.
6   J. C. Slater, *J. Chem. Phys.*, 1941, **9**(1), 16–33.
7   M. Komukae, T. Osaka, T. Kaneko and Y. Makita, *J. Phys. Soc. Jpn.*, 1985, **54**(9), 3401–3405.
8   L. H. Rashkovich, K. B. Meteva, Ya. É. Shevchik, V. G. Hoffman and A. V. Mishchenko, *Sov. Phys. Crystallogr.*, 1977, **22**(5), 613–615.
9   L. N. Rashkovich and K. B. Meteva, *Sov. Phys. Crystallogr.*, 1978, **23**(4), 447–449.
10   E. Rapoport, J. B. Clark and P. W. Richter, *J. Solid State Chem.*, 1978, **24**(3-4), 423–433.
11   B. Metcalfe and J. B. Clark, *Thermochim. Acta*, 1978, **24**(1), 149–153.
12   M. Wada, A. Sawada and Y. Ishibashi, *J. Phys. Soc. Jpn.*, 1979, **47**(5), 1571–1574.
13   L. C. Gupta, U. R. K. Rao, K. S. Venkateswarlu and B. R. Wani, *Thermochim. Acta*, 1980, **42**(1), 85–90.
14   B. M. Nirsha, E. N. Gudinitsa, A. A. Fakeev, V. A. Efremov, B. V. Zhadanov and V. A. Olikova, *Russ. J. Inorg. Chem.*, 1982, **27**(6), 770–772.
15   W. Bronowska and A. Pietraszko, *Solid State Commun.*, 1990, **76**(3), 293–298.
16   F. Romain and A. Novak, *J. Mol. Struct.*, 1991, **263**, 69–74.
17   R. A. Vargas and E. Torijano, *Solid State Ionics*, 1993, **59**(3–4), 321–324.
18   A. Preisinger, K. Mereiter and W. Bronowska, *Mater. Sci. Forum*, 1994, **166–169**, 511–516.
19   K. S. Lee, *J. Phys. Chem. Solids*, 1996, **57**(3), 333–342.
20   Y. Luspin, Y. Vaills and G. Hauret, *J. Phys. I*, 1997, **7**(6), 785–796.
21   E. Ortiz, R. A. Vargas and B. E. Mellander, *J. Chem. Phys.*, 1999, **110**(10), 4847–4853.
22   E. Ortiz, R. A. Vargas and B. E. Mellander, *Solid State Ionics*, 1999, **125**(1–4), 177–185.
23   W. Bronowska, *J. Chem. Phys.*, 2001, **114**(1), 611–612.
24   K. S. Lee, *Ferroelectrics*, 2002, **268**, 789–794.
25   J. Otomo, N. Minagawa, C. J. Wen, K. Eguchi and H. Takahashi, *Solid State Ionics*, 2003, **156**(3), 357–369.
26   D. A. Boysen, S. M. Haile, H. J. Liu and R. A. Secco, *Chem. Mater.*, 2003, **15**(3), 727–736.
27   J. H. Park, C. S. Kim, B. C. Choi, B. K. Moon and H. J. Seo, *J. Phys. Soc. Jpn.*, 2003, **72**(6), 1592–1593.

28 D. A. Boysen, T. Uda, C. R. I. Chisholm and S. M. Haile, *Science*, 2004, **303**(5654), 68–70.
29 J. H. Park, *Phys. Rev. B*, 2004, **69**(5).
30 K. Yamada, T. Sagara, Y. Yamane, H. Ohki and T. Okuda, *Solid State Ionics*, 2004, **175**(1–4), 557–562.
31 J. Otomo, T. Tamaki, S. Nishida, S. Q. Wang, M. Ogura, T. Kobayashi, C. J. Wen, H. Nagamoto and H. Takahashi, *J. Appl. Electrochem.*, 2005, **35**(9), 865–870.
32 A. I. Baranov, V. V. Grebenev, A. N. Khodan, V. V. Dolbinina and E. P. Efremova, *Solid State Ionics*, 2005, **176**(39–40), 2871–2874.
33 E. Ortiz, R. A. Vargas, B. E. Mellander and A. Lunden, *Pol. J. Chem.*, 1997, **71**(12), 1797–1802.
34 K. Gesi, *J. Phys. Soc. Jpn.*, 1992, **61**(1), 162–167.
35 J. Otomo, T. Tamaki, S. Nishida, S. Q. Wang, M. Ogura, T. Kobayashi, C. J. Wen, H. Nagamoto and H. Takahashi, *J. Appl. Electrochem.*, 2005, **35**(9), 865–870.
36 T. Uda and S. M. Haile, *Electrochem. Solid-State Lett.*, 2005, **8**(5), A245–A246.
37 H. Gharibi, R. A. Mirzaie, E. Shams, M. Zhiani and M. Khairmand, *J. Power Sources*, 2005, **139**(1–2), 61–66.
38 T. Uda, D. A. Boysen, C. R. I. Chisholm and S. M. Haile, *Electrochem. Solid-State Lett.*, 2006, **9**(6), A261–A264.
39 B. A. Peppley, J. C. Amphlett, L. M. Kearns and R. F. Mann, *Appl. Catal., A*, 1999, **179**(1–2), 21–29.
40 C. R. I. Chisholm and S. M. Haile, *Chem. Mater.*, 2006, submitted.
41 Z. Jirak, M. Dlouha, S. Vratislav, A. M. Balagurov, A. I. Beskrovnyi, V. I. Gordelii, I. D. Datt and L. A. Shuvalov, *Phys. Status Solidi A*, 1987, **100**(2), K117–K122.
42 C. R. I. Chisholm, California Institute of Technology, Pasadena, CA, 2002, http://resolver.caltech.edu/CaltechETD:etd-01292003-150309.
43 H. Matsunaga, K. Itoh and E. Nakamura, *J. Phys. Soc. Jpn.*, 1980, **48**(6), 2011–2014.
44 R. J. Nelmes and R. N. P. Choudhary, *Solid State Commun.*, 1978, **26**(11), 823–826.
45 K. Imai, *J. Phys. Soc. Jpn.*, 1983, **52**(11), 3960–3965.
46 C. R. I. Chisholm, R. B. Merle, D. A. Boysen and S. M. Haile, *Chem. Mater.*, 2002, **14**(9), 3889–3893.
47 J. D. Bernal and R. H. Fowler, *J. Chem. Phys.*, 1933, **1**(8), 515–548.
48 L. Pauling, *J. Am. Chem. Soc.*, 1935, **57**, 2680–2684.
49 W. F. Giauque and M. F. Ashley, *Phys. Rev.*, 1933, **43**(1), 81–82.
50 A. V. Belushkin, W. I. F. David, R. M. Ibberson and L. A. Shuvalov, *Acta Crystallogr., Sect. B*, 1991, **47**, 161–166.
51 B. V. Merinov, *Crystallogr. Rep.*, 1997, **42**(6), 906–917.
52 B. V. Merinov, A. I. Baranov, L. A. Shuvalov and B. A. Maksimov, *Sov. Phys. Crystallogr.*, 1987, **32**(1), 47–57.
53 Y. Yoshida, Y. Matsuo and S. Ikehata, *J. Phys. Soc. Jpn.*, 2003, **72**(6), 1590–1591.
54 R. H. Flowers, R. J. Gillespie, E. A. Robinson and C. Solomon, *J. Chem. Soc.*, 1960, 4327–4339.
55 J. S. Bass, R. J. Gillespie and E. A. Robinson, *J. Chem. Soc.*, 1960, 821–836.
56 E. Kanda, M. Yoshizawa, T. Yamakami and T. Fujimura, *J. Phys. C: Solid State Phys.*, 1982, **15**(33), 6823–6831.

Co-doping of scandia–zirconia electrolytes for SOFCs

John T. S. Irvine,[a] Jeremy W. L. Dobson,[a] Tatiana Politova,[a] Susana García Martín[b] and Atef Shenouda[a]

Received 27th March 2006, Accepted 9th May 2006
First published as an Advance Article on the web 20th July 2006
DOI: 10.1039/b604441g

Scandia stabilised zirconias offer much better electrical performance than conventional yttria stabilised materials; however, the limited availability and high cost of scandia have generally limited interest in its application in fuel cells. Political and economic changes over the last decade have significantly enhanced scandia's availability, rendering it worth considering for commercial application, even though there is still some uncertainty about its ultimate market price. A small addition of 2 mol% yttria to scandia stabilised zirconia results in stabilisation of the cubic phase and so avoids the major phase changes that occur on thermal cycling of scandia substituted zirconias, which might be expected to be detrimental to long term electrolyte stability. This addition of yttria does slightly impair the electrical conductivity of the scandia stabilised zirconia, although this can be reversed by further addition of ceria. Samples which are cubic throughout the studied temperature range basically show two linear conductivity regions in Arrhenius conductivity plots. A key observation is that the low temperature activation energy decreases and the high temperature activation energy increases as yttrium content increases and scandium content decreases. This correlates with the strength of short-range order as indicated by neutron and electron diffraction studies. Although scandia substitution increases conductivity and decreases high temperature activation energy, it also increases the tendency to short-range ordering at lower temperatures, resulting in a significant increase in activation energy for conduction. This is attributed to the ionic size of the Sc ion which favours a lower coordination number than that associated with ideal fluorite phases. It should also be realised that Zr, which has a similar size to Sc, also prefers a lower coordination number than is ideal for fluorite hence driving the tendency for short-range order in zirconia fluorites.

Introduction

Solid oxide fuel cells offer clean conversion of chemical to electrical energy at high efficiencies. Current developments are mainly based upon an yttria stabilised zirconia electrolyte, which functions very well at temperatures in the 850–1000 °C range for unsupported electrolytes and at temperatures as low as 700 °C for supported thin films. Two design concepts predominate, the more expensive tubular

[a] *University of St Andrews, School of Chemistry, St Andrews, Scotland, UK KY16 9ST*
[b] *Universidad Complutense Departamento de Química Inorgánica, Facultad de Ciencias Químicas, 28040, Madrid, Spain*

design and the simpler planar design. The major weakness of the planar concept relates to interconnect and sealing problems. If a lower temperature electrolyte could be achieved, then much cheaper materials could be utilised for interconnects (steel, in fact) and the cost effectiveness of the planar design, in particular, could be greatly enhanced. One other concern about the electrolyte that should be highlighted is stability; the tetragonal modification of yttria stabilised zirconia undergoes a catastrophic transformation under hydrothermal conditions at about 300 °C.[1] Most commercial cubic zirconias are actually prepared with the composition 8 mol% Y_2O_3/92% ZrO_2, which strictly is just inside the two-phase cubic/tetragonal zirconia phase field.[2] This means that on ageing these electrolytes at fuel cell operating conditions, tetragonal precipitates occur reducing conductivity.[3,4] Furthermore, on cycling this electrolyte between room temperature and operating temperature in the presence of water (a product of fuel cell operation) degradation and failure are highly likely due to the presence of the tetragonal form. Any compositional inhomogeneity means that failure will occur. These problems can be reduced or even avoided if slightly higher yttria compositions are utilised.[4]

Scandia stabilised zirconia offers much higher conductivity than yttria stabilised zirconia at 1000 °C (×3) with a larger enhancement at lower temperature due to the lower activation energy (0.65 vs. 0.95 eV);[5] however, it has not been viewed as a serious alternative due to cost until recently. This has now changed with the increase in availability of scandia from Russia and China; presently, scandia is perhaps only 4 times as expensive per gram or approaching 2 times as costly per mole as yttria and it does seem that prices could decrease further. Although scandia stabilised zirconias could be viewed as offering a route to even lower temperatures of operation than can be achieved using supported yttria stabilised thin film electrolyte designs, we see the primary advantages of scandia systems as a dramatic decrease in internal cell resistance and offering more robust designs for low temperature operation. It is widely recognised in the field that once a low enough temperature of operation has been achieved to allow low cost steel interconnects to be utilised and sealing problems to be minimised, then there is no advantage for a further decrease in operating temperatures as most design concepts benefit from high temperatures, e.g. SOFC–gas turbine, and at these temperatures the internal reforming of hydrocarbons remains an attractive possibility.

Here we seek to improve the stability of scandia stabilised zirconias by co-doping, attempting to minimise any loss in conductivity caused by the co-doping by judicious choice of co-dopant.

Experimental

Stoichiometric amounts of Sc_2O_3, Y_2O_3, and ZrO_2 and in some instances CeO_2 were mixed and ground together to prepare samples of composition $(CeO_2)_x$ $(Y_2O_3)_y(Sc_2O_3)_z(ZrO_2)_{1-x-y-z}$. Dried powders were pressed into pellets of diameter 13 mm and thickness 2–2.5 mm at 100–250 MPa. The pressed pellets were sintered at 1500–1600 °C for 16 h and then quenched to room temperature from 1000 °C. The relative densities were calculated from pellet geometry and mass and were in the range of 88–92%. A Stoe Stadi P X-ray diffractometer (10–90° 2θ, step size 0.02° 2θ, Cu Kα radiation) was used to determine crystalline structures and unit cell parameters. Ionic conductivities were measured using a Solartron 1260 frequency response analyser in the frequency range 6 MHz to 0.1 Hz at temperatures 300–1000 °C with Pt paste electrodes.

For neutron powder diffraction, experiments were performed at Studsvik on the instrument NPD. Data were collected at room temperature in runs averaging four hours. This was long enough to allow the maximum peak to reach ca. 10 000 counts. All data were refined using the FullPROF Rietveld refinement program.

For transmission electron microscopy the samples were ground in n-butyl alcohol and ultrasonically dispersed. A few drops of the resulting suspension were deposited in a carbon-coated grid. SAED studies were performed with an electron microscope JEOL 2000FX (double tilt ±45°) working at 200 kV.

EXAFS experiments were conducted at Daresbury laboratories, using the instruments 9.3, 8.1 and 7.1. All samples were ground as finely as possible using a steel percussion mortar and an agate pestle and mortar. For transmission experiments on instrument 9.3 the samples were intimately mixed in a ratio of 10 : 1 with a spectroscopically pure graphite binder and pressed into 13 mm discs using a uniaxial pellet press. For experiments on the instruments 7.1 and 8.1 samples were held as loose powder in metal holders using Kapton tape. For transmission experiments the samples were cut 10 : 1 with graphite but undiluted samples were used for reflectance geometry experiments. Data handling was performed using the Daresbury suite of software, namely Ecabs or Excalib for normalising and summing data, Exback and Exbrook (within Ecabs) for background subtraction. Refinements were performed using the Excurv98[6] software running on the Daresbury computer *xrsserv1*.

Ionic conductivity measurements

Scandia stabilised zirconia offers lower temperatures of SOFC operation than can be achieved using yttria stabilised zirconia especially at intermediate temperatures such as 700 °C. Compositions with Sc_2O_3 content from 8 to 12 mol% are members of a fluorite-type cubic solid solution. The best ionic conductivity was observed for the compositions with 10 and 11 mol% Sc_2O_3, which also exhibited the least degree of degradation on annealing at 800 °C for 1500 h. These compositions can readily be quenched with the cubic structure; however, there is a tendency to transform to the thermodynamically stable lower temperature rhombohedral (high Sc) or tetragonal (low Sc) forms.[7,8] This cubic to rhombohedral transformation manifests itself in conductivity measurements as a step that becomes more likely to be observed on repeated cycling,[9] see Fig. 1.

Co-doping of scandia zirconia with yttria can achieve stabilisation of the cubic phase.[10–12] Compositions in the ternary system $(Y_2O_3)_x(Sc_2O_3)_{(11-x)}(ZrO_2)_{89}$ (YxSc11 − xZr89, x = 0–11) have been investigated, with yttria addition being found to improve the phase stability of scandia stabilised zirconia. Even 1 mol% Y_2O_3 addition eliminates the rhombohedral phase ($Sc_2Zr_7O_{17}$, the beta-phase) and stabilises the cubic structure at room temperature as evidenced by the step change observed in conductivity, Fig. 1.

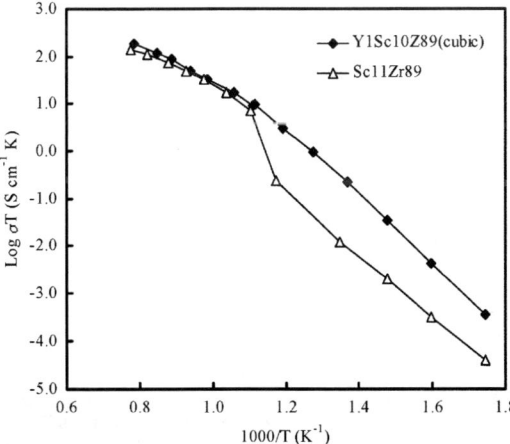

Fig. 1 Temperature dependences of conductivity of the Sc11Zr89 and Y1Sc10Zr89 compositions sintered at 1600 °C for 16 h.

Although the best overall ionic conductivity is observed for quenched yttria free compositions, the best conductivity over all temperatures for cycled samples is observed for samples containing 1 or 2 mol% Y_2O_3 composition. For higher yttria contents conductivity decreases significantly. Except where there is evidence of a phase change in the average crystal structure, two linear regions are observed in the Arrhenius conductivity plots joined by a curved region over a limited temperature range. In scandia rich compositions, the lower temperature region exhibited an activation energy of 1.3 eV and the high temperature region 0.65 eV. The difference between these activation energies in the wider series $(Y_2O_3)_x(Sc_2O_3)_{(11-y)}$ $(ZrO_2)_{100-x-y}$ decreases as yttria content decreases, as shown in Fig. 2 which presents activation energies as a function of Y : Sc ratio.

On close inspection of Fig. 2 it can be seen that increasing the proportion of scandia relative to the total yttria and scandia content increases the low temperature activation energy but does not greatly affect the high temperature activation energy in these Sc rich compositions. On comparison with scandia free compositions, such as 8 mol% yttria stabilised zirconia, which has a high temperature activation energy of 0.93 eV and a low temperature activation energy of 1.08 eV,[13] it seems that yttria substitution in zirconia yields a lower activation energy at low temperatures and a higher activation energy at high temperatures than scandia substitution. Furthermore a small degree of replacement of zirconia by ceria seems to decrease activation energy in both high and low temperature regimes, Fig. 2, and so additional Ce substitution can be seen to compensate for any loss in conductivity caused by partial replacement of Sc by Y in zirconias.

The widely accepted model for this change in activation energy in zirconia system conductivities on heating is that defect association breaks down and hence activation energy decreases.[14] This association may be interpreted as electrostatic and/or strain driven. Whilst the physical origins of this model clearly play an important role in determining activation energies in these systems, this model is not consistent with the trends in activation energy observed here. Scandia is better matched to the zirconia host lattice than yttria and would be anticipated to have a smaller association energy than yttria, which is the opposite to that observed. The high temperature activation energy is lower in the presence of scandia than yttria, which seems to indicate that the strain and/or electrostatically induced association is still present and that this may only break down at very much higher temperatures. The high activation energy observed at low temperatures is related to an additional factor, short-range order of defects, something that is strongly supported by various structural studies as described below. Thus the following model may be considered to explain the

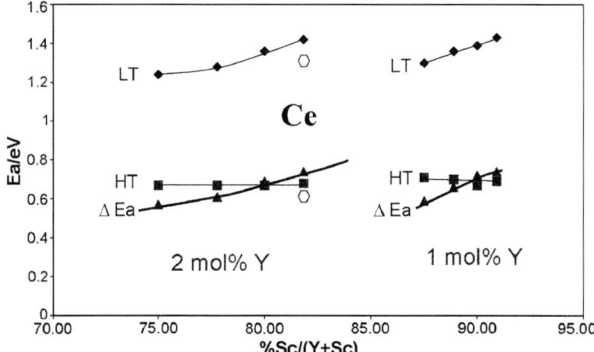

Fig. 2 Comparison of activation energies for ionic conduction plotted as a function of the ratio of Y to Y + Sc in the series $(Y_2O_3)_x(Sc_2O_3)_{(11-y)}(ZrO_2)_{100-x-y}$ for $x = 1, 2$ and $x + y = 8, 9, 10, 11$. Activation energies are shown for both temperatures above 700 °C (HT) and below 650 °C (LT) as well as the difference between these values (ΔEa). Also shown for comparison are values for a typical cerium-doped composition $(CeO_2)(Y_2O_3)_2(Sc_2O_3)_9(ZrO_2)_{88}$ ()

observed conductivity transition in zirconia fluorites, where E_a relates to the overall activation energy, with the following contributions to the activation energy: E_m—migration, E_{assoc}—defect association and E_{order}—short-range order of defects associates.

$$\text{Below} \sim 650 \quad E_a = (E_m + E_{assoc}) + E_{order} \tag{1}$$

$$\text{Above} \sim 700 \quad E_a = (E_m + E_{assoc}) \tag{2}$$

Structural studies

The clearest evidence for the influence of short-range ordering of defects upon this conductivity transition comes from *in situ* neutron diffraction studies of yttria stabilised zirconia where diffuse peaks relating to short-range ordering disappear at the same temperature as the transition in activation energy.[13] Diffuse scattering or a modulated diffuse background are characteristic features of both neutron and electron diffraction patterns obtained for zirconia based compounds stabilised in the cubic defect fluorite phase. The short-range order itself has been well studied using electron diffraction techniques, where the presence of an ordered oxygen sublattice is well documented as having resemblance to tetragonal zirconia,[15] C-type structure[16,17] and distorted pyrochlore,[16,18] depending upon composition, *e.g.* Fig. 3. Usually the average structure, as evidenced by X-ray powder diffraction, is cubic fluorite for all of these systems.

Introduction of scandia into the lattice in place of zirconia might be expected to cause less short-range ordering; however the opposite seems true. The diffuse peaks are much stronger in scandia containing samples, Fig. 4.

These strong diffuse features are entirely consistent with the higher activation energy observed in scandia containing compositions. Two types of scattering are observed: narrow peaks arising from the fluorite Bragg reflections and broad peaks that were observed at some forbidden lattice positions, *i.e.* hkl: $h + k, h + l, k + l = 2n$; $0kl$: $k, l = 2n$; hhl: $h + 1 = 2n$ and $h00$: $h = 2n$. The diffuse scattering is hence likely to arise from localised tetragonal distortions. The magnitude of diffuse scattering also increases with increasing scandium dopant content.

The unit cell parameters derived from the average fluorite structure change linearly with Sc-content, Fig. 5; however the temperature factors also show strong evidence of short-range order. As has been previously discussed[19,20] the temperature factors in an average structure compromise of two types of displacement, the thermal displacement and the static displacement due to atoms not being centred on their perfect crystallographic site. In these fluorite systems, the static displacement largely originates from short-range order of the sublattices, except at very low temperatures.[19] In previous studies, without scandia, the oxygen sublattice exhibited a much larger static displacement than the cation sublattice; however, in the presence

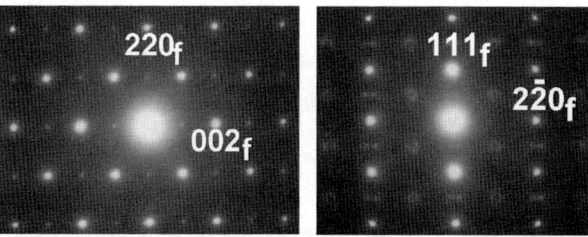

Fig. 3 Electron diffraction patterns of left $(Y_2O_3)_{0.08}(ZrO_2)_{0.92}$ showing diffuse tetragonal scattering peaks and right $(Y_2O_3)_{0.41}(ZrO_2)_{0.37}(Nb_2O_5)_{0.12}$ showing distorted pyrochlore type diffuse peaks.

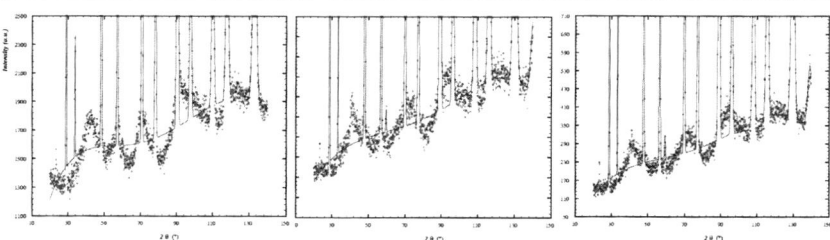

Fig. 4 Neutron diffraction patterns for from left to right, $(Sc_2O_3)_{0.05}(Y_2O_3)_{0.05}(ZrO_2)_{0.9}$, $(Sc_2O_3)_{0.075}(Y_2O_3)_{0.025}(ZrO_2)_{0.9}$ and $(Y_2O_3)_{0.1}$ $(ZrO_2)_{0.9}$ with baseline showing a clear modulation. The black lines are the Rietveld fit and the grey points are data.

of scandia, both cation and anion sublattices exhibit large static contributions. The isotropic temperature factors obtained for both anion and cation sites are presented in Fig. 6. The oxygen ITF values are very large and fairly independent of Sc content, except for the Y-free composition which has significantly higher values as might be expected considering the tendency to transform from cubic in the Y-free compositions. In all compositions, therefore, the oxygen sublattice exhibits considerable static displacement associated with short-range ordering of the anion sublattice. The cation sublattice exhibits much smaller static displacement but there is a clear increase in ITF with degree of Sc-substitution, thus whilst the cations in yttria stabilised zirconia are essentially centred on the their ideal positions, in the presence of scandium at least some cations are significantly displaced.

Coordination

The foregoing sections all point to the Sc^{3+} ion giving rise to much stronger short-range order than Y^{3+} ions, despite the scandium ion being much better matched to the host zirconia lattice in terms of size. In the high temperature region where short-range order has broken down, it was observed that scandia substitution decreases activation energy, as would be expected for a better match of dopant to lattice. It is therefore important to consider the atomic environments of the cations in the lattice. We have therefore considered these materials using the EXAFS technique comparing with previous literature and directly investigating.

Early EXAFS work on the YSZ system by Catlow et al.[21] indicated that the oxygen vacancies were located preferentially next to the zirconium ions at room temperature, with size effects being dominant. This appeared to be counterintuitive with respect to previous assumptions that the vacancies, [with an effective positive

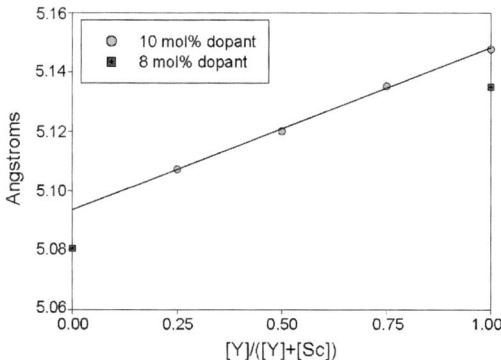

Fig. 5 Unit cell parameter as a function of composition for differing total levels of yttria and scandia doping for $(Sc_2O_3)_{0.1-y}(Y_2O_3)_y(ZrO_2)_{1.9}$ and $(Sc_2O_3)_{0.08-y}(Y_2O_3)_y(ZrO_2)_{1.92}$.

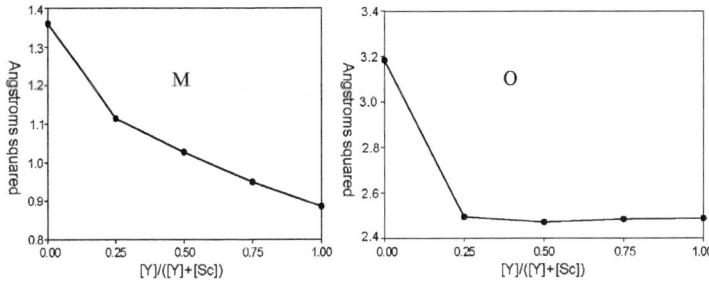

Fig. 6 Isotropic temperature parameters, B_{iso} as functions of composition for different total levels of yttria and scandia doping, left cation ITF and right oxygen ITF for $(Sc_2O_3)_{0.1-y}(Y_2O_3)_y(ZrO_2)_{1.9}$.

charge ($V_O^{\bullet\bullet}$)], would be preferentially sited adjacent to the Y^{3+} rather than the Zr^{4+}, in order to minimise charge density variations. Further work on the subject by Tuilier *et al.*[22] cautiously contradicted this result, claiming the vacancies *were* preferentially located next to the dopants. This work also pointed to considerable disorder in the anion sublattice, describing a "glass of anions" to account for oxygen displacements.

Other work has also produced similarly conflicting results. Goldman *et al.*,[23] Morikawa *et al.*[24] and Shan *et al.*[25] all considered the vacancies to be located next to the dopant species, whereas Komyoji *et al.*[26] and Veal *et al.*[27] reported the vacancies to be associated with the zirconium ion.

The matter appears to have been settled with the extensive works of Li *et al.*[28–31] They concluded that size was the most important factor determining the siting of vacancies, with vacancies being located next to the zirconium ions when large dopants such as yttrium or gadolinium were employed, but adjacent to the dopant ion for small dopant species such as iron or gallium. This latter point was also made in the computer simulations of Zacate *et al.*[32] whose results indicated that oxygen lattice relaxation was the principal consideration.

Studies on the related system of yttria doped ceria has shown that this too behaves in a similar manner, with vacancies clustering around the cerium and not the larger dopant. Several complex defect structures were proposed by Yamazaki *et al.*[33] for the location of vacancies in doped ceria, which may have analogues in zirconia systems.

Here, a range of yttria stabilised zirconia and scandia stabilised zirconias were investigated looking at the yttrium, scandium and zirconium edges. Various models were investigated and it was found that yttrium tended to have a coordination of 8 and zirconium tended to approach 6, as discussed above. More interesting was that scandium was also found to exhibit a coordination number approaching 6. As EXAFS models coordination environments by approximating the coordination environment as a thin spherical shell, coordination numbers are known to often be unrealistic especially for distorted environments. Thus one should only apply such coordination numbers with a degree of caution. A much more reliable parameter is the metal oxygen bond distance as this can be refined with precision and gives a very reliable indication of coordination, as larger coordination numbers must be associated with larger metal–oxygen distances.

In these studies the Y–O bond distance is always close to 2.30 Å, whereas the Sc–O and Zr–O distances are close to 2.15 Å, Fig. 7. This is also the case for M–O distances in co-doped samples. These results therefore clearly indicate that whilst yttrium has a strong preference for eightfold coordination, scandium and zirconium prefer sixfold coordination. This presents an apparent conflict as the fluorite structure is based upon essentially eightfold coordination, which is why there is

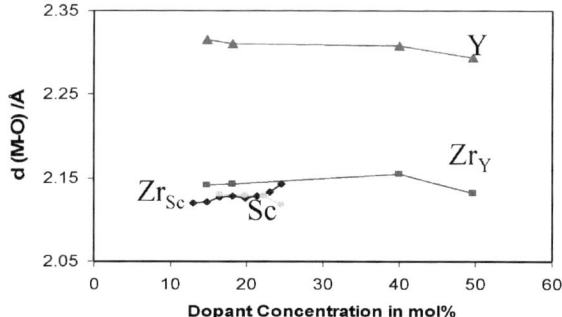

Fig. 7 Metal–oxygen bond distances determined from EXAFS studies of $(Sc_2O_3)_x(ZrO_2)_{1-x}$ and $(Y_2O_3)_y(ZrO_2)_{1-y}$.

such as strong tendency for short-range ordering in zirconias and especially scandia zirconias.

MO_2 stoichiometry does not require eightfold coordination, *e.g.* rutile is sixfold. The preferred structure type is well known to be a function of ion size and it is important to note that PbO_2, which has a larger cation radius than zirconium, does not prefer fluorite coordination. Thus Zr and Sc ions are at the limit of stability in eightfold coordination in oxide systems and will tend to short-range order.

Conclusions

It is argued in this article that the low temperature high activation energy behaviour in zirconia fluorites arises from short-range ordering driven by crystal chemical considerations. Thus at high temperatures in the not too concentrated regimes that are relevant to technological applications, an average fluorite structure with disordered cations and oxide vacancies is prevalent, eqn (3) (note that /4 implies 4 cations coordinated to each O or V_O). At low temperatures the preferred low coordination demands of species such as Zr or Sc drive an ordering of the oxygen vacancies to promote low coordination environments such as in eqn (4) for yttria zirconia. The true situation probably involves ordered and disordered regions in the lattice as presented in eqn (5).

$$MO_{8-y/4} \tag{3}$$

$$[YO_{8/4}]_y[ZrO_{6/4}V_{2/4}]_y[``ZrO_{6/3}'']_{1-2y} \tag{4}$$

$$([YO_{8/4}]_y[ZrO_{6/4}V_{2/4}]_y[``ZrO_{6/3}'']_{1-2y})_{1-\delta}[MO_{8-y/4}])_\delta \tag{5}$$

Acknowledgements

The authors would like to thank EPSRC for financial support (GR/R05772/01, EP/C002601/1 and GR/S62185) and for provision of synchrotron beamtime at Daresbury, the EU SOFC600 integrated project (20089) for support, Studsvik Research Centre for allocation of beamtime on NPD (experiment 396) and Dr Angela Kruth for useful discussions.

References

1   A. Baumard and P. Abelard, *Adv. Ceram.*, 1985, **12**, 555.
2   (*a*) H. G. Scott, *J. Mater. Sci.*, 1975, **10**, 1527; (*b*) I. R. Gibson, PhD thesis, University of Aberdeen, 1995.
3   F. T. Ciacchi, K. M. Crane and S. P. S. Badwal, *Solid State Ionics*, 1994, **74**, 49.

4   I. R. Gibson, G. P. Dransfield and J. T. S. Irvine, *J. Eur. Ceram. Soc.*, 1998, **18**, 661–667.
5   S. P. S. Badwal, *J. Mater. Sci.*, 1987, **22**, 4125.
6   N. Binsted, EXCURV98: CCLRC Daresbury Laboratory computer program, 1998.
7   R. Ruh, H. J. Garrett, R. F. Domagala and V. A. Patel, *J. Am. Ceram. Soc.*, 1977, **60**, 399.
8   R. L. Mangunov, G. L. Shkylyar and V. F. Katridi, *Inorg. Mater.*, 1989, **25**, 1035.
9   T. Politova and J. T. S. Irvine, *Processing and Microstructure-Property Relations in SOFC Components and Ceramic Gas-Separation Membranes*, ed. J. R. Jurado, CSIC, Madrid, 2003, pp. 149–157.
10 T. I. Politova and John T. S. Irvine, *Solid State Ionics*, 2004, **168**, 153–165.
11 F. T. Ciacchi and S. P. S. Badwal, *J. Eur. Ceram. Soc.*, 1991, **7**, 185–195–197–206.
12 S. P. S. Badwal, F. T. Ciacchi, S. Rachendran and J. Drennan, *Solid State Ionics*, 2000, **109**, 167.
13 I. R. Gibson and J. T. S. Irvine, *J. Mater. Chem.*, 1996, **6**, 895–898.
14 (*a*) A. S. Nowick and D. S. Park, in *Superionic Conductors*, ed. G. Mahan and W. Roth, Plenum Press, New York, 1976, pp. 395–412; (*b*) A. S. Nowick, D. Y. Wang, D. S. Park and J. Griffith, in *Fast Ion Transport in Solids*, ed. P. Vashishta, J. N. Mundy and G. K. Shenoy, North Holland, Amsterdam, 1979, pp. 673–679; (*c*) J. A. Kilner and C. D. Waters, *Solid State Ionics*, 1982, **6**, 253–259; (*d*) J. A. Kilner and B. C. H. Steele, in *Nonstoichiometric Oxides*, ed. O. T. Sørensen, Academic Press, New York, 1981pp. 233–269.
15 (*a*) Susana Garcia-Martin, Miguel A. Alario-Franco, Duncan P. Fagg and John T. S. Irvine, *J. Mater. Chem.*, 2005, **15**, 1903–1907; (*b*) S. Suzuki, M. Tanaka and M. Ishigame, *Jpn. J. Appl. Phys.*, 1985, **24**(4), 401; (*c*) V. Lanteri, R. Chaim and A. H. Heuer, *J. Am. Ceram. Soc.*, 1986, **69**(10), C258; (*d*) M. Yashima, S. Saski and M. Kakihana, *Acta Crystallogr., Sect. B*, 1994, **50**, 66.
16 S. García-Martín, M. A. Alario-Franco, D. P. Fagg, A. J. Feighery and J. T. S. Irvine, *Chem. Mater.*, 2000, **12**, 1729–1737.
17 S. Suzuki, M. Tanaka and M. Ishigame, *Jpn. J. Appl. Phys.*, 1985, **24**(4), 401.
18 R. L. Withers, J. G. Thompson, P. J. Barlow and J. C. Barry, *Aust. J. Chem.*, 1992, **45**, 1375.
19 J. T. S. Irvine, A. J. Feighery, D. P. Fagg and S. García-Martín, *Solid State Ionics*, 2000, **136/137**, 879–885.
20 D. N. Argyriou, *J. Appl. Crystallogr.*, 1994, **27**, 155.
21 C. R. A. Catlow, A. V. Chadwick, G. N. Greaves and L. M. Moroney, *J. Am. Ceram. Soc.*, 1986, **69**, 272.
22 M. H. Tuilier, J. Dexpert-Ghys, H. Dexpert and P. Lagarde, *J. Solid State Chem.*, 1987, **69**, 153.
23 A. I. Goldman, E. Canova, Y. H. Kao, W. L. Roth and R. Wang, in *EXAFS and Near Edge Structure III*, ed. K. O. Hodgson, B. Hedman and J. E. Penner-Hahn, Springer Verlag, New York, 1984, p. 442.
24 H. Morikawa, Y. Shimizugawa, F. Marumo, T. Harasawa, H. Ikawa, K. Tohji and Y. Udagawa, *J. Jpn. Ceram. Soc.*, 1988, **96**, 253.
25 Z. J. Shen, T. K. Li, K. Q. Lu and Y. Q. Zhao, *J. Chin. Silic. Soc.*, 1988, **16**, 270.
26 D. Komyoji, A. Yoshiasa, T. Moriga, S. Emura, F. Kanamaru and K. Koto, *Solid State Ionics*, 1992, **50**, 291.
27 B. W. Veal, A. G. McKale, A. P. Paulikas, S. J. Rothman and L. J. Nowicki, *Physica B*, 1988, **150**, 234.
28 P. Li, I.-W. Chen and J. E. Penner-Hahn, *Phys. Rev. B*, 1993, **48**, 10063.
29 P. Li, I.-W. Chen and J. E. Penner-Hahn, *Phys. Rev. B*, 1993, **48**, 10074.
30 P. Li, I.-W. Chen and J. E. Penner-Hahn, *Phys. Rev. B*, 1993, **48**, 10082.
31 P. Li, I.-W. Chen and J. E. Penner-Hahn, *J. Am. Ceram. Soc.*, 1994, **77**, 118.
32 M. O. Zacate, L. Minervini, D. J. Bradfield, R. W. Grimes and K. E. Sickafus, *Solid State Ionics*, 2000, **128**, 243.
33 S. Yamazaki, T. Matsui, T. Ohashi and Y. Arita, *Solid State Ionics*, 2000, **136–137**, 913.

Mass storage in space charge regions of nano-sized systems (Nano-ionics. Part V)†‡

J. Maier

Received 9th March 2006, Accepted 9th May 2006
First published as an Advance Article on the web 3rd August 2006
DOI: 10.1039/b603559k

The consideration of space charge effects at interfaces results in the prediction of considerable storage anomalies in two-phase materials. They are particularly expected in nano-sized systems building a bridge between chemical and electrostatic storage. These effects are systematically treated in the model of an abrupt planar contact. Several examples of fundamental and/or practical interest are analysed in greater detail.

1. Introduction

Previous papers on nano-ionics mainly dealt with questions of thermodynamics of ionic charge carriers and focused on transport properties. This contribution is primarily concerned with the possibility of excess storage in nanocrystalline or nano-composite materials owing to the necessarily occurring space charge effects. The examples considered in ref. 1–3, addressed local non-stoichiometries caused by ion redistribution which did not result in a global excess storage; this contribution explicitly deals with net overall storage (see also ref. 4). To give an example, if a given phase is contacted by a second one (see Fig. 1) which provokes an ionic charge transfer, excess ions are formed in one phase at the expense of a deficiency in the other. Consequently, a local but not a global storage of material is achieved in this way. In an ionic system excess storage of an element—and, in order to be specific, let us consider excess storage of the parent metal M—can only occur on the level of the electronic carrier concentrations. In order for a storage of the neutral component M to occur, an ion excess or deficiency of M^+ must be compensated by an electronic excess or deficiency (excess electron or hole concentration). Naturally, this M excess has to be imported from the ambient. Since in the space charge regions, owing to electrostatics (see also below, eqn (9)), the respective ion (M^+) excess must be locally accompanied by an electron (e^-) deficiency and an M^+ deficiency by an e^- excess, a compensation that leads to a significant M-storage, must occur heterogeneously. A characteristic example was mentioned in ref. 4–7: at the contact of Li_2O and a noble metal (as electron acceptor); Li can be heterogeneously stored in that Li^+ occupies interstitial sites in Li_2O close to the boundary compensated by the e^- sitting on the noble metal surface sites (see Fig. 2).

This paper gives a systematic treatment of the defect chemistry of storage in binary compounds starting with the bulk, then treating semi-infinite boundaries in the

Max-Planck-Institut für Festkörperforschung, Heisenbergstr. 1, D-70569, Stuttgart, Germany.
E-mail: s.weiglein@fkf.mpg.de

† For Part IV see ref. 4.
‡ Dedicated to Professor Hans-Joachim Queisser on the occasion of his 75th birthday.

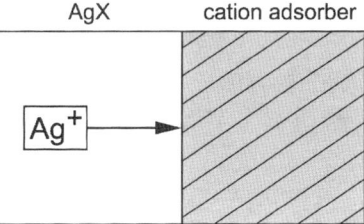

Fig. 1 Silver adsorption to a cation adsorbing second phase, the adsorbed charge being compensated by vacancies.

abrupt planar contact model.[8,9] Subsequently it will deal with the mesoscopic situation. There the number density is extremely high but possible overlap of space charge layers reduces the local charge storage by not allowing the material to internally compensate as much interfacial charge as possible in the semi-infinite system. Thermodynamically, the treatment of storage is related to the understanding of how space charge profiles depend on the component chemical potential (*e.g.* μ_M).[10,11] Such points are addressed in particular along with describing special examples.

2. Bulk stoichiometry

Before we discuss non-stoichiometries at boundary regions let us consider the well understood bulk situation.[12,13] Without significant loss of generality we consider a mixed conductor M^+X^-, neglect disorder in the X-sublattice and assume simple defect chemistry. Hence the defects under concern are metal ion interstitials or vacancies as well as excess electrons and electron holes (*i.e.* in Kröger–Vink notation $M_i^{\cdot}$, V_M', e', $h^{\cdot}$) the molar concentrations (number of moles per volume) of which we abbreviate by i, v, n, p. More complex defect chemistry (X-defects, associates, higher-valent ions) leads to a higher mathematical complexity but not to conceptually different situations, as long as we stick to low concentrations.

The non-stoichiometry δ in $M_{1+\delta}X$ (for a comprehensive treatment of defect chemistry see, *e.g.* ref. 14 and 15) is given by

$$\delta \cdot c^o = i - v = n - p \tag{1}$$

(c^o: concentration of regular M-lattice sites = inverse molar volume of MX ≡ $1/V_{MX}$; δc^o = excess M-concentration ≡ c^{ex}; throughout the paper we will assume monovalent defects.) This can easily be generalised to impure samples by including the concentration of the impurity (*e.g.* N^{2+} substituting for M^+) ion in the definition of c^{ex} (such as $i - v + [N_M^{\cdot}]$, *e.g.* in $((Ag_{1-2x+\delta}Cd_x)Cl)$. The far r.h.s. of eqn (1) follows from electroneutrality. Again without loss of conceptual generality it is sufficient to consider 3 situations: (i) $M_i^{\cdot}$ and V_M' are majority defects, *i.e.* $i \simeq v \simeq$

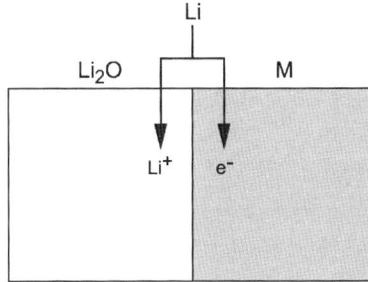

Fig. 2 Excess lithium storage in a composite of a phase that dissolves Li^+ and a phase that accommodates e^-.

$K_F^{1/2}$ (K_F: mass action constant of the Frenkel reaction), (ii) both ionic defects are minority effects, $n \simeq p \simeq K_B^{1/2}$ (K_B: mass action constant for the electron transfer from valence to conduction band), (iii) one ionic charge carrier (we take $M_i^{\cdot}$) and one electronic charge carrier (consequently e') are majority defects, *i.e.* $i \simeq n$. As an increased storage of M is achieved by increasing the chemical potential of M, the dependence of δ on μ_M is of particular importance.

In the last case (case (iii))

$$\delta \cdot c^{\circ} \simeq i \simeq n \simeq K_M^{1/2} a_M^{1/2} = K_M^{1/2} \exp \frac{1}{2} \frac{\mu_M - \mu_M^{\circ}}{RT} \qquad (2)$$

(a_M: M-activity defined by $\mu_M \equiv \mu_M^{\circ} + RT \ln a_M$ where $\mu_M^{\circ} = \mu_M$ (pure metal); K_M is the mass action constant of metal incorporation: $M + V_i \leftrightarrows M_i^{\cdot} + e'$).

In the second case (case (ii))

$$\delta \cdot c^{\circ} = i - v = (K_M/K_B^{1/2})a_M - K_F (K_M/K_B^{1/2})^{-1} a_M^{-1}. \qquad (3)$$

As the defect regime concerned also includes the intrinsic point at which δ vanishes exactly, it is convenient to express the r.h.s. by the intrinsic values ($c_{v,i}^* \equiv i^* = v^* = K_F^{1/2}$; a_M^*).[13] Then the well-known titration curve follows:

$$\delta \cdot c^{\circ} = c_{i,v}^* \left[\left(\frac{a_M}{a_M^*}\right) - \left(\frac{a_M}{a_M^*}\right)^{-1} \right] = 2K_F^{1/2} \sinh\left\{\frac{\mu_M - \mu_M^*}{RT}\right\}. \qquad (4)$$

For valencies other than ± 1 the absolute value of the pre-factor of the sinh-argument (*i.e.* the power of (a_M/a_M^*)) deviates from unity.

Lastly, in case (i), $\delta \cdot c^{\circ} = i - v$ the approximation $i \simeq v$ leads to zero, which expresses the small value of δ which is of the order of the minority species (as in case (ii)). A useful solution is obtained by using the identity $i - v = n - p$, and then by proceeding similarly as in the previous case, with the result

$$\delta \cdot c^{\circ} = (K_M/K_F^{1/2})a_M - K_B(K_M/K_F^{1/2})^{-1} a_M^{-1} \qquad (5)$$

or

$$\delta \cdot c^{\circ} = 2c_{n,p}^* \sinh\left\{\frac{\mu_M - \mu_M^*}{RT}\right\} = 2K_B^{1/2} \sinh\left\{\frac{\mu_M - \mu_M^*}{RT}\right\} \qquad (6)$$

where $c_{n,p} \equiv n^* = p^* \equiv p(\delta=0) = n(\delta=0) = K_B^{1/2}$. Again, for valencies other than ± 1, the argument contains a pre-factor (which is different from case (ii)).

3. Stoichiometry in the space charge zones

In the space charge zones all concentrations can severely differ due to electrical fields. Pronounced deviations from the "Dalton composition" (M_1X_1) occur (($M_{1+\delta(x)}X)^{\gamma+}$) at the expense of a local space charge (note that generally γ, which is proportional to the space charge density ($\gamma = \rho/F$), will deviate from δ owing to variations in the electronic concentrations). Let us first consider the local non-stoichiometry $\delta(x)$ for which

$$\delta(x)c^{\circ} = i(x) - v(x) = n(x) - p(x) + \rho/F \qquad (7)$$

(x measures the distance from the interface, F denotes Faraday's constant). The charge density ρ is given according to Poisson's equation

$$\rho = -\varepsilon d^2\phi/dx^2 \equiv -\varepsilon \Delta\phi'' \qquad (8)$$

(ε: dielectric constant, ϕ: electrical potential, $\Delta\phi \equiv \phi - \phi_{\infty}$, ϕ_{∞}: bulk value of ϕ) being a complicated function of the positional coordinate x and the component potential μ_M.

The constancy of the electrochemical potential of the mobile defects (v, i, n, p) ensures that

$$i(x)/i_\infty = p(x)/p_\infty = v_\infty/v(x) = n_\infty/n(x) = \exp -\frac{F\Delta\phi(x)}{RT} \equiv \mathrm{æ}(x) \qquad (9)$$

as long as Boltzmann distribution can be assumed (∞ denotes bulk values). The expression for the local non-stoichiometry reads (dash is indicating the positional derivative)

$$\delta(x)c^\circ = i_\infty \mathrm{æ}(x) - v_\infty/\mathrm{æ}(x) = n_\infty/\mathrm{æ}(x) - p_\infty \mathrm{æ}(x) - (\varepsilon/F)\,\Delta\phi''. \qquad (10)$$

The relation can be very much simplified by considering the three master cases.

In case (iii), where the accumulation or depletion of the ionic majority defect determines δ, we can write

$$\delta(x)c^\circ = K_M^{1/2} a_M^{1/2} \mathrm{æ}(x) = K_M^{1/2} \exp\left\{\frac{(\mu_M - \mu_M^\circ)}{2RT}\right\} \exp -\frac{F\Delta\phi}{RT}. \qquad (11)$$

It is important to realise the magnitude of the effects. Even though case (iii) is the case in which severe bulk non-stoichiometries may occur, $\delta(x)$ is additionally enhanced by $\mathrm{æ}(x)$ (compare eqn (11) and (2)), a factor which can amount to several orders of magnitude. Eqn (11) also describes depletion as long as $0 < \delta < \delta_\infty$.

Case (ii), in which both i and v have to be considered, results in

$$\delta(x)c^\circ = c_{i,v}^* \left[\left(\frac{a_M}{a_M^*}\mathrm{æ}\right) - \left(\frac{a_M}{a_M^*}\mathrm{æ}\right)^{-1}\right] = 2K_F^{1/2} \sinh\left\{\frac{\mu_M - \mu_M^* - F\Delta\phi}{RT}\right\}. \qquad (12)$$

Note again that $\Delta\phi$ is a function of μ_M, and hence eqn (11) and (12) do not explicitly describe δ as a function of μ_M (and T) (see below).

Case (i) requires a different consideration from the previous section, as now $i - v \neq n - p$. Unlike in the bulk case where $i \simeq v$ and $|i - v| \ll i$, v, the non-stoichiometry is—for cases of interest—given by the field effect on the majority level, so that $\delta(x) = i(x) - v(x)$ can be directly evaluated:

$$\delta(x)c^\circ = c_{i,v\infty}[\mathrm{æ}(x) - \mathrm{æ}^{-1}(x)] = 2K_F^{1/2} \sinh -\frac{F\Delta\phi}{RT}. \qquad (13)$$

Because of the severe changes in the space charge regions, the ranking of the carrier concentrations can change and inversion situations may lead to further cases to be considered. Furthermore situations also need to be considered in which mobile or immobile impurities are of importance. Only the latter (*i.e.* the Mott–Schottky case) is of particular interest;[16] there a frozen impurity (let us assume N_M') sets a constant charge density since the counter carrier (let us assume V_M') is depleted. Then obviously

$$\delta(x)c^\circ = m\left(1 - \exp\frac{F\Delta\phi}{RT}\right) \qquad (14)$$

with $\Delta\phi < 0$, $m \equiv [N_M']$ and $\delta \cdot c^\circ \equiv i - v + m \simeq m - v$. In order not to lose sight of our objective we will now turn to the overall excess stoichiometry. There the explicit dependence on x will be exploited (Gouy–Chapman[17] or Mott–Schottky[16] profiles) with the boundary concentration or potential appearing as the parameter whose dependence on control parameters such as μ_M remains to be elucidated.

4. Global phase stoichiometry

4.1. Semi-infinite samples

In (one-dimensionally) inhomogeneous situations, as it is the case if interfacial effects come into play, we replace δ by the arithmetic mean value δ_m (cf. $M_{1+\delta_m}X$). For purely interfacial effects we can split δ_m into the bulk value δ_∞ and the deviation $\Delta\delta_m$ from it, which is due to the boundaries. In the formulation $M_{1+\delta_\infty+\Delta\delta_m}X$, the value of $\Delta\delta_m$ depends on the thickness of MX (even if the boundary profiles decay to a bulk value within the phase). Hence we also introduce the quantities $\langle\delta\rangle_\ell$ and $\langle\Delta\delta\rangle_\ell$ which refer to the mean δ value and the mean excess value within a layer of characteristic thickness ℓ and not to the total sample thickness L. Because of additivity $\langle\Delta\delta\rangle_\ell = \frac{L}{\ell}\Delta\delta_m$ where $\Delta\delta_m c^\circ = \int c^{ex}\,dx/L$ and integration comprises the total sample. (With regard to the calculation of $\Delta\delta_m$ and then the definition of c° and L we have different possibilities. In a composite, e.g. of a matrix phase (see Ag_2Z example 1 (Section 6.1)) and an insulating phase (e.g. Al_2O_3 in example 1) it is sensible to relate c°, L to the Ag_2Z-phase rather than conceiving δ as an average over the two-phase system. Also, integration should include the adsorption layer which we thus attribute to Ag_2Z solely. In the case of the contact of two phases MX and MX′ in both of which δ differs, it is sensible to consider δ and c°, L separately for the two phases, otherwise appropriate mean values are to be used. As the integrated charge density in both phases has to be the same, this point should not cause conceptual problems. Note that the product $\Delta\delta_m L c^\circ$ gives the excess mole number.) Introducing the respective volume fractions (φ), δ_m can obviously be expressed equivalently as $\Delta\delta_m + \delta_\infty$, as $\varphi_\ell\langle\Delta\delta\rangle_\ell + \delta_\infty$ or as $\varphi_\ell\langle\delta\rangle_\ell + \varphi_\infty\delta_\infty$, whereby $\langle\delta\rangle_\ell = \langle\Delta\delta\rangle_\ell + \delta_\infty$ represents the mean non-stoichiometry in the boundary. The weighting factor φ_ℓ is the volume fraction of the layer (here ℓ/L); $\varphi_\infty = 1 - \varphi_\ell$ is the volume fraction of the bulk.

As we refer to quasi-one-dimensional systems, treatment of different cases means integration of Gouy–Chapman or Mott–Schottky profiles which for a variety of cases have been given in the literature (see e.g. ref. 10,15,18–21).

A few cases of relevance will be considered now. In order to refer to interesting situations we presuppose large space charge potentials, i.e. large concentration variations (for simplicity, however, we still use Boltzmann approximation). As in ref. 8 and 22 it is useful to refer to the maximum boundary concentration enhancement for the given space charge potential which occurs at $x = 0$, i.e. in the first MX layer adjacent to the interfacial core:

$$\zeta_{j_0} \equiv c_j\,(x=0)/c_{j\infty} \tag{15}$$

or to the "degree of influence"[8] ($-1 \leq \vartheta \leq 1$)

$$\vartheta_j \equiv \frac{\zeta_{j_0}^{1/2} - 1}{\zeta_{j_0}^{1/2} + 1}. \tag{16}$$

Let us (following ref. 8 and 22) refer to the averaged deviation from bulk stoichiometry as given by

$$\Delta\delta_m c^\circ = \frac{1}{L}\left[\int (c_i - c_{i\infty})\,dx - \int (c_v - c_{v\infty})\,dx\right] \tag{17}$$

As we may face asymmetric boundaries at $x = 0$ and $x = L$, we consider only a single interface even though integrating from 0 to L. The other boundary effect has then to be superimposed appropriately. In order to indicate this, we use the upper index 1. For symmetric boundary conditions (at $x = 0$ and $x = L$) simply

$\Delta\delta_m = 2\Delta\delta_m^1$. For Gouy–Chapman profiles the result is

$$\Delta\delta_m^1 c^\circ = \frac{2(2\lambda)}{L}\left[\frac{\vartheta_i i_\infty}{1-\vartheta_i} - \frac{\vartheta_v v_\infty}{1-\vartheta_v}\right] = \frac{(2\lambda)}{L}\vartheta_i\left[\frac{2i_\infty}{1-\vartheta_i} + \frac{2v_\infty}{1+\vartheta_i}\right]. \tag{18}$$

The Debye length λ is given by

$$\lambda = \sqrt{\frac{\varepsilon RT}{2F^2 c_\infty}} \tag{19}$$

where c_∞ is the bulk majority concentration.

Let us, without loss of generality, consider M_i to be enriched; then the following cases have to be distinguished. If both interstitials and vacancies are majority carriers in the bulk, the first term in eqn (18) predominates and for $\vartheta \to 1$ it yields

$$\Delta\delta_m^1 c^\circ = (2\lambda/L)(i_\infty i_0)^{1/2} = \sqrt{2\varepsilon\varepsilon_0 RTi_0}/(FL) \tag{20}$$

which could have been directly derived from the appropriately simplified profile $c \propto$ (const $+ x)^{-2}$.[15] If both are minority carriers, but $(n_\infty \gg)\ i_\infty > v_\infty$, the same equation is valid, however the r.h.s. rearrangement is not ($\lambda^{-1} \propto n_\infty^{1/2} \neq i_\infty^{1/2}$):

$$\Delta\delta_m^1 c^\circ = (2\lambda/L)(i_\infty i_0)^{1/2} = \sqrt{2\varepsilon\varepsilon_0 RTi_0}\sqrt{i_\infty/n_\infty}/(FL). \tag{21}$$

If vacancies—even though depleted—are present in higher concentrations than interstitials, it follows that

$$\Delta\delta_m^1 \cdot c^\circ = (2\lambda/L)(v_\infty - \sqrt{v_0 v_\infty})$$

$$\simeq (2\lambda/L)v_\infty + \begin{cases} -\sqrt{2\varepsilon\varepsilon_0 RTv_\infty}/(FL) \\ \text{if } V_M' \text{ is majority carrier} \\ -\sqrt{2\varepsilon\varepsilon_0 RTv_\infty}\sqrt{v_\infty/n_\infty}/(FL) \\ \text{if not.} \end{cases} \tag{22}$$

(Rather than correcting v_∞ by $-\sqrt{v_0 v_\infty}$ it may be more important to consider inversion effects due to interstitials leading to the correction term $-\sqrt{i_0 i_\infty}$.)

In Mott–Schottky situations in which the counter-carrier to the dopant (e.g. V_M') is depleted, severe non-stoichiometry effects appear, too. As pointed out in the previous section in such Mott–Schottky layers the dopant (of concentration n) is assumed to be immobile, this means $m' = 0$ if we can ignore frozen-in profiles. For simplicity we again assume the valencies to be ± 1. (If impurities are mobile, Gouy–Chapman profiles are established close to the boundary, and the situations are closely related to the ones described above.[18]) Owing to $\Delta\phi'' \propto \rho \propto m$, $\Delta\phi'(x = \lambda^*) = 0$, $\phi(x=0) = \Delta\phi_0$, $\Delta\phi(x = \lambda^*) = 0$, a parabolic profile is established between $x = 0$ and $x = \lambda^*$ and λ^* given by $\lambda^* = 2\lambda\sqrt{\frac{F}{RT}|\Delta\phi_0|}$ (see e.g. ref. 19). If $[M_i]$ becomes higher than $[V_M']$ in the space charge regions (but lower than m), δ is dominated by i and approximately

$$\Delta\delta_m^1 c^\circ = \frac{1}{L}\frac{1}{v_\infty^{1/2}}\frac{i_0}{[\ln i_0/i_\infty]^{1/2}} = \frac{(\lambda^*/L)i_0}{2\ln(i_0/i_\infty)} = \frac{(\lambda/L)i_0}{(\ln(i_0/i_\infty))^{1/2}} \tag{23}$$

(As $m' = 0$, differences in the definition of δ between pure and impure samples nullify in $\Delta\delta_m$.) In the derivation of eqn (23) the quadratic term in the electrical potential profile has been neglected. This is justified as the depletion effects are most pronounced for $x \ll \lambda^*$.[18,19] If i is lower than v even in the space charge regions, then $\Delta\delta_m$ is dominated by the vacancy depletion with the result

$$\Delta\delta_m^1 c^\circ = \frac{(\lambda^*/L)m^2}{2v_0 \ln(v_\infty/v_0)} = \frac{\lambda}{L}\frac{m^2}{v_0}\frac{1}{(\ln(v_\infty/v_0))^{1/2}} \tag{24}$$

If on the other hand i is so greatly enhanced that it surpasses the impurity level, we face a mixed situation that, owing to the steep profiles, may be described by a Gouy–Chapman solution.[18,23]

In all these relations the remaining parameter is the defect concentration value for $x = 0$, whose dependence on the control parameters (see μ_M) has to be elucidated (see below).

4.2. Finite (mesoscopic) samples

Now we discuss the relations for mesoscopic situations in which the films (particles) are so thin that a bulk value is not attained anywhere in the sample. Notwithstanding this, eqn (11)–(13) are still valid. In order to verify this, one best figures a contact of the mesoscopic phase with the extended phase. Given sufficient mobility of the M-defect to guarantee zero gradients $\tilde{\mu}'_i = 0 = -\tilde{\mu}'_v = \tilde{\mu}'_{M^+}$, and given sufficient immobility of the X-defect to warrant a metastable contact, these equations follow from the (spatial) equilibrium conditions.

The simplest situation is met for Gouy–Chapman conditions, in which $M_i^·$ and V'_M are mobile majority defects (we follow ref. 10 and 22). If we again assume $M_i^·$ to be the enriched defect, first integration of Poisson's equation demands

$$\Delta \delta_m c^o \propto 2\varepsilon \phi'_{x=0}. \tag{25}$$

As in the mesoscopic case, ϕ' disappears in the sample center (symmetric boundary conditions assumed). The boundary field has been given in ref. 21. The result is

$$\Delta \delta_m c^o = \frac{(4\lambda)}{L} c_\infty \left[\left(\frac{i_0}{i_\infty} + \frac{i_\infty}{i_0} \right) - \left(\frac{i_{1/2}}{i_\infty} + \frac{i_\infty}{i_{1/2}} \right) \right]^{1/2}. \tag{26}$$

with i_∞ as the value in the bulk of the extended phase and $i_{1/2}$ the value in the sample centre. Because of symmetry, it is sensible here to include both space charge regions in $\Delta \delta_m$ and φ (i.e. $\Delta \delta_m = 2\Delta \delta_m^1$). This is automatically ensured by integrating over half the sample.

As we consider large effects we can further neglect each second term in both brackets leading to

$$\Delta \delta_m c^o = 2 \left[2RT\varepsilon \, (i_0 - i_{1/2}) \right]^{1/2} / (FL). \tag{27}$$

The parameter $i_{1/2}$ is correlated with i_0, i_∞ and λ, L via elliptical integrals of the first kind

$$L/(2\lambda) = 2\sqrt{i_\infty/i_{1/2}} \times [E(i_\infty/i_{1/2}; \pi/2) - E(i_\infty/i_{1/2}; \arcsin \sqrt{i_{1/2}/i_0})] \tag{28}$$

where $E(k; \chi) \equiv \int_0^\chi d\alpha \, (1 - k^2 \sin^2 \alpha)^{-1/2}$. Eqn (28) implies that $i_{1/2} \to i_0$ for $L \to 0$.

As the influence of V'_M has disappeared, this relation also holds for cases in which $i_\infty > v_\infty$. Other cases (both i, v not majority carriers, or depletion of the majority M-defect and accumulation of minority M-defect) have to be evaluated separately.

If we deal with mesoscopic Mott–Schottky layers (charge density given by a constant impurity content, m, the counter-carrier being depleted) the conditions $\Delta\phi'' \propto \rho = Fm$, $\Delta\phi(x=0) = \Delta\phi(x=L) = \Delta\phi_0$, $\Delta\phi'(x=L/2) = 0$ lead to the simple result

$$\Delta\phi = -\frac{mF}{2\varepsilon} \left[x^2 - Lx - \Delta\phi_0 \frac{2\varepsilon}{Fm} \right] = -\frac{mF}{2\varepsilon} [x^2 - Lx + \lambda^{*2}]. \tag{29}$$

As before, we assume an impurity with valence $+1$ and metal ion vacancies (V'_M) as the defects that compensate the impurities in the bulk; near the boundary they are depleted.

If the accumulated interstitials are of no influence for the non-stoichiometry, the result is (integrating from $x = 0$ to $L/2$)

$$\int v \, dx \simeq \frac{4\lambda^2}{L} v_0 \left\{ \exp\left(\frac{L}{4\lambda}\right)^2 \left[1 + \left(\frac{L}{4\lambda}\right)^2\right] - 1 \right\}. \tag{30}$$

For the solution the quadratic term of the parabolic potential has been neglected between $x = 0$ and $x = L/4$ (note that the approximated $\Delta\phi$ intersects the precise $\Delta\phi$ at $x = L/4$) while between $x = L/4$ and $x = L/2$ the value at $x = L/2$ has been taken. If also the linear term is neglected, the result is

$$\int v \, dx = v_0 L/2. \tag{31}$$

If the accumulated interstitials become more important for the non-stoichiometry than the vacancies but a Mott–Schottky solution ($\rho \propto m$) is still retained, it follows that

$$\int i \, dx = \frac{4\lambda^2}{L} i_0 \left\{ 1 - \exp - \left(\frac{L}{4\lambda}\right)^2 \left(1 - \left(\frac{L}{4\lambda}\right)^2\right) \right\} \simeq \frac{i_0 L}{2}. \tag{32}$$

Again the far r.h.s. corresponds to a neglect of both quadratic and linear terms in $\Delta\phi$, or neglect of quadratic terms in the development of the l.h.s, which is the same. In the accumulated case we expect the same half which is close to the boundary to be dominant ($i = i_0$), while in the depleted case the centre part (between $L/4$ and $3L/4$) is expected to be predominant ($v = v_{1/2}$). The fact that in both cases the boundary values i_0, v_0 appear, is due to the fact that $v_{1/2} \to v_0$ in this extreme limit,[24] as is also the case for Gouy–Chapman profiles. As pointed out by Jamnik[25] the thickness dependencies of $\int c \, dx$ in Mott–Schottky and Gouy–Chapman cases can be strikingly different.

5. Extra storage at boundaries

In ref. 10 a variety of examples are considered in which ionic space charge effects occur, hence local non-stoichiometries and also global non-stoichiometries as far as a given phase is concerned. Let us repeat the major point and consider the contact AgCl/AgBr or the contact AgCl/Al$_2$O$_3$. In the first case a net silver ion transfer from one phase to the other occurs and in the second an internal silver ion adsorption to the Al$_2$O$_3$ surface (see Fig. 3). Hence in both cases there is no overall storage if we perform an integration of the excess Ag$^+$ concentration over the total two phase system ($\int \Delta\delta \, dx = 0$). What is missing in one phase can be found in the other. This can be different, if the Ag$^+$ excess is compensated by an ionic (Y$^-$) excess in the counter-phase (see Fig. 4). Most interesting is the case Y$^- \equiv e^-$, which results in elemental excess non-stoichiometries and directly connects with the question of the dependence of carrier concentrations on the chemical potential of the respective

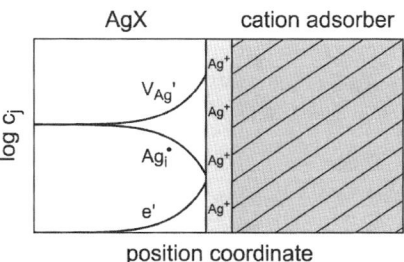

Fig. 3   Defect concentrations at the contact of a silver halide to an Ag$^+$-adsorbing second phase. There is no significant overall Ag storage.

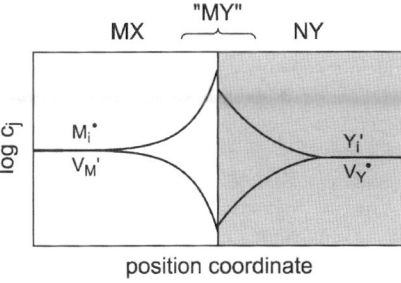

Fig. 4 Storage of MY at the interface of a phase MX and a neighbouring phase dissolving Y^-.

component. Some principal points have already been addressed in ref. 15 in the context of transport (Kröger–Vink diagram of boundary regions[11,26]). Here, however, we will focus on storage directly and consider cases in which distinct changes in the electron concentration occur in one of the phases compensating the defect metal ion charges in the other, a phenomenon that is expected to be of considerable practical interest. The major difficulty is the understanding of the boundary concentration ($x = 0$) or the integrated charge density as a function of the chemical potential of the element (*e.g.* dependence of oxygen or hydrogen storage on P_{O_2} or P_{H_2} or the storage of Li as a function of the cell voltage in a Li battery). For the sake of clarity we will perform the more detailed discussion along with experimental examples.

The storage capacity is primarily a chemical capacity (C_δ).[27,28] An appropriate differential definition is *via* $d\delta/\delta\mu \propto (d\ln\delta/d\ln a)\delta$ whereby the inverse of the bracketed factor is a simple integer in many cases (see N in example 4 (Section 6.4)). For practical purposes by storage capacity the integral quantity $\bar{C}_\delta = \int C_\delta d\mu / \int d\mu$ is meant. In most cases $\int d\mu$ covers a certain constant range, and C_δ is simply identified with $\Delta\delta$ itself. Note that *e.g.* in Li-batteries μ_{Li} is determined by the cell voltage which directly relates C_δ to an electrical storage quantity.

Even though an investigation of the dependence of the defect concentrations and hence of δ on the component potential suffers from a lack of knowledge with respect to interfacial chemistry, a few remarks are to be made (see also Section 6.3). For this purpose we refer to oxides and the dependence on P_{O_2}. There are two points that make the P_{O_2} dependence different from the bulk situation. One refers to the fact that both mean defect concentration in the boundary and the proportion of the boundary zones are P_{O_2} dependent resulting in a generally different N-value ($N \equiv \partial \ln c/\partial \ln P$) from the bulk value ($N_\infty$) even if the N value at $x = 0$ (N_0) does not differ from N_∞. The second reason stems from the fact that many situations demand N_0 to differ from the bulk value. Let us consider both points (see also ref. 26). If the profile is of Gouy–Chapman type and the defect is accumulated, the integral excess charge is proportional to $\lambda\sqrt{c_0 c_\infty}$ which simplifies to $\text{const}\sqrt{c_0}$ if the defect is a majority carrier in the bulk and to $\sqrt{c_0 c_\infty/c_\infty^*}$ if not (c_∞^* characterising the bulk majority level). The overall excess concentration depends on P_{O_2} with $N_{eff} = N_0/2$ or $N_0/2 + (N_\infty - N_\infty^*)/2$ differing from N_∞ even if $N_0 = N_\infty$. Similarly for a depletion effect the excess overall concentration is $\propto \lambda c_\infty$ and hence N_{eff} ($N_\infty/2$ or $N_\infty - N_\infty^*/2$, respectively) differs from N_∞.

Understanding of N_0 requires a detailed knowledge of the boundary defect chemistry. Simple conditions for which $N_0 \neq N_\infty$ or $N_0 = N_\infty$ are discussed in ref. 10 and 15. An important case is the following: If concentration changes introduced by P-change are small against the absolute c-value, then simply $N \simeq 0$. There are situations in which this is not fulfilled in the bulk (*e.g.* $N_\infty = \pm1/6$ in native, non-stoichiometric oxides), while well fulfilled in the boundary zones simply owing to accumulation ($N_0 \simeq 0$). A more detailed discussion was given for ionic and electronic defect concentrations in AgCl as a function of μ_{Ag} or P_{Cl_2}.[11]

In the case of storage in the abrupt model bulk-like properties up to the contact ($\nabla \mu^\circ = 0$) have been assumed. However, close to the interface μ° ("energy level") may also be different from the bulk value. Among others, possible reasons can be elastic fields due to mismatch. (Moreover, in nano-crystals μ° may differ from the value for a macroscopic sample everywhere in the crystallite due to the capillary pressure.) In these cases excess storage phenomena may also occur without space charges requiring a more individual treatment. The point that at the boundary core ($x = s$) μ° is different from the value in the material, which is fulfilled in any of these models and caused the excess charges in the above treatment, can also lead to more or less neutral storage; gas storage *via* storage by physi- or chemisorption belongs to this category (also in the limit the much debated H_2 storage in carbon nanotubes[29]).

More complex phenomena are addressed in the context of three-dimensionality of profiles,[30] in the case of partial charge transfer or in the case of non-idealities in the carrier distributions. Under-potential deposition is a phenomenon to be mentioned in this context which is well-studied in electrochemistry,[31] and in the limit of a total charge separation, similar to processes discussed here.

6. Examples

6.1. Example 1: Metal storage in mixed conductor/insulator composites: interfacial silver storage in chalcogenides

It is well-established that surface active insulator inclusions can tremendously improve conductivity properties of ion conductors (heterogeneous doping). Typically in $AgCl/Al_2O_3$, Ag^+ ions can be adsorbed at the Al_2O_3 surface, as a consequence of which Ag^+ is enriched in the space charge regions of AgCl (see Fig. 3).[8] The overall excess Ag^+ particle number is zero. This would be different if significant e^- concentrations occurred (see Fig. 5). Ag_2S is a mixed conductor and indeed significant anomalous non-stoichiometries have been found by Petuskey[32] in his careful coulometric titration study of Ag_2S/Al_2O_3 composites. Since according to the bulk defect chemistry the ionic disorder is even greater than electronic concentration variations, the magnitude of the observed non-stoichiometry suggests thermodynamic variations (μ° changes) as a consequence of interfacial interactions. Indeed DFT calculations have shown that a silver layer can be energetically favourably attached to a corundum surface if it is oxygen terminated.[33] As expected an interaction of Ag^+ with the surface takes place and, rather surprisingly, electrons are donated to the alumina's conduction band. If Al_2O_3 was a good electronic conductor, this mechanism would be analogous to under-potential deposition. A significant excess stoichiometry has also been found for Ag_2S grain boundaries and dislocations by Stubican *et al.*[34]

Candidates for large anomalies due to pure space charge effects would be materials in which conduction electrons are majority carriers. Ag_2Se is such a material in which $[e']_\infty \simeq [h^\bullet]_\infty$ and for $a_{Ag} \simeq 1$ $[Ag_i^\bullet]_\infty$ is also of similar

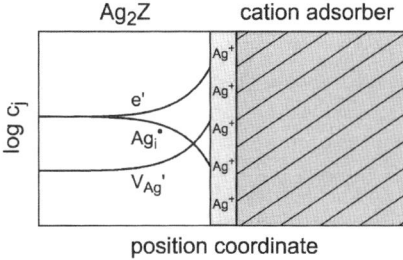

Fig. 5 Excess Ag-storage at a contact of a silver chalcogenide with a cation adsorber. The excess charge is compensated by e^- in Ag_2Z leading to significant Ag storage.

magnitude.[35] The extent to which recently reported magneto-resistance effects[36–38] are connected to this is not clear.[37] Coulometric titration shows significant silver excess which might also be due to supersaturation. According to Janek *et al.*[37] subtle nano-structures form which are supposed to be of decisive influence for the magneto-resistance. Again thermodynamic inhomogeneities and/or metastabilities seem to be of importance.

6.2. Example 2: Metal storage in composites of ion conductors with an electronically conducting phase: interfacial lithium storage

An example of practical interest is the aforementioned possibility of storing Li (of the order of 1 layer per boundary corresponding to a boundary volume percentage of several tens in nano-sized systems) in composites in which each single phase is not able to store Li (*i.e.* Li^+ and e^-).[4,5] Such a combination is Li_2O|noble metal (see Fig. 6). Li_2O exhibits plenty of accessible interstitial sites in which Li^+ can be accommodated but does not exhibit a redox-active element that could take up a lot of electrons. (Nonetheless, similar to Al_2O_3 above, DFT calculations[5] show that at least thin films of Li_2O have a remarkable affinity towards electrons.) According to DFT calculations[5] this mechanism is even feasible in the case of a base metal such as Ti. (On the other hand if Li_2O is Li deficient, Ti acts as an electron donor[5]). Of course, depending on the interfacial structure the excess Li^+ can also reside directly in the Li_2O/metal interface while the e^- is in the metal surface. The interest in such composites is a fundamental one, as they represent the bridge between a super-capacitor and electrode. This difference is blurred if Li_2O particles are so small that the space charge zones overlap (Fig. 7). The application interest lies in the possibility of optimising the combination of the two competing properties, storage capacity and storage rate.[39]

A central question in this context is the dependence of the excess storage on Li potential as this decides upon the related reversible cell voltage. It is helpful to conceive the two phases as two mixed (Li^+, e^-) conductors (see Fig. 6 and 8). In order to decide upon the variation of the space charge potential with μ_{Li} (Fig. 8), it is not sufficient to know the mass action constant K_{Li} which describes the incorporation of Li^+ and e^- as in the corresponding bulk problem. The individual redistribution constants (Li^+(phase 1) $\rightleftharpoons Li^+$(phase 2); e^-(phase 1) $\rightleftharpoons e^-$(phase 2)) also have to be considered, of which one is redundant owing to $K_{Li} = a_{Li}a_{e^-}/a_{Li}$. As, however, quantitative data are missing, further conclusions are difficult.

One point is worth pointing out: Even though boundary storage and bulk storage occur simultaneously if transport equilibrium is achieved, their μ_{Li} dependencies will generally differ, so that significant boundary storage may occur earlier or later than bulk storage if μ_{Li} is varied with time. Ref. 40 provides an example of predominant boundary storage at high μ_{Li}, while ref. 41 refers to predominant boundary storage at low μ_{Li}.

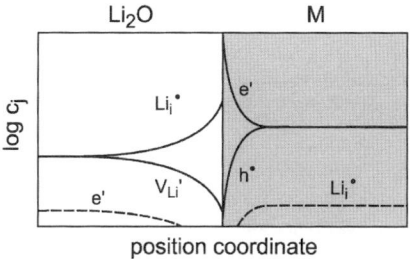

Fig. 6 The defect concentrations at Li_2O/M contact. For clarity a small solubility of e^- in Li_2O and of Li^+ in M has been assumed.

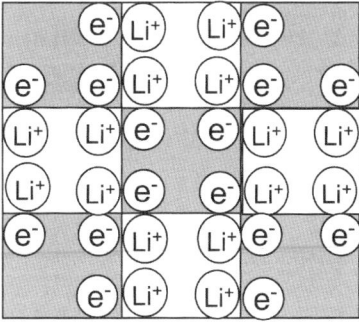

Fig. 7 The Li_2O/M composite forms the bridge between an electrostatic composite and a battery electrode if the space charge profiles overlap.

6.3. Example 3: Salt storage in ion conductor composites

The storage of metals in composites can be generalised to the storage of a salt M^+Y^- in composites in which one phase (MX) can store M^+ but not Y^- excess and the other (NY) can store Y^- but not N^+ excess (see Fig. 4). A useful candidate may be the combination Li_2O/CaF_2 for the storage of LiF. Also, a negative storage seems possible by formation of anion and cation vacancies separately leading to, *e.g.*, MY deficiency.

6.4. Example 4: Surface storage: nano-sized ceria powder

The following example—even though referring to such a "negative mass storage", *i.e.* to a depletion of a neutral component—is instructive in terms of understanding storage effects in terms of the chemical potential of the neutral component (here oxygen partial pressure).

It is well-known that slightly acceptor-doped nanocrystalline ceria exhibits an overall increased *n*-type conductivity. It has also be proven that simultaneously the overall ion transport ($V_{\ddot{O}}$) is hindered. In ref. 42 a detailed analysis showed that due to positively charged grain boundaries, conduction electrons are enriched in the space charge zones (acting as short-circuits for electron conduction) while oxygen vacancies are severely depleted (acting blocking for ion transport). As in the case of $SrTiO_3$ we use the same space charge model for surfaces. This is reasonable as surfaces also offer the possibility of dopant segregation (*e.g.* Gd^{3+}, Y^{3+}) as well as a lower metal to oxygen ratio ("segregation of oxygen vacancies") (Fig. 9). Model calculations suggest a higher concentration of oxygen vacancies but with lower

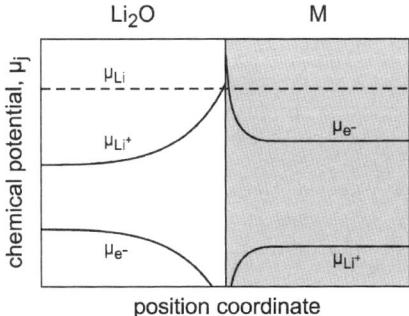

Fig. 8 The relation between μ_{Li^+}, μ_{e^-} and the component potential μ_{Li} at the Li_2O/M contact. The chemical Li-potential also relates to the cell voltage of an Li_2O/M electrode with respect to Li if a Li^+ conductor is inserted.

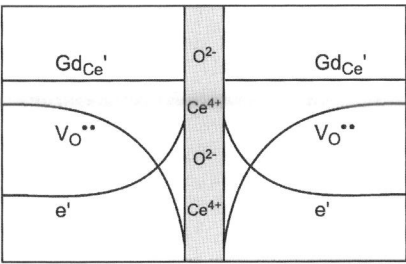

Fig. 9 Excess O-deficiency at CeO_2 grain boundaries. In the case considered the excess core charge is due to an enhanced Ce^{4+}/O^{2-} ratio.

valence.[43] A lowered oxygen stoichiometry was suggested for the grain boundary core of $SrTiO_3$.[44-46] For ceria powder, it was evidenced by thermogravimetry that the oxygen stoichiometry is a function of size simply because oxygen variations essentially occur at the surface effects.[47] The same reason accounts for the size dependence of the effective lattice parameter.[48] All these points can be understood by simple space charge considerations which also shed light on the nature of the positive surface charge.

Accordingly, in the space charge zones we face an e'-enrichment and a vacancy depletion, the charge being compensated by the surface core charge (constituted by oxygen vacancies $V_{Os}^{\varepsilon+}$ with effective charge $\varepsilon+$ and/or segregated dopant ions denoted by $A_s^{a\cdot}$). Information for a further analysis concerning the core charge stems from thermogravimetric results, in particular from the partial pressure dependence of the weight changes. If we assume the surface effects to be of comparable magnitude to the grain boundary effects, it follows from conductivity measurements that the core charge is much larger than the changes in mole number on P_{O_2} variation (if converted *via* molar charge F). Even though more detailed information is lacking, the following will show how to extract more useful information on the nature of the core defect. Let us first assume the core charge to be constituted by vacancy excess (Fig. 9). As now Σ/F determines δ and $\Sigma/F \gg \Delta\delta$ is presupposed, it follows that $\Delta \ln \delta \simeq 0$ in the P-range considered (*i.e.* negligible relative change in oxygen content), a result which is inconsistent with the measured values. Rather one has found $\Delta \ln \delta/\Delta \ln P_{O_2}$ to be within 0.1 and 0.5.[47,49] Hence the core charge should be significantly influenced by impurity segregation.

The other extreme situation is that the core charge is predominantly given by impurity segregation, *i.e.* $\Sigma/F \approx a[A_s^{a\cdot}] \gg [V_{Os}^{\varepsilon\cdot}]$. In this case we can proceed as follows. For the incorporation reaction

$$\frac{1}{2}O_2 + V_{Os}^{\varepsilon\bullet} + \varepsilon e'_\infty \rightleftharpoons O_O \qquad (33)$$

in which bulk electrons have been consumed, a (heterogeneous) mass action law holds:

$$P_{O_2}^{1/2}[V_{Os}^{\varepsilon\cdot}][e'_\infty]^\varepsilon = K_{s\infty} \exp[(-\varepsilon F/RT)\Delta\phi_{s\infty}] \qquad (34)$$

where $K_{s\infty}$ is the chemical constant, and the electrical potential difference between surface core and bulk has to be taken into account through $\Delta\phi_\infty$. As now $\Delta\phi_{s\infty}$ is not very dependent on redox variations, the r.h.s. of eqn (34) must be independent of P_{O_2}. Hence

$$-\partial \ln [V_{Os}^{\varepsilon\cdot}]/\partial \ln P_{O_2} = \partial \ln \delta/\partial \ln P_{O_2} = 1/2 - \varepsilon/4. \qquad (35)$$

Now let us interpret the values of 0.11 and 0.28 measured in ref. 47. Obviously in this model the observable value of 0.28 means a predominant ionisation degree of +1 ($\varepsilon = 0.9$, *i.e.* $V_O^{\bullet}$ with admixtures of $V_O^{\times}$), while the value of 0.11 points towards a mixture of V_O^x and $V_O^\cdot$ ($\varepsilon = 1.5$). The value of 0.5 that was observed by Porat *et al.*[49] corresponds to the limit of the predominant presence of neutral oxygen vacancies ($\varepsilon \simeq 0$). The fact that the oxygen charge is distinctly less than -2 is a well-established fact for most surfaces, the detailed values depending sensitively on the conditions. Note that the constancy of $\Delta\phi_{s\infty}$ and hence of its predominant contribution, namely $\Delta\phi_{0\infty}$, has been measured for nanoceramics. This and the similar $\delta(P_{O_2})$ behaviour found in ref. 50 corroborates the assumption of a similar behaviour for grain boundaries and surfaces. In turn, the constancy of the space charge potential delivers a direct argument for the non-validity of the first case (O-deficiency as major source of boundary charge): If $V_{Os}^\cdot$ constitutes the core charge, $[V_{Os}^\cdot]$ should—because of the electrical potential—be constant as well (in Mott–Schottky layers Σ is determined by $[A^{a\cdot}]\lambda^* \propto [A^{a\cdot}]\sqrt{-\Delta\phi_{0\infty}}$; eqn (35) which is also valid in this case, would then predict $\partial \ln [e']_\infty / \partial \ln P = 0$ which is definitely incorrect (correct value is $-1/4$). In summary, the core is still so significant that its variation with P_{O_2} determines the overall stoichiometry changes in the thermogravimetric experiments in nano-sized ceria.

Storing oxygen excess in this way, rather than realising a deficiency such as here, would presuppose roomy interstitial structures and probably metal terminated grain surfaces. Another possibility would be increasing oxygen interstitial concentrations (in other oxides of course) in the space charge regions which are compensated by holes or appropriate impurities in the grain boundary. With regard to hydrogen storage, interstitial accommodation is simpler to achieve. (In fact the proton is so small that it can dive into the O^{2-};[51] it is then accommodated as $OH_O^\cdot$. The occurrence of the latter is shown by a vast amount of experimental results of H_2 (or H_2O) dissolution in oxides.[51–53] In contrast to water storage, the important case of H_2 storage demands a pronounced redox variability.) Yet, with regard to interfacial storage, in the case of H one expects mostly covalent storage rather than a storage in dissociated form. Generally for optimisation of interfacial storage, composites appear to be more suitable than just polycrystalline samples.

References

1 J. Maier, *Solid State Ionics*, 2002, **154–155**, 291.
2 J. Maier, *Solid State Ionics*, 2003, **157**, 327.
3 J. Maier, *Solid State Ionics*, 2002, **148**, 367.
4 J. Jamnik and J. Maier, *Phys. Chem. Chem. Phys.*, 2003, **5**, 5215.
5 Y. F. Zhukovskii, P. Balaya, E. A. Kotomin and J. Maier, *Phys. Rev. Lett.*, 2006, **96**, 058302–1.
6 P. Balaya, H. Li, L. Kienle and J. Maier, *Adv. Funct. Mater.*, 2003, **13**, 621.
7 H. Li, P. Balaya and J. Maier, *J. Electrochem. Soc.*, 2004, **151**(11), A1878.
8 J. Maier, *J. Electrochem. Soc.*, 1987, **134**, 1524.
9 J. Jamnik, J. Maier and S. Pejovnik, *Solid State Ionics*, 1995, **75**, 51.
10 J. Maier, *Prog. Solid State Chem.*, 1995, **23**, 171.
11 J. Maier, *Solid State Ionics*, 1989, **32/33**, 727.
12 C. Wagner and W. Schottky, *Z. Phys. Chem.*, 1930, **B11**, 163.
13 N. Valverde, *Z. Phys. Chem., Neue Folge*, 1971, **74**, 146.
14 F. A. Kröger, *Chemistry of Imperfect Crystals*, North-Holland, Amsterdam, 1964.
15 J. Maier, *Physical Chemistry of Ionic Materials. Ions and Electrons in Solids*, John Wiley & Sons, Ltd, Chichester, 2004.
16 (a) N. F. Mott, *Proc. R. Soc. London, Ser. A*, 1939, **171**, 27; (b) W. Schottky, *Z. Phys.*, 1939, **113**, 367; (c) W. Schottky, *Z. Phys.*, 1942, **118**, 359.
17 (a) G. Gouy, *J. Phys.*, 1910, **9**, 457; (b) D. L. Chapman, *Philos. Mag.*, 1913, **25**, 475.
18 S. Kim, J. Fleig and J. Maier, *Phys. Chem. Chem. Phys.*, 2003, **5**, 2268.
19 H. K. Henisch, *Semiconductor Contacts*, Clarendon Press, Oxford, 1984.
20 S. M. Sze, *Semiconductor Devices*, John Wiley & Sons, Ltd, New York, 1985.
21 J. Th. G. Overbeek, in *Colloid Science*, ed. H. R. Kruyt, Elsevier, Amsterdam, 1952, vol. I.
22 J. Maier, *Solid State Ionics*, 1987, **23**, 59.
23 Note that an inversion effect leads to a different majority carrier (no longer $N_M^\cdot$). Then the screening and hence the Debye length is no longer controlled by m^{18}.

24   In greater detail:

$$i_{1/2} \to i_0 \left(1 - \left(\frac{L}{4\lambda}\right)^2\right) \simeq i_0,$$

$$v_{1/2} \to v_0 \left(1 + \left(\frac{L}{4\lambda}\right)^2\right) \simeq v_0$$

25   J. Jamnik, *Solid State Ionics*, (Proc. SSI-15, Baden-Baden, 2005), in press.
26   In ref. 11 the fact that both mean boundary concentration and boundary length are important has been used to explain the quite small P_{O_2} (~ -0.15) exponent (N) of the excess n-type conductance of ZnO when compared to the bulk (1/4). While this is in principle correct, the value $-1/8 = -0.13 \simeq -0.15$ only follows—under the assumption $N_0 = N_\infty = -1/4$—in a Mott–Schottky inversion situation (see below; this was not consistently set out in ref. 11); note that a bulk value of 1/4 means that electrons are in minority there. If we refer to Gouy–Chapman profiles, the electrons are necessarily still in minority also in the boundary zones (with respect to either a native ionic defect or an impurity defect) which results in $N_{eff} = (N_0 + N_\infty)/2$ rather than $N_\infty/2$. In such a case the measured exponent of $\sim 1/8$ is reproduced if $N_0 = 0$, which is, *e.g.* the case for a frozen positive surface charge. Such a frozen surface charge may be realised by strong impurity segregation or by a structurally determined O-deficiency (as verified for $SrTiO_3$ grain boundaries[44–46]). If, however, electrons are in majority in the bulk, indeed $N_\infty/2$ follows, but then $N_\infty = -1/6$. If we refer to a Mott–Schottky situation in which electrons are enriched but not to a level that they become majority carriers (no inversion),[18] $N_0 - N_\infty^*/2 \simeq N_0$ would follow as the bulk majority level is independent of P_{O_2}, *i.e.* as $N_\infty^* \simeq 0$ (while $N_0/2$ is the result if they represent the major carriers close to the interface, see above). Then a smaller local exponent has to be invoked in order to explain the value of -0.15. A similar point refers to the depletion in SnO_2,[11] there $N_0/2$ follows a Gouy–Chapman situation only if $N_\infty = -1/6$, while $N_\infty = -1/4$ means that λ is not dominated by the electron concentration (again this was not consistently set out in ref. 11).
27   J. Maier, *Solid State Phenomena*, 1994, **39–40**, 35.
28   J. Jamnik and J. Maier, *Phys. Chem. Chem. Phys.*, 2001, **3**, 1668.
29   (*a*) A. C. Dillon, K. M. Jones, T. A. Bekkedahl, C. H. Kiang, D. S. Bethune and M. J. Heben, *Nature*, 1997, **386**, 377; (*b*) M. Hirscher, M. Becher, M. Haluska, U. Dettlaff-Weglikowska, A. Quintel, G. S. Duesberg, Y.-M. Choi, P. Downes, M. Hulman, S. Roth, I. Stepanek and P. Bernier, *Appl. Phys. A*, 2001, **72**, 129; (*c*) G. Seifert, *Appl. Phys. A*, 2004, **168**, 265.
30   I. Lubomirsky, J. Fleig and J. Maier, *J. Appl. Phys.*, 2002, **92**, 6819.
31   J. O'M. Bockris and A. K. V. Reddy, *Modern Electrochemistry*, Plenum Press, New York, 1970.
32   W. T. Petuskey, *Solid State Ionics*, 1986, **21**, 117.
33   E. A. Kotomin, R. A. Evarestov, Yu. A. Mastrikov and J. Maier, *Phys. Chem. Chem. Phys.*, 2005, **7**, 2346.
34   (*a*) V. S. Stubican, C. M. Lin and E. Macey, *Adv. Ceram.*, 1987, **23**, 97; (*b*) V. Stubican, *Philos. Mag.*, 1993, **68**, 809.
35   U. von Oehsen and H. Schmalzried, *Ber. Bunsen-Ges. Phys. Chem.*, 1981, **85**, 7.
36   R. Xu, A. Husmann, T. F. Rosenbaum, M.-L. Saboungi and B. L. Littlewood, *Nature*, 1997, **390**, 57.
37   J. Janek, B. Mogwitz, G. Beck, M. Kreutzbruck, L. Kienle and C. Korte, *Prog. Solid State Chem.*, 2004, **32**, 179.
38   G. Beck and J. Janek, *Physica B*, 2001, **308–310**, 1086.
39   From the standpoint of transport, composites of metals and mixed conductors or electron conductors have already been considered by C. Wagner[54] and M. P. Setter and J. B. Wagner[55].
40   H. Li, G. Richter and J. Maier, *Adv. Mater.*, 2003, **15**, 736.
41   Y.-S. Hu, L. Kienle, Y.-G. Guo and J. Maier, *Adv. Mater.*, 2006, **18**, 1421.
42   S. Kim and J. Maier, *J. Electrochem. Soc.*, 2002, **149**, J73.
43   T. X. T. Sayle, S. C. Parker and C. R. A. Catlow, *Surf. Sci.*, 1994, **316**, 329.
44   Z. Zhang, W. Sigle, F. Phillipp and M. Rühle, *Science*, 2003, **302**, 846.
45   R. A. De Souza, J. Fleig, J. Maier, O. Kienzle, Z. Zhang, W. Sigle and M. Rühle, *J. Am. Ceram. Soc.*, 2003, **86**, 922.
46   Z. Zhang, W. Sigle and M. Rühle, *Phys. Rev. B*, 2002, **66**, 094108.
47   S. Kim, R. Merkle and J. Maier, *Surf. Sci.*, 2004, **549**, 196.
48   J. P. Nair, E. Wachtel, I. Lubomirsky, J. Fleig and J. Maier, *Adv. Mater.*, 2003, **15**, 2077.

49 O. Porat, H. L. Tuller, E. B. Lavik, Y.-M. Chiang, in *Nanophase and Nanocomposite Materials II*, eds. S. Komarneni, J. Parker and H. Wollenberger, Materials Research Society, Pittsburgh, PA, 1997, p. 99.
50 P. Knauth and H. L. Tuller, *J. Eur. Ceram. Soc.*, 1999, **19**, 831.
51 K.-D. Kreuer, *Chem. Mater.*, 1996, **8**, 610.
52 H. Iwahara, T. Esaka, H. Uchida and N. Maeda, *Solid State Ionics*, 1981, **314**, 259.
53 (*a*) Y. Larring and T. Norby, *Solid State Ionics*, 1995, **77**, 147; (*b*) A. S. Nowick, *Solid State Ionics*, 1995, **77**, 137.
54 C. Wagner, *J. Phys. Chem. Solids*, 1972, **33**, 1051.
55 M. P. Setter and J. B. Wagner, *Solid State Ionics*, 1988, **28/30**, 1579.

NMR and impedance studies of nanocrystalline and amorphous ion conductors: lithium niobate as a model system

Paul Heitjans, Muayad Masoud, Armin Feldhoff and Martin Wilkening

Received 28th February 2006, Accepted 31st March 2006
First published as an Advance Article on the web 24th July 2006
DOI: 10.1039/b602887j

Lithium niobate has been chosen as a model system for spectroscopic studies of the influence of different structural forms and preparation routes of an ionic conductor on its ion transport properties. The Li diffusivity in nanocrystalline $LiNbO_3$, prepared either mechanically by high energy ball milling or chemically by a sol-gel route, was studied by means of impedance and solid state ^7Li NMR spectroscopy. The Li diffusivity turned out to be strongly correlated with the different grain boundary microstructures of the two nanocrystalline samples and with the degree of disorder introduced during preparation, as seen especially by HRTEM and EXAFS. Although in both samples nanostructuring yields an enhancement of the Li diffusivity compared to that in coarse grained $LiNbO_3$, the Li diffusivity in ball milled $LiNbO_3$ is much higher than in chemically prepared nanocrystalline $LiNbO_3$. The former $LiNbO_3$ sample has a large volume fraction of highly disordered interfacial regions which seem to be responsible for fast Li diffusion and to have a structure very similar to that of the amorphous form. This is in contrast to the chemically prepared sample where these regions have a smaller volume fraction.

1 Introduction

Solid materials with a high ionic diffusivity are of vital interest in materials science due to their potential applications as solid electrolytes, *e.g.*, in secondary Li ion batteries.[1-4] Therefore, one of the fundamental aims in materials science is to understand the chemical and physical principles determining the transport characteristics of solids in order to design new materials with tailored diffusion properties.

A way to change the diffusion parameters of a given polycrystalline ionic conductor is to reduce its grain size down to the nanometre range, *i.e.*, to increase the volume fraction of the interfacial regions and to introduce a large number of grain boundaries.[5] As the transport properties of these regions can deviate considerably from those of the bulk material, this procedure may have a remarkable impact on the overall ionic diffusivity of the material.[5] Often grain boundaries in

University of Hannover, Institute of Physical Chemistry and Electrochemistry, Callinstr. 3-3a, 30167, Hannover, Germany. E-mail: heitjans@pci.uni-hannover.de; Fax: +49 511 762 19121; Tel: +49 511 762 3187

nanocrystalline materials provide fast diffusion pathways for small cations and anions like Li^+ and F^-, or even larger anions like O^{2-}, so that an enhancement of the diffusivity is observed.[6–18] Sometimes, however, the interfacial regions have a blocking effect resulting in a reduction of the ionic conductivity.[19] This aspect is of considerable interest especially in the case of yttria-stabilized ZrO_2.[20]

Up to now, it is not completely clear whether the grain boundary regions of nanocrystalline ceramics prepared by different routes have similar local structures. Because of the strong relationship between the transport properties and the microscopic structure, it is of great interest to study systematically ionic diffusion in nanocrystalline materials of the same chemical composition but prepared by different routes. In the present study, lithium niobate, $LiNbO_3$, serves as a model substance to elucidate the correlations between cation dynamics and the structural features of a nanocrystalline ceramic.

Solid-state nuclear magnetic resonance (NMR) techniques[21,22] in combination with impedance spectroscopy measurements[5,23] were used to probe microscopic as well as macroscopic Li diffusion parameters of $LiNbO_3$. In its single crystalline form, lithium niobate is a poor Li conductor with a very small Li diffusion coefficient.[6,23,24] The same result also holds for the microcrystalline form, i.e., a polycrystalline powder sample with an average particle size in the μm range.[6,23] Interestingly, the amorphous or glassy form of $LiNbO_3$ revealed an enhancement in Li diffusivity by several orders of magnitude.[23,25,26] The activation energy of Li conductivity, reflecting long-range Li diffusion, is reduced from about 1.2 eV in the case of single crystalline $LiNbO_3$ to about 0.6 eV in disordered $LiNbO_3$.[23] By means of ^7Li spin–lattice relaxation NMR, probing short-range Li motion, an activation energy of about 0.3 eV was found for amorphous $LiNbO_3$.[27] This is about one third of the value obtained by ^7Li NMR for Li diffusion in the microcrystalline form. In nanocrystalline $LiNbO_3$, Li diffusion is also drastically enhanced compared to its coarse grained counterpart.[23,27,28] The latter result holds at least for a nanocrystalline powder prepared by ball milling with an average particle size of about 20 nm.[23,27] In both samples, amorphous and nanocrystalline, the Li diffusion process can be described by practically the same set of parameters, as we have shown by an NMR relaxation and line shape study.[31] Thus it was indirectly concluded that the interfacial regions of the nanocrystalline material prepared by ball milling seem to have an amorphous structure. The enhanced diffusivity in amorphous $LiNbO_3$ was tentatively ascribed to the 'free volume' enclosed in its disordered structure leading to a distribution of energy barriers and thus to diffusion pathways with low thermal activation.

The present study is aimed at the question of whether the latter observation is a common feature of amorphous and nanocrystalline materials or at least characteristic of lithium niobate. This includes the question of whether the particle size of a nanocrystalline material is the main parameter determining the transport properties or whether structural properties of the grain boundaries play an important role too. For this purpose we have investigated two nanocrystalline samples with the same grain size but prepared by two different techniques, viz. chemically via a sol-gel method[29] and mechanically using high energy ball milling.[30] By the first preparation route the nanocrystallites are formed from smaller building units, similar to other techniques like inert gas condensation,[31] chemical vapor deposition,[32] or pulsed electro-deposition.[33] In contrast to that approach, nanocrystalline materials can also be obtained by reducing the grain size of their coarse grained counterparts, e.g., by sputtering with a radio frequency field or with heavy ions[34] or just by high energy ball milling.[5,30]

In addition to a detailed structural characterization of the samples with X-ray diffraction (XRD), high-resolution transmission electron microscopy (HRTEM), and extended X-ray absorption fine structure (EXAFS) spectroscopy, the samples were investigated by differential scanning calorimetry (DSC) and thermogravimetry (TG) as well as by Raman spectroscopy.

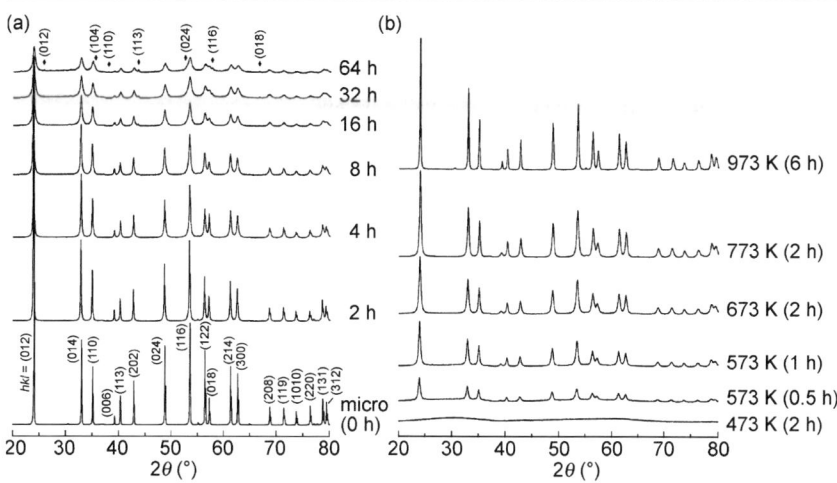

Fig. 1 (a) Effect of ball milling for various milling times, ranging from 2–64 h, on the XRD patterns of LiNbO$_3$. Increasing the milling time results in a broadening of XRD lines due to the decrease of particle size. The *hkl* indices of α-Al$_2$O$_3$ are shown in the XRD pattern of the sample milled for 64 h. After 16 h ball milling an average particle size of 23 nm is reached. (b) Heat treatment of amorphous LiNbO$_3$ prepared by complete hydrolysis of the lithium niobium double alkoxide. The XRD patterns were recorded after different periods of calcination at the indicated temperatures (in a progressive way). The XRD pattern for 773 K (1 h) represents sol-gel prepared nanocrystalline LiNbO$_3$ with an average particle size of about 27 nm.

2 Sample preparation and characterization

2.1 Preparation and XRD analysis of nanocrystalline and amorphous LiNbO$_3$

Phase-pure nanocrystalline LiNbO$_3$ was prepared from the microcrystalline source material which was obtained from Alfa Aesar (99.9995%). The coarse grained LiNbO$_3$ consists of irregularly shaped crystallites with an average particle diameter of some tens of microns. A SPEX 8000 ball mill equipped with an α-Al$_2$O$_3$ vial set and a ball of about 4 g made of the same material was used for ball milling. The ball-to-powder ratio was chosen to be 1 : 1, ref. 23. Although LiNbO$_3$ is stable in air, the whole preparation was done under an Ar atmosphere to avoid reaction of the hygroscopic nanocrystalline material with water vapour or CO$_2$. After 16 h of ball milling, for instance, the average crystallite size was 23 nm. This value was calculated from the broadening of the XRD lines using the Scherrer equation.[35] The XRD profiles were measured with a Philips PW 1800 diffractometer (Bragg–Brentano geometry) using Cu Kα radiation. Prior to the determination of the line broadening, the Kα_1 and Kα_2 contributions were separated from each other using the correction procedure introduced by Rachinger.[36] The effect of ball milling on the XRD profiles is shown in Fig. 1(a). A Rietveld structure refinement showed that the sample milled for 16 h contains at most 5% of crystalline alumina due to abrasion of the vial set and the ball during the milling procedure. As compared to the influence of the different structural forms of LiNbO$_3$ (*cf.* Fig. 5) this small admixture of Al$_2$O$_3$ has only a negligible effect on the ionic conductivity. For milling times equal or longer than 16 h the broadened XRD lines are superimposed on some broad background humps being characteristic of amorphous LiNbO$_3$ (see below and XRD pattern at 473 K of Fig. 1(b)). Thus, the XRD results already indicate that this nanocrystalline LiNbO$_3$ sample seems to be a heterogeneous mixture of a crystalline and an amorphous phase.

For the preparation of the other nanocrystalline sample using a sol-gel technique, amorphous LiNbO$_3$ was synthesized by the double alkoxide route.[29] Equimolar ratios of lithium ethoxide (Aldrich, 1 M solution in abs. ethanol) and niobium ethoxide (Alfa Aesar, neat liquid, 99.999%) were dissolved in abs. ethanol (0.2 M each). The solution was refluxed for 24 h at 352 K to prepare lithium niobium double alkoxide. Up to this point the whole preparation was done under an inert gas atmosphere. Next, the double alkoxide was completely hydrolyzed with 7.5 equiv. of deionized water dissolved in abs. ethanol (2 M). The lithium niobium hydroxide hydrated gel obtained was filtered and dried. Calcination for 2 h at 473 K under O_2 flow yields X-ray amorphous $LiNbO_3$ (see Fig. 1(b), XRD pattern at bottom). The pattern shows no sharp XRD lines; the broad background humps are characteristic of amorphous $LiNbO_3$.

Another amorphous $LiNbO_3$ sample which was used for the EXAFS investigations (see section 2.4) was prepared in a similar way, [27] but, instead of being completely hydrolyzed, the double alkoxide solution was only partially (1/3) hydrolyzed. The resultant gel was dried for several hours in an oxygen atmosphere at 470 K, followed by a brief calcination for 5 min at 620 K in order to burn alkyl residues. The sample consists of irregularly shaped particles with diameters between 1–20 μm and was previously used for a detailed ^7Li NMR relaxation study.[27]

Further calcination of the X-ray amorphous sample which was prepared by complete hydrolysis is shown in Fig. 1(b). After calcination for 30 min at 573 K the XRD pattern of $LiNbO_3$ emerges. The average particle size after heat treatment at 773 K is 27 nm when calculated from the XRD broadening using the Scherrer equation. In the following paragraphs this nanocrystalline sample is called the sol-gel nanocrystalline $LiNbO_3$. Note that the characteristic background contribution for an amorphous phase (see above) is missing in the XRD pattern for this chemically prepared sample. Thus, we conclude that the amount of amorphous regions is much less than in the mechanically prepared one. This conclusion is corroborated by a detailed analysis of the TEM micrographs of these two nanocrystalline samples (see section 2.3). Furthermore, nanocrystalline $LiNbO_3$ prepared by the sol-gel technique gives rise to relatively sharp XRD lines as compared to the milled samples.

2.2 Sample characterization by Raman spectroscopy, DSC and TG analysis

Raman spectra of microcrystalline and nanocrystalline $LiNbO_3$, recorded with a Bruker optics RFS 100/S spectrometer, are shown in Fig. 2. The effect of grain size reduction can be clearly seen in Fig. 2(a). The intensity of all Raman bands decreases with increasing milling time. Fig. 2(b) shows the evolution of the Raman bands of $LiNbO_3$ upon heating the hydrated double alkoxide precursor. The prepared gel shows very broad diffuse bands indicating a wide distribution of bond angles. For details we refer to ref. 37.

DSC curves and TG analyses were performed using a SETSYS evolution analyzer (SETARAM). The DSC curve of the prepared gel shows a characteristic endothermic peak around 400 K associated with a first TG weight loss step which can be attributed to the removal of water, alcohol and loosely bonded organic groups. At 600 K a strong exothermic peak due to the completion of pyrolysis is detected as confirmed by a second weight loss step. After that no weight loss and no significant DSC peaks are detected. It is estimated that the amorphous $LiNbO_3$ prepared by calcination at 473 K for 2 h will continue to consist of about 4% organic residuals. The DSC/TG curves of the samples calcined at 773 and 973 K did not show any DSC or weight loss peaks up to 1530 K, which is the melting point of $LiNbO_3$.

The micro- and single-crystalline $LiNbO_3$ samples show no thermal changes or weight losses up to the melting point. However, the 16 h ball milled $LiNbO_3$ reveals a broad endothermic signal at temperatures extending from 350–500 K. This is associated with a weight loss peak of about 3% followed by a second characteristic broad peak between 540 and 700 K which is absent in the case of microcrystalline $LiNbO_3$. The first peak is attributed to the loss of water. Hence, contact with water and air was carefully avoided during sample preparation for impedance and NMR measurements (see above). The second peak is attributed to grain growth of the

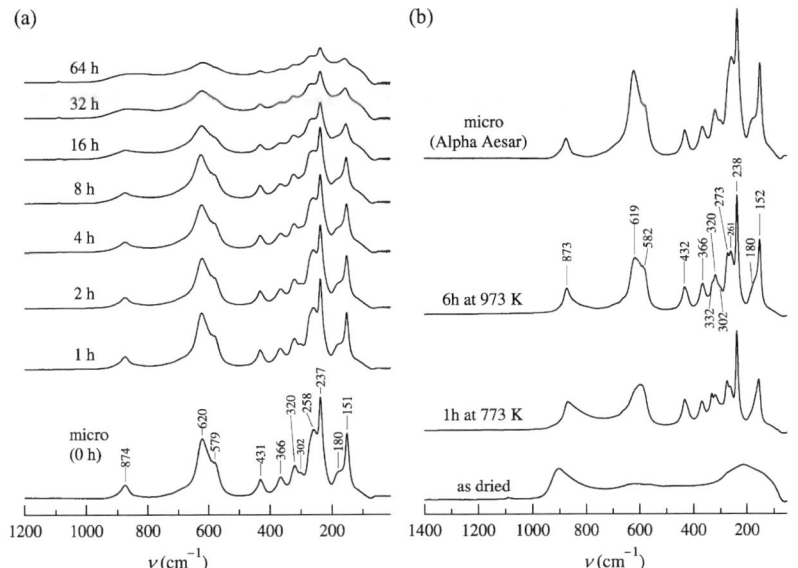

Fig. 2 (a) The effect of grain size reduction with increasing milling time on the Raman spectra of LiNbO$_3$. (b) Raman spectra of the sol-gel precursor illustrating the evolution of nanocrystalline LiNbO$_3$ upon heating. The Raman spectrum of the commercially available microcrystalline LiNbO$_3$ is shown for comparison.

nanocrystalline particles. Such a peak is absent for the coarse grained material and, furthermore, also absent in the case of the sol-gel nanocrystalline material. The latter observation implies that the interfacial regions differ according to the preparation route. In order to avoid any grain growth a temperature of 450 K was chosen as an upper limit for all NMR and impedance measurements.

2.3   Characterization of nanocrystalline LiNbO$_3$ by means of HRTEM

HRTEM micrographs were taken at 200 kV using a JEOL JEM-2100F-UHR field-emission microscope that provides a point resolution better than 0.19 nm. In Fig. 3 HRTEM images of two nanocrystalline samples prepared by ball milling for 32 h (Fig. 3(a)) and chemically *via* the sol-gel method (Fig. 3(b)) are shown. The average crystallite size of the mechanically prepared nanocrystalline sample (*cf.* Fig. 1(a)) is about 20 nm, thus, it is similar to the one milled for only 16 h.

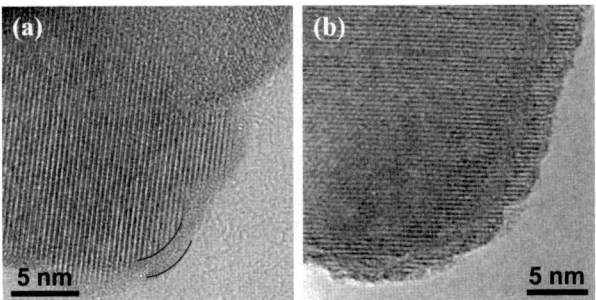

Fig. 3 (a) HRTEM micrograph of nanocrystalline LiNbO$_3$ prepared by ball milling for 32 h, with an average particle size of about 20 nm. A large amount of amorphous LiNbO$_3$ can be seen. (b) HRTEM micrograph of nanocrystalline LiNbO$_3$ prepared chemically *via* complete hydrolysis of the lithium niobium double alkoxide. Although the grain boundary regions are somewhat disordered, not much amorphous material can be detected.

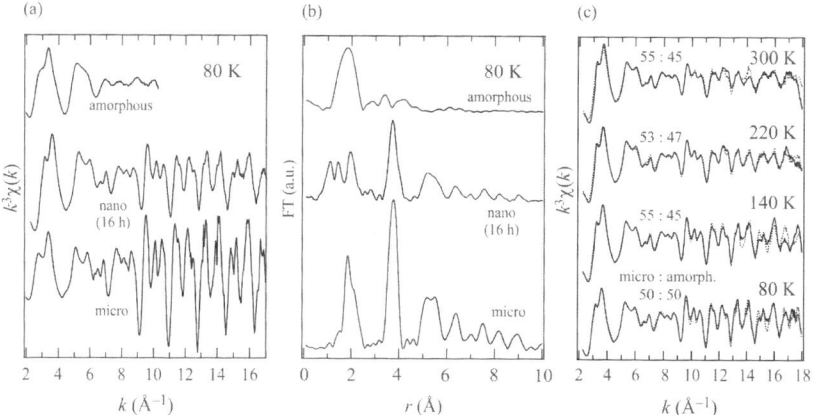

Fig. 4 (a) Nb K-edge EXAFS spectra for microcrystalline, mechanically prepared nanocrystalline and amorphous LiNbO₃ at T = 80 K. (b) Fourier transforms of the EXAFS spectra of the three forms of LiNbO₃. (c) EXAFS spectra of the nanocrystalline sample. Data were recorded at different temperatures. Dotted curves represent the fits using a linear combination of the EXAFS spectra of microcrystalline and amorphous LiNbO₃ at the same temperature. Nanocrystalline LiNbO₃ can be described by a linear combination with a ratio of approximately 1 : 1 of microcrystalline and amorphous LiNbO₃.

Fig. 3(a) shows clearly the heterogeneous nature of ball milled LiNbO₃ consisting of a crystalline and an amorphous phase of LiNbO₃. The nanocrystallites are surrounded by an amorphous grain boundary region of about 2 nm thickness (indicated in Fig. 3(a) by two solid lines). Contrary to this, the chemically prepared nanocrystalline sample of LiNbO₃ (Fig. 3(b)) shows a much smaller amount of amorphous LiNbO₃. The particles seem to be highly crystalline. Their grain boundary regions are somewhat disordered, however, and they are much thinner compared to those of the milled sample. Some terraces and surface steps are visible.

2.4 EXAFS measurements

In addition to TEM measurements, a series of EXAFS experiments were carried out using station 9.2 at the Daresbury synchrotron radiation source in order to explore the nature of the grain boundaries of a mechanically prepared sample *via* high energy ball milling for 16 h which we used previously for an NMR study.[6] All the measurements were done at the Nb K-edge and experimental data were collected in transmission mode. The minimum beam current was 150 mA at 2 GeV and the beam size was 10 mm × 0.6 mm. The energy selection was accomplished by a double crystal Si(220) monochromator. Standard ion chambers were used as detectors. The EXAFS data were analyzed using the Daresbury suite of EXAFS programs. For further details see ref. 37 and 38.

EXAFS data of amorphous LiNbO₃ being prepared *via* partial hydrolysis and results of microcrystalline LiNbO₃ (Aldrich, 99.999%, see ref. 6) were used as internal references for analyzing the experimental EXAFS spectra of the ball milled LiNbO₃ at various temperatures. The data obtained at 80 K are presented in Fig. 4(a) and their Fourier transforms are shown in Fig. 4(b). A linear combination of EXAFS spectra of microcrystalline and amorphous LiNbO₃ is fitted to the spectra of the nanocrystalline sample (*cf.* Fig. 4(a) and (c)). The fit shows that the nanocrystalline material consists of about equal shares of crystalline and amorphous LiNbO₃. Thus, the findings obtained by XRD and TEM measurements are corroborated by the EXAFS data. Obviously, the ball milled samples of LiNbO₃ can be regarded as structurally heterogeneous materials. The nanocrystallites seem to be embedded in an amorphous matrix of $LiNbO_3$. Similar results have been obtained recently by Pooley and Chadwick from an EXAFS measurement at 300 K.[39]

3 Experimental

3.1 Impedance spectroscopy

Most of the impedance measurements were carried out in the frequency range from $\nu = 5$ Hz to 13 MHz using an RF impedance analyzer HP 4192A. The impedance detection range was 0.1 mΩ–1 MΩ. A home-built impedance cell was used for the measurements. It was placed in a horizontal-tube furnace controlled by an Eurotherm 818 programmable temperature unit. The temperature was measured 3 mm away from the sample and its accuracy is better than 0.5 K.

Due to the low conductivity of single crystalline $LiNbO_3$ an Alpha Novocontrol high-resolution dielectric analyzer was utilised which works in the frequency range from $\nu = 3$ μHz to 10 MHz and in a large impedance range, i.e., from 10^{-2} to 10^{14} Ω. A connection head (BDS 1200) served as the standard sample cell. The temperature was monitored with an accuracy of 0.3 K by means of a PT100 temperature unit. It is either adjusted with freshly evaporated nitrogen or with an electrical heater controlled by WinDETA (Novocontrol).

Pellets of the powder samples were prepared by applying an uniaxial pressure of about 0.75 GPa. The thickness ranged between 0.2–0.4 mm. In the case of single-crystalline $LiNbO_3$, plates with a thickness of 0.45 mm and an area of 15×12 mm^2 were used for the measurements. The surfaces of the plates were polished and sputtered with gold.

All of the experiments were done either in inert gas or in a dry oxygen atmosphere. The results proved to be highly reproducible, no hysteresis behaviour of the conductivity was observed when excluding any influence of water vapour on the experiments. Complex plane impedance plots were analysed by electrical equivalent circuit software.

3.2 NMR measurements

^{7}Li spin–lattice relaxation NMR rates $1/T_1$ as a function of inverse temperature were recorded using a modified MSL 100 Bruker spectrometer at a Larmor frequency $\omega_0/2\pi$ of 77.72 MHz. The MSL 100 is connected to a tunable Oxford cryomagnet and equipped with a Kalmus 400 W power amplifier. The 90° pulse length using a commercial Bruker broadband probe was about 5 μs. Relaxation rates in the laboratory frame were measured with the standard saturation recovery pulse sequence,[40] $n \times 90° -$ delay time $t_i - 90° -$ acq., where a train of ($n = 10$) 90° pulses destroys any longitudinal magnetization M_z so that the subsequent recovery of M_z can be detected via the last 90° pulse as a function of thirty different delay times t_i ranging between $t_1 > 0$ and $t_{30} > 5T_1$. Up to 32 scans were accumulated for each value of t_i. Magnetization transients $M_z(t_i)$ were obtained by integration of the free induction decays.

^{7}Li NMR spectra were recorded using the solid-echo pulse sequence, $\phi - t_e - 90° - t_e -$ echo, modified for a spin-3/2 nucleus, i.e., $\phi = 64°$. The time-domain data were converted into the frequency-domain by Fourier transformation starting from the echo top at $t = t_e$. The interpulse delay t_e ranged between 10 and 100 μs. It was chosen such that t_e was smaller than the transverse relaxation time T_2.

4 Results and discussion

4.1 Impedance spectroscopy

The conductivity spectra $\sigma'(\nu)$ of all the different forms of $LiNbO_3$ investigated, viz. single-crystalline, microcrystalline, nanocrystalline as well as amorphous $LiNbO_3$,

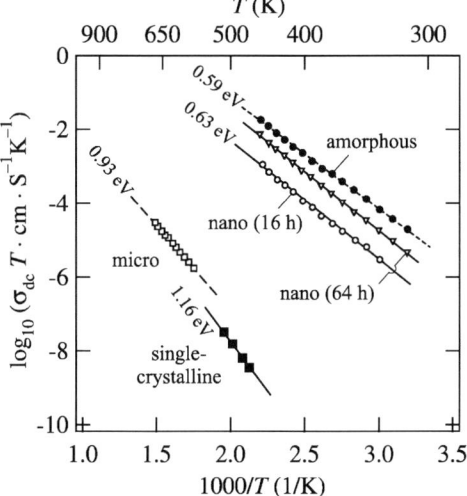

Fig. 5 Temperature dependence of σ_{dc} for amorphous (•) and ball milled nanocrystalline LiNbO$_3$ (milling time: 16 h (○), 64 h (▽)). The results are compared to those obtained for commercially available microcrystalline (□, Alfa Aesar) and single-crystalline (■) LiNbO$_3$. The microcrystalline LiNbO$_3$ was used as source material to prepare the nanocrystalline samples. Dashed and solid lines show fits according to the Arrhenius relation of eqn (2).

consist of well-defined σ_{dc} plateaus at low frequencies and a typical dispersive regime in the high frequency range due to correlated Li jumps on shorter time scales. The spectra can be described empirically by the power law.

$$\sigma'(\nu) = \sigma_{dc} + A(\nu 2\pi)^s \qquad (1)$$

$$T\sigma_{dc} = A'\exp(-E_{A,\sigma}/k_B T) \qquad (2)$$

with the exponent s ranging between 0.55 and 0.77, ref. 23. The temperature dependence of σ_{dc} is shown in Fig. 5. Each sample shows single Arrhenius behaviour over the respective measured temperature ranges so we assume that no change in the conduction mechanism occurs. The lowest Li conductivity is found for single-crystalline and microcrystalline LiNbO$_3$. It is increased by seven orders of magnitude for amorphous LiNbO$_3$. A drastic enhancement of the Li conductivity, though somewhat less than that for amorphous LiNbO$_3$, is also found for the nanocrystalline material prepared *via* 16 h high energy ball milling. Whereas the diffusion process in single-crystalline and microcrystalline LiNbO$_3$ is determined by an activation energy $E_{A,\sigma}$ of about 1 eV, the corresponding values of $E_{A,\sigma}$ for the nanocrystalline (16 h ball milled, 23 nm particle size) and amorphous forms are reduced by a factor of two resulting in activation energies of about 0.59 eV. $E_{A,\sigma}$ refers to the long-range transport process of the charge carriers. Interestingly, both samples follow an Arrhenius relation (eqn (2)) with the same activation energy, thus reflecting similar conduction pathways in amorphous and mechanically prepared nanocrystalline LiNbO$_3$. With increasing milling time the data points of the nanocrystalline material are shifted towards the Arrhenius line of the amorphous sample, so that the difference between the $T\sigma_{dc}(1/T)$ values of the two materials becomes smaller than one order of magnitude.

Contrary to these observations, the behaviour of the sol-gel prepared nanocrystalline sample with an average particle size of about 27 nm is completely different. Fig. 6(a) compares the σ_{dc} temperature dependence of the sol-gel nanocrystalline LiNbO$_3$ with that of the amorphous sample prepared from the double alkoxide calcined for 2 h at 473 K and with that of a sample heat treated at 973 K. The latter has an average

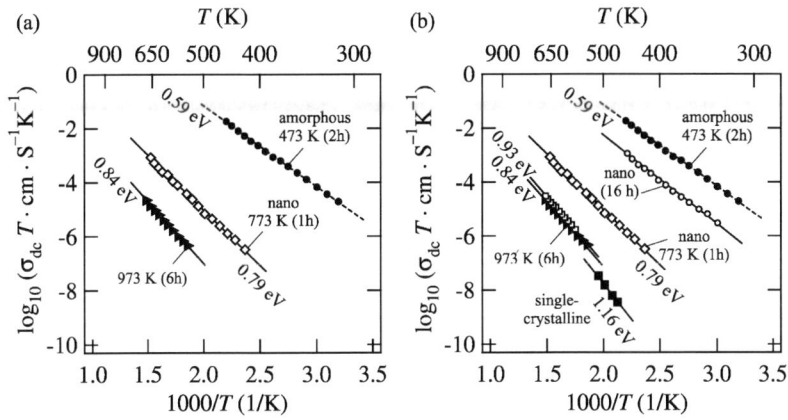

Fig. 6 Temperature dependence of the dc conductivity of amorphous LiNbO₃ in comparison with those of sol-gel prepared nanocrystalline LiNbO₃ (773 K, 1 h, *cf.* Fig. 1) and with a sample heat-treated at much higher temperature (973 K, *cf.* Fig. 1 for the corresponding XRD pattern).

particle size of about 100 nm. Its conductivity behaviour is nearly the same as that obtained for microcrystalline LiNbO₃ which was used as the source material for high energy ball milling (see section 2.1). The data points coincide and the activation energies are very similar, *viz.* 0.84 and 0.93 eV, respectively (*cf.* Fig. 6(b)). Although the conductivity for the sol-gel prepared nanocrystalline LiNbO₃ is higher by two orders of magnitude when compared with the results of the sample prepared at 973 K and of microcrystalline LiNbO₃, it is much smaller than that of the sample prepared by milling and of amorphous LiNbO₃, respectively. The chemically prepared nanocrystalline sample takes a medial position between the single-crystalline and microcrystalline form on the one hand and the ball milled nanocrystalline and amorphous material on the other hand. The activation energy of about 0.79 eV comes closer to the values $E_{A,\sigma}$ obtained for the coarser grained materials. Starting with the amorphous sample, $E_{A,\sigma}$ first increases from 0.59 eV by about 0.2 eV to 0.79 eV for the chemically prepared nanocrystalline material (27 nm) and then to 0.84 eV for the sample heat-treated at 973 K with a particle size around 100 nm. The latter change in activation energy can be attributed to the onset of grain growth leading to a reduction of interfacial regions in the material.

The difference in transport properties of the two nanocrystalline samples cannot be explained by different particle sizes as the average diameter of the nanometer sized grains are very similar, namely 23 and 27 nm. The conductivity results imply that the nature of the interfacial regions of the chemically prepared sample are indisputably different from those present in the mechanically prepared one. Provided the microstructure of the grain boundaries is the relevant difference between both materials, it is obvious from the very similar conductivity behaviour of ball milled nanocrystalline and amorphous LiNbO₃ that the interfacial regions of the mechanically prepared material have an amorphous-like structure.

This finding is in agreement with the XRD, TEM and EXAFS results uncovering the amorphous structure of the interfacial regions. The same result was also concluded indirectly from measurements probing local Li diffusion *via* ^{7}Li spin–lattice relaxation NMR, previously performed in our laboratory.[27] The amorphous grain boundaries, which dominate the Li transport in the ball milled sample, represent fast conduction pathways resulting in a high Li conductivity with a low activation energy. However, these fast diffusion pathways seem to be absent in chemically prepared nanocrystalline LiNbO₃. The XRD results indicate, and the TEM images show rather clearly, that the chemically prepared sample is composed of nanometer sized grains with thin grain boundary regions. Large amounts of an

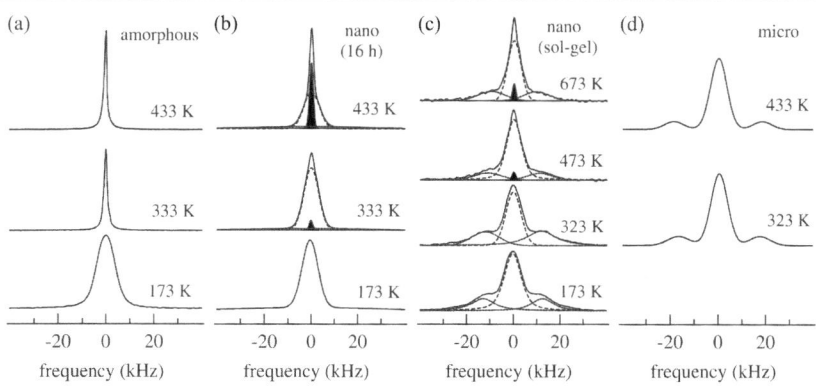

Fig. 7 Temperature dependent ^{7}Li NMR spectra ($\omega_0/2\pi = 77.7$ MHz) of (a) amorphous, (b) mechanically prepared nanocrystalline, (c) chemically (sol-gel) prepared nanocrystalline, and (d) microcrystalline LiNbO$_3$ delivered by Alfa Aesar. Narrow contributions A_f of the respective nanocrystalline samples are shown by filled areas, whereas the broad contributions are indicated by dashed lines. Quadrupole contributions to the NMR line of the sol-gel nanocrystalline material are represented by solid lines.

amorphous phase, and thus fast diffusion pathways, visible in the case of ball milled LiNbO$_3$, are missing. Thus, it can be easily understood why the chemically prepared nanocrystalline sample shows a relatively low conductivity, comparable to that of the microcrystalline material. In sol-gel prepared nanocrystalline samples there is no way to enhance the ionic conductivity in the same way as is easily possible for the ball milled samples.

It should be noted that sol-gel synthesis proves to be a highly suitable technique for the synthesis of nanosized crystallites with relatively thin and ordered interfacial regions. Quite recently, Niederberger *et al.*[41] reported similar results for LiNbO$_3$ prepared by a soft-chemistry route.

4.2 Solid-echo and relaxation NMR measurements

4.2.1 ^{7}Li NMR lineshape analysis. Information about structural aspects and Li dynamics can be obtained by recording ^{7}Li NMR spectra of the different forms of LiNbO$_3$. In Fig. 7 static ^{7}Li NMR spectra at selected temperatures between 153 and 673 K are shown for the amorphous, two nanocrystalline and microcrystalline samples. As ^{7}Li is a quadrupole nucleus with a spin quantum number of 3/2, it is expected that the Li spectra are composed of a central NMR line and two satellite contributions with minor intensity shifted to lower and higher resonance frequencies. The splitting of the central line into three NMR lines is simply due to the interaction of the quadrupole moment of the nuclei with a non-vanishing electric field gradient produced by the electric charge distributions in the neighbourhood of the Li site.

In the case of amorphous LiNbO$_3$ (Fig. 7(a)) the central transition is mainly observed when the NMR signal is recorded *via* the solid-echo pulse sequence. With increasing temperature the central line starts to narrow due to the onset of Li motions with jump rates in the kHz range. Its shape can be described with a Gaussian line at low temperatures, *i.e.*, in the rigid lattice regime below 200 K, and with a Lorentzian line at higher temperatures above, *e.g.*, 300 K.

The same description is valid also for the central transition of the microcrystalline sample (Fig. 7(d)). The onset of motional narrowing for microcrystalline LiNbO$_3$ starts at about 650 K, thus being shifted by some hundreds of degrees to higher temperatures due to the much lower Li diffusivity as compared to that in the highly disordered sample (see below and *cf.* Fig. 9). In the powder spectra of coarse grained

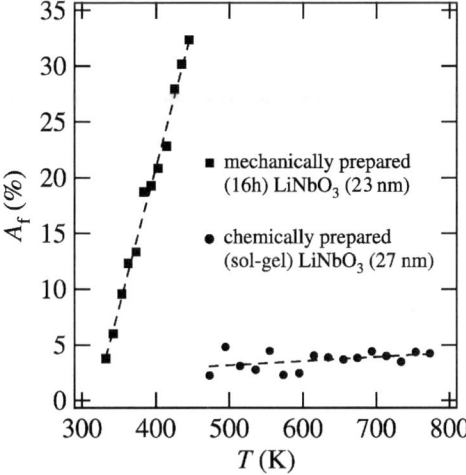

Fig. 8 Area fraction A_f (%) of the narrow ^{7}Li NMR contribution to the total area of the NMR line as a function of temperature for the two nanocrystalline materials prepared either mechanically by ball milling for 16 h (■) or chemically *via* the sol-gel technique (●).

LiNbO$_3$ a pair of well-defined quadrupole satellites appears next to the central NMR transition. This quadrupole splitting was also observed for single-crystalline LiNbO$_3$. The observation of satellite transitions is related to the well-defined crystalline, less defective structure of coarse grained LiNbO$_3$ as compared to that of amorphous LiNbO$_3$. In the latter, because of the disordered structure, the satellite lines are broadened or smeared out due to a relatively large distribution of electric field gradients and thus quadrupole coupling constants. In amorphous LiNbO$_3$ the quadrupole contributions can hardly be seen, forming only a broad background.

The same holds true for the nanocrystalline sample which was prepared by ball milling and which consists of a large amount of amorphous regions (Fig. 7(b)). Contrary to that, the sol-gel prepared nanocrystalline sample shows a higher degree of crystallinity so the satellite transitions can be clearly seen (Fig. 7(c)).

Thus, the degree of disorder can be probed by studying satellite contributions to the respective spectra. The two nanocrystalline samples take a medial position between amorphous and microcrystalline LiNbO$_3$; whereas the chemically prepared one is structurally more related to the microcrystalline sample and the ball milled material is similar to the amorphous one.

In addition to structural characteristics dynamical aspects can also be studied very well *via* recording ^{7}Li NMR spectra.[6,21,22,27] Contrary to the amorphous and microcrystalline material, the central line of the 16 h high energy ball milled sample shows two contributions at the same frequency when a temperature of about 330 K is attained. The narrow contribution, which can be described with a Lorentzian shape, is attributed to the fast subset of Li ions located in the structurally disordered interfacial regions. The broad Gaussian contribution represents the Li ions, which are still immobile at this temperature.[6] Similar results were obtained for the nanocrystalline anionic conductor CaF$_2$.[22,42] The NMR results confirm that this nanocrystalline LiNbO$_3$ sample is characterized by a heterogeneous, *i.e.*, two-component structure of crystalline grains and amorphous grain boundaries.

The fraction of mobile ions is given by the area of the narrow contribution relative to the area of the total resonance line. Fig. 8 shows the relative contribution A_f of the narrow NMR line as a function of temperature. At about 450 K the fraction of mobile ions, dominating the Li transport process in this material, reaches a value of

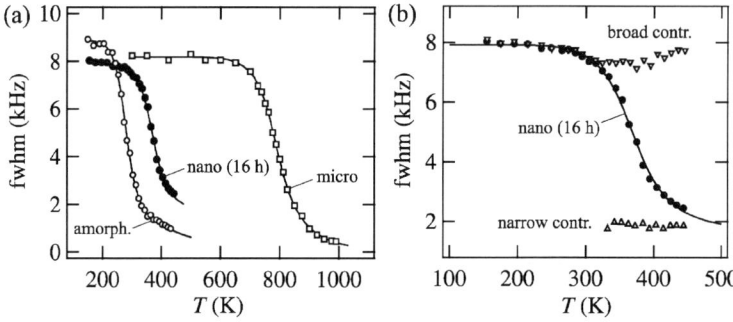

Fig. 9 (a) ^{7}Li NMR linewidths (fwhm: full width at half maximum of the overall NMR line) as a function of temperature for amorphous (○) and microcrystalline (□) LiNbO$_3$, as well as for the nanocrystalline materials prepared by high energy ball milling of the microcrystalline material for 16 h (●). The linewidths were obtained from ^{7}Li spectra which were acquired at a resonance frequency of 77.7 MHz and using the solid-echo pulse sequence. (b) Decomposition of the ^{7}Li NMR line of nanocrystalline LiNbO$_3$ prepared by ball milling for 16 h into a broad (∇) and narrow contribution (△). The linewidth of the entire NMR line (●) of Fig. 9(a) is shown for comparison. In both figures solid lines are to guide the eye. The data of microcrystalline LiNbO$_3$ are taken from ref. 6.

nearly 35% which is fairly consistent with the results from EXAFS measurements (see above and ref. 38). To avoid any grain growth the NMR experiments had to be restricted to an upper temperature limit of approximately 450 K (see above, section 2.2).

In contrast to this result, the heterogeneous motional narrowing of the sol-gel prepared sample is much less pronounced than it is in the case of the milled sample. A narrow contribution cannot be separated before a temperature of about 470 K is reached. This narrow NMR line shows a constant relative contribution of about 4%, indicating once more the relatively thin grain boundary regions of sol-gel prepared nanocrystalline LiNbO$_3$.

In order to ascertain the Li dynamics in a more quantitative manner, the linewidths of the different forms of LiNbO$_3$ were determined as a function of temperature. Fig. 9 shows this temperature dependence. For amorphous LiNbO$_3$ the NMR line starts to narrow at about 225 K, whereas the onset of motional narrowing of the microcrystalline material is shifted to about 700 K. The onset temperature T_{onset} is related *via* the empirical expression of Waugh and Fedin[43]

$$E_A^{MN}/eV = 1.617 \times 10^{-3} T_{onset}/K \qquad (3)$$

with the activation energy E_A^{MN} of the Li hopping process. The activation energy is estimated to be about 0.32 eV for amorphous LiNbO$_3$ and about 0.46 eV for the ball milled sample, whereas it is much higher for the microcrystalline source material, 1.07 eV.[6,27,44] Motional narrowing of the ball milled material is more comparable to that of amorphous LiNbO$_3$, as has also been reported quite recently by Chadwick *et al.*[28] In the range below 470 K the sol-gel prepared nanocrystalline material reveals no sharp decrease in linewidth with increasing temperature, like the other samples do, since its Li diffusivity is not dominated by fast diffusion pathways located in the interfacial regions. Thus, the same trend of Li diffusivity is found by NMR solid-echo experiments as was obtained by conductivity measurements (see section 4.1 and ref. 23). In order to analyze the two-component shape of the ^{7}Li NMR spectra of the ball milled nanocrystalline sample, the linewidths of the two distinct contributions are shown separately in Fig. 9(b) as a function of temperature. The emergence of the narrow contribution and the narrowing of the total line occurs simultaneously at about 330 K. At this temperature the NMR linewidth of the corresponding amorphous sample is already extremely narrowed.

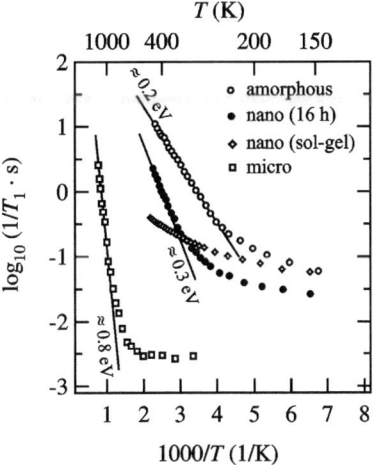

Fig. 10 ^{7}Li NMR spin–lattice relaxation rates vs. reciprocal temperature for amorphous (○) and nanocrystalline LiNbO$_3$ which was either prepared by ball milling for 16 h (●) or by sol-gel synthesis (◊). The data of commercially available microcrystalline LiNbO$_3$ (□) are shown for comparison. The data of microcrystalline LiNbO$_3$ are taken from ref. 6.

Irrespective of grain growth we have also intentionally performed some high temperature measurements up to 750 K on the sol-gel prepared nanocrystalline sample (not shown here, see ref. 37 for details). The narrow contribution started to be separable only above 470 K. It has a fwhm comparable to that of the amorphous form. As expected, it showed only a minor effect on the narrowing of the total line due to its nearly temperature independent small area fraction A_f (Fig. 8).

4.2.2 ^{7}**Li NMR spin–lattice relaxation.** In addition to the analysis of solid-echo NMR spectra, the effect of grain size and microstructure on the Li diffusion process was also investigated by ^{7}Li NMR spin–lattice relaxation measurements. In Fig. 10 the ^{7}Li NMR spin–lattice relaxation rates $T_1^{-1}(T^{-1})$ of the amorphous, two nanocrystalline and microcrystalline samples are shown. The data were recorded at a resonance frequency of 77.7 MHz. Except for the nanocrystalline sample prepared by the sol-gel route, the plots of T_1^{-1} vs. reciprocal temperature consist of a superposition of a diffusion induced and a background contribution:[27]

$$T_1^{-1} = T_{1\mathrm{diff}}^{-1} + T_{1\mathrm{bgr}}^{-1} \tag{4}$$

While the diffusion induced spin–lattice relaxation rates T_1^{-1} follow Arrhenius behaviour,

$$T_{1\,\mathrm{diff}}^{-1}(T^{-1}) = A''\exp(E_{A,\mathrm{SLR}}/k_B T), \tag{5}$$

which represents the low temperature flank of the characteristic $T_{1\mathrm{diff}}^{-1}(T_1^{-1})$ peak,[22] the background contribution $T_{1\mathrm{bgr}}^{-1}$ shows a weaker-than-activated temperature behaviour which can usually be described with a simple power law $T_{1\mathrm{bgr}}^{-1}(T^{-1}) \propto T^k$, see ref. 27. The background relaxation is mostly caused by lattice vibrations and/or paramagnetic impurities. However, in the case of microcrystalline LiNbO$_3$, this background contribution, which shows up below $T = 500$ K, seems to be temperature independent. Above 800 K the relaxation rate is predominantly induced by diffusion of Li ions with an activation energy of about 0.8 eV. The diffusion induced relaxation rate is proportional to the mean jump rate τ^{-1} of the Li ions. The corresponding diffusion induced $T_{1\mathrm{diff}}^{-1}(T^{-1})$ flank of the amorphous material is shifted to much lower temperatures, verifying much faster Li diffusion in this material. Above room temperature the relaxation rates are mainly induced by Li diffusion, the activation energy is drastically reduced and turned out to be about 0.22 eV. Recently, we also reported ^{7}Li NMR relaxation measurements for the amorphous sample which was prepared by partial hydrolysis of the lithium niobium double alkoxide (cf. section. 2.1 and ref. 27). For that material an activation energy of about 0.27 eV was found by means of ^{7}Li NMR relaxation.[27] Such small differences in activation energies are expected since both materials are prepared by slightly different preparation routes (cf. section 2.1), and will not be discussed further in the present paper.

Similar to the results for amorphous LiNbO$_3$, the diffusion induced flank of the heterogeneously structured, ball milled nanocrystalline LiNbO$_3$ sample is also shifted to lower temperatures in comparison with that of the coarse grained (microcrystalline) material. The activation energy of about 0.3 eV for the nanocrystalline material is comparable to that obtained for the completely disordered sample.

As expected, the spin–lattice relaxation behaviour of the chemically prepared nanocrystalline sample is different from that of the mechanically prepared one. In the investigated temperature range up to $T = 460$ K no sharp increase in $T_1^{-1}(T^{-1})$ is observed. Instead, only a non-diffusive relaxation background is detected whose temperature dependence is similar to that observed for the nanocrystalline sample prepared by ball milling for 16 h. Thus, ^{7}Li NMR spin–lattice relaxation measurements confirm once more that the overall Li diffusivity in the sol-gel prepared nanocrystalline material is much lower than that in the homogeneously or heterogeneously disordered samples, i.e. the amorphous and the nanocrystalline sample prepared by high energy ball milling, respectively.

5 Conclusion

Li transport in two differently prepared nanocrystalline LiNbO$_3$ samples with nearly the same average particle size in the range between 20–30 nm was extensively investigated by means of impedance spectroscopy and solid-state ^{7}Li NMR spectroscopy. The results were compared with those obtained for the micro- and single-crystalline as well as the amorphous form of LiNbO$_3$. Nanocrystalline LiNbO$_3$ was either prepared mechanically by high energy ball milling or chemically via the sol-gel technique. In both cases, nanostructuring results in enhanced Li diffusivity as compared to the cation transport in coarse grained or single-crystalline lithium niobate.

It was unambiguously shown by HRTEM images and EXAFS measurements that the grain boundaries of nanocrystalline LiNbO$_3$ prepared by ball milling have an amorphous structure. These results are fully corroborated by the data obtained from impedance spectroscopy and NMR relaxation experiments on the four different forms of LiNbO$_3$. Li ions which reside in the disordered interfacial regions have access to fast diffusion pathways with low activation energies. As shown by dc-conductivity measurements and ^{7}Li NMR, the overall Li transport is dominated by these fast charge carriers. However, although Li diffusivity in sol-gel prepared nanocrystalline is also enhanced, its Li transport behaviour is more related to that of the coarse grained material than to amorphous LiNbO$_3$. HRTEM images as well as XRD results show that the grain boundaries are thinner and have a much more ordered structure than in the ball milled samples. Thus, we could clearly show that transport properties are strongly related to the nature of the preparation-dependent grain boundary microstructure.

The results concerning the microstructures of interfacial regions in LiNbO$_3$ are in agreement with recent conclusions from EXAFS measurements on other, but Li free, nanocrystalline oxides such as sol-gel prepared SnO$_2$, ref. 45 and 46, sol-gel and mechanically prepared ZrO$_2$, ref. 47, as well as Y-stabilised cubic ZrO$_2$, ref. 48, prepared by polymer spin coating.

The sol-gel technique, like other soft-chemistry routes, gives access to nanocrystalline materials with grain boundaries similar to those in bulk materials, whereas ball

 This journal is © The Royal Society of Chemistry 2006

milling leads to oxides with a higher level of disorder in the interfacial regions. In cases where larger amounts of a highly conducting material are needed, the latter preparative route[30] may be advantageous. It also allows easy extension to composite nanocrystalline ceramics.[5,7,9]

Acknowledgements

The authors wish to thank C. Rüscher, H. Behrens and T. Gesing (Institute of Mineralogy, Hannover University) as well as M. Klüppel, J. Meier and B. Huneke (German Institute for Rubber Technology, Hannover) for their help and valuable technical support. The cooperation of D. Bork, R. Winter, D.-M. Fischer, N. Greaves and A. Dent in the EXAFS project is highly appreciated. M. M. would like to acknowledge a grant from the Center of Solid State Chemistry and New Materials (Hannover University). P. H. is grateful to G. E. Murch and I. V. Belova for their hospitality at the University of Newcastle (Callaghan, Australia), where part of this article was written.

References

1   S. Chandra, *Superionic Solids, Principles and Applications*, North-Holland, Amsterdam, 1981.
2   *Lithium Ion Batteries*, ed. M. Wakihara and O. Yamamoto, Wiley-VCH, Weinheim, 1998.
3   M. S. Whittingham, *Chem. Rev.*, 2004, **104**, 4271.
4   J.-M. Tarascon and M. Armand, *Nature*, 2001, **414**, 359.
5   P. Heitjans and S. Indris, *J. Phys.: Condens. Matter*, 2003, **15**, R1257.
6   (*a*) D. Bork and P. Heitjans, *J. Phys. Chem. B*, 1998, **102**, 7303; (*b*) D. Bork and P. Heitjans, *J. Phys. Chem. B*, 2001, **105**, 9162.
7   S. Indris, P. Heitjans, H. E. Roman and A. Bunde, *Phys. Rev. Lett.*, 2000, **84**, 2889.
8   S. Indris and P. Heitjans, *J. Non-Cryst. Solids*, 2000, **307**, 555.
9   M. Wilkening, S. Indris and P. Heitjans, *Phys. Chem. Chem. Phys.*, 2003, **5**, 2225.
10   (*a*) W. Puin and P. Heitjans, *Nanostruct. Mater.*, 1995, **6**, 885; (*b*) W. Puin, S. Rodewald, R. Ramlau, P. Heitjans and J. Maier, *Solid State Ionics*, 2000, **131**, 159.
11   U. Brossmann, R. Würschum, U. Södervall and H.-E. Schaefer, *Nanostruct. Mater.*, 1999, **12**, 871.
12   P. Knauth, *J. Solid State Electrochem.*, 2002, **147**, 115.
13   P. Knauth and H. L. Tuller, *Solid State Ionics*, 2000, **136–137**, 1215.
14   Y.-M. Chiang, E. B. Lavik, I. Kosacki, H. L. Tuller and J. Y. Ying, *J. Electroceram.*, 1997, **1**, 7.
15   A. Tschöpe, E. Sommer R. and Birringer, *Solid State Ionics*, 2001, **139**, 255.
16   J. Lee, J. H. Hwang, J. J. Mashek, F. O. Mason, A. E. Miller and R. W. Siegel, *J. Mater. Res.*, 1995, **10**, 2295.
17   C.-W. Nan, A. Tschöpe, S. Holten, H. Kliem and R. Birringer, *J. Appl. Phys.*, 1999, **85**, 7735.
18   G. Li, L. Li, S. Feng, M. Wang, L. Zhang and X. Yao, *Adv. Mater.*, 1999, **11**, 146.
19   P. Mondal, A. Klein, W. Jaegermann and H. Hahn, *Solid State Ionics*, 1999, **118**, 331.
20   U. Brossmann, G. Knöner, H.-E. Schaefer and R. Würschum, *Rev. Adv. Mater. Sci.*, 2004, **6**, 7.
21   P. Heitjans, S. Indris and M. Wilkening, in *Solid-State Diffusion and NMR in: Diffusion Fundamentals*, ed. J. Kärger, F. Grinberg and P. Heitjans, Leipziger Universitätsverlag, Leipzig, 2005, pp. 226–245.
22   P. Heitjans, A. Schirmer, S. Indris, in *NMR and β-NMR Studies of Diffusion in Interface-Dominated and Disordered Solids in: Diffusion in Condensed Matter-Methods, Materials, Models*, ed. P. Heitjans and J. Kärger, Springer, Berlin/Heidelberg, 2005, pp. 367–415.
23   M. Masoud and P. Heitjans, *Defect Diffus. Forum*, 2005, **237–240**, 1016.
24   S. Lanfredi and A. Rodrigues, *J. Appl. Phys.*, 1999, **86**, 2215.
25   A. Glass, K. Nassau and T. Negran, *J. Appl. Phys.*, 1978, **49**, 4808.
26   K. Nassau, *J. Non-Cryst. Solids*, 1980, **42**, 423.
27   M. Wilkening, D. Bork, S. Indris and P. Heitjans, *Phys. Chem. Chem. Phys.*, 2002, **4**, 3246.
28   A. V. Chadwick, M. J. Pooley and S. L. P. Savin, *Phys. Status Solidi C*, 2005, **2**, 302.
29   S. Ono, H. Mochizuki and S. Hirano, *J. Ceram. Soc. Jpn.*, 1996, **104**, 574.
30   S. Indris, D. Bork and P. Heitjans, *J. Mater. Synth. Process.*, 2000, **8**, 245.
31   H. Gleiter, *Prog. Mater. Sci.*, 1989, **33**, 223.

32 V. V. Srdic, M. Winterer, G. Miehe and H. Hahn, *Nanostruct. Mater.*, 1999, **12**, 95.

33 H. Natter, M. Schmelzer, S. Janßen and R. Hempelmann, *Ber. Bunsen-Ges. Phys. Chem.*, 1997, **101**, 1706.

34 *Nanophase Materials*, ed. G. Hadjypanayis and R. W. Siegel, Kluwer Academic Publishers, Netherlands, 1994.

35 (*a*) H. P. Klug and L. E. Alexander, *X-Ray Diffraction Procedures*, Wiley, New York, 1974; (*b*) P. Scherrer, *Göttinger Nachrichten*, 1918, **2**, 98.

36 (*a*) W. Rachinger, *J. Sci. Instrum.*, 1948, **25**, 254; (*b*) B. Warren, *X-Ray Diffraction*, Addison Wesley, New York, 1969.

37 M. Masoud, PhD thesis, Hannover University, 2005.

38 M. Masoud, P. Heitjans, D. Bork, R. Winter, N. Greaves, D.-M. Fischer and A. Dent, to be submitted.

39 M. J. Pooley and A. V. Chadwick, *Radiat. Eff. Defects Solids*, 2003, **158**, 197.

40 E. Fukushima and S. B. W. Roeder, *Experimental Pulse NMR*, Addison-Wesley, Reading, 1981.

41 M. Niederberger, N. Pinna, J. Polleux and M. Antonietti, *Angew. Chem., Int. Ed.*, 2004, **43**, 2270.

42 W. Puin, P. Heitjans, W. Dickenscheid and H. Gleiter, *Defects in Insulating Materials*, ed. O. Kanert and J. Spaeth, World Scientific, Singapore, 1993.

43 J. S. Waugh and E. I. Fedin, *Sov. Phys. Solid State*, 1963, **4**, 1633.

44 Y. Xia, N. Machida, X. Wu, C. Lakeman, L. van Wüllen, F. Lange, Ch. Levi and H. Eckert, *J. Phys. Chem. B*, 1997, **101**, 9901.

45 S. Davis, A. Chadwick and J. Wright, *J. Phys. Chem. B*, 1997, **101**, 9180.

46 A. V. Chadwick, *Radiat. Eff. Defects Solids*, 2003, **158**, 21.

47 A. V. Chadwick, M. J. Pooley, K. E. Rammutla, S. L. P. Savin and A. Rougier, *J. Phys.: Condens. Matter*, 2003, **15**, 431.

48 G. Rush, A. V. Chadwick, I. Kosacki and U. Anderson, *J. Phys. Chem. B*, 2000, **104**, 9597.

A ^{27}Al, ^{29}Si, ^{25}Mg and ^{17}O NMR investigation of alumina and silica Zener pinned, sol-gel prepared nanocrystalline ZrO$_2$ and MgO

L. A. O'Dell,*[a] S. L. P. Savin,[b] A. V. Chadwick[b] and M. E. Smith*[a]

Received 8th February 2006, Accepted 4th May 2006
First published as an Advance Article on the web 21st July 2006
DOI: 10.1039/b601928e

Alumina and silica Zener pinning particles in sol-gel prepared nanocrystalline ZrO$_2$ and MgO have been characterised using ^{27}Al and ^{29}Si MAS NMR after annealing at various temperatures up to 1200 °C. The structures of the pinning phases were found to differ not just between the two metal oxide systems but also depending on the exact method of manufacture. Three distinct transitional alumina phases have been observed in different alumina-pinned samples annealed at 1200 °C, one in particular identified by a peak at a shift of 95 ppm in the ^{27}Al NMR spectrum. Both the alumina and silica pinning phases reacted with the MgO nanocrystals, forming spinel in the case of alumina, and enstatite and forsterite in the case of silica. Despite reacting readily with the MgO, the silica pinning particles were effective at restricting grain growth, with 11 nm MgO nanocrystals remaining after annealing at 1000 °C.

1. Introduction

Materials consisting of crystals with one or more dimension measuring less than 100 nm are known as nanocrystalline.[1–3] When materials exist in this state, they often exhibit physical[4] and chemical[5] properties that are very significantly different to their coarse-grained or bulk equivalents. For example, nanocrystalline metal oxides can show ionic conductivity up to five orders of magnitude higher than their bulk counterparts.[6] Properties such as these have made nanocrystalline metal oxides the subject of intense study over the past decade due to their potential in applications such as electrochemistry,[7] catalysis,[8,9] biological detection[10] and many others.[11–15] ZrO$_2$ and MgO are two of the most thoroughly studied metal oxide systems, and their nanocrystalline forms are most commonly used in catalysis, with their high oxygen ion conductivity making them especially suited to redox reactions. Their highly reactive nature is thought to arise from a combination of their high surface area and an abundance of corner/edge sites due to the small size of the crystals.

Nanocrystalline ZrO$_2$ has already been shown to have extremely useful catalytic properties.[16–19] Zirconia can exist in several phases in the nanocrystalline state due to size-stabilization of the tetragonal phase in crystallites below 30 nm in diameter,[20] with typical ZrO$_2$ nanocrystal sizes in as-prepared samples varying between 1.5[19] and 36 nm.[16] This is due to the tetragonal phase having a lower free surface energy

[a] Department of Physics, University of Warwick, Coventry, UK CV4 7AL. E-mail: L.A.O-Dell@warwick.ac.uk; M.E.Smith.1@warwick.ac.uk
[b] School of Physical Sciences, University of Kent, Canterbury, Kent, UK CT2 7NR

than the monoclinic phase and hence being thermodynamically favoured at small crystallite sizes.[21] The tetragonal phase of zirconia is desirable in many catalytic applications, and is most commonly produced by doping with yttrium.

Nanocrystalline MgO has also shown great promise in catalytic applications.[22–25] It has been demonstrated that its reactivity per unit surface area is higher than for bulk MgO,[26] and that this reactivity is not significantly reduced by pressing the powder sample into a shaped pellet at moderate pressures.[27] Nanocrystalline MgO typically exists as irregular polyhedral crystallites[24] which are generally smaller than ZrO_2 nanocrystals (as small as 4 nm in diameter in some as-prepared samples[22,25,26]). This is in part due to its high ionicity and hence high lattice energy, making crystallite growth less favourable. However, in both systems rapid and significant crystallite growth can occur upon heating, and as-prepared crystal diameters can increase by an order of magnitude or more at temperatures below 1000 °C. Growth can be a serious problem because as the crystals grow above the nano regime the metal oxide will revert to its bulk properties. Since certain applications involve high operating temperatures and some manufacturing processes require heating of the system, a solution to this problem is desired.

Zirconia nanocrystals have been shown to grow by grain boundary diffusion,[28] and so one method that should be successful in restricting growth in this system is Zener pinning.[29] This involves a small amount of a second phase being introduced to the system, which in the final product exists as discrete particles in the nanocrystal interface region, pinning the grain boundaries in place. Zener pinning is advantageous over other methods of restricting crystallite growth such as doping[30] or surface treatment[31] since the metal oxide remains in a pure form with most of its surface unaffected. The presence of a second phase can also have advantageous effects on sample reactivity, for example a mixture of MgO and Al_2O_3 has been shown to have enhanced adsorptive properties over the two separate materials.[32,33] The Zener pinning method has already been studied and shown to be effective at restricting nanocrystal growth,[34–39] but almost nothing is known about the structure of the pinning particles themselves and their interactions with the nanocrystals. Since the pinning particles are thought to be at least an order of magnitude smaller than the nanocrystals themselves, techniques such as XRD and TEM cannot easily yield useful information about them. Nuclear magnetic resonance, however, is an extremely effective probe for this type of material and has recently been successfully used to study silica and alumina pinning particles in a nanocrystalline SnO_2 system.[40] ^{17}O, ^{25}Mg and ^{1}H NMR investigations have already been published on pure ZrO_2 and pure MgO nanocrystals,[41,42] and this paper presents an NMR investigation into silica and alumina pinning particles in the ZrO_2 and MgO systems.

A wide range of techniques have been used to manufacture nanocrystalline metal oxides, such as ball milling,[43] laser evaporation,[44] freeze-drying,[45] reflux-digestion,[46] emulsion precipitation,[15] hydrothermal decomposition[47] and ballistic deposition.[48] However, a commonly favoured method of synthesis is via the sol-gel route. This method provides the ability to control the homogeneity and physical characteristics of the final material by use of different precursor materials (e.g. a zirconium propoxide precursor has been shown to produce the tetragonal form of ZrO_2[17]), hydrolysis routes and heat treatments. It also allows a precise amount of the pinning phase to be introduced homogeneously throughout the sample, which is important since it has been demonstrated that a non-uniform distribution of pinning particles can inhibit their pinning ability.[49] The sol-gel method requires annealing at temperatures between 600 and 800 °C to obtain a fully dried and pure sample. Although a novel annealing method has been proposed to restrict growth without pinning,[50] the minimum particle size obtained by this method was 60 nm, which is significantly larger than the crystallite sizes desired for many applications, and so Zener pinning should be especially useful when manufacturing nanocrystalline materials this way.

2. Experimental

In previous work conducted by our group,[51] the optimum weight quantities for effective Zener pinning of metal oxide nanocrystals were found to vary between 3 and 15% depending on the pinning material and the metal oxide in question. In this investigation, for the sake of consistency, each of the pinned metal oxide systems have been manufactured to contain either 10% alumina or 15% silica by weight.

Three different alumina-pinned ZrO_2 samples were manufactured, hereafter designated CP (Co-hydrolysis method, Propoxide metal precursor), AP (Addition method, Propoxide metal precursor) and AC (Addition method, Chloride metal precursor). The CP-ZrO_2–Al_2O_3 was made by mixing 164 ml zirconium iso-propoxide (70 wt% in propanol, Aldrich Chemical Co.), 25 ml aluminium tri-sec-butoxide (Aldrich Chemical Co.) and 50 ml butan-2-ol (Aldrich Chemical Co.). This solution was removed from the glove box and stirred for 1 h and then water was added, with stirring, until the solution gelled. The solid was then dried at 80 °C. Sample AP-ZrO_2–Al_2O_3 was manufactured using 164 ml zirconium iso-propoxide, to which water was added until the solution gelled. The solid was then dried at 80 °C and ground using a pestle and mortar. 100 ml butan-2-ol was added to the white powder and mixed, followed by 25 ml aluminium tri-sec-butoxide. The subsequent slurry was dried at 80 °C overnight. Lastly, sample AC-ZrO_2–Al_2O_3 was made by dissolving 117.7 g zirconyl chloride octahydrate (Aldrich Chemical Co.) in approximately 150 ml of water. Ammonia was added drop-wise whilst stirring until the solution gelled. The white solid was then washed with dilute nitric acid to remove chloride ions, then dried at 100 °C and ground using a pestle and mortar. 75 ml of butan-2-ol was subsequently added to the white powder and mixed, followed by 25 ml aluminium tri-sec-butoxide, and the mixture was stirred until the solution gelled. The white solid was then dried at 100 °C. Each of these methods resulted in 10% alumina by weight in each sample.

The silica-pinned ZrO_2 was manufactured in a way similar to the CC method described above. 111.1 g of zirconyl chloride octahydrate was dissolved in approximately 150 ml of water. A separate solution of 27.8 ml tetraethylorthosilicate (TEOS, Aldrich Chemical Co.) in approximately 50 ml ethanol was prepared. These two solutions were mixed and stirred for 1 h. Ammonia was then added drop-wise whilst stirring until the solution gelled. The white solid was washed with dilute nitric acid to remove chloride ions, then dried at 100 °C overnight. This resulted in a sample containing 15% silica by weight. Another sample was manufactured using 70% ^{17}O enriched water, using a zirconium iso-propoxide precursor.

The alumina-pinned MgO was made by adding water to 1476 ml magnesium methoxide (6–10% in methanol, Aldrich Chemical Co.) whilst stirring until the solution gelled. The gel was then heated at 80 °C until fully dry, after which 200 ml butan-2-ol was added to the powder to form a slurry. 25 ml aluminium tri-sec-butoxide was then added to this and mixed until it gelled. The solid was then dried at 80 °C, resulting in 10% alumina by weight.

For the silica-pinned MgO, 27.8 ml TEOS was added to 1395 ml magnesium methoxide and the solution was stirred for 1 h. Water was subsequently added whilst stirring, until the solution gelled. The sample was then dried at 80 °C, resulting in 15% silica by weight in the final product.

After grinding each of the dry products with a pestle and mortar, portions of each sample were removed and annealed separately under air for 1 h at 400, 600, 800, 1000 and 1200 °C.

^{29}Si spectra were obtained using a 7.05 T magnet (giving a ^{29}Si Larmor frequency of 59.6 MHz) with a 7 mm MAS probe spinning the samples at 4 kHz. These were referenced using the signal from tetramethylsilane set to 0 ppm. The pulse width used was around 1.5 µs, which produced a tip angle of $\sim 30°$. The recycle delay was set to 20 s, although the average acquisition time for each sample was generally longer than for the other nuclei at 24–48 h due to the relatively low natural abundance of the ^{29}Si

isotope, 4.7%, and the small amount of silica in the samples, 15% by weight. Unless otherwise stated, all ^{27}Al experiments presented here were conducted in a 14.1 T magnetic field (^{27}Al Larmor frequency 156.4 MHz) with a 3.2 mm probe spinning samples at $\sim$20 kHz. These spectra were referenced against yttrium aluminium garnet (YAG), with the signal corresponding to the octahedrally coordinated aluminium site set to 0.7 ppm so that chemical shifts reported are relative to the primary shift reference of 1M $[Al(H_2O)_6]^{3+}$ at 0 ppm. A pulse width of 0.5 μs was used to give a small tip angle of $\sim$15° with a recycle delay of 1 s and an overall acquisition time of between 5 and 15 min for each sample. Certain ^{27}Al spectra were also recorded with similar parameters at 18.8 T using a 2.5 mm probe, and at 8.45 T using a 3.2 mm probe. ^{25}Mg experiments were carried out in a 14.1 T magnet, (at a ^{25}Mg Larmor frequency of 36.7 MHz) using a 4 mm probe spinning samples at 8–10 kHz. The spectra were referenced against bulk MgO, with the single, narrow peak from this set to 0 ppm. A pulse width of 2 μs was used (corresponding to a magnetisation tip angle of approximately 90°), with a recycle delay of 60 s and an overall acquisition time of between 30 min and 24 h for each sample depending on the signal strength. All MAS NMR spectra were recorded using a one-pulse sequence with the Spinsight program and Varian Chemagnetics CMX, Infinity or Infinity Plus spectrometers. The dmfit2003 program,[52] and Quad Fit[53] were used to simulate the spectra and fit the peaks.

Static ^{17}O NMR spectra were recorded using a custom-built probe (see Fig. 1) in a 7.05 T magnetic field, corresponding to a ^{17}O Larmor frequency of 40.7 MHz. The sample was held inside a quartz container which itself was held in place by a copper rf coil. Heating of the sample was achieved *via* a resistive nickel–chromium coil wound in a bifilar way with its axis parallel to the magnetic field and the rf coil at the centre. This coil was held in place by a quartz tube and ceramic cement, and insulated with silica fibre material. The coil had a resistance of approximately 22 Ω, and sample temperatures of up to 1000 °C were easily achievable from this using a 100 W DC power supply. Calibration of the probe was conducted with a thermocouple placed at the position of the rf coil. At any given power setting, temperatures were reproducible to within ±5 °C. After adjusting the power, one hour was allowed for the temperature of the sample to reach equilibrium. The heated section of the probe was surrounded by a water jacket to keep the outside of the probe at ambient temperature. The pulse width used was 8 μs which was determined to be approximately equal to a 90° tip angle. T_1 measurements were conducted using the saturation recovery technique, in which a train of 90° pulses was used to saturate the magnetisation, and subsequently another 90° pulse was applied after a delay time

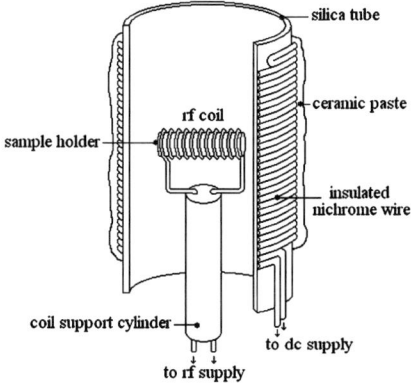

Fig. 1 A cut-away schematic diagram of the variable temperature NMR probe used for the static ^{17}O experiments.

τ, and the FID signal was measured in the usual way. The saturation pulses were separated by 500 µs (roughly the length of the FID), and values of τ ranged from 0.0001–7 s depending on the temperature of the sample. A decreasing signal to noise ratio with increasing temperature meant that a single T_1 measurement took up to 24 h to complete. H_2O was used as a reference, with the single sharp line from this set to 0 ppm. Since the ^{17}O isotope is only 0.037% naturally abundant, the sample made using the enriched water was used for all ^{17}O NMR experiments, and the signal arising from the quartz sample holder is negligible.

X-Ray diffraction experiments were conducted on a Philips PW 1720 diffractometer with Cu Kα radiation (1.514 Å) and the resultant data was analysed using the Traces v3.0 (Diffraction Technology, Pty) software, which also estimated the average metal oxide nanocrystal sizes via the Scherrer relation. Electron micrographs were obtained using a Zeiss SUPRA 55VP scanning electron microscope with a 10 keV electron beam, and a Jeol 2000FX transmission electron microscope with a 200 keV beam. Samples were prepared for the TEM by sonic dispersion in acetone and subsequent deposition on a carbon grid. Thermo-gravimetric analysis was performed on a small amount of the unannealed samples in air using a heating rate of 10 °C per minute using a TG-750 Stanton Redcroft instrument.

3. Results and discussion

3.1 ZrO_2–Al_2O_3

Fig. 2(a) shows the ^{27}Al MAS NMR spectra obtained for the alumina-pinned zirconia prepared by the co-hydrolysis route with the zirconium propoxide precursor (referred to as sample CP hereafter). Four, five and six-fold coordinated aluminium sites are present in the sample, giving signals at approximately δ = 70, 40, and 10 ppm, respectively. Fig. 3 illustrates the advantage of recording such spectra at more than one field. In Fig. 3(b), only three peaks are visible to the untrained eye, whereas at a higher field (Fig. 3(a)) the sixfold coordinated aluminium peak can be seen to consist of two distinguishable lines corresponding to two different octahedral aluminium environments. By simulating the aluminium spectra at two fields using a Gaussian distribution in the quadrupolar coupling constant C_Q for each peak, a more accurate estimation of the relative intensities (i.e. abundances), of each aluminium site can be made. Two-field simulation data are presented in Table 1.

A significant amount of fivefold coordinated aluminium exists in the CP samples annealed at temperatures up to 1000 °C. This fact, along with the asymmetric nature of each lineshape (due to a distribution in C_Q), indicates a high level of disorder. This disorder may arise either from the alumina existing as an amorphous phase or from alumina crystals that are extremely small and thus have a large surface to volume ratio, the surface being more disordered than the bulk. Most likely it is a combination of the two. The CP sample annealed at 1200 °C shows just 7% five-coordinated aluminium compared with 37% for the 1000 °C sample, suggesting crystallisation of the alumina between these temperatures. However, simulations of the lineshape suggest the presence of two tetragonal and two octahedral sites, indicating two types of alumina phases. From a qualitative consideration of these results, the presence of at least two transition aluminas[54–63] can be postulated. These rarer aluminas (η, ρ, χ, κ, δ and θ) can occur as intermediate states between the more stable γ and α alumina phases in the following sequences:

Bayerite/Boehmite → γ → δ → θ → α[56,58,61,62]
Gibbsite → χ → κ → α[59,63]

While the structures of θ[62] and κ[63] are known, the structures of χ and δ are not universally agreed upon since they contain a high degree of inherent disorder making them difficult to characterise.[62] The fact that they often co-exist in varying amounts

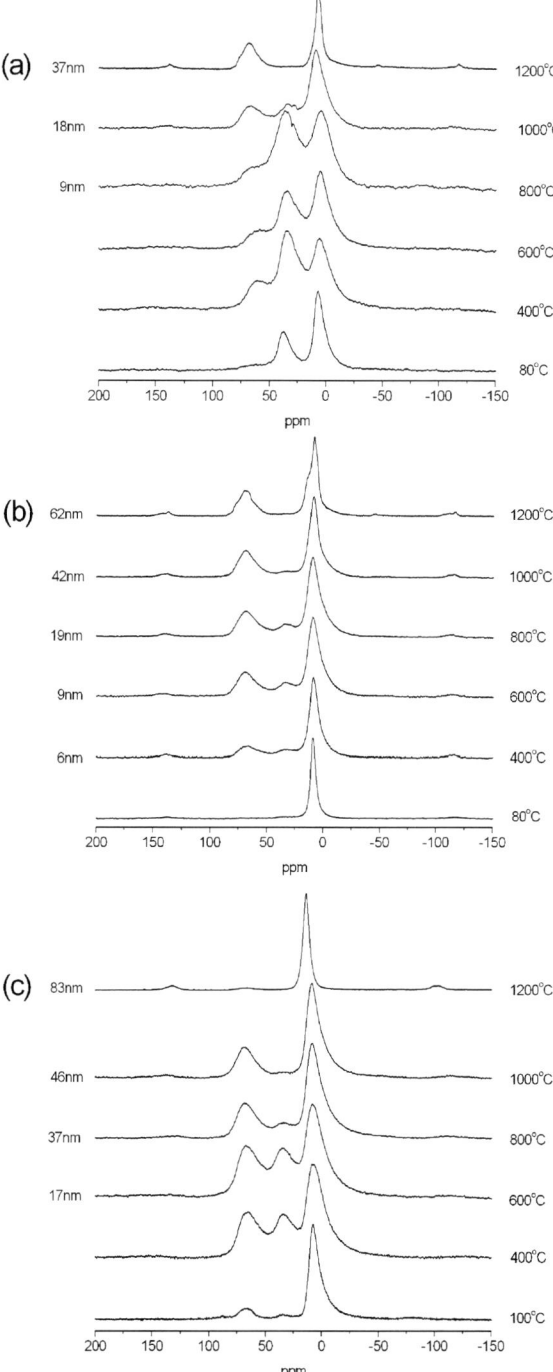

Fig. 2 ^{27}Al MAS NMR spectra obtained from the alumina-pinned zirconia samples, (a) CP-ZrO$_2$–Al$_2$O$_3$, (b) AP-ZrO$_2$–Al$_2$O$_3$ and, (c) AC-ZrO$_2$–Al$_2$O$_3$. Each sample was previously annealed at the temperatures indicated on the right. The lengths shown on the left of the spectra are the average metal oxide nanocrystal diameters as estimated by XRD using the Scherrer equation. Where these distances are not shown, the XRD spectra were too broad to enable reliable peak fitting. All NMR spectra will hereafter be presented in this format.

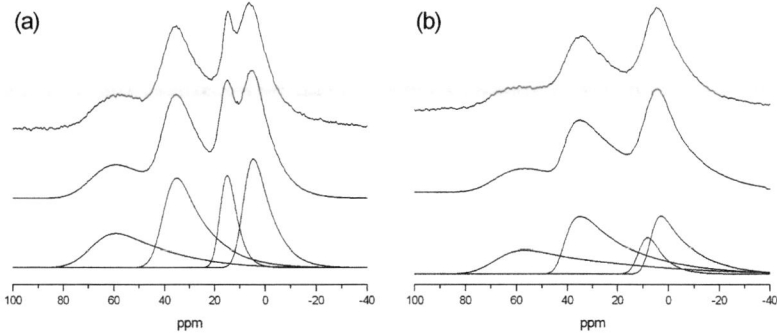

Fig. 3 Spectra and simulations of the ^{27}Al MAS NMR spectra of CP-ZrO$_2$–Al$_2$O$_3$ after annealing at 600 °C, for spectra obtained at (a) 18.8 T and (b) 14.1 T. For each plot, the top line is the experimental spectrum obtained, the middle line is the total simulated spectrum, and the bottom line shows the individual simulated peaks.

and also the fact that the temperatures at which they occur can vary with particle size[55] and heating rate[59] complicate matters further. It is thought that γ, δ and θ have defect spinel structures with oxygen anions arranged in a face-centred cubic (fcc) structure with aluminium cations occupying tetrahedral and octahedral sites, but the exact ratios of octahedral and tetrahedral cations in each phase is disputed in the literature.[54–63] The fact that the anion lattice remains fcc in transition aluminas means that the phase transitions do not show up in DTA experiments,[57] unlike the θ to α transition in which the anions rearrange into a hexagonal close packed structure *via* a nucleation mechanism.[58] Although bulk transition aluminas are expected not to contain any five-fold coordinated aluminium, this has been observed at the surface regions of small crystallites, and in phases ρ and δ.[55]

A likely candidate for one of the aluminas present in the CP 1200 °C sample is θ-Al$_2$O$_3$, since this phase is associated with small crystal sizes[54] and it has been observed that alumina doped with zirconia consists entirely of the θ phase after heat treatment at 1200 °C.[60] It is possible that one of the octahedral lines is due to α-Al$_2$O$_3$, which in the bulk phase shows a single, narrow peak at a shift of 14 ppm (this position is half-way between the two octahedral peaks observed, and so it is difficult to say which peak it is more likely to be). However, since there are two tetrahedral lines, it is more likely to be two transitional aluminas since all transition aluminas have both octahedral and tetrahedral aluminium sites (no transitional alumina phase has been reported to contain exclusively tetrahedral aluminium in these types of materials). The presence of more than one alumina phase is perhaps a result of a distribution in particle sizes. The relatively small amount of fivefold coordinated aluminium suggests that disorder due to surface effects plays only a small role, probably due to growth of the alumina particles with the zirconia nanocrystals (Zener pinning involves a coupled grain growth between the pinning particle and the pinned phase[49,64]). Zirconia has been shown to restrict the θ to α alumina phase transition by preventing α nucleation at the alumina surface, and Zr–O–Al bonds have been reported to stabilize transition alumina structures,[60] so this could explain the lack of any α-Al$_2$O$_3$ in sample CP.

Fig. 2(b) shows the spectra obtained from sample AP. The as-prepared 80 °C sample shows a single narrow peak at a chemical shift of 9 ppm. This is likely to be boehmite (γ-AlOOH), which the literature reports to produce a signal at a shift in the range 5–9 ppm.[65] Boehmite contains only octahedrally coordinated aluminium and is commonly formed after hydrolysis of aluminium tri-*sec*-butoxide. The presence of this particular alumina precursor provides further evidence for the identity of the transition aluminas above, restricting the choice to δ or θ, since these are the transition phases forming when boehmite undergoes dehydration during heat

Table 1 Data obtained from simulations of the ^{27}Al spectra obtained at multiple fields. Errors for the chemical shifts and relative intensities are given in brackets

Sample	Annealing Temperature /°C	Aluminium Coordination	δ/ppm	Relative Intensity (%)	Mean C_Q / MHz ± 1.0 MHz
AC	600	4	77 (2)	36 (4)	10.5
		5	43 (1)	23 (2)	9.0
		6	16 (1)	41 (2)	8.0
CP	600	4	75 (2)	25 (2)	13.0
		5	43 (1)	37 (2)	9.5
		6	15 (3)	13 (3)	5.0
		6	10 (1)	26 (2)	7.5
CP	1000	4	78 (2)	25 (3)	10.0
		5	37 (1)	17 (3)	5.0
		6	16 (1)	59 (1)	7.0
CP	1200	4	87 (3)	15 (2)	12.0
		4	73 (2)	19 (4)	4.0
		5	40 (3)	7 (2)	10.0
		6	20 (1)	34 (2)	9.0
		6	9 (1)	25 (2)	3.0
AP	600	4	78 (1)	24 (2)	8.5
		5	41 (1)	14 (3)	8.5
		6	15 (1)	62 (1)	7.0
AP	1200	4	76 (1)	29 (2)	6.0
		5	37 (1)	5 (2)	6.0
		6	19 (1)	47 (2)	3.0
		6	9 (1)	19 (1)	7.5
MgO–Al$_2$O$_3$	600	4	82 (3)	31 (7)	10.5
		5	41 (1)	13 (2)	10.0
		6	17 (1)	51 (3)	7.0
		6	15 (4)	6 (2)	2.0
MgO–Al$_2$O$_3$	1200	4	95 (1)	23 (2)	12.0
		4	72 (2)	6 (1)	4.0
		5	33 (2)	4 (1)	4.0
		6	17 (1)	63 (2)	5.0
		6	11 (1)	4 (1)	2.0

treatment. Since δ-Al$_2$O$_3$ is believed to contain much more disorder than θ-Al$_2$O$_3$, which would show up in the NMR spectrum as a fivefold coordinated aluminium peak, it would seem that θ-Al$_2$O$_3$ is the transition alumina phase being observed in sample AP, with δ-Al$_2$O$_3$ being the second alumina phase present in sample CP. The boehmite phase is eliminated after annealing at 400 °C, by which temperature most hydroxyl groups are removed from the system. It can be seen from Fig. 2 that sample AP contained much less five-coordinated aluminium than CP throughout the entire temperature range (14% compared with 37% for the two samples annealed at 600 °C), and linewidths were also narrower in this sample at all temperatures. The alumina in this sample is therefore much less disordered than in sample CP. This is probably because in sample AP the alumina pinning phase was added in a second stage of sample preparation after the zirconia nanocrystals were already partially formed, and so the alumina particles are likely to be less well distributed throughout the system than in sample CP where the zirconium and aluminium precursors were hydrolysed together. Hence the alumina in sample AP may exist as larger or agglomerated particles which would have a smaller surface area to volume ratio, and would crystallise more thoroughly than the much smaller particles in sample CP. This explanation is corroborated by the fact that the pinning effect of the alumina in sample AP is much less than in sample CP, with 62 nm diameter nanocrystals at

1200 °C in AP compared to 37 nm in CP. This would be expected if the distribution of pinning particles was inhomogeneous.[49] The 80 °C spectrum also suggests a fairly large particle size, with just a single narrow 9 ppm boehmite line. The 400 °C lineshape shows the presence of the γ-Al$_2$O$_3$ phase, with the relative intensities of the six- and four-fold coordinated peaks in the 3 : 1 ratio. Over the range of temperatures from 600–1000 °C the lineshape changes very little, with only a variation in the intensity of the fivefold coordinated aluminium peak. At 1200 °C there are two octahedral aluminium peaks but only one tetrahedral peak. This suggests a mixture of θ-Al$_2$O$_3$ and α-Al$_2$O$_3$, but with both octahedral peaks occurring at shifts 5 ppm away from the 14 ppm at which the bulk α phase peak normally occurs, it is difficult to identify which one is the α peak.

The spectra for sample AC are shown in Fig. 2(c). This sample was manufactured using the same sol-gel route as sample AP but with a zirconium chloride precursor rather than zirconium propoxide, and yet the alumina pinning particles show significantly different results. A fourfold coordinated aluminium peak is present in the 100 °C spectrum, probably as a result of the slightly higher drying temperature that has removed some hydroxyl groups from the boehmite phase. Much more fivefold coordinated aluminium is present in sample AC than in sample AP over the annealing temperature range 400–800 °C (23% in AC compared with 14% in AP at 600 °C), and the lineshapes are generally broader, indicating a more highly amorphous alumina. In the sample annealed at 1200 °C, the α phase of alumina has formed, producing a peak at 14 ppm), with a small amount (4%) of fourfold coordinated alumina still present, which could perhaps be accounted for by aluminium sites at the surface of the α-alumina particles. While the alumina had not fully transformed to α-Al$_2$O$_3$ in sample AP, here it seems the alumina particles have grown sufficiently large for the complete transition to occur. The minimum θ-Al$_2$O$_3$ particle size that can undergo transition to α has been reported as being approximately 20 nm,[58] and so the alumina particles in sample AC have probably grown along with the zirconia nanocrystals to something in the order of this size or larger.

The differences in alumina pinning particle evolution between samples AC and AP can be accounted for by the way the zirconia nanocrystals are formed after hydrolysis of the zirconia precursors. Since the metal precursors are different, the size, distribution in size, amount of disorder, phase, shape and agglomeration of the resultant zirconia nanocrystals may differ, and any one of these factors could have an effect on the second stage of the preparation route when the aluminium precursor is added and hydrolysed, thus affecting the initial state or location of the alumina pinning particles. What is certainly clear from the sizes of the zirconia nanocrystals given in Fig. 2 is that the pinning ability of the alumina is stronger after manufacturing by method AP than AC. However, method CP produced the smallest nanocrystals overall due to the more effective mixing of the alumina and zirconia phases during the co-hydrolysis stage of preparation. Fig. 4 shows an SEM image of sample AC (annealed at 1200 °C), in which two distinct regions are visible. Firstly, a region of large zirconia crystals which formed after the initial stage of the preparation method (region A). These aggregated before addition of the alumina, and hence are unpinned and have undergone significant growth during annealing. Surrounding this is the second region of much smaller zirconia crystals (region B). These occur on the surface of region A, and it is here that the alumina pinning particles would have formed after the second stage of the addition process, pinning the zirconia nanocrystals and restricting their growth during heat treatment. This illustrates that the addition sol-gel method results in a less homogeneous distribution of the pinning phase than co-hydrolysis.

It is interesting to compare the values of quadrupolar coupling constants C_Q, and their distribution widths ΔC_Q, obtained from these samples with those obtained from bulk amorphous alumina.[66] In general, both the mean C_Q values and distribution widths are larger by a factor of two in samples CP, AC and AP than

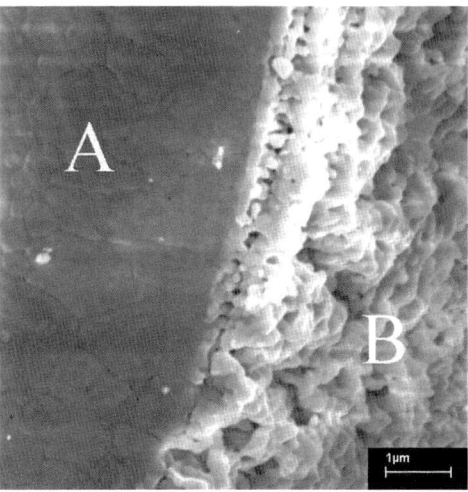

Fig. 4 An SEM image of sample AC after annealing at 1200 °C.

in bulk amorphous alumina. However, in both cases ΔC_Q is between 60 and 100% of the magnitude of C_Q. This indicates that while the alumina particles pinning the zirconia contain the same amount of inherent disorder as bulk amorphous alumina, the extent of the distortion of the aluminium environments is higher. This must be due to boundary strain in the pinning particles (due to their small size and confinement between the larger metal oxide nanocrystals) that prevents the structural relaxation that occurs in bulk-phase amorphous alumina.

It should be noted that no crystalline alumina phases were observed in any of the zirconia samples at any annealing temperature by XRD. This is likely to be due to the sizes of the alumina particles, which are expected to be approximately an order of magnitude smaller than the metal oxide nanocrystals and therefore the XRD peaks will be broadened significantly by the Scherrer effect. This is a clear illustration of the advantage of using NMR to characterise these types of nanocrystalline materials.

3.2 ZrO$_2$–SiO$_2$

The ^{29}Si MAS NMR results for the silica particles pinning the zirconia nanocrystals are shown in Fig. 5. As the annealing temperature increases, there is a steady evolution from lower n to higher n Q^n silicon environments. The 100 °C spectra can be most accurately simulated using three peaks representing Q^1, Q^2 and Q^3 species at chemical shifts of $\delta = -79$, -86 and -95 ppm, respectively. These are most likely silicon atoms with one or more hydroxyl groups left attached as remnants from the sol-gel preparative route. It is well known from TGA and ^{1}H NMR experiments that hydroxyl groups can remain in sol-gel prepared nanocrystalline materials at temperatures up to 600 °C.[67] It is also possible that some organic groups are present in the sample at the lower temperatures. By 800 °C, most OH groups have been removed and the silica network has become more fully connected, with only Q^3 and Q^4 species remaining. The 1200 °C spectrum can be fitted using only a single broad Q^4 peak centred at $\delta = -110$ ppm. The disappearance of the Q^3 peak between 800 and 1200 °C indicates that the pinning particles are undergoing significant growth in this temperature range, since larger particles would have a much lower fraction of Q^3 surface silicon environments (the interior of the particles will be dominated by Q^4 sites at these temperatures). Indeed, as can be seen in Fig. 5, the zirconia nanocrystals themselves have grown by a factor of five during this stage of annealing, and since Zener pinning involves a coupled grain growth mechanism between the nanocrystals and pinning particles,[49,64] larger silica particles at 1200 °C

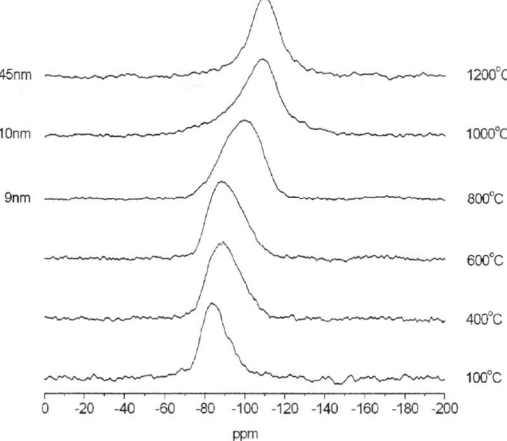

Fig. 5 ^{29}Si MAS NMR spectra obtained from the silica-pinned zirconia samples heated to various temperatures.

seem very likely. This structural evolution of the silica pinning particles is very similar to that previously observed in the silica-pinned nanocrystalline SnO_2 system, however in that system the tin oxide nanocrystals underwent less growth and a significant percentage of Q^3 silicon species remained after annealing at 1200 °C.[40]

The highly crystalline nature of the 1200 °C annealed silica-pinned zirconia nanocrystals is evident in Fig. 6, with three lattice fringes visible in one of the nanocrystals in the lower middle of the TEM image. The average zirconia nanocrystal diameter in this sample as measured using TEM was around 40 nm, which is in very good agreement with the 45 nm estimate from the XRD.

^{17}O static NMR spectra obtained from the oxygen enriched silica-pinned zirconia at various temperatures up to 285 °C are presented in Fig. 7. Two main peaks can be seen. The first peak centred at around 400 ppm arises due to oxygen in ZrO_3 and ZrO_4 environments, and also in Zr–O–Si bonds. The second, less intense peak centred at -100 ppm corresponds to water and hydroxyl groups which remain in the sample as a result of the sol-gel preparation process. This peak narrows between 20 and 65 °C due to increasing mobility of these groups, and at temperatures above 100 °C it decreases in intensity as the groups are removed from the sample (Fig. 8 shows that loss of water and hydroxyl groups is a process that occurs over the temperature range 20–600 °C, but that it mostly complete by 200 °C). In order to measure the mobility of the oxygen ions within the zirconia nanocrystals, only the first peak at 400 ppm was integrated in saturation recovery T_1 measurements. The measured T_1 values at each temperature are shown in Table 2. According to the BPP model,[68] the relaxation time T_1 is linked to the activation energy E_A for the mobility

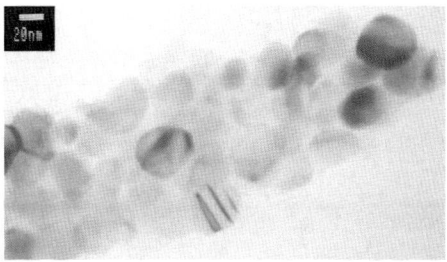

Fig. 6 A TEM image of the silica-pinned zirconia nanocrystals annealed at 1200 °C.

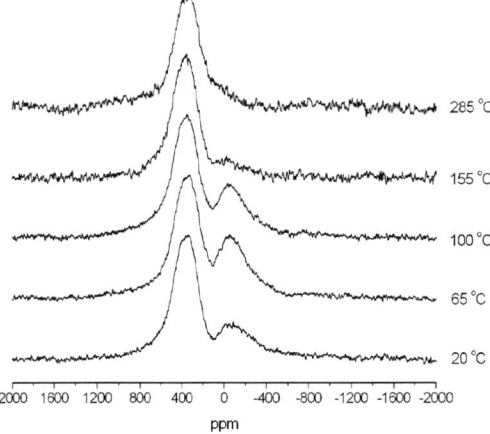

Fig. 7 The ^{17}O static NMR spectra obtained from the oxygen-enriched silica-pinned zirconia.

mechanism by the relationship

$$\ln T_1 \propto \frac{-E_A}{kT} \tag{1}$$

T is the temperature in Kelvin and k is Boltzmann's constant. It should be noted that eqn (1) and the T_1 measurements in Table 2 refer to the high temperature side of the T_1 minimum. From this relation, a value of 100 ± 20 meV was obtained for this sample, which is in agreement with the value of 80 ± 40 meV obtained using T_1 measurements previously recorded from unpinned nanocrystalline zirconia over a similar range of temperatures.[69]

3.3 MgO–Al₂O₃

In the ^{27}Al MAS NMR spectrum of the as-prepared MgO–Al$_2$O$_3$ sample, a narrow peak at $\delta = 9$ ppm is present corresponding to a well defined sixfold coordinated aluminium environment (see Fig. 9). This is boehmite, as seen previously in sample AP (Section 3.1). The spectra of the samples annealed at 400 and 600 °C show a second sixfold coordination aluminium site at a chemical shift of around 15 ppm,

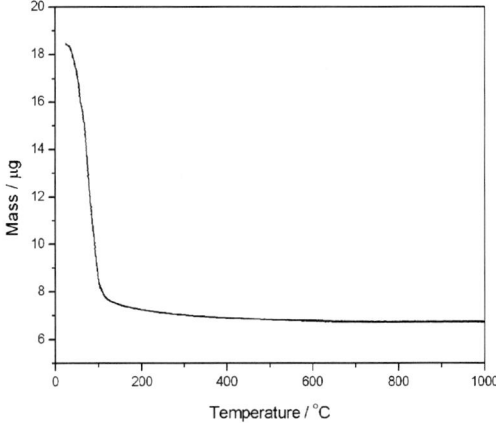

Fig. 8 A thermogravimetric analysis plot of the silica-pinned zirconia.

Table 2 The T_1 relaxation times obtained from the oxygen-enriched silica-pinned zirconia at various temperatures. Errors are given in brackets

Temperature/°C	T_1/s
20 (2)	0.19 (0.01)
65 (5)	0.27 (0.03)
100 (5)	0.37 (0.02)
155 (5)	0.58 (0.04)
285 (5)	1.20 (0.08)

visible as a sharp protrusion on the left side of the main sixfold coordination peak in the 600 °C spectrum in Fig. 9. The peak is relatively narrow, with a width of just 300 Hz and a relatively small C_Q distribution. This is unlikely to be α-Al_2O_3, which generally forms at a much higher temperature in these materials, and is probably spinel ($MgAl_2O_4$), which contains only octahedrally coordinated aluminium in the fully-ordered crystalline phase. This phase remains present throughout the range of annealing temperatures, however the peak becomes overlapped by another much broader sixfold coordinated aluminium peak which gradually moves to a more positive shift as the annealing temperature increases, and so the sharper peak cannot be clearly seen in the higher temperature NMR spectra in Fig. 9, although its presence can be seen in the simulations in Fig. 10. Small peaks corresponding to spinel were observed in the XRD spectrum of the 1200 °C sample. Significant fivefold coordinated aluminium appears in the 400 °C spectrum along with a doubling of the widths of the four- and sixfold coordination lines, indicating more disordered alumina. From 600 to 1000 °C the amount of fivefold coordination aluminium decreases to zero and the MgO nanocrystals undergo significant growth, suggesting that the alumina particles themselves are growing and crystallising. At 1200 °C the broad line for the fourfold coordinated aluminium can be simulated using two peaks, one of which remains at a shift of 76 ppm and another peak that has moved out to a shift of 95 ppm. This indicates the presence of another transitional alumina phase, which is unlikely to be the θ or δ phases that appeared in the CP- and AP-ZrO_2–Al_2O_3 samples (see Section 3.1) since these phases showed no feature at 95 ppm. This phase may therefore be χ, κ, ρ, or η alumina. A suggested candidate is η, in which "quasi-trihedral" aluminium has been reported to exist at the surface region.[57] The feature occurring here at 95 ppm is not likely to be three-coordinated aluminium (such a site would normally gain a hydroxyl group and become a more stable tetrahedral structure), but it may well be a highly distorted tetrahedral environment that is tending towards a trihedral-like shape.

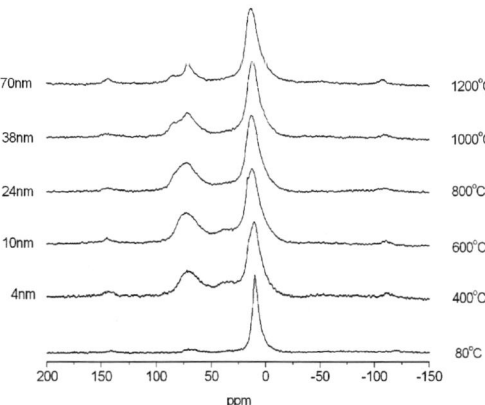

Fig. 9 The ^{27}Al MAS NMR spectra obtained for the alumina-pinned nanocrystalline MgO.

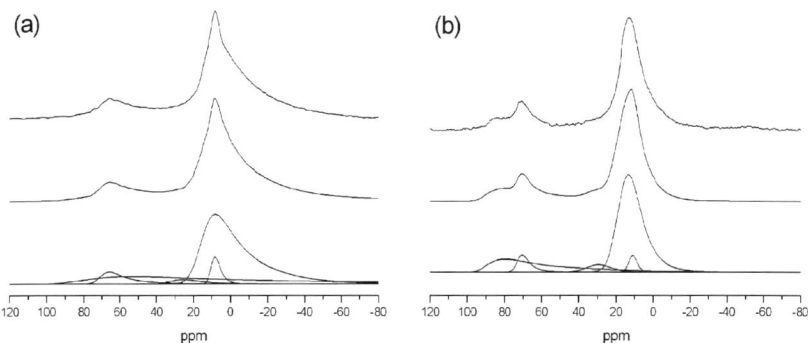

Fig. 10 Spectra and simulations of the ^{27}Al MAS NMR spectra of the alumina-pinned MgO annealed at 1200 °C at (a) 8.45 T and (b) 14.1 T. For each plot, the top line is the experimental spectrum obtained, the middle line is the total simulated spectrum, and the bottom line shows the individual simulated peaks.

3.4 MgO–SiO$_2$

Fig. 11 shows how the ^{29}Si MAS NMR spectra for the MgO–SiO$_2$ samples evolve with annealing temperature. In the as-prepared sample, three distinguishable peaks occur at shifts of −78, −85 and −91 ppm, positions that correspond to Q^0, Q^1 and Q^2 silicon environments, respectively. These are simply silicon atoms with hydroxyl groups and organic groups still attached as remnants of the sol-gel process. After annealing at 400 °C, these groups have been eliminated, but rather than producing a silica network with higher order Q^n species, a magnesium silicate glass has formed, with a broad peak centred at −72 ppm. The more positive shift demonstrates that Mg has entered the silica network and reduced the average Q^n speciation. This glass remains up to 1000 °C, however at 800 °C a narrower peak occurs at −93 ppm corresponding to a Q^2 silicon environment. This indicates the formation of enstatite (MgSiO$_3$) which contains pyroxene-type chains of (SiO$_3$)$_\infty$[70] and has been shown to form in these types of materials at around 780 °C.[71] At 1000 °C a sharp Q^0 peak occurs at −62 ppm. This is due to the formation of forsterite (Mg$_2$SiO$_4$), and by

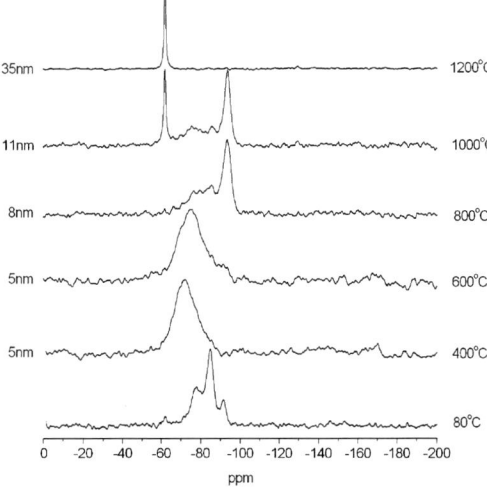

Fig. 11 ^{29}Si MAS NMR spectra obtained from the silica-pinned MgO samples annealed at various temperatures.

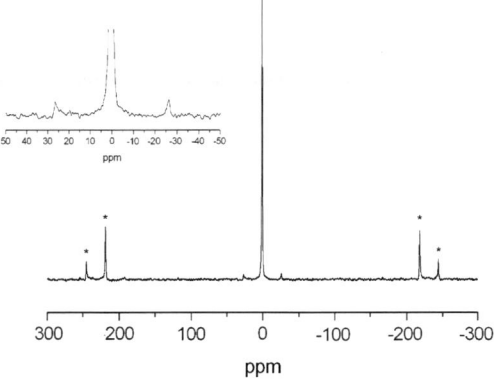

Fig. 12 ^{25}Mg MAS NMR spectrum of the silica-pinned MgO annealed at 1200 °C, with an expanded view inset on the left. Spinning sidebands are denoted with asterisks.

1200 °C only this phase remains. The forsterite was also observed with ^{25}Mg MAS NMR. Fig. 12 shows the main MgO peak at 0 ppm and the inset on the top left is an expansion of the area around the base of this peak, where the quadrupolar lineshape[72] for one of the two magnesium sites in forsterite is visible as two small peaks at $\delta = -27$ and 27 ppm. These sharp features match exactly with the singularities of a second-order quadrupolar central transition lineshape with $C_Q = 4.3$ MHz and an asymmetry parameter $\eta = 0.40$.[73] Simulation has shown the isotropic chemical shift of this line to be at $\delta = (77 \pm 2)$ ppm, the first time this shift has been reported. A second magnesium site present in forsterite has a C_Q of 5.0 MHz, but an asymmetry parameter of 0.96,[73] meaning that its main singularity is hidden under the MgO line in this spectrum, and so is not clearly visible. Given that all of the silica is fully transformed to forsterite, just 15% of the Mg atoms will exist in this phase, and combining this with the broad nature of the quadrupolar lineshapes, the intensity of the signal from forsterite is much less than that from the MgO. However, the presence of the forsterite has been clearly observed by both ^{29}Si and ^{25}Mg MAS NMR whilst none was detected at all by XRD techniques due to the phase existing as extremely small particles. This is another example of the advantage of NMR over XRD in investigating nanocrystalline materials. Fig. 13 shows how the linewidth of the MgO peak in the ^{25}Mg spectra decreased with increasing annealing temperature. This is simply due to the growth of the MgO

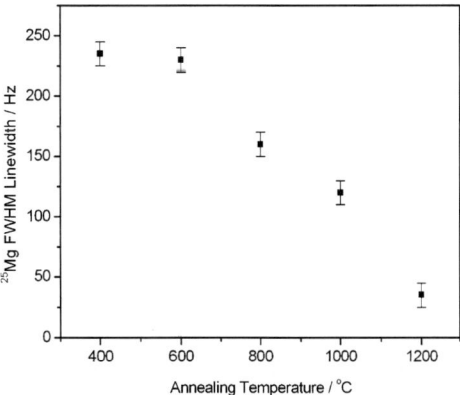

Fig. 13 A graph showing how the linewidths of the silica-pinned MgO samples varied with annealing temperature.

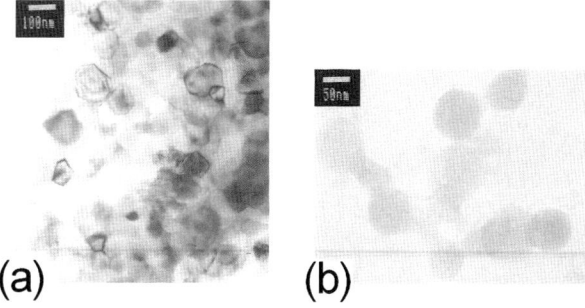

Fig. 14 TEM images of the silica-pinned nanocrystalline MgO after annealing at 1200 °C.

nanocrystals, and the corresponding reduction in their surface area to volume ratio. It should be noted here that these linewidths are narrower than those previously reported for sol-gel prepared nanocrystalline MgO over a comparable range of crystal diameters,[42] suggesting that the nanocrystals are more highly crystalline.

Fig. 14 shows two different nanocrystal morphologies observed in the silica-pinned MgO after annealing at 1200 °C. Fig. 14(a) shows single-crystal polyhedral MgO structures, with electron diffraction lines visible in some of the nanocrystals. The average diameter of these crystals is around 90 nm, significantly larger than the 35 nm value as estimated by XRD. This difference is accounted for by a second particle type shown in Fig. 14(b). These are spherical particles between 50 and 70 nm in diameter but consisting of much smaller domains measuring around 5 nm in diameter. These structures are expected to contain most of the pinning silica (transformed to forsterite in this particular sample), which would explain the small size of the individual domains. Perhaps due to their spherical morphology, these particles were significantly less aggregated than the polyhedral MgO nanocrystals.

It is interesting to note that despite the fact that the silica has reacted very readily with the MgO nanocrystals, the pinning effect is still significant and even after annealing at 1000 °C, the average diameter of the metal oxide nanocrystals is just 11 nm.

4. Conclusions

Alumina and silica pinning particles undergo thermal evolution in structure that depends heavily on their method of manufacture and on the metal oxide system that they are pinning. In fact, in the ZrO_2–Al_2O_3 system, the nature of the alumina pinning phase was even found to be dependent on the type of zirconium precursor. While both the co-hydrolysis and addition sol-gel routes were successfully employed to manufacture nanocrystalline ZrO_2–Al_2O_3, the co-hydrolysis route resulted in a more homogeneous distribution of the pinning phase and was therefore the superior method of preparing zirconia nanocrystals resistant to growth.

The alumina particles pinning ZrO_2 retained significant disorder up to 800 °C, which may have been due to their amorphous nature or their high surface area to volume ratio due to their extremely small size. Likely it was a combination of both factors. At higher temperatures, the particles showed more well-defined crystalline NMR lineshapes, having undergone significant growth along with the metal oxide nanocrystals that they were pinning.

Three distinct transitional alumina phases have been observed using ^{27}Al MAS NMR, two appearing at 1200 °C in the alumina particles pinning the zirconia (postulated to be θ and δ), the other pinning the MgO and identified by an aluminium peak at a chemical shift of $\delta = (95 \pm 1)$ ppm (suggested to be η-Al_2O_3). None of these phases were observed using XRD due to the extremely small particle size.

The MgO nanocrystals reacted very readily with both the alumina and silica pinning phases, forming spinel in the former case, and enstatite and forsterite in the latter. Despite this reactivity, the pinning effect of the silica was significant, with MgO nanocrystals of 11 nm average diameter remaining after annealing at 1000 °C.

The activation energy for oxygen ion mobility in the silica-pinned zirconia was observed to be the same as that for pure nanocrystalline zirconia. It therefore appears that the pinning phase does not have a significant effect on the oxygen ion conductivity within the nanocrystals.

Acknowledgements

The EPSRC are thanked for funding the work on nanocrystalline materials through grants (GR/S61881 and GR/S61898). MES thanks both EPSRC and the University of Warwick for partial funding of NMR equipment at Warwick. The authors would like to thank Dr T. J. Bastow, CSIRO Division of MIT, Melbourne, Australia, for advice on high temperature NMR and loan of a probe. Steve York is also thanked for his help with electron microscopy.

References

1 H. Gleiter, Nanostructured materials: basic concepts and microstructure, *Acta Mater.*, 2000, **48**, 1–29.
2 A. V. Chadwick, Small, but perfectly formed: the microstructure of nanocrystalline oxides, *Radiat. Eff. Defects Solids*, 2003, **158**, 21–30.
3 C. Suryanarayana and C. C. Koch, Nanocrystalline materials—current research and future directions, *Hyperfine Interact.*, 2000, **130**, 5–44.
4 A. I. Gusev, Effects of the nanocrystalline state in solids, *Phys. Uspekhi*, 1998, **41**, 49–76.
5 C. N. R. Rao, G. U. Kulkarni, P. J. Thomas and P. P. Edwards, Size dependent chemistry: properties of nanocrystals, *Chem. Eur. J.*, 2002, **8**, 28–35.
6 P. Heitjans and S. Indris, Diffusion and ionic conduction in nanocrystalline ceramics, *J. Phys.: Condens. Matter*, 2003, **15**, R1257–R1289.
7 C. W. Kwon, S. J. Hwang, M. H. Delville, C. Labrugère, A. Vadivel Murugan, B. B. K. Kale, K. Vijayamohanan and G. Campet, Electrochemistry of inorganic nanocrystalline electrode materials for lithium batteries, *Act. Passive Electron. Compon.*, 2003, **26**, 23–29.
8 K. Tennakone and J. Bandara, Photocatalytic activity of dye-sensitized tin(IV) oxide nanocrystalline particles attached to zinc oxide particles: long distance electron transfer *via* ballistic transport of electrons across nanocrystallites, *Appl. Catal., A*, 2001, **208**, 335–341.
9 J. Y. Ying and T. Sun, Research needs assessment on nanostructured catalysts 1, *J. Electroceram.*, 1997, **3**, 219–238.
10 P. Alivisatos, The use of nanocrystals in biological detection, *Nat. Biotechnol.*, 2004, **22**(1), 47–52.
11 F. Quaranta, R. Rella, P. Siciliano, S. Capone, C. Distante, M. Epifani and A. Taurino, Preparation and characterisation of nanostructured materials for an artificial olfactory sensing system, *Sens. Actuators, B*, 2002, **84**, 55–59.
12 M. E. McHenry and D. E. Laughlin, Nano-scale materials development for future magnetic applications, *Acta Mater.*, 2000, **48**, 223–238.
13 E. Lucas, S. Decker, A. Khaleel, A. Seitz, S. Fultz, A. Ponce, W. Li, C. Carnes and K. J. Klabunde, Nanocrystalline metal oxides as unique chemical reagents/sorbents, *Chem. Eur. J.*, 2001, **7**, 2505–2510.
14 S. Rajagopalan, O. Koper, S. Decker and K. J. Klabunde, Nanocrystalline metal oxides as destructive adsorbents for organophosphorus compounds at ambient temperatures, *Chem. Eur. J.*, 2002, **8**(11), 2602–2607.
15 F. C. M. Woudenberg, W. F. C. Sager, N. G. M. Sibelt and H. Verweij, Dense nanostructured t-ZrO$_2$ coatings at low temperatures *via* modified emulsion precipitation, *Adv. Mater.*, 2001, **13**, 514–516.
16 J. A. Wang, M. A. Valenzuela, J. Salmones, A. Vázquez, A. García-Ruiz and X. Bokhimi, Comparative study of nanocrystalline zirconia prepared by precipitation and sol-gel methods, *Catal. Today*, 2001, **68**, 21–30.
17 M. K. Mishra, B. Tyagi and R. J. Jasra, Effect of synthetic parameters on structural, textural, and catalytic properties of nanocrystalline sulfated zirconia prepared by sol-gel technique, *Ind. Eng. Chem. Res.*, 2003, **42**, 5727–5736.

18 S. Castillo, R. Gómez and M. Morán–Pineda, Effect of sol-gel derived Al_2O_3–ZrO_2 and Al_2O_3–TiO_2 oxides on the selectivity of NO reduction by CO under oxidizing conditions, *React. Kinet. Catal. Lett.*, 2003, **79**, 271–279.

19 X. M. Liu, G. Q. Lu and Z. F. Yan, Synthesis and stabilization of nanocrystalline zirconia with MSU mesostructure, *J. Phys. Chem. B*, 2004, **108**, 15523–15528.

20 B. L. Kirsch, A. E. Riley, A. F. Gross and S. H. Tolbert, Probing the effects of interfacial chemistry on the kinetics of phase transitions in amorphous and tetragonal zirconia nanocrystals, *Langmuir*, 2004, **20**, 11247–11254.

21 R. C. Garvie and M. F. Goss, Intrinsic size dependence of the phase transformation temperature in zirconia microcrystals, *J. Mater. Sci.*, 1986, **21**, 1253–1257.

22 I. V. Mishakov, A. F. Bedilo, R. M. Richards, V. V. Chesnokov, A. M. Volodin, V. I. Zaikovskii, R. A. Buyanov and K. J. Klabunde, Nanocrystalline MgO as a dehydro-halogenation catalyst, *J. Catal.*, 2002, **206**, 40–48.

23 S. Utamapanya, K. J. Klabunde and J. R. Schlup, Nanoscale metal oxide particles/clusters as chemical reagents. Synthesis and properties of ultrahigh surface area magnesium hydroxide and magnesium oxide, *Chem. Mater.*, 1991, **3**, 175–181.

24 P. Jeevanandam and K. J. Klabunde, A study on adsorption of surfactant molecules on magnesium oxide nanocrystals prepared by an aerogel route, *Langmuir*, 2002, **18**, 5309–5313.

25 J. V. Stark and K. J. Klabunde, Nanoscale metal oxide particles/clusters as chemical reagents. Adsorption of hydrogen halides, nitric oxide and sulfur trioxide on magnesium oxide nanocrystals and compared with microcrystals, *Chem. Mater.*, 1996, **8**, 1913–1918.

26 O. B. Koper, I. Lagadic, A. Volodin and K. J. Klabunde, Alkaline-earth oxide nanoparticles obtained by aerogel methods. Characterisation and rational for unexpectedly high surface chemical reactivities, *Chem. Mater.*, 1997, **9**, 2468–2480.

27 E. Lucas, S. Decker, A. Khaleel, A. Seitz, S. Fultz, A. Ponce, W. Li, C. Carnes and K. J. Klabunde, Nanocrystalline metal oxides as unique chemical reagents/sorbents, *Chem. Eur. J.*, 2001, **7**, 2505–2510.

28 B. L. Kirsch and S. H. Tolbert, Stabilization of isolated hydrous amorphous and tetragonal zirconia nanoparticles through the formation of a passivating alumina shell, *Adv. Funct. Mater.*, 2003, **13**, 281–288.

29 C. Zener, quoted in C. S. Smith, Grains, phases and interphases, an interpretation of microstructure, *Trans. Metall. Soc.*, 1948, **175**, 15–51.

30 Y. W. Zhang, Y. Yang, S. Jin, C. S. Liao and C. H. Yan, Doping effect on the grain size and microstrain in the sol-gel derived rare earth stabilized zirconia nanocrystalline thin films, *J. Mater. Sci. Lett.*, 2002, **21**, 943–946.

31 S. Shukla, S. Seal, R. Vij and S. Bandyopadhyay, Effect of HPC and water concentration on the evolution of size, aggregation and crystallisation of sol-gel nano zirconia, *J. Nanopart. Res.*, 2002, **4**, 553–559.

32 G. M. Medine, V. Zaikovskii and K. J. Klabunde, Synthesis and adsorption properties of intimately intermingled mixed metal Oxide Nanoparticles, *J. Mater. Chem.*, 2004, **14**, 757–763.

33 C. L. Carnes, P. N. Kapoor, K. J. Klabunde and J. Bonevich, Synthesis, characterisation, and adsorption studies of nanocrystalline aluminium oxide and a bimetallic nanocrystalline aluminium oxide/magnesium oxide, *Chem. Mater.*, 2002, **14**, 2922–2929.

34 S. L. P. Savin and A. V. Chadwick, Restricting the high temperature growth of nanocrystalline tin oxide, *Radiat. Eff. Defects Solids*, 2003, **158**, 73–76.

35 Y. Al-Angry, S. L. P. Savin, K. E. Rammutla, M. J. Pooley, E. R. H. Van Eck and A. V. Chadwick, The stabilization of metal oxide nanocrystals by the addition of alumina, *Radiat. Eff. Defects Solids*, 2003, **158**, 209–213.

36 S. L. P. Savin, A. V. Chadwick, L. A. O'Dell and M. E. Smith, EXAFS study of confined nanocrystalline oxides, *Phys. Status Solidi*, 2005, **2**(1), 661–664.

37 G. Couturier, C. Maurice and R. Fortunier, Three-dimensional finite-element simulation of Zener pinning dynamics, *Philos. Mag.*, 2003, **83**(30), 3387–3405.

38 N.-L. Wu, S.-Y. Wang and I. A. Rusakova, Inhibition of crystallite growth in the sol-gel synthesis of nanocrystalline metal oxides, *Science*, 1999, **285**, 1375–1377.

39 S. L. P. Savin, *Sol-Gel Derived Materials for Chemical Sensing*, PhD Thesis, University of Kent, 2003.

40 L. A. O'Dell, S. L. P. Savin, A. V. Chadwick and M. E. Smith, Structural studies of silica- and alumina-pinned nanocrystalline SnO_2, *Nanotechnology*, 2005, **16**, 1836–1843.

41 A. V. Chadwick, G. Mountjoy, V. M. Nield, I. J. F. Poplett, M. E. Smith, J. H. Strange and M. G. Tucker, Solid-state NMR and X-ray studies of the structural evolution of nanocrystalline zirconia, *Chem. Mater.*, 2001, **13**, 1219–1229.

42 A. V. Chadwick, I. J. F. Poplett, D. T. S. Maitland and M. E. Smith, Oxygen speciation in nanophase MgO from solid-state ^{17}O NMR, *Chem. Mater.*, 1998, **10**, 864–870.

This journal is © The Royal Society of Chemistry 2006

43 A. V. Chadwick, M. J. Pooley, K. E. Rammutla, S. L. P. Savin and A. Rougier, A comparison of the extended X-ray absorption fine structure of nanocrystalline ZrO_2 prepared by high energy ball milling and other methods, *J. Phys.: Condens. Matter*, 2003, **15**, 431–440.

44 G. Williams and G. S. V. Coles, Gas sensing properties of nanocrystalline metal oxide powders produced by a laser evaporation technique, *J. Mater. Chem.*, 1998, **8**, 1657–1664.

45 G. Baldinozzi, D. Simeone, D. Gosset and M. Dutheil, Neutron diffraction study of the size-induced tetragonal to monoclinic phase transition in zirconia nanocrystals, *Phys. Rev. Lett.*, 2003, **90**, 216103.

46 S. F. Yin and B. Q. Xu, On the preparation of high-surface-area nano-zirconia by reflux-figestion of hydrous zirconia gel in basic solution, *ChemPhysChem*, 2003, **3**, 277–281.

47 Y. Ding, G. Zhang, H. Wu, B. Hai, L. Wang and Y. Qian, Nanoscale magnesium hydroxide and magnesium oxide powders: control over size, shape, and structure *via* hydrothermal synthesis, *Chem. Mater.*, 2001, **13**, 435–440.

48 Z. Dohnálek, G. A. Kimmel, D. E. McCready, J. S. Young, A. Dohnálková, R. S. Smith and B. D. Kay, Structural and chemical characterization of aligned crystalline nanoporous MgO films grown *via* reactive ballistic deposition, *J. Phys. Chem. B*, 2002, **106**, 3526–3529.

49 A. Saha and D. C. Agrawal, Microstructure development in hybrid sol-gel prepared Al_2O_3–ZrO_2 composites, *J. Mater. Sci. Lett.*, 1998, **17**, 1333–1336.

50 I. W. Chen and X. H. Wang, Sintering dense nanocrystalline ceramics without final-stage grain growth, *Nature*, 2000, **404**, 168–171.

51 Y. Al-Angari, *Studies of Methods to Restrict the Grain Growth of Nanocrystalline Metal Oxides*, PhD Thesis, University of Kent, 2002.

52 D. Massiot, F. Fayon, M. Capron, I. King, S. Le Calvé, B. Alonso, J.-O. Durand, B. Bujoli, Z. Gan and G. Hoatson, Modelling one- and two-dimensional solid state NMR spectra, *Magn. Reson. Chem.*, 2002, **40**, 70–76.

53 T. F. Kemp, *High Field Solid State ^{27}Al NMR of Ceramics and Glasses*, Masters Thesis, University of Warwick, 2004.

54 C. Pecharromán, I. Sobrados, J. E. Iglesias, T. González-Carreño and J. Sanz, Thermal evolution of transitional aluminas followed by NMR and IR spectroscopies, *J. Phys. Chem. B*, 1999, **103**, 6160–6170.

55 I. Bennett and R. Stevens, Calcination and phase changes in alumina, *Br. Ceram. Trans.*, 1998, **97**, 117–125.

56 C. Wolverton and K. C. Hass, Phase stability and structure of spinel-based transition aluminas, *Phys. Rev. B*, 2000, **63**, 024102.

57 R. S. Zhou and R. L. Snyder, Structures and transformation mechanisms of the η,γ and θ transition aluminas, *Acta Crystallogr., Sect. B*, 1991, **47**, 617–630.

58 H. L. Wen and F. S. Yen, Growthcharacteristics of boehmite-derived ultrafine theta and alpha-alumina particles during phase transformation, *J. Cryst. Growth*, 2000, **208**, 696–708.

59 S. W. Jang, H. Y. Lee, S. M. Lee, S. W. Lee and K. B. Shim, Mechanical activation effect on the transition of gibbsite to α-alumina, *J. Mater. Sci. Lett.*, 2000, **19**, 507–510.

60 B. Djuričić, S. Pickering, P. Glaude, D. McGarry and P. Tambuyser, Thermal stability of transition phases in zirconia-doped alumina, *J. Mater. Sci.*, 1997, **32**, 589–601.

61 M. Nguefack, A. F. Popa, S. Rossignol and C. Kappenstein, Preparation of alumina through a sol-gel process. Synthesis,characterisation, thermal evolution and model of intermediate boehmite, *Phys. Chem. Chem. Phys.*, 2003, **5**, 4279–4289.

62 S. H. Cai, S. N. Rashkeev, S. T. Pantelides and K. Sohlberg, Phase transformation mechanism between γ- and θ-alumina, *Phys. Rev. B*, 2003, **67**, 224104.

63 B. Ollivier, R. Retoux, P. Lacorre, D. Massiot and G. Férey, Crystal structure of κ-Alumina: an X-ray powder diffraction, TEM and NMR study, *J. Mater. Chem.*, 1997, **7**, 1049–1056.

64 V. V. Srdić and D. I. Savić, Grain growth in sol-gel derived alumina–zirconia composites, *J. Mater. Sci.*, 1998, **33**, 2391–2396.

65 M. E. Smith, Application of Al-27 NMR techniques to structure determination in solids, *Appl. Magn. Reson.*, 1993, **4**, 1–64.

66 G. Kunath-Fandrei, T. J. Bastow, J. S. Hall, C. Jäger and M. E. Smith, Quantification of aluminium coordination in amorphous alumina by combined central and satellite transition NMR, *J. Phys. Chem.*, 1995, **99**, 15138–15141.

67 P. N. Gunawidjaja, M. A. Holland, G. Mountjoy, D. M. Pickup, R. J. Newport and M. E. Smith, The effects of different heat treatment and atmospheres on the NMR signal and structure of TiO_2–ZrO_2–SiO_2 sol-gel materials, *Solid State Nucl. Magn. Reson.*, 2003, **23**, 88–106.

68 N. Bloembergen, E. M. Purcell and R. V. Pound, Relaxation effects in nuclear magnetic resonance adsorption, *Phys. Rev.*, 1948, **73**, 679.

69 I. J. F. Poplett, M. E. Smith and J. H. Strange, A novel high temperature NMR probe design: application to ^{17}O studies of gel formation of zirconia, *Meas. Sci. Technol.*, 2000, **11**, 1703–1707.

70 J. V. Smith, The crystal structure of proto-enstatite, $MgSiO_3$, *Acta Crystallogr.*, 1959, **12**, 515–519.

71 K. J. D. MacKenzie and R. H. Meinhold, Thermal reactions of chrysotile revisited; a ^{29}Si and ^{25}Mg MAS NMR study, *Am. Mineral.*, 1994, **79**, 43–50.

72 M. E. Smith and E. R. H. van Eck, Recent advances in experimental solid state NMR methodology for half-integer spin quadrupolar nuclei, *Prog. Nucl. Magn. Reson. Spectrosc.*, 1999, **34**, 159–201.

73 B. Derighetti, S. Hafner, H. Marxer and H. Rager, NMR of Si-29 and Mg-25 in Mg_2SiO_4 with dynamic polarisation technique, *Phys. Lett.*, 1978, **66**, 150–152.

General Discussion

Professor Navrotsky opened the discussion of Professor Kilner's paper: LSGM is energetically more stable than perovskites with doping only in one sublattice, but the enthalpy varies linearly with dopant concentration. In doped zirconias there is a strong minimum (maximum stabilisation) with doping. These differences suggest that optimum size matching is important in both types of structures, but that short range order is more dominant in the fluorites. This is a theme to bear in mind throughout the discussions.

Professor Maier commented: It may be quite natural that transport along the boundary is fast while hindered laterally. The problem comes in the evaluation, as such a channelling effect for a tracer diffusing in through a grain boundary is not taken account of in the Le Claire evaluation.[1]

1   See D. Gryasnov, J. Fleig and J. Maier, *Solid State Ionics*, 2006, in press.

Professor Kilner said: I agree, and it would only be noticeable under certain experimental conditions, *i.e.*, small grain sizes and low bulk diffusion coefficients.

Professor Irvine asked: The LSM/YSZ composite data look very interesting. Have you checked the surface exchange and diffusion properties of oxygen in manganese doped zirconias?

Professor Kilner replied: We cannot discount the fact that the manganese may have diffused over the surface of and subsequently affected the surface exchange rate of the zirconia in the composite. Because we have to use dense materials for the isotopic exchange experiments, the composites have been sintered at fairly high temperatures (1350 °C for five hours).[1] We have measured the surface exchange rate in manganese doped material and this data is unpublished, to my best recollection the surface exchange was not markedly changed, however this does need confirmation. We are also intending to look at the manganese content of the zirconia surfaces in the composite materials by high resolution SIMS.

1   M. Dhallu and J. A. Kilner, Oxygen transport in YSZ/LSM composite materials, *J. Fuel Cell Sci. Technol.*, 2005, **2**(1), 29–33.

Professor Heitjans asked: You showed this remarkable correlation between the expansion coefficient and the diffusivity. You attributed it to vacancy concentration. This is a microscopic interpretation while expansion is normally ascribed to anharmonic lattice vibrations which represent a macroscopic aspect. Could you comment on how these different views come together?

Professor Kilner answered: The expansion coefficients shown in that correlation[1] are due to what is often termed the "chemical expansion" caused by a loss of oxygen in these materials. If you hold one of these perovskites at a constant temperature, particularly a high cobalt content material, you can cause the material to expand by reducing or contract by re-oxidising the material. This is entirely separate from, and in addition to, the anharmonic lattice vibration component you have already mentioned, and is caused by the change in cation radius on changing valence state, plus some contribution from the relaxations around the oxygen vacancies.

1 H. Ullmann, N. Trofimenko, F. Tietz, D. Stöver and A. Ahmad-Khanlou, Correlation between thermal expansion and oxide ion transport in mixed conducting perovskite-type oxides for SOFC cathodes, *Solid State Ionics*, 2000, **138**(1–2), 79–90.

Professor Haile commented: I think it's important to emphasize that the chemical expansion coefficient is technologically problematic in cases where the material is exposed to an oxygen concentration gradient. This includes fuel cell electrolytes, interconnects and oxygen permeation membranes. In a fuel cell cathode with high electrochemical activity, however, there should be no significant oxygen partial pressure gradient and therefore no gradient in chemical expansion.

Professor Irvine added: There are some quite important issues regarding chemical expansion of fuel cell cathodes also. Stresses induced during high temperature processing can seriously affect the stability of membranes used in fuel cells, especially over the longer term.

Professor Kilner said: I agree that under normal operating conditions this need not be a serious problem. However, as has been stated, it can be a serious problem during processing, particularly if any cycling of oxygen partial pressure is required.

Professor Haile said: One has to be careful about cause and effect here if one is going to design and implement rational solutions to the problem of fuel cell stability. Chemical expansion refers to the fact that variable valence oxides undergo expansion as a result of the lowering of the average oxidation state of the cations (*i.e.*, oxygen loss). This may happen as a result of heating, or, more typically, as a result of exposure to reducing atmospheres. Cathode materials are not generally exposed to reducing conditions, either during processing or during operation and therefore chemical expansion *per se* does not impact the long term stability of fuel cell cathodes. On the other hand, fuel cell interconnects and electrolytes are subjected to reducing conditions. More significantly, they are subjected to a chemical potential *gradient* (reducing conditions on one side, oxidizing conditions on the other) which, in principle, generates large internal stresses (due to the differential chemical expansion). There is some speculation that these stresses are relieved by creep as an explanation as to why the components do not more routinely fracture. In terms of cathode materials, long term stability is more typically impacted by deleterious chemical reaction with the electrolyte or possibly other components in the fuel cell.

Professor Beck opened the discussion of the papers by Professor Haile and Professor Irvine: Is there a volume change during the transition? If so, one could check the entropy change by looking at dT/dP because the entropies calculated from temperature dependent measurements of ΔH may be hampered by unknown enthalpic effects.

Professor Haile replied: There is indeed a volume change at the transition, and the P–T phase diagram has been measured, allowing, in principle, for comparison with the Clapeyron equation. From the phase diagram reported by Rapoport[1] dP/dT is $\sim 3.3 \times 10^8$ Pa K^{-1}. Combining our own (unpublished) measurements of the thermal expansion of monoclinic CsH_2PO_4 (determined by *in situ* X-ray diffraction) with the volume of the superprotonic cubic unit cell reported by Priesinger,[2] the volume change at the transition is ~ 1.0 Å^3 per formula unit (or 0.9 vol%). Together, these imply a transition entropy of 21 J K^{-1} mol^{-1}. This value is in very good agreement with the 23 J K^{-1} mol^{-1} that we have measured directly by thermal methods (measurements of ΔH) and reported in this manuscript. It must be noted, however, that Priesinger has reported a substantially larger volume change than what we have measured, which would imply an accordingly larger entropy change (by almost a factor of two). We believe that our thermal expansion measurements of the monoclinic phase, which were carried out for the critical purpose of evaluating the mechanical suitability of CsH_2PO_4 as a fuel cell electrolyte, are more accurate than the earlier results. The high temperature cell volume reported by Priesinger is comparable to that of similar sulfate–phosphate compounds in their cubic, super-protonic phases,[3] and therefore can be considered reliable for this analysis.

1   E. Rapoport, J. B. Clark and P. W. Richter, High-pressure phase relations of RbH_2PO_4, CsH_2PO_4 and KD_2PO_4, J. Solid State Chem., 1978, **24**, 423–433.
2   A. Preisinger, K. Mereiter and W. Bronowska, The phase transition of CsH_2PO_4 (CDP) at 505 K, Mater. Sci. Forum, 1994, **166–169**, 511–516.
3   C. R. I. Chisholm, R. B. Merle, D. A. Boysen and S. M. Haile, Superprotonic phase transition in $CsH(PO_3H)$, Chem. Mat., 2002, **14**, 3889–3893.

Professor Yashima asked: Did you try to study the nuclear/electron density distribution in order to investigate the disordered structure?

Professor Haile replied: No, we have not ourselves performed high temperature structure refinements of either $CsHSO_4$ or CsH_2PO_4. To the best of my knowledge there has not been a literature study of these compounds aimed at resolving the nuclear or electron density. In my group, we have carried out very preliminary structural studies of some superprotonic mixed sulfate–phosphates using conventional laboratory X-ray sources.

Professor Yashima then asked: Did you use a split-atom model?

Professor Haile replied: Yes, all refinements of these compounds to date have employed split-atom models (both our preliminary studies of mixed sulfate–phosphates and all the literature reports of $CsHSO_4$ and CsH_2PO_4).

Professor Yashima said: It is a good idea to use the maximum entropy method to study the disorder. We can collaborate on this if you wish.

Professor Haile answered: I would be delighted to collaborate with you on this.

Professor Maier asked:
(1) Are the hydrogen partial pressures at which you expect decomposition of CsH_2PO_4 ($Cs_3P + PH_3 + H_2O$) beyond that of fuel cell conditions?
(2) Even if the situation is thermodynamically stable, one would expect decomposition to occur as one removes the product (PH_3, H_2O) in a non-equilibrium flow system. What do long-time experiments show?

Professor Haile answered: Full thermodynamic data are not readily available for all of the relevant compounds, however, we can use the available data to consider the formation of H_xP compounds from solid phosphorous and gaseous hydrogen. First, we note that there are no known solid compounds in this binary system (as reported in the NIST database), so we consider gas phase species. At 200 °C, and at 1 atm H_2, the partial pressures in the possible gaseous species are: $p(P_4) = 3 \times 10^{-5}$ atm; $p(PH_3) = 1 \times 10^{-5}$ atm; $p(PH_2) = 3 \times 10^{-15}$ atm. These values are extremely low, and consistent with our experimental examination of CsH_2PO_4 under reducing conditions. Mass spectrometry analysis of the evolved gases showed the concentrations of these species to be below the detection limits.[1] Even under kinetic conditions in which evolved species are being removed from the system, they are unlikely, at such low concentrations, to play a role in long term stability. A more likely species is gaseous $(P_2O_3)_2$, but this also was not detected by mass spectrometry. Furthermore, we have indeed performed long term measurements directly confirming the stability of CsH_2PO_4-based fuel cells. With humidified H_2 supplied to the anode, humidified O_2 supplied to the cathode (both flowing at 50 sccm), a current density of 100 mA

cm^{-2} and a temperature of 235 °C, a fuel cell with 260 μm thick CsH_2PO_4 maintained an absolutely steady voltage for over one hundred hours, at which point the experiment was shut off.[1] Even longer term tests are under way, and all evidence supports the conclusion that CsH_2PO_4 displays sufficient thermodynamic stability for fuel cell applications.

1   D. A. Boysen, T. Uda, C. R. I. Chisholm and S. M. Haile, High performance solid acid fuel cells through humidity stabilization, *Science*, 2004, **303**, 68–70.

Professor Islam addressed Professor Haile:

(1) You use the Pauling model for ice to examine the configurational entropy. A sharp increase in the mobility of protonic defects is argued, rather than a sharp increase in the concentration of such defects. What are the implications for this?

(2) With a high concentration of mobile protons, do you see any evidence of a cooperative migration mechanism?

Professor Haile answered: In the Pauling model, the concentration of defects, defined by analogy to ice as species such as $H_3PO_4{}^0$, $HPO_4{}^-$, and doubly occupied hydrogen bonds, is small such that they do not contribute significantly to the configurational entropy. From a consideration of the electrostatic energy of such defects, it is reasonable to expect that the low and high temperature phases of CsH_2PO_4 would support comparable concentration of such defects. Furthermore, even in liquids such as H_2SO_4 and H_3PO_4, self-ionization leads to defect concentrations of only a fraction of a mol%. But this now begs the question, why does the conductivity rise so dramatically at the transition? If the concentration of defects is indeed relatively unchanged, the logical conclusion is that a dramatic increase in their mobility is responsible for the superprotonic behavior. Another interpretation, which we did not explicitly describe in the manuscript, is that *all* protons become mobile above the transition, whether classified as defects or not. If the mobility is so high that defect and normal species interchange at an extremely high rate, these two interpretations may, in fact, describe the same physical phenomenon. As of yet we have not observed significant evidence for cooperative motion.

Professor Beck said: I question whether there are really no cooperative effects which would not correlate with the model given for an evaluation of the entropy according to Pauling's model. Such caterpillar effects are known for many ionic conductors.

Professor Haile replied: The point we are making is that if one accepts the Pauling model for ice as being descriptive of disordered CsH_2PO_4, then the concentration of defects must be low. It may well be that for those few defects which are present, cooperative motion occurs. This level of detail in understanding the proton transport mechanism remains to be resolved.

Professor Irvine asked: The literature has suggested decomposition from $CsHPO_4$ to $CsPO_3$, which is a fully condensed phosphate. Have you considered partial decomposition to, say, a 2-phosphate unit such as pyrophosphate? This would suggest an equilibrium between $2HPO_4{}^{2-}$ and $P_2O_7{}^{4-}$.

Professor Haile responded: As written in the manuscript, we have explicitly considered dehydration according to $CsH_2PO_4(s) \rightarrow CsH_{2-2x}PO_{4-x}(s) + xH_2O(g)$, $(0 \leq x \leq 1) \rightarrow CsPO_3 + H_2O$ $(x = 1)$.

Indeed, it is reasonable to anticipate the existence of intermediate phases between CsH_2PO_4 and $CsPO_3$. In their high temperature X-ray diffraction study of CsH_2PO_4, Preisinger *et al.*[1] reported the occurrence of sharply crystalline $Cs_2H_2P_2O_7$ as a precursor to complete dehydration. In contrast, we see no evidence for this phase in our thermal gravimetric studies. Our (unpublished) results indicate that $Cs_2H_2P_2O_7$ is not thermodynamically stable under any condition, but that it may be metastable at low temperatures (below about 150 °C). Strictly speaking, therefore, there is no equilibrium between CsH_2PO_4 and $Cs_2H_2P_2O_7$. Alternatively, amorphous phases with variable stoichiometry (H_2O content) may form. In particular, we suspect that partially dehydrated amorphous phases can appear on the surface of CsH_2PO_4 particles and inhibit H_2O loss.

Another possibility that one might consider is that within the structure of CsH_2PO_4 pyrophosphate (e.g., $P_2O_7^{4-}$) defects occur. Such a situation, which has been suggested occasionally in the literature, is, in my opinion, highly unlikely. The phosphorous to phosphorous distance in CsH_2PO_4 is about 4.5 Å, whereas in a P_2O_7 unit it is less than 3 Å. The local strain that such a substitution would require of the structure would be enormous. Given that CsH_2PO_4 and related compounds are resistant even to replacement of $H_2PO_4^-$ by HSO_4^-, it is hard to imagine that they could accommodate replacement of two $H_2PO_4^-$ groups by a $H_2P_2O_7^{=}$ group. Thus, proposing a role for such species in the proton transport mechanism in crystalline CsH_2PO_4 is optimistic, at best.

1   A. Preisinger, K. Mereiter and W. Bronowska, The phase transition of CsH_2PO_4 (CDP) at 505 K, *Mater. Sci. Forum*, 1994, **166–169**, 511–516.

Professor Kilner opened the discussion of Professor Irvine's paper: It is well known that the mobility of the ions on the two sublattices are very different, the cation mobilities are much lower than the anion mobilities. Do you think that the thermal history, particularly at high temperature, will be important in determining the clustering? In other words, does the vacancy distribution adjust to a pre-existing cation distibution?

Professor Irvine responded: Yes, the vacancy distribution will adjust to the pre-existing cation distribution. It is important to note that on long term annealing a fairly random cation distribution will start to order and hence anion clustering will increase and indeed change in nature. This seems to be closely correlated with the degradation of ionic conduction properties on extended annealing.

Professor Kilner then asked: Do you think that the doped cerias will be different from the doped zirconia? In fact, will each material be different?

Professor Irvine answered: There are surprisingly quite strong similarities, but the underlying driver for extended defect order in the zirconias is that Zr is too small for 8-fold coordination which is not the case for Ce.

Professor Yashima asked: Did you deduce detailed defect structural models? Detailed models including clusters and their fractions have been proposed by, for example, Professor Ishizawa,[1] for YSZ through single crystal and EXAFS techniques.

1   N. Ishizawa, Y. Matsushima, M. Hayashi and M. Ueki, *Acta Crystallogr., Sect. B*, 1999, **55**, 726.

Professor Irvine responded: The main point of this work is to demonstrate that these defect clusters are organised into extended structures. Interesting studies such as those of Ishizawa[1] or Hull[2] probe the atomic structure around the constituent defects. We are seeking to understand the continuity and arrangement of such defects.

1 N. Ishizawa, Y. Matsushima, M. Hayashi and M. Ueki, *Acta Crystallogr., Sect. B*, 1999, **55**, 726–733.
2 J. P. Goff, W. Hayes, S. Hull, M. T. Hutchings and K. N. Clausen, *Phys. Rev. B*, 1999, **59**, 14202–14219.

Professor Drennan asked two questions:

(1) From your preparation method, are you confident that your materials are homogeneous?

(2) Do you believe that adding Y_2O_3 to the system affects the reaction of Sc_2O_3 with zirconia?

Professor Irvine replied:

(1) Yes. For materials that are carefully analysed for ordering effects we use repeated grinding and firing to ensure homogeneity, as confirmed by TEM analysis. This would not necessarily apply for ceramics optimised for technical performance where pinhole free structure is more important than homogeneity.

(2) Not so much the reaction as the product, but, yes, it will modify coordination significantly as evidenced by the loss of the rhombohedral phase.

Professor Haile asked: Could you clarify the distinction between defect trapping/ association energy and ordering energy? Are either of these connected to cation ordering?

Professor Irvine replied: Defect association is very short range, and is more or less related to point defects. Ordering energy relates to further association of the point defects into extended defects. Ordering is related to lattice ordering which includes the cation lattice; however, in these fluorites the oxygen sublattice is generally the dominant factor in lattice ordering due to the higher anion diffusion rate, allowing the anion lattice to equilibrate more easily.

Professor Navrotsky asked: Is cation ordering important in the long-term degradation of scandia doped zirconia, does it explain irreproducibility in differently prepared samples, and can one extract cation ordering kinetics from engineering ageing data (over weeks/months of operation)?

Professor Irvine replied: The answer is yes to all three questions, but it is still difficult to extract suitable quality reproducible data to refine good models.

Professor Kilner commented: We must be careful when interpreting some of the fuel cell ageing experiments as they measure total dc conductivity and there may well be a grain boundary component to the degradation.

Professor Irvine responded: This is a fair point, especially with less pure electrolyte powder sources. However, there is considerable evidence from ac experiments on ageing, which include a cool down phase in the cycles to allow the bulk and grain boundaries to be discriminated, that bulk ageing is a real and important phenomenon.

Professor Kilner said: The cations will also have a different mobility in the fluorite lattice. Thus different systems might react in different ways.

Professor Navrotsky opened the general discussion: A fundamental question is over what rate of ionic motion the idea of configurational entropy falls apart, in the sense that you can no longer count discrete sites and occupancies and must instead consider some sort of communal entropy? This is analogous to the same question relevant to electron hopping, *i.e.*, Fe^{2+}/Fe^{3+} in magnetite.

Professor Maier said: Given certain bond strengths and dielectric constants, at low temperature you find low defective situations exhibiting Arrhenius type of behaviour. As temperature increases, the defect concentration becomes so high that interactions influence the effective formation energies. If they are attractive, defect concentrations increase in an over-Arrhenius way, leading eventually to a transition to a partially or totally molten state, hence typically the complete disorder in the superionic phase is indicated in the low T phase (for a simple interpretation see Hainovsky and Maier[1]).

1   N. Hainovsky and J. Maier, *Phys. Rev. B*, 1995, **51**, 15789.

Professor Navrotsky replied: Such "sublattice melting" has been studied by Michael O'Keeffe's group in the 1980s and we have evidence for a similar transition in Y_2O_3, where half the normal entropy of fusion is associated with a disordering transition some 100 K below the melting point, with the remaining entropy (and enthalpy) of fusion anomalously low.

Dr Gray-Weale commented: In principle there need be no connection between mobility and disorder, very many ions could be 'mobile', and yet almost all close to their perfect-crystal sites. On a potential energy landscape, the configurational entropy is given by the number of potential energy minima; the diffusivity or conductivity by the heights of the energy barriers between minima. If you varied the heights of the barriers without varying the number of minima, transport would be drastically slowed or accelerated without a change in the configurational entropy. In work on glasses and supercooled liquids this argument is put against any connection between configurational entropy and diffusivity or electrical conductivity.

The problem with this argument is that in practice a connection between the two is often found.[1] The explanation is that, at least within a given class of materials, the number of minima and the heights of the barriers do not vary independently. The potential energy landscape is governed by real forces, and changes in the chemistry of a material will produce correlated changes in the number of minima and the heights of barriers.

1   A. Gray-Weale and P. A. Madden, *J. Phys. Chem. B*, 2004, **108**, 6624.

Professor Haile added: One can certainly envisage situations in which there is a great deal of disorder yet little mobility, implying that the fraction of mobile ions is low. This could again describe the situation encountered in glassy, amorphous materials, or even others with 'frozen-in' disorder. In contrast, it is much harder to envisage the opposite, where there is a high fraction of mobile ions yet the material is not disordered. In this context, what would a mobile species be, if not one residing, for example, in an interstitial site, or next to a vacant site?

Professor Irvine said: In response to the suggestion that the significant proportion of ions that are mobile might redefine thermodynamics, one should realise that unfortunately only a very small fraction of oxide ions are mobile in the materials so far discovered apart from perhaps the cobaltate perovskites.

Professor Haile said: Alex (Professor Navrotsky), you've identified a fundamental question in statistical thermodynamics; when does the idea of configurational entropy fall apart? As the mobility of ions increases, the potential energy landscape becomes shallower and shallower and it becomes more difficult to distinguish occupied sites from those which are not and therefore use them to evaluate the configurational entropy. We are trying to see how far the 'simple' picture can take us before going to more complex models.

Professor Bruce commented: Concerning entropic contributions, one must also remember that at high concentrations of mobile defects there will be considerable dynamic ordering such that an ion hop will be followed by a slower reorganisation of the surrounding mobile defects.

Professor Irvine replied: Yes, this is consistent with our thoughts on activation energies. If the ordering length scale is more extended a greater activation energy will be required to transport ions.

Professor Haile addressed Professor Irvine: Are the intensities of the super-structure peaks in the electron diffraction patterns consistent with oxygen ordering or would it require cation ordering?

Professor Irvine answered: Yes, we believe they are. These superstructure peaks are entirely consistent with diffuse neutron diffraction peaks which are absent from X-ray diffraction. This indicates clearly that the oxygen sublattice is largely responsible for the superstructure peaks. Although electron diffraction does relate to a shorter length scale, the diffuse neutron peaks do seem to relate to the same phenomenon.

Dr Slater asked: With the fluorite structure allowing the possibility of oxygen excess, is it possible that some of the conduction is due to oxygens displaced into interstitial sites? That is, if vacancies are trapped by dopants, can additional vacancies occur around the dopant to give a more favourable coordination generating interstitial oxygens?

Professor Irvine replied: It certainly is a possibility, however as we are working with large vacancy contents the defect equilibria will strongly mitigate against significant interstitial formation, apart from any transient occupation during ion motion.

Professor Navrotsky commented: The simultaneous existence of oxygen vacancies and interstitials could be explained in a system such as $UO_{2+x} \cdot YO_{1.5}$. Early vapour pressure measurements and recent calorimetry at Davis show surprisingly large stabilization.

Professor Bruce asked a general question: Can anyone offer some suggestions concerning how we might design better H^+ or O^{2-} ionic conductors? Professor Kilner may say grain boundaries but is there likely to be more cation transport in the grain boundaries of an oxide ion conductor that would result in more ageing effects?

Professor Beck answered: Professor Navrotsky gave a good hint as to how to proceed. She should try and combine effects to counterbalance the difficult problems, *e.g.*, use auxetic materials to balance thermal expansion effects which may be enhanced by a specific substitution.

Professor Nazar addressed Professor Haile: How important is it to develop solid solutions on the PO_4/SO_4 site, and hence on the acid sites, in order to improve proton conductivity at low temperatures?

Professor Haile replied: Initially, we believed this was quite important—that developing solid solutions would enable us to tune the concentration of proton interstitials and/or vacancies. For example, by introducing some phosphate groups into $CsHSO_4$ we hoped we could introduce proton interstitials. What we found, rather remarkably, is that these materials are extremely resistant to such doping. Instead of achieving solid-solution substitutions, we discovered a series of intermediate sulfate–phosphates along the pseudo-binary line between $CsHSO_4$ and CsH_2PO_4: β-$Cs_3(HSO_4)_2[H_{2-x}(P_{1-x},S_x)O_4]$ ($x \approx 0.5$),[1] α-$Cs_3(HSO_4)_2(H_2PO_4)$,[2] $Cs_2(HSO_4)(H_2PO_4)$,[3] and $Cs_5(HSO_4)_3(H_2PO_4)_2$.[4] Furthermore, all but the first of these four compounds exists over an immeasurably small $SO_4 : PO_4$ composition range. In hindsight, we can rationalize the resistance to doping in terms of the local crystal chemical features of the hydrogen bond. A typical $O \cdots O$ distance is about 2.7 Å, with the proton residing essentially within the electron cloud of the oxygen atoms. The proton pulls these ostensibly unfriendly oxygen atoms to distances they would not otherwise accept. Creating a proton vacancy at this site leaves the structure in an unfavorable, high energy configuration, which apparently cannot be accommodated by local structural relaxation. Instead, the structure arranges the dopant ion and unoccupied hydrogen bond sites in an ordered manner, resulting in the generation of a new structure type. Indeed, the structure of α-$Cs_3(HSO_4)_2(H_2PO_4)$ can be easily understood as $CsHSO_4$ with every third S in the HSO_4 chain replaced by P and new hydrogen bonds introduced in a regular manner. That said, I don't think that it would be possible, even if solid solutions could be created, to obtain in an ordered structure the kind of conductivity we observe in the superprotonic, disordered phases. Doping simply could not be expected to gain us five orders of magnitude (or even three, or even one). Ultimately, our exploratory synthesis yielded far more important information than we were initially seeking. All of the new structures exhibit superprotonic transitions despite their distinct low temperature structures. Their behavior thus demonstrates that such transitions are not limited to any particular structure type, but are rather a more general characteristic of compounds with large counter-cations and extended hydrogen bond networks.

1 S. M. Haile, P. M. Calkins and D. Boysen, Superprotonic conductivity in β-$Cs_3(HSO_4)_2(H_x(P,S)O_4)$, *Solid State Ionics*, 1997, **97**, 145–151.
2 S. M. Haile, G. Lentz, K.-D. Kreuer and J. Maier, Superprotonic conductivity in $Cs_3(HSO_4)_2(H_2PO_4)$, *Solid State Ionics*, 1995, **77**, 128–134.
3 C. R. I. Chisholm and S. M. Haile, Structure and thermal behavior of the new superprotonic conductor $Cs_2(HSO_4)(H_2PO_4)$, *Acta Crystallogr., Sect. B*, 1999, **55**, 937–946.
4 S. M. Haile and P. M. Calkins, X-Ray diffraction study of $Cs_5(HSO_4)_3(H_2PO_4)_2$, a new solid acid with a unique hydrogen-bond network, *J. Solid State Chem.*, 1998, **140**, 251–265.

Professor Haile opened the discussion of Professor Maier's paper: Has there been sufficient consideration of space charge effects and the possibly negative influence of grain boundaries in very thin and nanocrystalline ceria and zirconia electrolytes for fuel cell electrolytes?

Professor Maier answered: I think people are becoming more and more aware of it. Yet it has to be noted that space charge effects are most prominent in systems with low carrier concentrations, but not so much in highly doped systems used in fuel cells.

Professor Islam asked: You mentioned surface storage and nano-sized ceria powder. Do you find evidence of a greater degree of segregation of acceptor dopants at the surfaces of such nano-sized systems?

Professor Maier replied: In many cases we found lesser space charge potentials in nano-sized grains.

Dr Gray-Weale asked: When you observe a non-trivial nanoscale effect, such as when neighbouring space charge layers overlap, and the local effect on material properties is not merely the sum of effects from the overlapping layers, what is the source of the nonlinearity?

Professor Maier answered: It is the non-linearity of the Poisson-Boltzmann equation.

Professor Kilner asked: Given that these composite structures you have examined look interesting, can they be made practically?

Professor Maier responded: The Li_2O/Ru nanocomposites can be easily formed by electrochemical reduction of RuO_2 by lithium. We have a paper in press[1] that elaborates on this preparation method.

1 Y.-S. Hu, Y.-G. Guo, W. Sigle, S. Hore, P. Balaya and J. Maier, *Nat. Mater.*, 2006, **5**, 713.

Professor Nazar said: The problem of charge-storage is not only one of Li^+ storage, but of the necessity of transport of O^{2-} (in RuO_2/Li, for example) for reversible storage. How do you see overcoming this problem—and have you conducted studies on materials with high oxide ion conductivity in their reduced state to see if the polarization is reduced?

Professor Maier answered: In Ru_2O, Li^+, O^{2-} and e^- are all fairly mobile. In IrO_2 where at least O^{2-} is not so mobile, the reversibility is far less.

Professor Heitjans commented: Concerning salt storage in ion conductors, the system Li_2O/CaF_2 for the storage of LiF is rather attractive for NMR studies, and we intend to perform corresponding measurements.

Professor Maier replied: At the moment we are thinking about how to realise this experimentally.

Dr Zhukovskii said: I'd like to make a supporting comment with regard to Professor Maier's paper. It would be important to involve electronic structure calculations on comparatively simple models of metal–metal oxide or metal–metal fluoride interfaces to explain effects of electronic charge transfer through the interface. We have simulated Ti/Li_2O (111) and Ag and Cu over LiF (001) interfaces. Their 2D slab models could explain both enhanced Li storage inside the interface and charge transfer towards the metal, thus explaining the Me^-/Li^+ model.

Mr O'Dell opened the discussion of Professor Heitjans' paper: Where is the 5% alumina located in the ball milled sample, and what is its effect on the Li ion conductivity?

Professor Heitjans responded: As I showed in my talk, when going from the microcrystalline to the nanocrystalline sample prepared by ball milling the conductivity is increased by several orders of magnitude (see Fig. 5 of the Discussion paper). Compared to that, the admixture of 5% alumina has a neglible effect and does not at all influence the conclusions drawn.

As a matter of fact, we quite recently measured the dc conductivity of the nanocrystalline composite $(1 - x)LiNbO_3:xAl_2O_3$ explicitly as a function of the alumina content x in the range from $x = 0$ to 0.7.[1] For $x = 0.05$ an increase in the conductivity by a factor of 1.3 over the value for $x = 0$ was found at, *e.g.*, 423 K. At this temperature the conductivity of nanocrystalline $LiNbO_3$ (milled for 16 h) is larger than that of microcrystalline $LiNbO_3$ by a factor of about 10^5, as can be estimated from the data in Fig. 5.

1   R. Amade W. Iwaniak and P. Heitjans, 2006, unpublished results.

Dr Andreev said: Your model for the ball milled sample is that this material is heterogeneous, consisting of a crystalline core surrounded by an amorphous layer. Could there be an alternative model, given the fact that ball milling is a high-impact treatment which may lead to microstrains? So, basically you have a homogeneous particle with a high degree of microstrain which is most severe on the surface. This model would also explain the absence of glass transition on DSC traces.

Professor Heitjans replied: The heterogeneous nature of our ball milled $LiNbO_3$ samples was confirmed in particular by our HRTEM and EXAFS results and by the fact that the conductivity parameters as well as the NMR linewidth and relaxation results of the milled samples are close to those of the purely amorphous material. The similarity of the nanocrystalline with the purely amorphous material increases with increasing milling time. Concerning DSC traces, we presently have only data for the sample milled for 16 h, and the glass transition of the corresponding amorphous portion may be masked by other processes at elevated temperatures like grain growth and grain-boundary relaxation. Another early example from my group, where the picture of nanocrystalline grains embedded in amorphous grain boundaries applies, is α-$LiBO_2$.[1] DSC curves showed the recrystallization of the amorphous part being determined by the milling time.

Further systems supporting the heterogeneous model have been studied by, *e.g.*, Alan Chadwick and coworkers,[2] as you know.

1   C. H. Rüscher, E. Tobschall and P. Heitjans, Amorphization of α-$LiBO_2$ by ball milling: investigation by XRD, IR-spectroscopy and DTA, in *Applied Mineralogy*, ed. D. Rammlmair, J. Mederer, Th. Oberthür, R. B. Heimann and H. Pentinghaus, Balkema Publishers, Rotterdam, 2000, p. 221.
2   A. V. Chadwick, M. J. Pooley, K. E. Rammutla, S. L. P. Savin and A. Rougier, A comparison of the extended X-ray absorption fine structure of nanocrystalline ZrO_2 prepared by high-energy ball milling and other methods, *J. Phys.: Condens. Matter*, 2003, **15**, 431; see also A. V. Chadwick, *Diffusion Fundamentals*, 2005, **2**, 44.

Professor Navrotsky said: I am still concerned about the effect of alumina from ball milling, particularly on the amorphous phase. Could the amorphous fractions contain alumina, and on what distance scale—nanocomposite or true solid solution, for example?

A disordered $LiNbO_3$ can be thought of as a corundum structure and the disorder could be stabilised by solid solution with $Al_2O_3{}^-$ corundum.

Professor Heitjans answered: The amorphous grain boundaries may contain a small amount of alumina. According to our XRD patterns, where in the case of the sample with the longest milling time (64 h) the lines for α-Al_2O_3 are just visible besides those of $LiNbO_3$, we have a composite rather than a solid solution. As I explained in my response to Mr O'Dell, the effect of the alumina on conductivity is very small.

Professor Islam asked: You have a general finding that ball milling leads to a higher level of disorder (and amorphous material) in the interfacial regions. Is this due to mechanical factors or are other issues involved?

Professor Heitjans replied: The highly disordered or amorphous structure of the interfaces of ball milled samples indeed results from the high mechanical energy exerted on the material. Details of the process depend on the type of mill, the materials of the milling vials and balls, the milling time, the temperature and other parameters, and are not completely understood yet. This is the subject of current research within mechanochemistry, which has become a very active field in recent years.

Professor Beck said: Could the mobility effects discussed be brought about by the same mechanisms that Professor Maier described in his paper? Professor Maier spoke of space charge effects which enhanced the mobility of ions in one phase by the adjacent crystals of a second one, *e.g.*, alumina. I would like to mention earlier work on electrostatic potentials and isopotential surfaces ranging even out of the surface of a crystal which have been calculated and have been proven to disturb the space potentials in an adjacent crystal lattice, thereby creating structural distortions and defects which would also enhance the mobility of ions within a surface layer.[1]

1 H. P. Beck and T. Beyer, *Z. Kristallogr.*, 1998, **213**, 501.

Professor Heitjans responded: As I pointed out in my response to Mr O'Dell, in the present case of ball milled $LiNbO_3$, "heterogeneous doping" by alumina due to the abrasion of the milling vial set cannot explain the measured increase in the conductivity by several orders of magnitude. Here the main effect is related to the quasi-amorphous structure of the grain boundaries. The earlier theoretical work you refer to is very interesting but does not seem to apply to our case.

Dr Wilson asked: We saw in the first session (Professor Irvine's paper) how differences in ion conductivity may be understood in terms of the local ionic coordination environments. Do you have a feel for the differences in the local environment between the crystalline and amorphous $LiNbO_3$ systems? Put more simply, why is the conductivity so much higher in the amorphous state compared to the crystalline state (at comparable temperatures)?

Professor Heitjans answered: Clearly we now no longer talk about the effect of heterogenous doping on the conductivity, but deal with the variation of local crystalline structure (brought about by homogenous doping in the case you refer to) and its influence on ionic conductivity. First, we established a strong correlation between the conductivity and overall structural disorder increasing in the sequence: single crystal, microcrystalline, nanocrystalline (sol–gel), nanocrystalline (ball milled), amorphous $LiNbO_3$. Second, concerning the local structure in the amorphous state with respect to that in the crystalline state, we found from our EXAFS measurements that amorphous $LiNbO_3$ consists of smaller and less distorted NbO_6 octahedra, which are connected to fewer and farther situated NbO_6 by a smaller number of lithium–oxygen polyhedra. This result supports the idea that in the amorphous state the short-range order is similar to that in the crystalline one, consisting of the same building units, whereas the medium- and long-range order is different, and that the amorphous state contains homogeneously distributed "free volume", which may be the main reason for the enhanced conductivity and diffusivity of the bulk amorphous sample.

Of course, the amorphous structure is not unique by definition, and there may be differences between the conductivities of amorphous samples prepared in different ways which, however, are small compared to those with respect to the crystalline state.

Professor Yashima asked:
(1) What do you think the effect of OH^-, H_2O and/or CO_2^{3-} (carbon) was on the surface/bulk states of ball milled and sol–gel samples?
(2) Although the ball milling is carried out under Ar, the concentration of water is not zero and can affect the results. You could check this using IR/Raman spectra.

Professor Heitjans replied: As stated in the paper, the materials were handled under dry and inert conditions in a glove-box with argon atmosphere. Milling was performed using a vial set of alumina, which was additionally placed in an airtight container made of steel. Although great care was taken in order to avoid reaction of the nanocrystalline material with water vapour or CO_2, an influence on the results cannot be excluded *a priori*. However, when cycling the temperature of the final samples in the range up to about 450 K, which was not exceeded in order to exclude grain growth, no hysteresis effect in the conductivity was observed. Another argument is the following. Due to the absence of large volume fractions of interfacial regions in amorphous $LiNbO_3$, the influence of water *etc.* should be strongly reduced. Note that, according to Fig. 5 of our paper, the conductivity of ball milled $LiNbO_3$ (nearly) reaches that of the amorphous sample but is in no case higher than this, which would be expected if a large amount of water was responsible for the high conductivity.

Apart from the effect of grain-size reduction with increasing milling time, our Raman spectra agree very well with the literature data for $LiNbO_3$ and there are no indications for the presence of water in our samples.

Mr O'Dell commented: Thermogravimetric analysis shows complete loss of organic and hydroxyl groups from sol–gel prepared nanocrystalline samples by 300 °C.

Professor Yashima replied: But $\sim$ 300 °C heat treatments are not high enough to remove impurities. Even if we heat treat at 500 °C, it is not enough to remove them, particularly the carbonates and carbon. This is well known.

Professor Heitjans commented: DSC/TG curves of samples calcined at 500 °C (like at 700 °C) did not show any DSC or weight loss peaks up to the melting point.

Professor Yashima then asked: Did you analyse the local structure? From EXAFS spectra, you can obtain interatomic distances, the coordination number and thermal parameters. If you can determine the local structure around Li, you can determine the mechanism of the Li diffusion.

Professor Heitjans responded: Yes, we did the EXAFS analysis, as referred to in the paper and indicated in my response to Dr Wilson. In addition, we studied the local structure by ^7Li NMR and ^{93}Nb MAS-NMR (not mentioned in the paper). Nevertheless, we are not at the stage of understanding the Li diffusion mechanisms in this complex system in detail. The question of the exact mechanism is not even settled yet for $LiNbO_3$ single crystals in the literature.

Professor Irvine asked: Do the ferroelectric properties of $LiNbO_3$ change with the degree of (nano)crystallinity as observed from either impedance spectroscopy (capacitive) or SSNMR?

Professor Heitjans replied: Bulk crystalline $LiNbO_3$ has a Curie temperature of about 1500 K. I expect this to decrease with grain size decreasing to the nanometer range. We did not explicitly study the ferroelectric properties of our nanocrystalline samples, which have average grain sizes of about 20 nm. Please note that in our measurements the temperature did not exceed about 450 K in order to avoid grain growth.

Professor Drennan opened the discussion of Mr O'Dell's paper:
(1) With these very reactive materials (*i.e.*, sol–gel mixtures), why do you not expect to see some formulation of $ZrSiO_4$ at low temperatures?
(2) You report the formation of $MgAl_2O_4$—why doesn't this interfere with your observations?
(3) Your electron microscopy is very selective. How do you know where your SiO_2 and Al_2O_3 are in the structure?

Mr O'Dell answered:
(1) If it were present, it would certainly be visible in the ^{29}Si MAS NMR spectrum as a narrow peak at a shift of around 60 ppm, representing the isolated tetrahedral Q^0 silicon environment.

(2) The $MgAl_2O_4$ is present in the ^{27}Al MAS NMR spectra across the range of annealing temperatures, and seems to remain separate from the rest of the alumina. Although another octahedral aluminium peak overlaps it at certain temperatures, simulations at multiple fields have allowed the two peaks to be distinguished unambiguously. $MgAl_2O_4$ was also observed by XRD, as stated in the paper.

(3) Due to the extremely small sizes of the pinning particles, it is often very difficult to image them successfully using standard TEM. Our efforts at this are improving, and recently we have managed to obtain images of the pinning phases in other systems (CeO_2, TiO_2). The pinning phases are not visible in the images in the paper (with the exception of Fig. 4 which clearly shows the alumina phase in the ZrO_2 system), and so their exact location in the structure remains unknown. However, NMR in combination with other techniques (XRD, EXAFS) can still provide structural information that can be used to postulate their location relative to the metal oxide nanocrystals, and their morphology. Of course, HRTEM could unambiguously answer these questions, and we are planning to use this technique in the near future.

Dr Kendrick commented: Zircon formation occurs at 1600–1700 °C even when sol–gel and coprecipitation techniques are used for the synthetic methods. Mineralisers are required, and NaF/LiCl or transition metal oxides are used to lower the temperature of formation.

Professor Beck said: It is known in the literature (*e.g.*, by M. Veith *et al.*, University of Saarland[1,2]) that *via* the sol–gel $MgAl_2O_4$ synthesis route, inverse spinel is formed first due to the reaction path defined by the precursors. This should be taken into account when interpreting your NMR data.

1   M. Veith, A. Atherr and H. Wolfanger, *Chem. Vap. Deposition*, 1999, **5**, 87.
2   F. Meyer, R. Hempelmann, S. Mathur and M. Veith, *J. Mater. Chem.*, 1999, **9**, 1755.

Mr O'Dell responded: This is an interesting point of which I was unaware, and I shall certainly bear it in mind when interpreting future work. However, the spinel observed in this paper showed only a sharp octahedral peak in the ^{27}Al MAS NMR spectrum, and no sharp tetrahedral aluminium site was observed, indicating that this is the regular spinel, not the inverse form. XRD also showed that only the regular spinel structure was present.

Professor Drennan addressed Professor Maier: How confident are you that the ionic measurement techniques that presently exist can deal with the new nano-materials?

Professor Maier replied: There is still need for improvement. We achieved this progress by developing microelectrode impedance, with the help of which one can arrive at space resolution. In cases of extremely good ionic conduction we could even address single nanoparticles by the impedance spectrometer.

Professor Heitjans said: I would like to stress the importance of studying the same material by different techniques in order to get a more complete and consistent picture of the atomic transport. As an example, I mention our early measurements on nanocrystalline CaF_2 both by NMR[1,2] and impedance spectroscopy,[3] which appear to be the first combined study of this kind in a nanocrystalline ion conductor. NMR allowed us to differentiate between slow F^- ions in the grains and fast ones in the grain boundaries. The considerably enhanced overall diffusivity and conductivity were ascribed to F^- transport parallel to space charge layers at the grain boundaries. The measured conductivity enhancement was then also quantitatively explained in terms of the large fraction of interface regions in nanocrystalline as compared to microcrystalline CaF_2.[4] Thus, complementary and consistent information on the fast ion diffusion was obtained by impedance and NMR spectroscopies on a nanocrystalline material, which this time was prepared by the inert gas condensation technique. For the NMR measurements the as-prepared powder was used, while it was compacted for the conductivity measurements. Generally, no grain growth was observed below 500 K and the results showed no hysteresis effects. CaF_2 has thus become another model system for the influence of interfaces on ion transport. Contrary to the case of $LiNbO_3$, dealt with in our *Faraday Discussion* paper, the strong conductivity increase in nanocrystalline CaF_2 is not due to "amorphization" of the structure of the diffusion pathways.

1   W. Puin, P. Heitjans, W. Dickenscheid and H. Gleiter, in *Defects in Insulating Materials*, ed. O. Kanert and J.-M. Spaeth, World Scientific, Singapore, 1993, p. 137
2   P. Heitjans, A. Schirmer and S. Indris, in *Diffusion in Condensed Matter - Methods, Materials, Models*, ed. P. Heitjans and J. Kärger, Springer, Berlin/Heidelberg, 2005, p. 367.
3   W. Puin and P. Heitjans, *Nanostruct. Mater.*, 1995, **6**, 885.
4   W. Puin, S. Rodewald, R. Ramlau, P. Heitjans and J. Maier, *Solid State Ionics*, 2000, **131**, 159.

Dr Ruiz-Trejo asked a general question: Can we have a guess of what would happen to the transport properties of ceramic proton conductors if we can prepare samples with nanograins? Can we expect electronic conductivity in $BaCeO_3$ or $SrZrO_3$?

Professor Maier answered: Again the high doping level suggests that these effects are not significant. More important are the increased blocking effects by insulating boundaries, *e.g.*, in $BaZrO_3$, the most promising material for intermediate fuel cells.

Professor Ishihara addressed Professor Maier: Since ionic mobility has not been measured clearly up to now, there has been no detailed discussion about the effects of nanosize on the mobility of ions. However, I wonder whether a change in mobility occurs in the case of nanosized particles. Do you think that the change in mobility occurs and causes the change in conductivity?

Professor Maier replied: The interface can influence both mobility and defect concentration. As a rule of thumb: in materials with high carrier mobility in the bulk, the interfacial structure is likely to lead to lower mobilities therein. Then, particularly in systems with low carrier concentration, space charge effects are very important. With regard to materials with very low mobilities, interfaces are likely to provide pathways of higher mobility.

Professor Bruce asked: Considering the system with an ionic conductor in contact with, say, Al_2O_3: You said that the greatest effect is on the system with fewest defects. Therefore do you not expect a proportionally greater increase of mobile defects for the immobile sublattice and hence the possibility of material transport at the interface, and ageing effects?

Professor Maier responded: What eventually counts is the conductivity. If the mobility of the sublattice is very low, point defects are not even formed at the contact. Should they be formed, their effect on the overall conductivity would be negligible. In AgCl, for example, the effects on the Ag^+ sublattice are dominant, the Cl^- sublattice offers a negligible concentration of defects.

Dr Nastar said: High mobility of species is required just to reach the equilibrium state. Another way to induce grain boundary segregation is to irradiate the system. Mobility might be enhanced by the creation of point defects, but segregation profiles will also be controlled by the mobility coefficients.

Professor Maier replied: It may be a good idea in some cases to "catalyse" boundary effects by irradiation. However, after that, the carriers may freeze again and are then of no use for conductivity (but may be for other properties such as catalysis).

Professor Haile commented: There appears to be some evidence in the literature that composites of proton conducting oxides and metals lead to higher proton incorporation in the oxide than there is in the single component oxide.

Dr De Souza addressed Professor Heitjans: If one has an amorphous shell of material surrounding a crystalline core of the same material, would you expect a space-charge zone at the interface, since the standard defect chemical potentials will be different for the two phases? Do you think this effect will be large?

Professor Heitjans replied: In principle, in the ball milled nanocrystalline $LiNbO_3$ samples space charge effects at the grain boundaries should also influence the conductivity. However, due to the fact that the conductivity of the ball milled samples is rather similar to that of the bulk amorphous one, I think that in this case the space charge contribution turns out to be relatively small. Ascribing the conductivity of the sol–gel prepared nanocrystalline sample, whose grain boundaries are essentially not amorphous (see Fig. 3 of our Discussion paper), to space charge effects, it may be estimated that the latter appear to be responsible for only about 0.1% of the enhancement in the ball milled sample (16 h) at, e.g., 450 K (cf. Fig. 6b). This would mean that, nevertheless, space charge effects lead to an enhancement of the conductivity in sol–gel prepared nanocrystalline $LiNbO_3$ by a factor of 10^2 with respect to microcrystalline $LiNbO_3$. As I mentioned in my comment to Professor Drennan, nanocrystalline CaF_2, which was prepared by noble gas condensation, is another example of a chemically homogeneous system we have studied, where the grain boundaries are not amorphous and the enhancement can be ascribed to space charge effects.

Nanostructured materials for lithium-ion batteries: Surface conductivity *vs.* bulk ion/electron transport

B. Ellis,[a] P. Subramanya Herle,[a] Y.-H. Rho,[a] L. F. Nazar,*[a] R. Dunlap,[b] Laura K. Perry[c] and D. H. Ryan[c]

Received 22nd February 2006, Accepted 9th May 2006
First published as an Advance Article on the web 7th August 2006
DOI: 10.1039/b602698b

Lithium metal phosphates are amongst the most promising cathode materials for high capacity lithium-ion batteries. Owing to their inherently low electronic conductivity, it is essential to optimize their properties to minimize defect concentration and crystallite size (down to the submicron level), control morphology, and to decorate the crystallite surfaces with conductive nanostructures that act as conduits to deliver electrons to the bulk lattice. Here, we discuss factors relating to doping and defects in olivine phosphates $LiMPO_4$ (M = Fe, Mn, Co, Ni) and describe methods by which *in situ* nanophase composites with conductivities ranging from 10^{-4}–10^{-2} S cm^{-1} can be prepared. These utilize surface reactivity to produce intergranular nitrides, phosphides, and/or phosphocarbides at temperatures as low as 600 °C that maximize the accessibility of the bulk for Li de/insertion. Surface modification can only address the transport problem in part, however. A key issue in these materials is also to unravel the factors governing ion and electron transport within the lattice. Lithium de/insertion in the phosphates is accompanied by two-phase transitions owing to poor solubility of the single phase compositions, where low mobility of the phase boundary limits the rate characteristics. Here we discuss concerted mobility of the charge carriers. Using Mössbauer spectroscopy to pinpoint the temperature at which the solid solution forms, we directly probe small polaron hopping in the solid solution Li_xFePO_4 phases formed at elevated temperature, and give evidence for a strong correlation between electron and lithium delocalization events that suggests they are coupled.

Introduction

The creation of redox-active transition metal framework structures that host mobile interstitial Li^+ ions is crucial in developing high capacity lithium-ion batteries. Lithium transition metal phosphates such as $LiFePO_4$,[1] $LiMnPO_4$,[2] $Li_3V_2(PO_4)_3$[3] and $LiVPO_4F$[4] have been recognized as promising positive electrodes for these systems because of their energy storage capacity combined with electrochemical and thermal stability. These are related either to fast-ion conducting phases, or minerals

[a] *University of Waterloo, Department of Chemistry, Waterloo, Ontario Canada N2L 3G1*
[b] *Department of Physics, Dalhousie University, Halifax, Canada*
[c] *Department of Physics, McGill University, Montreal, Canada*

such as olivine and tavorite. Owing to the inherently low electronic conductivity in the bulk it is absolutely essential to optimize their properties to obtain good electrochemical characteristics. This includes modification of crystal growth to minimize lattice defects and particle size (hence reducing the path length for electron and lithium ion transport), and modification of the crystallite surface to create conductive species that can act as "electronic wires" to feed electrons into the lattice without blocking access of lithium. Critical factors include how to design and tailor the ideal nanostructure and determine what its optimum morphology would be, and the nature of the interface.

A group of materials that serve as a good model for developing these concepts are the $LiMPO_4$ phosphates (M = Fe, Co, Ni, Mn), the most prominent members of a family of the polyanion compounds.[1,5] Their promise is due to the "inductive effect" of the XO_4 (X = Si, S, P) polyanion, which elevates the M^{2+}/M^{3+} redox couple by about 1.5–2 V for X = P.[6] A redox potential of 3.45 V $vs.$ Li/Li^+ results in the case of $LiFePO_4$, making it a particularly appealing material for hybrid energy systems where cost and safety are of major concern. Both the Fe and Mn compositions are also attractive owing to their low environmental impact. The $Mn^{2+/3+}$ couple is raised to a very desirable potential of 4.1 V in the phosphate framework; however, extraction of Li from this material is both slow, and incomplete due to a combination of factors that are not fully understood. These pertain to the Jahn–Teller distortion experienced by Mn^{3+} which creates mechanical stress in the lattice at high levels of oxidation.[7,8] Efforts to overcome this drawback have been successful to varying degrees.[9] Consideration of the inductive effect and preliminary electrochemical studies show both $LiCoPO_4$ and $LiNiPO_4$ will have very high redox couples (4.8 and >5.0 V, respectively), making them suitable only in the presence of an electrolyte with very high oxidation stability.[10]

The $LiMPO_4$ family adopts a common structure displayed by silicate minerals such as olivine, $MgFeSiO_4$, that constitute a large fraction of the earth's crust. The lattice comprises a network of MO_6^{n+} octahedra interwoven with XO_4^{n-} tetrahedra. The mobile alkali ions, Li^+ in the case of $LiMPO_4$, form one-dimensional chains in the structure that run parallel to planes of corner-shared MO_6 octahedra, and along the [010] direction. Calculations of "free" ion transport in the absence of interactions with localized electron sites in the lattice suggest that the ion mobility along the chain direction is high,[11] but the material does not appear to be a fast ion conductor.[12] The electronically insulating effect of the tetrahedral XO_4^{n-} groups on which the inductive effect relies gives rise to isolation of the redox centers within the lattice. Correspondingly, incorporation of a XO_4^{n-} polyanion, such as phosphate, increases the band gap $vis a vis$ the oxide to values that are in the range of 3.7 eV in $LiFePO_4$ based on calculations and experiment.[13] Electron transport in this very poor semiconductor ($\sigma \approx 10^{-9}$ S cm^{-1}) is dependent upon small polaron hopping of Fe^{3+} holes within the lattice. Recent calculations predict an activation energy of 0.185 eV for a "free polaron" carrier in the absence of ionic interactions.[14]

The consequence of electronic transport limitation has led to immense efforts to overcome it, including methods to coat the phosphate particles with carbon,[15] embed them in a carbon matrix,[16] and lay down metal particles to form a composite.[17] The latter have all resulted in an increase in the working capacity of the material to approach theoretical capacity at relatively fast rates of electron extraction and insertion in the material. Another recently explored avenue is doping the framework with a supervalent ion to render it inherently conductive. The proposition was made that doping induced formation of a mixed valent state at the iron centers of Fe^{2+}/Fe^{3+}. Doped compositions $Li_{0.99}M_{0.01}FePO_4$ (M = Nb, Zr, Mg, Ti) were reported to be black p-type semiconductors with conductivities as high as $\sim 10^{-2}$ S cm^{-1} at room temperature.[18] Hole conductivity was suggested to arise from the formation of minority type Fe^{3+} carriers within the lattice. Since the stoichiometry as presented in fact suggests a sub-valent <2+ state of Fe in the case of Nb, Zr and Ti doping, loss of lithium is necessary to account for such an effect. Interest in this report stemmed not only from the good performance of the material as a Li-ion electrode, but also from the 8-fold order of magnitude increase in conductivity. The origin and reproducibility of the conductivity have since been the source of extensive controversy (and many studies on doping) with speculation arising because the results violate commonly held concepts of electron mobility in this class of materials.

Another property of metal phosphates is their propensity to undergo carbothermal reduction. The reaction with carbon and $LiFePO_4$ results in oxidation of the carbon to CO or CO_2, and reduction of the neighboring Fe and P ions in the lattice to form Fe_2P and/or Fe_3P. This well known reaction has been recently used to reduce Fe^{3+} to Fe^{2+} to form $LiFePO_4$ from Fe^{3+} precursors, where control of the oxygen partial pressure and temperature is necessary to inhibit iron phosphide formation.[19] Higher temperatures result in reduction to Fe_2P whose presence in the bulk at temperatures above 850 °C has been implicated.[20] Unequivocal evidence for Fe_2P formation has been observed by X-ray diffraction, when sufficiently large amounts of carbon are present.[21] Careful control of such a reaction can result in intergranular conductivity in nanophase composites, and as we have demonstrated in a preliminary communication,[22] is responsible for the enormous enhancement of conductivity observed in carbon-containing $LiFePO_4$. As we show here, the reaction is general and even more favoured in metal phosphates of higher oxidizing potential such as $LiCoPO_4$, and $LiNiPO_4$, but not $LiMnPO_4$, and is responsible for the previously reported enhanced conductivity in Cr-doped $LiFePO_4$.[23] The use of lower reduction temperatures in more reducing atmospheres allows the reaction to be sustained at temperatures that maximize the surface area and accessibility of the material for Li insertion reactions. Nitridation can also be accomplished through the use of reactive gases that produce Fe_2N, and results in an increase in conductivity.

Since such surface structures can only partially solve the transport problem, a key issue in these materials is to furthermore disentangle the factors governing ion and electron transport within the lattice. Important to this is the creation of solid solutions over a wide lithium concentration range to facilitate coupled ion and electron transport. Such solid solutions were shown to form in $Li_{3-x}V_2(PO_4)_3$, for example, by disorder of V^{4+}/V^{5+} ions in the oxidized lattice that drives the delocalization of lithium ions.[24] In the $LiMPO_4$ family of materials, extraction of lithium forms a two-phase $LiMPO_4/MPO_4$ mixture that is in part driven by volume change between the structures. However, it was recently demonstrated that a transition to a Li_xFePO_4 solid solution (SS) phase occurs at about or above 485 K, where lithium occupation is random within the lattice.[25] Here we show using Mössbauer spectroscopy that electron delocalization in the solid solution phases is due to rapid small polaron hopping. We also show that the onset temperature of *electron* delocalization is correlated to the state of lithium disorder, suggesting the two transport mechanisms are coupled. Thus, the transport is limited by neither carrier alone, but by their concerted mobility through the lattice. This provides insight into the transport mechanism not only in $LiFePO_4$, but in an ever increasing family of phosphate, fluorophosphate and silicate materials being considered as the new generation of lithium-ion cathodes.

Experimental

Synthesis

Several solid state methods have now been described in the literature for the formation of $LiFePO_4$, triphylite, including (a) reaction of precursors such as $FeC_2O_4 \cdot 2H_2O/NH_4H_2PO_4/Li_2CO_3$ followed by treatment in various gases; (b) reaction of iron(III) precursors such as $FePO_4$ with lithium sources followed by treatment in reducing gases; and (c) precipitation of vivianite $Fe_3(PO_4)_2 \cdot 8H_2O$ and reaction with Li_3PO_4 followed by treatment in inert gases. All three methods were used in this study. Oxalates of nickel, cobalt and manganese were used to synthesize the corresponding phospho-olivines, according to procedure (a). A typical process for the synthesis of triphylite involves the rigorous ball milling of the solid precursors in stoichiometric amounts, followed by sintering at 350 °C for 6 h. Final sintering of the powder occurs at 600–700 °C under various atmospheres (Ar, 7% H_2 in N_2, NH_3) for 2–10 h. Further treatment involved the pressing of a circular pellet (12 mm diameter under 3 tons of pressure) and sintering in flowing Ar at 800–1000 °C for various compositions.

XRD, STEM and SEM

X-Ray diffraction was performed on a Bruker D8-Advantage powder diffractometer using Cu Kα radiation (λ = 1.5405 Å) from 2θ = 10–80° at a scan rate of 6 s per step of 0.01°. X-Ray data sets were refined by conventional Rietveld methods using the GSAS package with the EXPGUI interface.[26] The background, scale factor, zero point, lattice parameters, atomic positions, and coefficients for the peak shape function were iteratively refined until convergence was achieved. TEM analysis was carried out by embedding a small portion of the sintered pellet in epoxy resin, and slicing the sample with an ultramicrotome. The slice was supported on a 200 mesh Cu grid. STEM imaging and EDX spot elemental analysis was performed using a Hitachi S5200 operating at 30 kV in STEM mode to determine the Fe : P ratios. SEM samples were gold coated and examined in a LEO 1530 field emission scanning electron microscope (FESEM) instrument equipped with an energy dispersive X-ray spectroscopy (EDX) attachment. Images were recorded at 15 kV with a secondary electron detector.

X-Ray photoelectron spectroscopy

Materials were analyzed with a VG Scientific XPS Microprobe ESCA Lab 250 using focused monochromatic Al Kα radiation (1486.6 eV). Samples were deposited on a Cu substrate with an irradiated area of 0.4 × 1 mm^2, and loaded in the chamber at a pressure of less than 10^{-10} mbar.

Conductivity measurements

Pellet surfaces were polished prior to variable temperature conductivity measurements that were performed using four-point d.c. methods. Electrode contacts were affixed using silver or gold paste in linear geometry on a thin section of a pellet of approximate dimensions: 1 mm × 1 mm × 5 mm.

Electrochemical measurements

Electrochemical evaluation of the materials were carried out in coin cells using a commercial (MacPile™) multichannel galvanostat/potentiostat operating in galvanostatic mode. Typical cathode loadings were in the range of 5–6 mg cm^{-2} and an electrode diameter of 10 mm was used throughout. The positive electrodes comprised 80 wt% active material, 10% Super S carbon and 10 wt% PVdF binder. The electrolyte was composed of a 1 M $LiPF_6$ solution in 1 : 1 EC–DMC; and the anode consisted of lithium metal supported on a stainless steel disc.

Results and discussion

(a) Cation occupation, doping and substoichiometry

The olivine structure represented by $LiMPO_4$ (i.e., $ABXO_4$, where X = Si, P, B, Be) contains two crystallographic sites occupied by the A and B cations, as seen in Fig. 1. Both sites (known as M1 and M2) can be described as having slightly distorted octahedral co-ordination. In iron–magnesium silicate olivines that form a solid

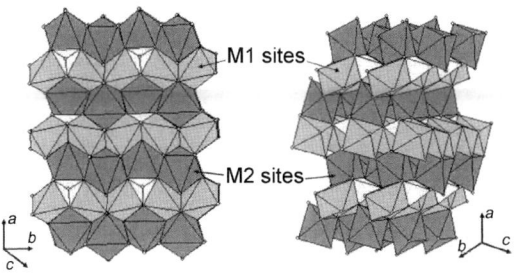

Fig. 1 The olivine structure, adopted by several minerals including $LiFePO_4$. The octahedral M1 (Li) and M2 (Fe) cation sites are labelled; the bridging phosphate groups are shown as tetrahedra.

solution composition between fayalite (Fe_2SiO_4) and forsterite (Mg_2SiO_4), there is disorder of the 2+ cations on the M1 and M2 sites. When the metal cations differ in charge however (such as the case of $LiFePO_4$), there is generally strict ordering of ions in these metal sites based on size and energy preferences: the M2 site houses the cation of greater charge. However, low temperature syntheses at 130 °C based on hydrothermal chemistry have been reported to yield disorder as high as 7%.[27]

A starting point for our investigations was to examine the consequence of Li substoichiometry and doping of cations on the M1 and M2 sites. The reason was two-fold: first, evidence from Fe-substituted nickel phosphates such as $Li_{1-3x}Fe_x$-$NiPO_4$ suggest that olivine phosphates can be synthesized with sustainable cation vacancies on the M1 site,[28] thus the direct synthesis of any lithium deficient iron phosphates that are able to sustain solid solution behaviour at room temperature would indeed be novel. Second, as mentioned above, compositions with very low dopant levels such as $Li_{0.99}M_{0.01}FePO_4$ (where M = Mg^{2+}, Al^{3+}, Zr^{4+}, Nb^{5+}) have been reported.[12] Since the nominal stoichiometries imply a sub-valent state of Fe ($<Fe^{2+}$) on the basis of charge balance (*i.e.*, for $Li_{0.99}Zr_{0.01}FePO_4$, the formal oxidation state of iron would be +1.97), loss of lithium during processing would have to occur to account for Fe^{3+} hole carrier formation. The dopant could act as a stabilizer. The precise site occupancy of these dopants was not established, however. These results have stimulated considerable debate about the precise defect properties. Cation doping of $LiFePO_4$ raises key questions as to the favored substitution site (M1 *versus* M2), the type of compensating defect, and whether the doping process is favorable on energetic grounds. Our preliminary work that examined the compositions $Li_xZr_{0.01}FePO_4$ (x = 0.87 to 0.99) showed that dopants were not essential, and suggested that percolating "nano-networks" of metal-rich phosphides within the grain boundaries of $LiFePO_4$ crystallites are responsible for the enhanced electronic conductivity.[22] Recent structural and electrochemical studies of Delacourt *et al.* were also unsuccessful in Nb doping of $LiFePO_4$; instead, they showed that crystalline α-$NbOPO_4$ and/or an amorphous (Nb, Fe, C, O, P) coating was formed around $LiFePO_4$ particles, which is believed to be responsible for the superior electrochemical activity.[29] In accord with these experimental results are recent calculations reported by Islam *et al.*, that examined a range of dopants including divalent (*e.g.*, Mg, Mn, Co), trivalent (*e.g.*, Al, Ga, Y), tetravalent (*e.g.*, Zr, Ti), and pentavalent (*e.g.*, Nb, Ta) ions.[30] Their calculations reveal that low favorable energies are found only for divalent dopants on the Fe (M2) site (such as Mg and Mn). On energetic grounds, $LiFePO_4$ is not tolerant to aliovalent doping (*e.g.*, Al, Ga, Zr, Ti, Nb, Ta) on either Li (M1) or Fe (M2) sites.

Our attempts to dope the $LiFePO_4$ lattice are consistent with these findings. We chose Zr^{4+} as a target dopant since the calculated effective ionic radii of Li^+ and Zr^{4+} (0.76 and 0.72, respectively) are sufficiently similar that Zr could easily substitute onto a lithium site. Table 1 summarizes refined lattice parameters for various stoichiometries, synthesized with iron oxalate, with both lithium and iron or

Table 1 Refined lattice parameters of olivine phases after addition of supervalent cations to probe the presence of doping and possible site preferences for the dopant

Dopant target site(s)	Stoichiometry	Refined olivine lattice parameters			Detectable impurities
		$a/Å$	$b/Å$	$c/Å$	
None	$LiFePO_4$	10.3203(2)	6.0045(1)	4.6934(2)	None
M1	$Li_{0.96}Zr_{0.01}FePO_4$	10.3203(3)	6.0041(2)	4.6957(3)	Below detection limit
M1	$Li_{0.88}Zr_{0.03}FePO_4$	10.3260(2)	6.0047(1)	4.6948(1)	NASICON-structured phosphate
M1	$Li_{0.99}Zr_{0.01}FePO_4$	10.3221(3)	6.0049(1)	4.6924(2)	None
M1	$Li_{0.91}Cr_{0.03}FePO_4$	10.3262(2)	6.0050(1)	4.6935(1)	Cr_2O_3, $Fe_2P_2O_7$
M1/M2	$Li_{0.94}Al_{0.06}Fe_{0.94}PO_4$	10.3210(2)	6.0052(2)	4.6941(1)	$AlPO_4$
M1/M2	$Li_{0.94}Y_{0.06}Fe_{0.94}PO_4$	10.3183(3)	6.0040(1)	4.6925(1)	YPO_4

simply lithium deficiencies. The data for the former does not indicate that doping of the olivine structure has taken place, owing to the minimal changes in the lattice parameter for pristine triphylite and the considerable presence of impurity phases produced which account for the preponderance of the supervalent cation population. To determine the extent of Zr doping in the lattice, $Li_{1-4x}Zr_xFePO_4$ compositions ($0 < x < 0.05$) were heated to 600 °C under an inert atmosphere in microcrystalline form. Fig. 2 shows X-ray diffraction patterns of the materials. The composition $Li_{0.96}Zr_{0.01}FePO_4$ fits well with the database pattern for $LiFePO_4$, suggesting that a small degree of lithium non-stoichiometry may be sustained within the lattice, as suggested by a very recent report.[31] However the XRD pattern for $Li_{0.88}Zr_{0.03}FePO_4$ clearly shows the presence of a new phase that is likely a mixed Li–Zr–Fe phosphate with a NASICON structure, as the lines index well to those of $LiZr_2(PO_4)_3$.[32] Determination of the lattice parameters of the olivine phase by Rietveld analysis did not clearly indicate that Zr substitution on the M1 site in the olivine occurred, as outlined in Table 1. It is expected, based on previous studies of Fe substituted $LiNiPO_4$ phases ($Li_{1-3x}Fe_xNiPO_4$), that the a and b lattice parameters would experience a small decrease with Zr substitution owing to the presence of vacancies on M1, and that c would undergo a slight increase.[28] This is observed for $Li_{0.96}Zr_{0.01}FePO_4$, but the change ($\pm0.15\%$) is barely significant. In the case of $Li_{0.88}Zr_{0.03}FePO_4$, the change in lattice parameter is minute and opposite to that

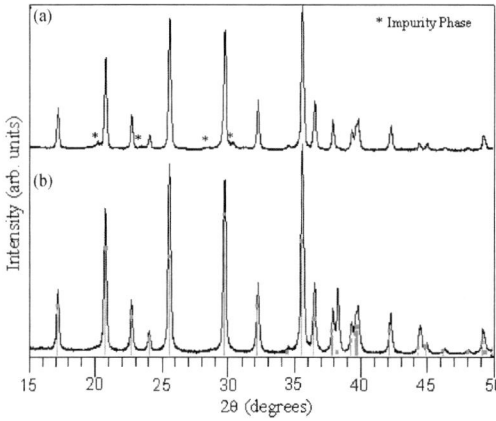

Fig. 2 X-Ray diffractograms of Zr doped materials (a) $Li_{0.88}Zr_{0.03}FePO_4$ and (b) $Li_{0.96}Zr_{0.01}$-$FePO_4$ prepared at 600 °C. A NASICON-structured impurity phase becomes evident at high lithium non-stoichiometry, indicated by the asterisks in the diffraction pattern of (a).

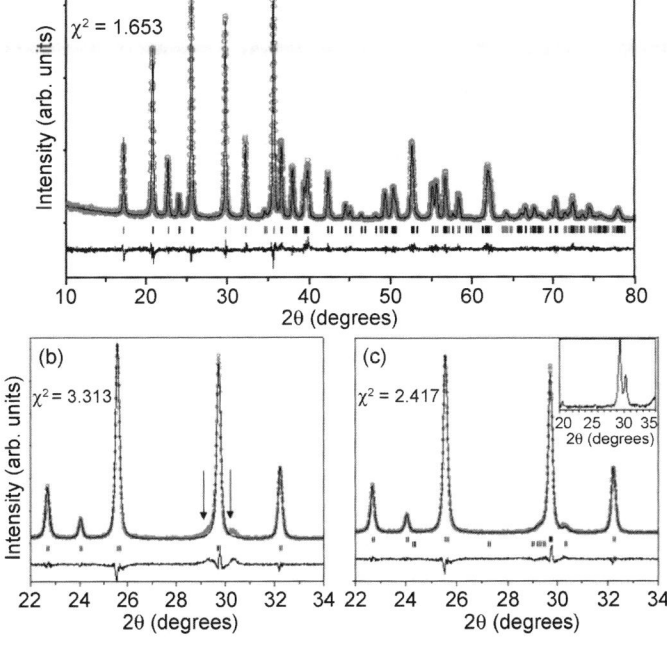

Fig. 3 Experimental X-ray ($\bigcirc$) and calculated diffraction patterns (-) based on Rietveld refinement together with the (*hkl*) reflections (|) and the difference curve of: (a) pristine olivine LiFePO$_4$ (refined cell: a = 10.3172(4) Å; b = 6.0018(2) Å; c = 4.6906(1) Å); (b) a representative sample of a lithium-deficient material, Li$_{0.91}$FePO$_4$ refined with only LiFePO$_4$, showing the contribution of iron pyrophosphate (arrows); (c) refinement using a two-phase mixture of LiFePO$_4$ and Fe$_2$P$_2$O$_7$ (lower phase tags in (c); refined cell: a = 5.5032(2) Å; b = 5.2759(2) Å; c = 4.4760(2) Å; α = 98.353(3)°; β = 98.539(3)°; γ = 104.009(4)°), indicating the significant improvement in the refinement. The XRD pattern of Fe$_2$P$_2$O$_7$ alone is shown in the inset in (c). The phase mixture fraction of all materials based on Rietveld analysis are listed in Table 2.

expected for a and b. Hence we conclude that if doping is sustainable, it is only at very low levels (*i.e.*, <3%).

We endeavoured to determine whether lithium deficient triphylite materials could be synthesized directly and uncover the possible stabilizing effects of very low amounts of Zr-doping. Following the original procedure, Li$_x$Zr$_{0.01}$FePO$_4$ powdered materials (0.99 < x < 0.87) were heated to 600 °C under inert atmosphere. Lithium deficient samples of Li$_y$FePO$_4$ (0.88 < y < 1) were also prepared without the addition of the zirconium isopropoxide dopant, chosen to be "valent-equivalent" to the Zr-doped samples. The resultant XRD patterns for representative crystalline powders are shown in Fig. 3. The materials prepared are almost entirely pure: compositions that are close to stoichiometric (such as Li$_{0.97}$FePO$_4$ or Li$_{0.96}$Zr$_{0.01}$-FePO$_4$) show only reflections due to single-phase LiFePO$_4$ (Fig. 3a). Closer inspection reveals that as the Li content decreases, a slight broadening in the (020) reflection at 29.7 and 30.3° in 2θ becomes evident, which can ultimately be resolved as very weak satellite lines attributable to Fe$_2$P$_2$O$_7$, the diffraction pattern of which is shown in Fig. 3c (inset). The agreement factors in the Rietveld fit increase markedly with the addition of this impurity phase to the refinement, as seen for Li$_{0.91}$FePO$_4$ (Fig. 3b and c). The weight and molar percentage of each phase present are summarized in Table 2. The calculated quantity of lithium in each sample closely matches the original pyrophosphate impurity, which increases linearly with lithium substoichiometry for both the Zr-containing and undoped compositions. Thus, it is

Table 2 Composition of two-phase mixtures formed from heat treatment of Li-deficient stoichiometries heat treated at 600 °C

Compound	$LiFePO_4$/mol%	$Fe_2P_2O_7$/mol%
$LiFePO_4$	100.0	—
$Li_{0.97}FePO_4$	98.1	1.9
$Li_{0.94}FePO_4$	95.7	4.3
$Li_{0.91}FePO_4$	91.8	8.2
$Li_{0.88}FePO_4$	88.8	11.2
$Li_{0.99}Zr_{0.01}FePO_4$	99.5	—
$Li_{0.96}Zr_{0.01}FePO_4$	97.5	2.5
$Li_{0.93}Zr_{0.01}FePO_4$	92.1	7.9
$Li_{0.90}Zr_{0.01}FePO_4$	89.1	10.9

conceivable that a metastable substoichiometric $Li_{(1-x)}FePO_4$ phase initially forms, with a stability regime between $\sim$200–400 °C as recently reported.[18] This phase then decomposes to $(1-x)$ $LiFePO_4$ + $x/2$ $Fe_2P_2O_7$ at temperatures above 600 °C. In turn, the pyrophosphate undergoes carbothermal reduction to iron phosphide at 800 °C, a lower temperature than that for $LiFePO_4$.

Further proof that substoichiometric Li compositions decompose to $Fe_2P_2O_7$ and fail to form Fe^{3+} hole carriers as a method of charge compensation was attained from direct analysis of the Fe^{3+} content by Mössbauer spectroscopy. The spectrum of a highly conductive $Li_{0.90}Zr_{0.01}FePO_4$ sample pellet is shown in Fig. 4. The fitted parameters are: isomer shift (IS) 1.217 mm s^{-1}; quadrupole splitting (QS) 2.905 mm s^{-1} and width 0.19 mm s^{-1} which are typical of octahedrally coordinated Fe^{2+} in $LiFePO_4$. The impurity phase $Fe_2P_2O_7$ is not visible in the Mössbauer spectrum. As iron is in the same oxidation state and similar coordination in this material, its Mössbauer parameters are similar to those for $LiFePO_4$ and this component lies under the signal for the majority phase $LiFePO_4$. The spectrum for a Fe^{+3} ion in octahedral coordination features a symmetric doublet; the quadrupole splitting for orthorhombic $FePO_4$ is 1.53 mm s^{-1}.[2] A least-squares fit to the spectrum that

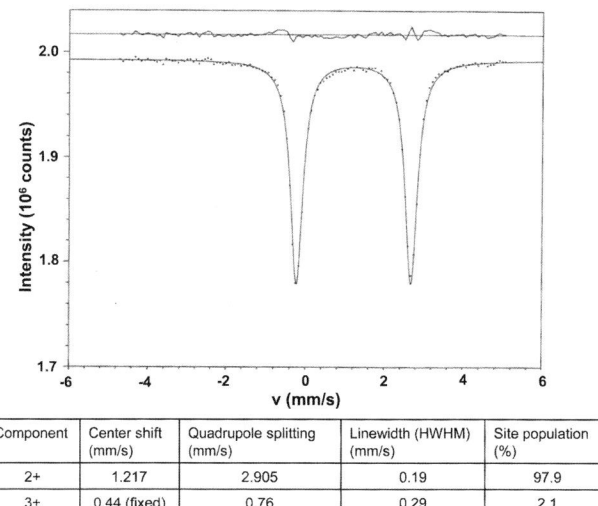

Component	Center shift (mm/s)	Quadrupole splitting (mm/s)	Linewidth (HWHM) (mm/s)	Site population (%)
2+	1.217	2.905	0.19	97.9
3+	0.44 (fixed)	0.76	0.29	2.1

Fig. 4 Mössbauer spectrum of a highly conductive sample $Li_{0.90}Zr_{0.01}FePO_4$ sintered at 800 °C. The fitted parameters are typical of Fe^{2+} in $LiFePO_4$. The other minor Fe^{2+} contributor, $Fe_2P_2O_7$, that is visible in the XRD pattern lies under this Fe^{2+} component.

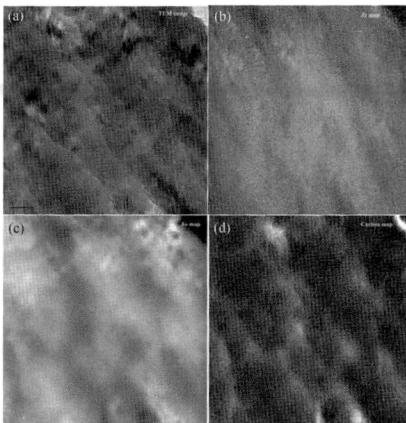

Fig. 5 TEM image of $Li_{0.90}Zr_{0.01}FePO_4$ pellet sintered at 800 °C showing (a) image; (b) Zr map; (c) Fe map, (d) C map.

included a second contribution from Fe^{3+} (with a "typical" fixed isomer shift of 0.44 mm s^{-1}) failed to yield any statistically significant evidence for its presence. At most, the Fe^{3+} contribution was 2%, and the quadrupole splitting did not match that of $FePO_4$. Thus, even with a 10% deficiency in Li content from pure $LiFePO_4$, the additional conductivity found in these composites cannot be predominately due to Fe^{+3} hole conductivity.

The total of the above observations strongly suggests that (a) Zr does not act as an internal dopant to stabilize lithium substoichiometry to any large extent; and (b) the Zr is likely primarily located on the surface of the particles. The latter is not surprising as the dopants were added as an alkoxide [*i.e.*, $Zr(OC_3H_7)_4 \cdot C_3H_7OH$], and the precursors were subjected to extensive ball-milling. This step effectively disperses the Zr (and more importantly, carbon from the alkoxide) but minimal Zr is incorporated into the olivine lattice. Indeed, EDX mapping of crystallites (Fig. 5) shows a uniform distribution of Zr both on the surface and within the grain boundaries of the crystallites. Finally, in the context of using dopants to increase electrochemical performance, it is worth noting that aliovalent doping of the M1 site (*i.e.* Mg^{2+}, Zr^{4+}, *etc.*) *could* potentially induce lithium vacancy formation, *i.e.*, $Li_{1-2x}Mg_xFePO_4$ at low levels. However transport of the Li^+ ions would be inhibited by the immobility of the dopant within the one-dimensional tunnels.

(b) Carbothermal reduction

Fig. 6 shows SEM images of the $LiFePO_4$ crystallites obtained after milling carbonaceous and non-carbonaceous precursors at 600 °C under flowing argon. The particle morphology of the former consists of aggregrates of primary nanoparticles <300 nm in dimension. This suggests, not surprisingly, that the carbon (whether from the iron oxalate, dopant alkoxide or both) binds strongly to the surface of the precursors, thus simultaneously coating the particle with carbon containing reagent, and restricting growth of the particles. Without the alkoxide carbon contribution, materials produced from vivianite yield a block-like morphology in large crystallite form.

Furthermore, the presence of carbon, either from the additional alkoxide or from the oxalate in the solid state route, is significant to the properties of the materials owing to the importance of carbothermal chemistry that occurs at high temperatures in these phosphates. Carbothermal reduction (CTR) is used extensively in industry to reduce metal oxides, such as ZnO, FeO, Cr_2O_3 and Al_2O_3 to pure metals.[33] Using

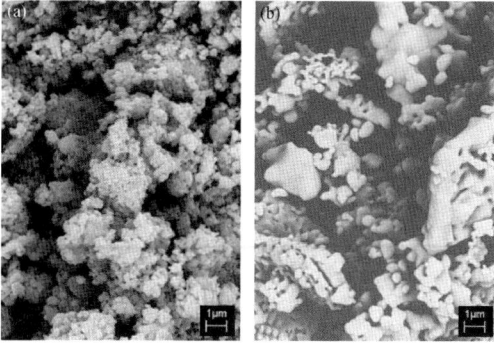

Fig. 6 SEM image of polycrystalline LiFePO$_4$ powders sintered at 600 °C prepared from (a) iron oxalate and (b) vivianite. Carbon present in the former controls particle size and morphology; particles average 300 nm in diameter.

the system of binary oxides as an example, the ease of reduction of a particular binary metal oxide can be described by the standard free energy of formation of the oxide, which is a measure of the affinity of the metal to be in an oxide lattice. The reduction can take place *via* two different carbon oxidation reactions:

$$C + O_2 \leftrightarrow CO_2 \tag{1}$$

$$2C + O_2 \leftrightarrow 2CO \tag{2}$$

The formation of carbon dioxide in eqn (1) represents a minimal volume change, and thus a change in entropy of almost zero. As a result, the standard free energy of formation of CO$_2$ is almost unchanged (-390 kJ mol^{-1}) regardless of temperature. However, the formation of carbon monoxide in eqn (2) involves an increase in entropy through an increase in volume of the system. Therefore, the standard free energy of formation of CO becomes increasingly negative as temperature increases. At approximately 700 °C, the formation of CO becomes more favourable than the formation of CO$_2$, resulting in stronger reducing conditions at higher temperatures. Slower kinetics and less reductive conditions exist at lower temperatures. As a result, it is theoretically possible to reduce any oxygen-containing mineral with carbon assuming a critical temperature is reached. A typical method of showing this information is an Ellingham plot of the standard free energy of compound formation *vs.* temperature for various metal oxide pairs; Table 3 summarizes the data of the Ellingham plot for oxide formation. The enthalpy of reduction increases with stability of the oxide so that very high temperatures are required (>1400 °C) to reduce very stable oxides such as MnO or MgO.

CTR has been previously reported as a solid state synthetic method for making lithium-containing battery materials, including LiFePO$_4$.[19] In that process, an excess of carbon is added to Fe$_2$O$_3$ and LiH$_2$PO$_4$ and the mixture is heated to 750 °C under argon. This reaction allows selective reduction (of only Fe^{+3}), simultaneous lithium incorporation into the lattice and provides a small excess of carbon at the conclusion

Table 3 Thermodynamic data for carbothermal reduction of selected binary oxides

Compounds	$-\Delta G°$ at 600 °C/kJ mol O$_2^{-1}$	Minimum Temperature for CTR/°C
Mg/MgO	1040	1850
Mn/MnO	645	1420
Fe/FeO	420	710
Co/CoO	340	240
Ni/NiO	335	280

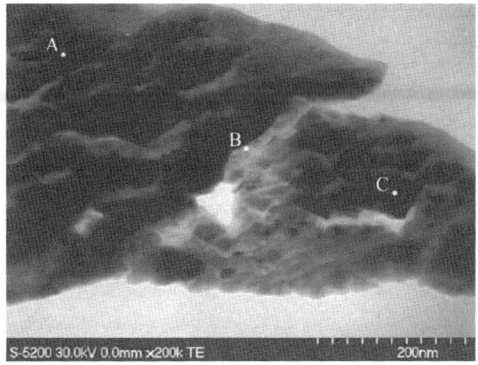

Region of Micrograph	Fe (atom%)	P (atom%)
Region A, bulk	49.4	50.6
Region B, grain boundary	62.3	37.7
Region C, bulk	53.3	46.7

Fig. 7 STEM image of sintered $Li_{0.90}Zr_{0.01}FePO_4$ collected using a Hitachi S5200 STEM operating at 30 kV EDX. EDX spot elemental analysis was used to determine the Fe : P ratios at the grain boundary and in the bulk, confirming the presence of iron phosphides at the grain boundaries.

of the reaction which is beneficial to electrode preparation. However, the carbothermal reduction of the olivines themselves has not been vigorously studied. Many synthetic methods of $LiFePO_4$ contain carbon, whether from decomposition of oxalates or acetates, or simply from the incorporation of additional carbon into the synthesis. These $LiFePO_4$–C composites are usually heated to temperatures near or above the minimum temperature for reducing iron oxide. Using the CTR data for oxides as a guide, it is clear that these conditions should result in reduction of the olivine phosphates to produce metal phosphide compounds (such as FeP, Fe_2P, Fe_3P in the case of $LiFePO_4$), many of which are metallic or semiconducting. If these compounds are produced in small quantities, they provide a conductive solid network to improve the overall conductivity of the olivine compounds, which was previously only possible by the addition of carbon (which decreases tap density).[34]

We have previously implicated the role of CTR in the increase in conductivity of $LiMPO_4$ (M = Fe, Ni) compounds. This enhancement is due to formation of a "nano-network" of electronically conductive species which forms at high temperatures, imaged by TEM and EELS mapping to reveal the presence of Fe_2P and carbon in the grain boundaries of the solid.[22] The onset of the enhanced conductivity commences near 800 °C for lithium-substoichiometric compounds, where we have observed the reduction of the $Fe_2P_2O_7$ present in these materials. Additional evidence for phosphides is shown in the STEM image (Fig. 7) of granules of $Li_{0.90}Zr_{0.01}FePO_4$. EDX analysis shows that the bulk of the crystallite (regions A, C) yields an iron to phosphorus ratio of roughly 1 : 1 as expected. However, the edge of the crystallite (region B) is significantly iron-rich with an Fe to P ratio of 2 : 1, indicating the presence of conductive Fe_2P. In combination with the carbon in the grain boundaries, this creates the nano-network which percolates through the entire sample. It is also likely that the amorphous metallic glass phase $Fe_{75}P_{20}C_{15}$ is formed at these temperatures by reaction of iron phosphide with carbon, which may serve to wet the grain boundary.

As departure from the original $LiFePO_4$ stoichiometry increases (and so does the $Fe_2P_7O_7$ content), the conductivity of the densified pellets increases. A summary of the conductivities of various starting stoichiometries is shown in Table 4. This value reaches a maximum near $x(Li) = 0.90$, after which the pellets are less dense, have a

Table 4 Room temperature conductivity values of lithium deficient $LiFePO_4$ composite pellets sintered at 800 °C

Stoichiometry	Conductivity @ 20 °C/S cm^{-1}
$LiFePO_4$	$<10^{-7}$
$Li_{0.97}FePO_4$	2.0×10^{-3}
$Li_{0.94}FePO_4$	6.5×10^{-3}
$Li_{0.91}FePO_4$	1.1×10^{-2}
$Li_{0.88}FePO_4$	$<10^{-7}$
$Li_{0.99}Zr_{0.01}FePO_4$	$<10^{-7}$
$Li_{0.93}Zr_{0.01}FePO_4$	5.1×10^{-4}
$Li_{0.90}Zr_{0.01}FePO_4$	5.3×10^{-3}
$Li_{0.87}Zr_{0.01}FePO_4$	$<10^{-7}$

more brittle texture and are no longer a deep black colour throughout. It is likely that the CTR reaction with $Fe_2P_2O_7$ has gone to completion, consuming the entire quantity of carbon in the grain boundaries, and leaving behind amorphous $Fe_x(P,C)_y$ without the connecting carbon infrastructure. The conductive nano-network is thus dismantled. In addition, dense pellets of carbon-free Li_xFePO_4 were prepared from vivianite ($x = 0.91, 0.94, 1.00$). These samples were each light grey and non-conductive ($\sigma < 10^{-7}$ S cm^{-1}), in contrast to $LiFePO_4$ prepared from iron oxalate at the same temperature, further demonstrating the role of carbon in the enhancement of conductivity in these materials.

Although electron transport in $LiFePO_4$ is a factor in poor electrochemical performance, the increase in conductivity *via* high temperature CTR does not benefit the electrochemistry of these materials. Fig. 8 shows the electrochemistry and conductivity plot of a conductive pellet of pristine $LiFePO_4$ sintered at 850 °C. Compared to $LiFePO_4$ powder sintered at 600 °C, the sample heated to higher

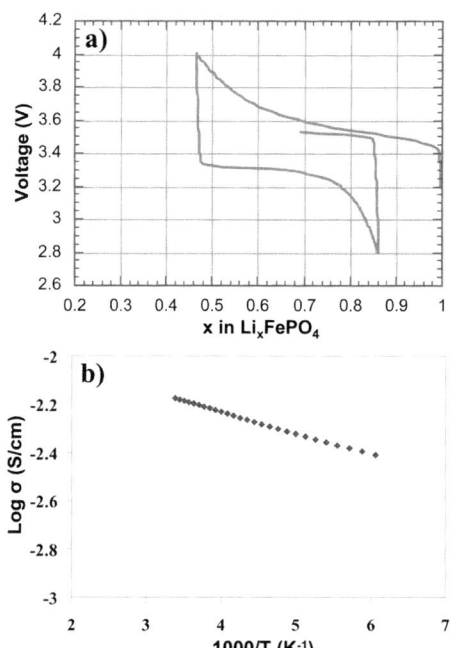

Fig. 8 (a) Electrochemistry and (b) electronic conductivity of a pelletized sample of $LiFePO_4$ sintered at 850 °C.

temperature has poor reversibility and a high polarization as a result of having a much larger particle size as a consequence of high temperature sintering.

(c) Carbothermal reduction of LiMPO$_4$ (M = Ni, Co, and Mn)

The concept of carbothermal reduction in LiFePO$_4$ was extended to other lithium metal phosphates including LiNiPO$_4$, which has a higher oxidation potential than LiFePO$_4$. Samples of Li$_x$NiPO$_4$ (x = 0.91–1.00) were prepared similarly to the corresponding iron compounds with the exception of atmosphere. Since LiNiPO$_4$ is stable at high temperatures under air and oxygen, it can be synthesized in oxidizing conditions which would promote formation of a mixed valent Ni$^{+2/+3}$ state for the lithium deficient stoichiometries while removing the possibility of phosphide formation. These oxidizing conditions can not be used to synthesize the iron compound, as LiFePO$_4$ oxidizes to Li$_3$Fe$_2$(PO$_4$)$_3$ in air above 400 °C.

Fig. 9 depicts X-ray diffraction patterns of powders and sintered pellets of Li$_{1-x}$NiPO$_4$ compositions prepared from carbon containing precursors. Single phase LiNiPO$_4$ is formed at 600 °C in air in both stoichiometric and slightly Li-substoichiometric compositions, resulting in bright yellow solids for all compositions. All showed no measurable gain in conductivity, showing values $<10^{-7}$ S cm^{-1}. Conversely, sintering of the carbon-containing nickel phosphate under inert gas at 600 °C yielded black powders. LiNiPO$_4$ is formed, along with other phosphate impurities such as Li$_4$P$_2$O$_7$, and Li$_2$Ni$_3$(P$_2$O$_7$)$_2$, as well as substantial quantities of Ni$_3$P, indicating that extensive carbothermal reduction of the phosphate has commenced prior to reaching this temperature. As a result, a pressed pellet sintered at 650 °C showed a significant (10^6) gain in conductivity (Fig. 10). The overall value is lower than for LiFePO$_4$ in part because the pellets are less densified at a lower temperature, and also because of the different nature and quantities of the phosphides within the grain boundary. Heat treatment of LiNiPO$_4$ composite pellets at temperatures above 650 °C produced pellets that are expanded and have a foamed texture (similar to that of pumice rocks), indicative of gas formation inside the solid. Clearly vigorous reduction resulting in the consumption of carbon and the subsequent release of CO and CO$_2$ takes place at these higher temperatures. A similar reaction also occurs in the case of LiFePO$_4$ made from iron oxalate; pellets sintered at temperatures >925 °C also expand and trap gas inside. These samples, however, are grey instead of black and are non-conductive, even though Fe$_2$P can be seen in the XRD pattern. This is a result of the phosphide crystallizing out of the grain

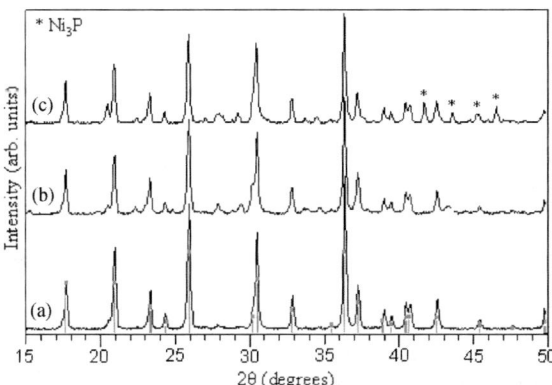

Fig. 9 Effect of sintering atmosphere on LiNiPO$_4$ composites: (a) LiNiPO$_4$ (600 °C, air); (b) Li$_{0.94}$NiPO$_4$ (600 °C, air); (c) Li$_{0.94}$NiPO$_4$ (600 °C, Ar) indicating the presence of substantial secondary phases including Ni$_3$P.

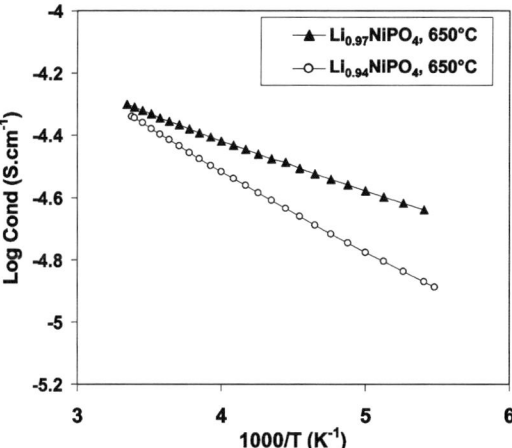

Fig. 10 Electronic conductivity plots of lithium-deficient Ni compounds in the presence of Ni_3P.

boundaries to form large particles, which, when combined with the consumption of carbon during this process, disconnects the conductive nano-network. We note that the temperature window between when iron phosphide/phosphocarbide begins to first form (~ 800 °C) and when it is extruded from the grain boundary as a result of Fe_2P crystallite growth (*i.e.*, 925 °C), is much broader than in the case of $LiNiPO_4$. Any iron phosphocarbide that is also formed (see above) may contribute to this wider stability range.

The mechanism for phosphide formation in the nickel compound differs from that for $LiFePO_4$. XRD patterns of the carbon-containing $LiNiPO_4$ precursors after the initial heat treatment at 350 °C show low intensity, poorly crystalline peaks which correspond to $LiNiPO_4$ and Li_3PO_4, and two high intensity peaks of metallic Ni. Thus the carbon from the nickel oxalate precursor reduces a portion of the metal to the elemental state. As the temperature of the reaction is increased to 600 °C, $LiNiPO_4$ crystallizes, along with Ni_3P and the other phosphate impurities. The peaks corresponding to nickel decrease and become very minor at this stage. We conclude that nickel metal is directly consumed in the production of Ni_3P or, due to the catalytic properties and reactivity of nickel nanoparticles, it is viable that it acts also as a catalyst, aiding the formation of Ni_3P by catalyzing the carbothermal reduction of $LiNiPO_4$.

In contrast, the remainder of the transition metal series of $LiMPO_4$ compounds (M = Mn, Fe, Co) do not undergo this early carbothermal reduction step. Each phosphate in this series is partially crystalline after firing at 350 °C under an inert atmosphere in the presence of trace amounts of carbon from the respective metal oxalate precursors. XRD reveals no traces of the respective metals. As a result, pure $LiMPO_4$ powders at 600 °C are produced; it is heat treatment at elevated temperatures that initiates the carbothermal reduction. In the case of $LiCoPO_4$, this process commences below 700 °C, as Co_2P and $Li_4P_2O_7$ are clearly seen in XRD patterns for powders fired at 700 °C. As stated previously for $LiFePO_4$, iron phosphide (and potentially) phosphocarbide formation occurs due to reduction at the grain boundaries at temperatures over 800 °C.

In the case of $LiMnPO_4$, the temperature for reduction is much higher than that for the other compounds, owing to the high thermodynamic stability of oxygen-containing Mn(II) compounds. We were unable to obtain any indications of reduction of carbon-containing $LiMnPO_4$ made from manganese oxalate at temperatures up to 1000 °C. Further attempts for reduction of $LiMnPO_4$, including the addition of 10 wt% carbon and firing in reducing atmospheres (in both NH_3 and a

This journal is © The Royal Society of Chemistry 2006

mixture of 7% H_2 in N_2) at 1000 °C, also failed to show any trace of reduced species after sintering. These results are consistent with the high stability of MnO, which will only undergo CTR at temperatures in excess of 1400 °C based on the Ellingham curves. In theory, phosphide and phosphocarbide formation as a result of CTR can occur in any of the carbon-coated $LiMPO_4$ compounds; the carbon, which deposits in the grain boundaries between the crystallites, can initiate reduction once the onset temperature is reached.

(d) Synthesis in highly reducing atmospheres

Although the generation of phosphides and phosphocarbides has been shown to benefit electrical conductivity of bulk materials, the temperatures required to generate these species result in excessive particle growth and, as a result, are detrimental to electrochemical performance. However, it is well known that reduction of iron(III) precursors to $LiFePO_4$ using hydrogen gas as a reducing agent can be carried out at lower temperatures compared to some CTR reactions. This synthesis method has been used in various recent synthetic routes,[35] and was used in the attempted doping of $LiFePO_4$ with chromium. Band structure calculations suggested that the Fermi energy of the latter is at the edge of the Fe valence band and high electronic conductivity was measured in the Cr-doped materials.[23]

A sample of $Li_{0.91}Cr_{0.03}FePO_4$ was prepared exactly according to the cited literature method.[26] Contrary to the report, the powder, upon sintering at 700 °C for ten hours under a 7% H_2–N_2 atmosphere, reduced to a large fraction of Fe_2P and Li_3PO_4. As a result, the starting powder was re-fired at 700 °C under argon gas to limit the severity of the reducing conditions. The powder XRD pattern is shown in Fig. 11. As is the case with all lithium-deficient compounds, $Fe_2P_2O_7$ is prominent in the diffractogram. Also clearly seen in the diffraction pattern are lines which match Cr_2O_3, the original source of Cr, which indicates there is clearly an excess of Cr_2O_3 that was not doped into the lattice. Refined lattice parameters for the olivine phase are: $a = 10.326(1)$ Å, $b = 6.0050(9)$ Å, $c = 4.6934(8)$ Å, all of which represent <0.1% differences in the respective parameters from pristine $LiFePO_4$, and imply that the olivine lattice remains undoped by chromium.

Although this synthetic method does not indicate the presence of doping, the reduction of $LiFePO_4$ to Fe_2P at this lower temperature with hydrogen (as opposed to 850 °C *via* CTR) allows the formation of small quantities of a conductive additive

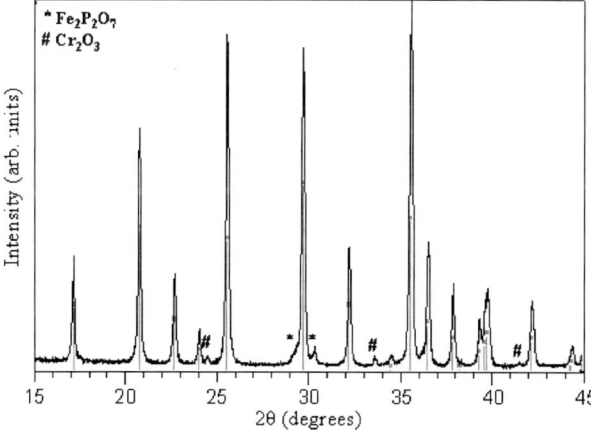

Fig. 11 A Cr-doped triphylite composition, $Li_{0.91}Cr_{0.03}FePO_4$, after firing at 600 °C for 15 h under Ar. The diffractogram shows the presence of Cr_2O_3 not incorporated into the olivine lattice and $Fe_2P_2O_7$, present in all lithium-deficient samples.

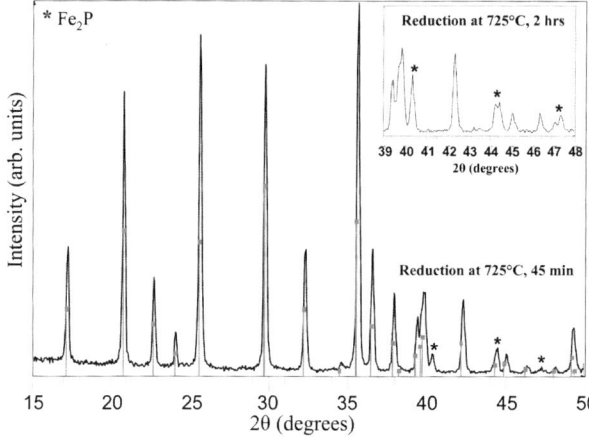

Fig. 12 XRD pattern of LiFePO$_4$ (from iron–oxalate precursors) subjected to further treatment at 725 °C in 7% H$_2$–N$_2$ for 45 min, causing a minor degree of reduction to Fe$_2$P. Iron phosphide is very evident after treatment for 2 h (inset).

without excessive growth in particle size. Samples of carbon-containing LiFePO$_4$ powder sintered at 600 °C were subjected to heat treatment at 725 °C under a 7% H$_2$–N$_2$ mixture for various lengths of time. Fig. 12 shows X-ray diffractograms after 45 min and 2 h. The presence of Fe$_2$P (inset) is clearly implicated after these relatively short heating times; Li$_3$PO$_4$ is also produced according to the reactions:

$$6LiFePO_4 + 13.5H_2 \rightarrow 3Fe_2P + 2Li_3PO_4 + H_3PO_4\uparrow + 12H_2O\uparrow$$

or

$$6LiFePO_4 + 16H_2 \rightarrow 2Fe_2P + 2Li_3PO_4 + 2FeP + 16H_2O\uparrow$$

and/or (in the presence of carbon from the precursor)

$$6LiFePO_4 + 8C \rightarrow 3Fe_2P + 2Li_3PO_4 + P\uparrow + 8CO_2\uparrow$$

We note that FeP is difficult to detect in the diffraction pattern, although it is visible by Mössbauer spectroscopy and Rietveld analysis of other materials produced by sol–gel methods, starting from iron(III) citrate.[36] The mechanism of hydrogen reduction of LiFePO$_4$ should be similar to that of carbon reduction; the LiFePO$_4$ surface groups which contact the hydrogen gas are most susceptible.

This is illustrated by a scanning electron microscopy image (Fig. 13) of a sample that has been reduced for only 45 min. Rutherford backscattering images taken from the SEM indicate the presence of phases which have different average atomic number (AAN); compounds of higher AAN will appear in brighter contrast in this mode. As a result, we can deduce that the large grey blocks in the micrograph are LiFePO$_4$ whose surface is speckled with tiny embedded crystals of Fe$_2$P (bright spots). As a result, the electrochemistry (Fig. 14) is vastly improved over LiFePO$_4$ powders sintered at 600 °C; the polarization is <0.2 V and the reversible capacity is 135 mA h g^{-1} at C/5. This brief reduction step allows the formation of a conductive surface layer of Fe$_2$P at a temperature low enough not to consume carbon or trigger large particle agglomeration, which in turn substantially improves the electrochemistry of LiFePO$_4$ prepared by a solid state route. Conversely, reduction for greater than one hour under 7% hydrogen at 725 °C has an adverse effect on battery performance; the material had poor reversibility (only 30% of theoretical capacity reached). This is likely to be a result of the phosphide covering the LiFePO$_4$ grain

Fig. 13 SEM micrograph portraying Fe_2P (black dots) on the surface of $LiFePO_4$ after sintering the material at 725 °C in 7% H_2-N_2 for 45 min.

surface enough to limit Li transport in and out of the bulk grains, effectively smothering the $LiFePO_4$ crystallites.

The synthesis of $LiFePO_4$ under flowing ammonia produces conductive surface species at even lower temperatures (600 °C). This reaction produces both Fe_2P and metallic Fe_2N, detectable by XRD after only two hours of reaction time (Fig. 15), with Li_3PO_4 as a by-product. The formation of the nitride after shorter periods of time is apparent in SEM images that nicely illustrate 30–50 nm Fe_2N nanocrystallites littering the surface of the $LiFePO_4$ (Fig. 16). Selective EDX analysis of these regions (not quantitative) revealed atomic N : Fe ratios of 1 : 1. This is in contrast to $LiFePO_4$ treated in flowing Ar that showed no visual evidence for Fe_2N nanodots (inset Fig. 6, right) or any surface nitrogen species by EDX. XPS studies also confirm the presence of Fe_2N in the NH_3-treated materials, as shown by the N1s spectrum (insert, Fig. 16). The N1s core peak at 397.4 eV is in close accord with that reported for Fe_2N (399 eV). We note that the mechanism for the formation of these species is complex, since $LiFePO_4$ begins to crystallize at close to the same temperature that iron oxalate is nitrided to produce Fe_2N. As the temperature increases, the Fe_2N may react with $NH_4H_2PO_4$ to produce Fe_2P or Li_3PO_4 to produce $LiFePO_4$, or it may remain intact. Examination of the XRD pattern shows triphylite peaks that are broader compared with samples fired under argon or hydrogen at similar temperatures, indicating that the sample is poorly crystalline. As a result, electrochemical results for this compound have yet to be optimized.

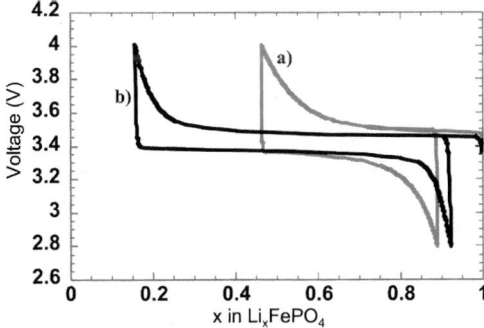

Fig. 14 Electrochemistry of (a) polycrystalline $LiFePO_4$ (600 °C, 15 h, Ar) and (b) the reduced triphylite composite (725 °C, 45 min, 7% H_2-N_2).

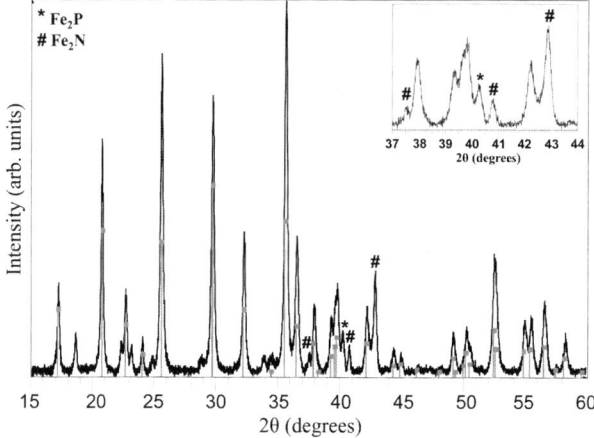

Fig. 15 X-Ray diffractogram of iron oxalate $LiFePO_4$ precursors, sintered under NH_3 (600°, 2 h). Both iron nitride (Fe_2N) and iron phosphide (Fe_2P) are present (inset).

(e) Small polaron hopping: correlation of electron mobility with lithium ion disorder

It is well known that extraction of lithium from $LiFePO_4$ results in formation of a two phase $LiFePO_4$–$FePO_4$ mixture owing, in part, to the 6% volume change between the phases. However, it was recently demonstrated that such two phase mixtures undergo a transition to a Li_xFePO_4 ($0 < x < 1$) solid solution at about 200 °C, where lithium ions and lithium vacancies randomly occupy the M1 sites within the lattice.[25] The presence of a mixed $Fe^{2+/3+}$ state on the M2 sites is implied, although only indirect evidence for it was derived from bond-sum data. This raises questions about the electron hopping rate between iron sites, and whether the electron delocalization which gives rise to this averaged valence state is coincident with lithium disordering. The latter phenomena are of importance, since electron transport in poor semiconductors such as $LiFePO_4$ takes place by small polaron migration, generated by either hole or electron carriers.[37] We chose to use variable temperature Mössbauer spectroscopy to probe the changes that occur as the material undergoes the phase transition. This technique can readily distinguish between Fe^{3+} and Fe^{2+} sites based on the difference between both isomer shift and quadrupole

Fig. 16 SEM micrograph of $LiFePO_4$ treated at 600 °C under NH_3, showing surface Fe_2P and Fe_2N as nanodots. Inset left: XPS N1s spectrum; inset right, SEM micrograph at the same magnification of the same material treated under Ar at 600 °C instead of NH_3.

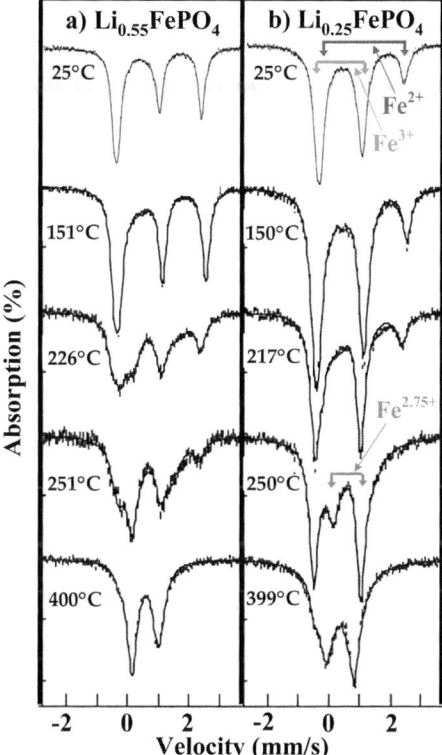

Fig. 17 Mössbauer spectra of chemically oxidized stoichiometries at room temperature and elevated temperatures illustrating the phase separation of the Fe^{2+}/Fe^{3+} components and the onset of solid solution behaviour, for (a) $Li_{0.55}FePO_4$ and (b) $Li_{0.25}FePO_4$.

splitting; moreover, it is sensitive to dynamics on timescales comparable to the Larmor precession time of the ^{57}Fe nucleus (about 10^{-8} s to about 10^{-6} s).[38] Rapid small polaron hopping between the Fe^{2+} and Fe^{3+} ions (*i.e.* $\ll 10^{-8}$ s) will show an averaged iron valence state, whereas a static situation or very slow hopping ($\gg 10^{-6}$ s) will distinguish between them. Thus, the onset of electron delocalization can be pinpointed and correlated to the lithium disordering temperature.

Samples of $LiFePO_4$ were oxidized using $NOBF_4$ to give two phase mixtures with nominal stoichiometries of $Li_{0.55}FePO_4$, $Li_{0.25}FePO_4$, and $FePO_4$. Mössbauer spectra recorded at room temperature of these two phase mixtures (Fig. 17a) contained contributions of the localised Fe^{2+} and Fe^{3+} components characterized by two doublets. They were fit with Mössbauer parameters of IS = 1.2 mm s^{-1}, QS = 3.0 mm s^{-1} and IS = 0.45 mm s^{-1}, QS = 1.5 mm s^{-1}, typical of Fe^{2+} and Fe^{3+} in these materials.[2] The relative two areas of the two components were in accord with the target starting stoichiometries. Mössbauer spectra recorded at temperatures from 130 to 430 °C illustrate the evolution from the initial two-phase composition to the solid solutions $Li_xFe^{2+/3+}PO_4$ as a function of temperature (Fig. 17b and c). The spectra of the samples at 150 °C still show the major features of the parent phases. Increases in temperature leads to the appearance of a new phase that grows in intensity as the parent phases diminish. Transformation to the single phase regime is essentially complete by 400 °C as shown by the collapse of the spectra to one doublet. The temperature dependence of the fitted parameters give the best illustration of the spectral changes as a function of heat treatment. The variation in the isomer shift of the Fe^{2+}/Fe^{3+} components and the solid solution phases in the

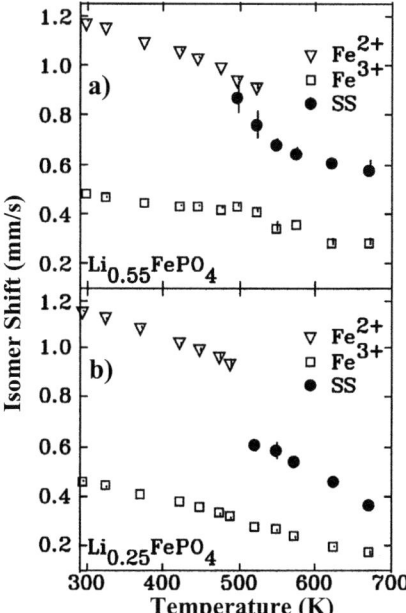

Fig. 18 Iron isomer shifts measured by Mössbauer spectroscopy for the Fe^{2+}, Fe^{3+} and solid solution phases of (a) $Li_{0.55}FePO_4$ and (b) $Li_{0.25}FePO_4$ at various temperatures.

phases $Li_{0.55}FePO_4$ (Fig. 18a) and $Li_{0.25}FePO_4$ (Fig. 18b) clearly show that the isomer shift of the solid solution phase lies between those of the parent Fe^{2+} and Fe^{3+} phases in both cases. Thus the $Li_{0.55}FePO_4$ phase displays a single quadrupole doublet with an isomer shift of 0.9 mm s^{-1} at 250 °C, midway between that of the parent phases. This demonstrates that the iron valence is intermediate between 2+ and 3+ in this rapid hopping regime, where the valence states are averaged on the timescale of the Mössbauer window. We can identify the exact temperature of the phase transition by the appearance of the averaged signal.

Both samples show an onset of rapid electron hopping behavior at about 225 °C, which is the same temperature as the onset of the transformation of two-phase mixtures of $LiFePO_4$–$FePO_4$ into a solid solution reported in the literature.[25,39] These two XRD studies demonstrated that the patterns for the two parent phases merge into one single phase Li_xFePO_4 that displays intermediate lattice parameters. The temperature at which this occurs is relatively invariant as a function of composition "x". Subsequent neutron diffraction studies have also revealed that the lithium ions are fully disordered in the solid solutions.[40] Our Mössbauer studies take us one step further in understanding the formation of the solid solution. Namely, the spectrum would not change during the phase transition if the delocalization of the lithium within the lattice led to a static delocalized distribution of the Fe^{2+} and Fe^{3+}. The clear evolution of the spectra within the Mössbauer "window" in which changes are evident shows that the electrons are dynamically delocalized in this rapid small polaron hopping regime. It is this behavior that is correlated with the random lithium population within the solid solution.

We believe that the averaging of the iron oxidation state induces disorder of the Li^+, rather than the converse. This supposition is based on comparison to our previous work on complex phase transitions driven by lithium de/intercalation in $Li_{3-x}V_2(PO_4)_3$, where a combination of charge ordering on the vanadium sites and lithium ordering/disordering amongst lattice sites was shown to be responsible for the features in the electrochemical curve.[24,41] Combined neutron diffraction, ^7Li

NMR and electrochemical studies revealed that two-phase transition behavior between single phase compositions corresponding to $x = 0, 0.5, 1.0, 2.0$ and 3.0 was adopted on lithium extraction. The single phases are characterized by highly localized V^{3+}/V^{4+} valence populations, and specific Li sites. However, solid solution electrochemical behavior was observed on insertion of Li^+ (and e^-), and was correlated with the delocalization of both lithium and the V^{n+} valence states as seen by diffraction and NMR. The driving force here was proposed to arise from disorder of the mixed V^{4+}/V^{5+} state in $V_2(PO_4)_3$ formed on emptying the lattice that then gave rise to random lithium siting on re-insertion. That is, lithium insertion results in disorder (in the absence of V^{n+} ordering to drive Li^+ ordering). The results suggested that Li-site ordering and the electron ordering are coupled. It also explains why valence substitution can drive the formation of solid solution regimes. For example, the solid solution γ-phase of $Li_3V_2(PO_4)_3$ that is only accessible above 450 K, is stabilized at room temperature by the addition of a M^{4+} dopant such as Zr^{4+} to form $Li_{3-x}(V_{1-x}Zr_x)_2(PO_4)_3$.

Conclusions

The tailoring of carrier transport in poorly conductive lithium metal phosphates hinges on the ability to precisely control bulk crystal chemistry, particle morphology, and surface chemistry, and the understanding of cooperative ion/electron transport in these highly localized systems. Our studies of the olivine phosphates show that aliovalent dopants do not contribute to the high conductivity observed in olivine phosphates treated at high temperatures under carbothermal or reducing conditions. There is also no evidence for significant dopant concentration in the olivine lattice— nor its stabilization of stable, substoichiometric lithium phases on treatment at processing temperatures that lead to high conductivity. The latter can be obtained in composite materials, however, through exploiting the reactivity of the phosphate surface. Partial reduction, either through carbothermal reduction and/or treatment in H_2 or treatment in ammonia forms metallic surface species (phosphides or nitrides respectively). Along with the residual carbon, these nanophase composites are highly conductive. This explains the properties of Cr-doped $LiFePO_4$. Ellingham curves can be used to determine the temperature at which carbothermal reduction occurs, to give an estimate of the reducibility of the phosphate. Therefore, although Fe, Co and Ni olivines are highly susceptible to reduction, Mn is not, and such species cannot be readily formed at temperatures $<1000\ °C$. Establishing control of the concentration, growth and placement of such metallic species is crucial to nanostructure optimization. Understanding the interface, and how electrons are delivered to the bulk, is also of vital importance. New surface modification methods can be anticipated in the future, which will provide even better nanophase conductivity in insulating or poorly conductive materials.

With respect to bulk transport properties, all lithium metal phosphates studied to date undergo two-phase transformations upon extraction of lithium from the lattice. The movement of the phase boundary represents a migration of both carriers: the electrons or holes (i.e. small polarons) and the lithium ions. The formation of solid solutions at elevated temperature can in principle enhance the mobility. In this regime in Li_xFePO_4, the rapid hopping of the small polarons can be probed by Mössbauer spectroscopy, which provides an accurate measure of when the $Fe^{2+/3+}$ sites become averaged on the Mössbauer time scale, and also provides an estimate of the hop frequency. The onset of lithium ion disorder is precisely correlated with the onset of rapid small polaron hopping, indicating that the two transport mechanisms are coupled. The transport is limited by neither carrier alone, but by their concerted mobility through the lattice, i.e., the migration of the phase boundary. This is expected to be a general phenomenon for lithium metal phosphates. Consideration of these factors may lead to methods by which solid solutions can be induced to form at room temperature through manipulation of the lattice energetics.

Acknowledgements

We gratefully acknowledge funding from NSERC (Canada) through its Discovery Grant Program. LFN especially thanks NSERC for support from the Canada Research Chair program.

References

1 (a) A. K. Padhi, K. S. Nanjundaswamy and J. B. Goodenough, *J. Electrochem. Soc.*, 1997, **144**, 1188–1194; (b) J. B. Goodenough, A. K. Padhi, C. Masquelier and K. S. Nanjundaswamy, US Patent 08/840,523, 1997.
2 G. Li, H. Azuma and M. Tohda, *Electrochem. Solid-State Lett.*, 2002, **5**, A135.
3 (a) H. Huang, S.-C. Yin, T. Kerr and L. F. Nazar, *Adv. Mater.*, 2002, **14**, 1525; (b) J. Barker and M. Y. Saidi, US Patent, 5,871,866, 1999; (c) M. Y. Saïdi, J. Barker, H. Huang and G. Adamson, *Electrochem. Solid-State Lett.*, 2002, **5**, A149; (d) S. C. Yin, P. Strobel, M. Anne and L. F. Nazar, *J. Am. Chem. Soc.*, 2003, **125**, 10402.
4 (a) J. Barker, M. Y. Saidi and J. L. Swoyer, *J. Electrochem. Soc.*, 2003, **150**, A1394; (b) J. Barker, M. Y. Saidi and J. Swoyer, US Patent 6,387,568, 2002; (c) J. Barker, M. Y. Saidi and J. L. Swoyer, *Electrochem. Solid-State Lett.*, 2003, **6**, A1.
5 A. Yamada, S. C. Chung and K. Hinokuma, *J. Electrochem. Soc.*, 2001, **148**, A224.
6 (a) K. S. Nanjundaswamy, A. K. Padhi, J. B. Goodenough, S. Okada, H. Ohtsuka, H. Arai and J. Yamaki, *Solid State Ionics*, 1996, **92**, 1; (b) A. K. Padhi, K. S. V. Manivannan and J. B. Goodenough, *J. Electrochem. Soc.*, 1998, **145**, 1518; (c) C. Masquelier, A. K. Padhi, K. S. Nanjundaswamy and J. B. Goodenough, *J. Solid State Chem.*, 1998, **135**, 228.
7 A. Yamada and S.-C. Chung, *J. Electrochem. Soc.*, 2001, **148**, A960.
8 M. Yonemura, A. Yamada, Y. Takei, N. Sonoyama and R. Kanno, *J. Electrochem. Soc.*, 2004, **151**, A1352.
9 C. Delacourt, P. Poizot, M. Morcrette, J.-M. Tarascon and C. Masquelier, *Chem. Mater.*, 2004, **16**, 93.
10 K. Amine, H. Yasuda and M. Yamachi, *Electrochem. Solid-State Lett.*, 2000, **3**, 178.
11 D. Morgan, A. Van der Ven and G. Ceder, *Electrochem. Solid-State Lett.*, 2004, **7**, A30.
12 M. S. Whittingham, Y. Song, S. Lutta, P. Y. Zavalij and N. A. Chernova, *J. Mater. Chem.*, 2005, **15**, 3362.
13 F. Zhou, K. Kang, T. Maxisch, G. Ceder and D. Morgan, *Solid State Commun.*, 2004, **132**, 181.
14 T. Maxisch, F. Zhou and G. Ceder, *Phys. Rev. B*, 2006, **73**, 104301.
15 (a) N. Ravet, Y. Chouinard, J. F. Magnan, S. Besner, M. Gauthier and M. Armand, *J. Power Sources*, 2001, **97**, 503; (b) N. Ravet and M. Armand, US Patent Application 0195591A1, 2002.
16 H. Huang, S. C. Yin and L. F. Nazar, *Electrochem. Solid-State Lett.*, 2001, **4**, A170.
17 F. Croce, A. D'Epifanio, J. Hassoun, A. Deptula, T. Olczac and B. Scrosati, *Electrochem. Solid-State Lett.*, 2002, **5**, A47.
18 S.-Y. Chung, J. T. Bloking and Y.-M. Chiang, *Nat. Mater.*, 2002, **123**, 1.
19 J. Barker, M. Y. Saidi and J. L. Swoyer, *Electrochem. Solid-State Lett.*, 2003, **6**, A53.
20 S.-Y. Chung and Y.-M. Chiang, *Electrochem. Solid-State Lett.*, 2003, **6**, A278.
21 B. Ellis, P. S. Herle and L. F. Nazar, Abstract #1074, *203rd Electrochemical Society Spring Meeting*, Paris, April, 2003.
22 P. S. Herle, B. Ellis and L. F. Nazar, *Nat. Mater.*, 2004, **3**, 147.
23 S. Q. Shi, L. J. Liu, C. Ouyang, D. S. Wang, Z. X. Wang, L. Q. Chen and X. J. Huang, *Phys. Rev. B*, 2003, **68**, 195108.
24 S. C. Yin, H. Grondey, P. Strobel, M. Anne and L. F. Nazar, *J. Am. Chem. Soc.*, 2003, **125**, 10402.
25 C. Delacourt, P. Poizot, J.-M. Tarascon and C. Masquelier, *Nat. Mater.*, 2005, **4**, 254.
26 B. H. Toby, *EXPGUI*, a graphical user interface for *GSAS*, *J. Appl. Crystallogr.*, 2001, **34**, 210–213.
27 S. Yang, Y. Song, P. Y. Zavalij and M. S. Whittingham, *Electrochem. Commun.*, 2002, **4**, 239.
28 A. Goni, L. Lezama, M. I. Arriortura, G. E. Barberis and T. Rojo, *J. Mater. Chem.*, 2000, **10**, 423.
29 C. Delacourt, C. Wurm, L. Laffont, J.-B. Leriche and C. Masquelier, *Solid State Ionics*, 2006, **177**, 333.
30 M. Saiful Islam, D. J. Driscoll, C. A. J. Fisher and P. R. Slater, *Chem. Mater.*, 2005, **17**, 5085.

31 A. Yamada, H. Koizumi, S.-I. Nishimura, N. Sonoyama, R. Kanno, M. Yonemura, T. Nakamura and Y. Kobayashi, *Nat. Mater.*, 2006, **5**, 357.

32 R. Brochu, A. Lamzibri, A. Aadane, S. Arsalane and M. Ziyad, *Eur. J. Solid State Inorg. Chem.*, 1991, **28**, 253.

33 J. W. Evans and L. C. DeJonghe, *The Production of Inorganic Materials*, Macmillan, New York, 1991, p. 64.

34 M. Doeff, Y. Hu, F. McLarnon and R. Kostecki, *Electrochem. Solid-State Lett.*, 2003, **6**, A207.

35 (*a*) M. Armand, M. Gauthier, J.-F. Magnan and N. Ravet, 2002, PCT WO 02/27824; (*b*) R. Dominko, J. M. Goupil, M. Bele, M. Gaberscek, M. Remskar, D. Hanzel and J. Jamnik, *J. Electrochem. Soc.*, 2005, **152**, A858.

36 Y.-H. Rho, L. F. Nazar, L. Perry and D. Ryan, *J. Electrochem. Soc.*, in press.

37 K. M. Rosso, D. M. A. Smith and M. Dupuis, *J. Chem. Phys.*, 2003, **118**, 6455.

38 S. Dattagupta, in *Mössbauer Spectroscopy*, ed. D. P. E. Dickson and F. J. Berry, Cambridge University Press, Cambridge, 1986, p. 198.

39 J. L. Dodd, R. Yazami and B. Fultz, *Electrochem. Solid-State Lett.*, 2006, **9**, 131.

40 C. Delacourt, J. Rodríguez-Carvajal, B. Schmitt, J.-M. Tarascon and C. Masquelier, *Solid State Sci.*, 2005, **7**, 1506.

41 S.-C. Yin, H. Grondey, P. Strobel and L. F. Nazar, *J. Am. Chem. Soc.*, 2003, **125**, 327.

Factors influencing the conductivity of crystalline polymer electrolytes

Edward Staunton, Yuri G. Andreev and Peter G. Bruce*

Received 9th February 2006, Accepted 28th March 2006
First published as an Advance Article on the web 3rd August 2006
DOI: 10.1039/b601945e

Crystalline polymer electrolytes conduct, in contrast to the established view for 30 years. The crystalline polymer poly(ethylene oxide)$_6$:LiXF$_6$, X = P, As, Sb is composed of tunnels formed from pairs of (CH$_2$–CH$_2$–O)$_n$ chains, within which the Li$^+$ ions reside and along which they may migrate. The anions are located outside the tunnels. PEO$_6$:LiXF$_6$ formed from PEO of average molecular weight 1000 Da has an average chain length of 40 Å compared with a typical crystallite size of 2500 Å, hence low molecular weight materials have many chain ends within a crystallite. More chain ends increase conductivity. Materials composed of polydispersed PEO (chains of different lengths) of average molecular weight 1000 Da exhibit a conductivity one order of magnitude greater than monodispersed materials of the same molecular weight. Replacing the –OCH$_3$ groups on the chain ends with –OC$_2$H$_5$ increases the conductivity by a further order of magnitude. Conductivity may also be increased by isovalent or aliovalent doping of the 6 : 1 complexes in which XF$_6^-$ is replaced by N(SO$_2$CF$_3$)$_2^-$ or SiF$_6^{2-}$, respectively.

Background

Ion transport in the solid-state was first demonstrated by Michael Faraday in the mid 1800s, when he reported the observation of ionic conductivity in PbF$_2$ and Ag$_2$S.[1] It took a further 150 years for an entirely new class of solid ionic conductor to be discovered, namely solid polymer electrolytes. First reported by Fenton, Parker and Wright in 1973, these materials consist of a salt, *e.g.* LiI, dissolved in a solid coordinating polymer, *e.g.* poly(ethylene oxide) (CH$_2$CH$_2$O)$_n$.[2] The metal to ether oxygen interactions are sufficiently strong to overcome the lattice energy of the salt, thus promoting dissolution.[3,4] The ions move through the polymer membrane giving rise to ionic conductivity. Polymer electrolytes also represent a new class of coordination compounds that are the direct solid-state analogues of the crown-ether and cryptand complexes, so well known in solution chemistry.[5]

In 1979 Armand *et al.* recognised the immense potential of these new solid electrolytes that are unique in combining solid yet flexible mechanical properties with ionic conductivity.[6] They are an ideal replacement for liquid electrolytes in electrochemical cells since, unlike ceramic electrolytes, they preserve good contact with solid electrodes as the latter change their dimensions with temperature and on charge/discharge. They are also processable and may be formed into various shapes. Solid polymer electrolytes hold the key to a new generation of all-solid-state electrochemical devices, including rechargeable lithium batteries. Indeed, they may represent the only option for the use of lithium metal electrodes in lithium batteries because of their ability to inhibit dendrite formation on charging. Lithium metal has the highest energy density of any anode for lithium batteries.

School of Chemistry, University of St Andrews, St Andrews, Fife, Scotland, KY16 9ST

Polymer electrolytes may be prepared as crystalline phases at certain discrete ether oxygen to cation ratios, but can be formed as amorphous phases over a wide composition range. Crystalline polymer electrolytes have long been considered insulators, with ionic conductivity being confined exclusively to the amorphous state above the glass transition temperature, T_g. In this state, local segmental (liquid-like) motion of the polymer chains facilitates ion transport by the dynamic formation of free volume. As the local segmental motion randomly generates suitable coordination sites adjacent to a given ion, the ion can hop into this newly created site, where it waits for the polymer chain reorganisation to generate a further site, enabling another hop. The degree of local polymer chain dynamics controls ion transport. Much effort has been expended over the last 30 years in forming highly amorphous polymer electrolytes with low T_g, thus maximising chain dynamics and hence conductivity. It is even possible to increase the salt content to a point where it becomes the dominant phase, forming so-called polymer-in-salt materials. However, the highest conductivities are exhibited by the polymer-in-salt electrolytes based on perchlorates, which are too unsafe for applications.[7] The conductivities of several amorphous polymer electrolytes are presented in Fig. 1. Despite the ingenuity of many synthesis chemists, the maximum level of conductivity obtained from amorphous polymer electrolytes at room temperature has remained too low for many applications. Gel polymer electrolytes have been prepared and these yield higher conductivities, however, they exhibit many of the disadvantages of liquid systems.[8] There is no doubt that true solid polymer electrolytes with high conductivities represent an important scientific and technological goal.

When such limits are reached in science a new direction is required. The temperature dependent conductivity of one of the best lithium-ion conducting solids at room temperature is presented in Fig. 1. Its conductivity is about one order of magnitude higher than the best amorphous polymer, yet it is a highly crystalline solid. As a ceramic, problems are encountered interfacing it to electrodes, precluding its use. However, if crystalline ceramics can support such high levels of conductivity why should crystalline polymers be insulators? Such a view had been entertained by Michel Armand many years earlier but then rejected in favour of ion transport in the amorphous state.[6] We embarked on the investigation of crystalline polymer electrolytes despite the widespread belief that such materials were insulators.

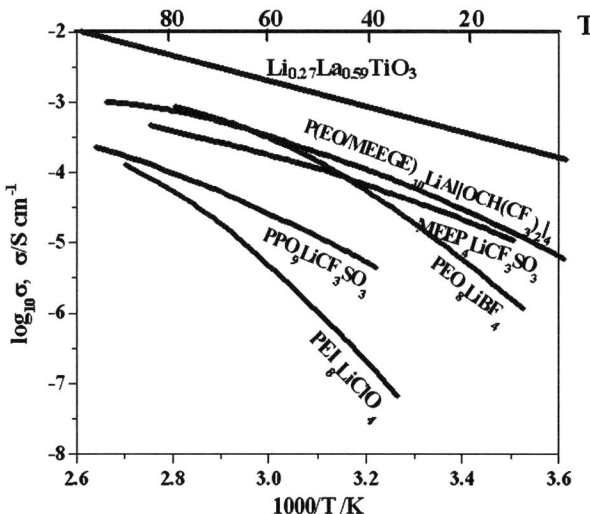

Fig. 1 Variation of conductivity with temperature for several amorphous polymer electrolytes and a crystalline ceramic.

Much of modern chemistry is built on a knowledge of structure, such knowledge is an essential pre-requisite for understanding function. There was a profound lack of information concerning the structure of polymer electrolytes due, in large measure, to difficulties in solving their structures by single crystal methods or any of the established powder diffraction approaches. Yet knowledge of the crystal structures was vital if we were to explore ion transport. By developing a new approach to *ab initio* structure solution from powder diffraction data we were able, for the first time, to solve the structures of a number of polymer electrolytes. This work led ultimately to the discovery of ionic conductivity in the crystalline polymer electrolytes $PEO_6:LiXF_6$, where X = P, As, Sb.[9-12] Important studies by other groups, including the use of molecular dynamics (MD) simulations, have also served to highlight that organisation and order can increase the conductivity of polymer electrolytes.[13,14]

We have now prepared crystalline polymer electrolytes with conductivities comparable to the best amorphous polymers. In this paper we shall focus our discussion on the factors that influence the ionic conductivity of crystalline polymer electrolytes. In particular we shall consider the influence of structure, chain length, polydispersity, chain end groups and doping. By controlling these factors, conductivities may be raised by up to two orders of magnitude compared with the original 6 : 1 complexes. Crystalline polymer electrolytes represent a new class of solid electrolyte, distinct in their mechanism of conduction from conventional amorphous polymer electrolytes and distinct in their mechanical properties from crystalline ceramic ionic conductors. They offer a new direction in research into ion transport in the solid-state and in the search for high ionic conductivity in polymeric systems.

Experimental

All manipulations of air sensitive materials were carried out in an argon filled high integrity glove box (MBraun) in which H_2O and O_2 levels were maintained below 1 ppm.

To prepare di-ethoxy end capped PEO, 64 g of KOH (Fischer Scientific, pellets, lab reagent grade) were powdered and added to 80 ml of chlorobenzene. The resulting slurry was stirred overnight. To this a solution of 25 g of iodoethane (Aldrich, 99.9%) in 150 ml of chlorobenzene (Acros Organics, 99%) was made up and cooled to 0 °C in an ice bath and purged with N_2. 10 g of OH terminated poly(ethylene glycol) Mw = 1000 (Fluka, 98%) were dissolved in 80 ml of chlorobenzene and added drop wise from a burette to the iodoethane containing solution over 15 min. The PEO solution was added slowly to an 8-fold excess of iodoethane to avoid the β-hydrogen elimination side reaction that could otherwise occur. The resulting slurry was filtered through a glass sinter and the filtrate collected and rotary evaporated at 40 °C to yield a crude product. Unreacted di-hydroxy terminated PEO and mono-ethylated PEO present in the mixture were separated by recrystallisation. The crude product was recrystallised twice by dissolution in warm toluene (Riedel de Haën, purum) and precipitated by adding iso-octane (Riedel de Haën, 99%) to yield di-ethoxy terminated PEO (yield = 40%). ^{1}H NMR showed that the peak normally associated with the hydroxyl group was absent, mass spectroscopy also showed no evidence of a peak associated with either di-hydroxy PEO or mono ethylated PEO. The resulting polymer was then dried at 35 °C under dynamic vacuum.

$LiPF_6$ (Stella SC hemita electrochemical grade for 99.99%) was used as received. $LiAsF_6$ (ABCR, 99.8%) and $LiSbF_6$ (Strem, 98%) were dried at 50 °C for 24 h under dynamic vacuum. Methoxy end capped poly(ethylene oxide) (Fluka, 98%) with average molar masses of 1000, 1500 and 2000 Da were dried for 4 days at 30 °C, also under dynamic vacuum. Monodispersed methoxy end capped poly(ethylene oxide) with a molecular mass of 1015 Da (Polypure, >97% molecularly pure), obtained by state-of-the-art preparative scale reverse phase HPLC, was used as received.

Masses of salt and the appropriate polymer suitable for the formation of a 6 : 1 complex were weighed out and dissolved together in dry acetonitrile. Following complete dissolution, the acetonitrile was permitted to evaporate slowly. The resulting white powders were dried overnight under dynamic vacuum at 35 °C. IR spectroscopy (FTIR spectrometer Nicolet 860) confirmed the absence of H_2O and CH_3CN from the powders. For conductivity measurements the powders were pressed into self supporting disks at room temperature.

Powder X-ray diffraction was carried out using a Stoe STAD/P powder diffractometer with Cu $K\alpha_1$ radiation operating in transmission mode and employing a small angle position sensitive detector (PSD). Data were collected with a step width of 0.02° in 2θ. To avoid contact with air the polymer electrolyte samples were sealed in Lindemann (glass) capillaries or between Mylar films, depending on whether the samples were in the form of a powder or a film.

Differential scanning calorimetry was carried out using a Netzch DSC 204 Phoenix with heating and cooling rates of 5 or $10°$ min^{-1}.

Conductivity data were obtained using AC impedance measurements carried out with a Solatron 1255 frequency response analyser and 1287 electrochemical interface, both under the control of a PC. A polarising potential of 25 mV was employed and data were collected over the frequency range 10^{-1} to 10^5 Hz. The polymer electrolyte disks were sandwiched between 2 stainless steel plates in a 2-electrode cell, which was itself located within an argon filled stainless steel chamber. The chamber was placed in a thermostatic bath in order to control the temperature of the cells.

Results and discussion

As part of our extensive studies concentrating on the structural chemistry of crystalline polymer electrolytes, we solved the structure of the 6 : 1 complexes PEO_6:LiXF$_6$, X = P, As, Sb [PEO=CH$_3$O(CH$_2$CH$_2$O)$_n$CH$_3$, where n is on average 22, corresponding to a molecular weight of 1000 Da], Fig. 2.[10] Their structures are similar and are composed of pairs of PEO chains that individually fold to form a half cylinder, which in turn interlocked, forming tunnels within which the Li$^+$ ions reside, coordinated by 3 ether oxygens from 1 chain and 2 from the other chain. The anions do not coordinate to the cations and are located outside the chains. The structure remains the same at higher molecular weights. This structure immediately suggested the possibility of Li$^+$ transport along the tunnels, which proved to be so.

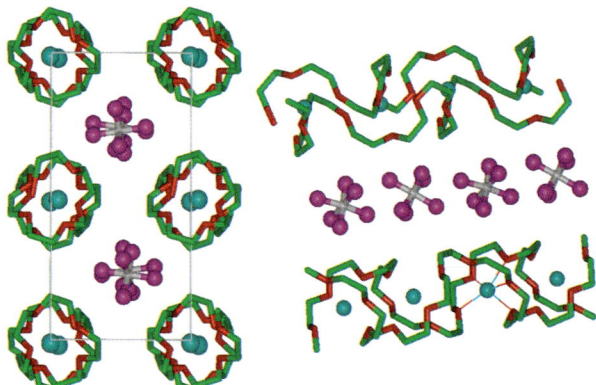

Fig. 2 The structure of PEO_6:LiAsF$_6$. (Left) View of the structure along the polymer chain axis showing rows of Li$^+$ ions perpendicular to the page. (Right) View of the structure showing the positions and the conformation of the chains. Light blue, lithium; green, carbon; red, oxygen; white, arsenic; magenta, fluorine; hydrogens not shown. Thin lines indicate coordination around Li$^+$ ions.

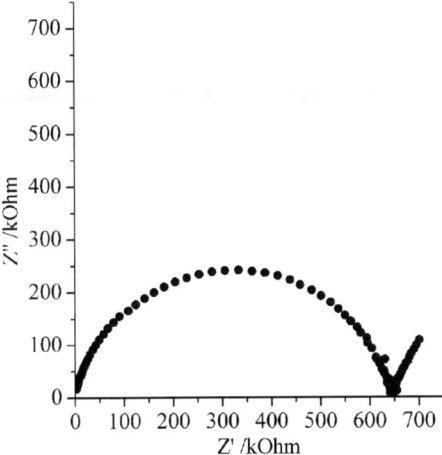

Fig. 3 Complex impedance of PEO_6:$LiAsF_6$ at room temperature. Capacitance associated with the semicircle is 1.9 pF cm^{-1}.

This discovery represented the first report of conductivity in crystalline polymer–salt complexes. NMR data indicated that ion transport was dominated by the motion of the Li^+ ions, something that is consistent with the crystal structure, although not in agreement with MD simulations that predict T_+ values lower than unity for short chains.[15] One of the potential benefits of crystalline polymer electrolytes is that they impose selectivity on ion transport, whereas amorphous materials above T_g, much like conventional liquids, are not selective, and in fact are dominated by anion transport. The role of the crystal structure in controlling conductivity was confirmed recently by our discovery of the first example of polymorphism in polymer electrolytes. We reported a new structural form of the 6 : 1 complex, named β-PEO_6:-$LiAsF_6$. This material does not possess a tunnel structure and its conductivity is significantly lower than that of the original, α, polymorph.[16]

The conducting 6 : 1 complexes are soft powders. AC impedance has confirmed that pressed disks and films of such materials do not exhibit grain boundary resistances and that the DC resistance of the films corresponds to the intra-granular resistance of the materials, Fig. 3. This is undoubtedly an advantage arising from the soft nature of the solids since ceramic materials, the grains of which cannot deform except under more extreme pressures, tend to lead to bottlenecks for ion transport between grains and hence to grain boundary resistances that are significantly higher than that of the bulk.[17] What other factors might influence the level of conductivity in these crystalline polymer electrolytes?

Influence of molecular weight

Above an average molar mass of 3200 Da, the PEO chains become entangled, inhibiting crystallisation. In order to maximise crystallinity, low molar masses are employed. NMR data have shown no evidence of an amorphous phase below the 3200 Da entanglement limit, hence conductivity below this limit is due to the crystalline 6 : 1 complex. Throughout most of this work, we have used PEO with an average molar mass of 1000 Da. However, varying the average molar mass, even below the entanglement limit, results in a variation in conductivity. This implies changes in the conductivity of the crystalline phase, Fig. 4. The conductivity increases on lowering the molar mass. At first sight this seems strange since ion transport in the crystalline state involves Li^+ ions migrating along the polymer tunnels, and one might expect that the longer chains associated with higher molar mass, and hence fewer chain ends, would yield more continuous pathways and therefore

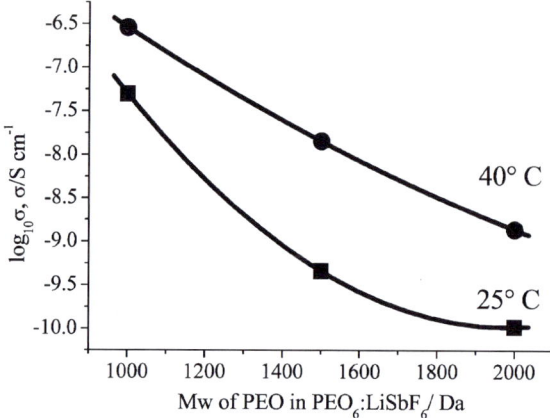

Fig. 4 Conductivity isotherms as a function of molecular weight of PEO in $PEO_6:LiSbF_6$.

higher conductivity. Analysis of the peak widths in the powder X-ray diffraction patterns reveals that the crystallite size increases with decreasing molar mass, Fig. 5, from 2000 Å at 2000 Da to 2500 Å at 1000 Da and is smaller than the 1–20 µm size of the particles as demonstrated by scanning electron microscopy. Hence each particle is composed of a mosaic of misaligned crystallites. This suggests that the misalignment of polymer tunnels that must occur between adjacent crystallites acts to impede ion transport and that the larger the crystallites, and therefore the fewer such inter-crystallite boundaries, the higher the conductivity. Confirmation that crystallite size has an influence on the conductivity comes from varying the conditions used to prepare the 6 : 1 complex. By removing the solvent rapidly, during the solvent casting process, smaller average crystallite sizes are obtained (1700 Å), as indicated by X-ray diffraction patterns with broader peaks (Fig. 5), and these materials show significantly lower conductivities. It should also be noted that in addition to the increase in crystallite size, the increase in the concentration of chain ends can increase the number of defects responsible for ion transport, see the next section.

Monodispersed poly(ethylene) oxide

Polydispersity is ubiquitous in polymers. However, a distribution of chain lengths often complicates an understanding of polymer properties. Although the

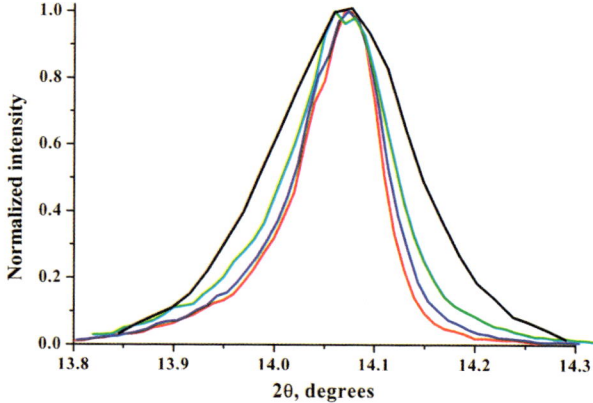

Fig. 5 The 201 diffraction peak of $PEO_6:LiSbF_6$ synthesized with Mw of PEO 1000 (red), 1500 (blue), 2000 (green), 2000 fast cast (black).

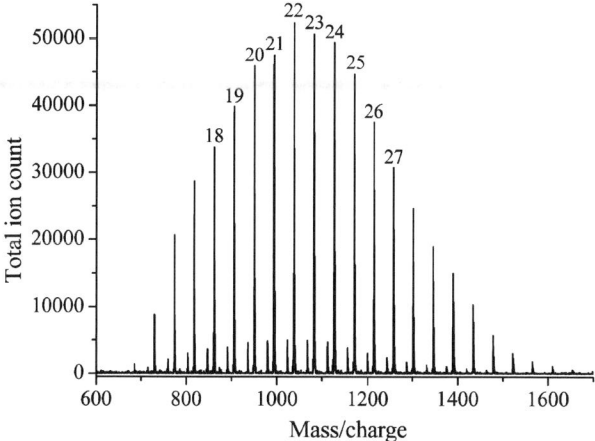

Fig. 6 Mass spectrum of polydispersed PEO, average molar mass 1000 Da. As sodium formate was used as the calibrant, all masses correspond to the ethylene oxide chain and one Na. Numbers above the peaks represent the number of EO units in the corresponding chains.

polydispersity of the PEOs used here is not severe, Fig. 6, there is a distribution in the chain lengths. As a result, the occurrence of chain ends along a continuous tunnel in the 6 : 1 structure is random, as shown schematically in Fig. 7a. What might the effect be of forming the 6 : 1 complex with PEO chains of identical length? Clearly, the occurrence of the chain ends along one of the two strands of the polymer forming a tunnel must then be regular, but the chain ends along both strands may also be coincident across the tunnels as shown in Fig. 7b. How might this influence the conductivity?

The vast majority of studies carried out on polymers involve the use of poly-dispersed materials, because the very nature of polymer synthesis leads to such polydispersity. The synthesis of monodispersed materials generally presents a formidable challenge, since conventional chain reactions must be avoided. A collaboration with Polypure in Norway has allowed the synthesis of monodispersed poly(ethylene) oxide with a molar mass of 1015 Da (see Experimental). The monodispersity of the poly(ethylene) oxide has been demonstrated by MALDI mass spectrometry. The monodispersed material corresponded to 22 ethylene oxide units.

Poly- and monodispersed $PEO_6:LiPF_6$ were prepared using polydispersed PEO of average molar mass 1000 and monodispersed PEO of molar mass 1015 Da, respectively. The temperature dependent conductivities of the two materials are presented in Fig. 8. The conductivity of the monodispersed material is approxi-mately one order of magnitude lower than the equivalent polydispersed $PEO_6:LiPF_6$, demonstrating that the dispersity in chain length influences signifi-cantly the conductivity. The activation energies for the two electrolytes are very similar. How might these results be explained?

The powder X-ray diffraction patterns for the poly- and monodispersed $PEO_6:LiPF_6$ materials are shown in Fig. 9. In both cases, the data may be fitted by the crystal structure for the $PEO_6:LiPF_6$ compound described above. The powder diffraction data do not observe directly the chain ends and are thus not sensitive to the discreteness of the chains. Instead, the model of the crystal structure is that of an infinite chain, in which the translational symmetry maps ethylene oxide units in one discrete chain onto those in the next chain along the tunnel. This is despite the fact that the chain length is ~ 40 Å (average in the case of the polydispersed material) within crystallites that are ~ 2500 Å, resulting in many chain ends within each crystallite, and in contrast to the usual situation in a long chain polymer where the chains are longer than the crystalline size of the regions. In the case of the

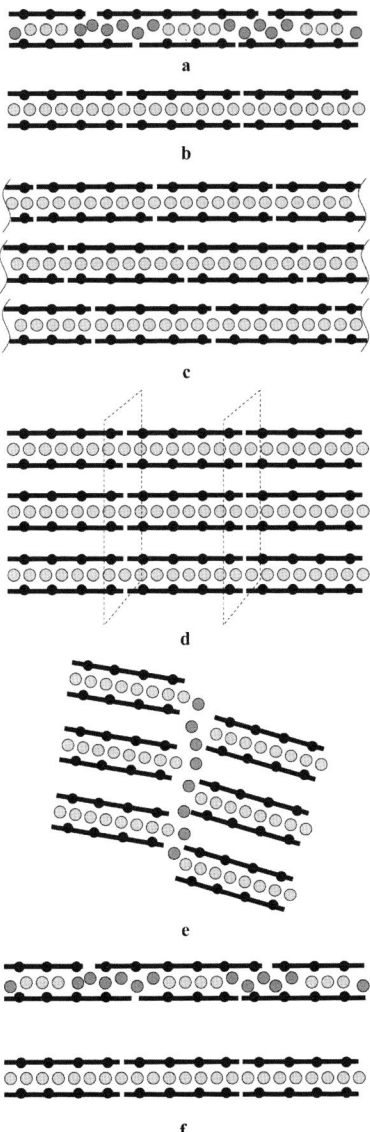

Fig. 7 (a) Schematic representation of part of the $PEO_6:LiXF_6$ crystal structure. Three PEO tunnels with their associated Li^+ ions are shown. The PEO chains are represented by the solid lines and are of different length, resulting in the random occurrence of chain ends. Li^+ ions are represented by circles, with the disordered Li^+ ions, near chain ends, being represented by darker shading. Anions are not shown. (b) A single tunnel formed from monodispersed PEO and showing the coincidence of chain ends in each of the two strands of the tunnel. (c) Model based on monodispersed PEO in which the chain ends coincide within a tunnel but NOT between tunnels. (d) Model based on monodispersed PEO in which the chain ends not only coincide within AND between tunnels such that the ends are located in a plane perpendicular to the tunnel axis. (e) Model based on monodispersed PEO in which the tunnels are canted and displaced. (f) Model indicating greater number of defects per unit length in polydispersed $PEO_6:LiXF_6$.

polydispersed material the non-observance of chain ends is not too surprising since the polydispersity ensures that the chain ends are randomly distributed throughout the crystal structure and hence not directly observed by diffraction (Fig. 7a). The

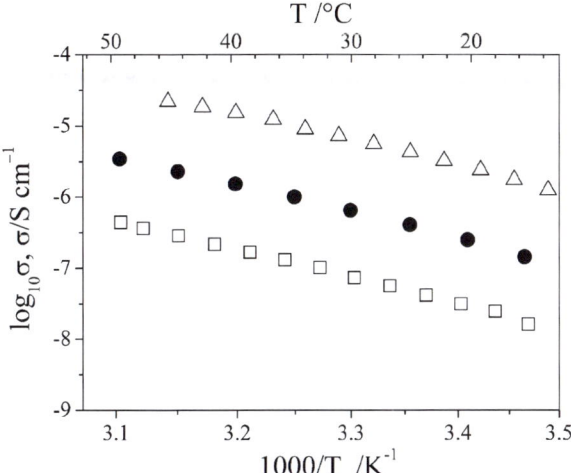

Fig. 8 Ionic conductivity σ (S cm^{-1}) of crystalline PEO$_6$:LiPF$_6$ complexes prepared with – OCH$_3$ terminated monodispersed PEO (open squares), –OCH$_3$ terminated polydispersed PEO (solid circles) and –OC$_2$H$_5$ terminated polydispersed PEO (open triangles).

invisibility of the chain ends in the diffraction data for the monodispersed material, in which the chains are of identical length, is perhaps less obvious and needs further consideration.

As can be seen in Fig. 9, the peak positions for the mono- and polydispersed materials are slightly different and therefore the change in dispersity is reflected in the powder diffraction data. The lattice parameters for the two phases are: for polydispersed PEO$_6$:LiPF$_6$, $a = 11.924(4)$, $b = 17.336(6)$ and $c = 9.202(3)$ Å, $\beta = 108.83(3)°$; and for monodispersed PEO$_6$:LiPF$_6$, $a = 11.801(4)$, $b = 17.419(2)$, $c = 9.19(4)$ Å, $\beta = 108.72(3)°$. These data show that the a lattice parameter contracts by 0.123(6) Å and b expands by 0.083(6) Å on going from the poly- to the mono-dispersed materials. The a lattice parameter lies along the axis of the polymer tunnels and this indicates that the monodispersed material involves a contraction along the tunnel axis. Since the average number of chain ends is the same for the poly- and monodispersed materials (same average molar mass), these changes must reflect differences in the distribution of the polymer chain ends. It is difficult to see why such

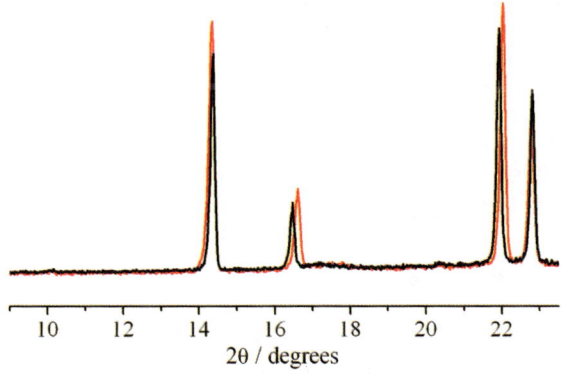

Fig. 9 Powder X-ray diffraction patterns of PEO$_6$:LiPF$_6$ prepared from monodispersed PEO (red) and polydisperse PEO (black).

changes should occur simply because of a regular arrangement of chain ends along one PEO strand of a tunnel, therefore we propose that the chain ends along both strands of the tunnels must be coincident, resulting in the structure shown schematically in Fig. 7b, *i.e.* the chains in each strand terminate at the same location. The changes in *a* and *b* suggest that the terminal $-OCH_3$ groups twist further out from the tunnels in the case of the monodispersed material, permitting the adjacent chains along the tunnels to approach more closely, whilst at the same time pushing the tunnels slightly farther apart.

So far we have only considered a single tunnel in monodispersed $PEO_6:LiXF_6$. There are two ways in which such tunnels may be arranged into an extended crystal structure, Fig. 7c and d. In the first, Fig. 7c, the chain ends in neighbouring tunnels are not coincident, whereas in the second, Fig. 7d, the tunnels are arranged such that the chain ends are all located in a sheet perpendicular to the tunnel axis. The second model would introduce additional long-range order and hence symmetry, due to the regularity of the chain ends. Since we have no evidence so far for such additional symmetry in the diffraction data we cannot assume this additional degree of order and hence favour the model depicted in Fig. 7c. Interestingly, Brandell *et al.* have explored the model in Fig. 7d further using MD simulations (referred to by the authors as the smectic model).[18] They concluded that the tunnels are unlikely to be aligned across a sheet, but instead are distorted and canted with respect to each other, Fig. 7e. If such a model were to exist across the entire crystal then it would destroy the translational symmetry on a length scale >40 Å (length of a PEO chain), *i.e.* the regions of long-range order would be confined to ~ 40 Å and this is not consistent with the diffraction data. However, it is possible that small segments of the structure shown in Fig. 7e could exist as extended defects within the otherwise more regular crystal structure. It is interesting to note that both diffraction and simulations agree that the chain ends in each strand of a tunnel are coincident, Fig. 7b.

The linear conductivity plots in Fig. 8 are described by an Arrhenius expression. The similar activation energies for the $-OCH_3$ end capped poly- and monodispersed materials suggest that the activation barrier for ion hopping along the tunnels is the same regardless of the dispersity. The differences in conductivity occur in the pre-exponential factors of the Arrhenius expression and hence would seem to reflect differences in the concentration of charge carriers. It is important to remember that the crystal structure obtained from the diffraction data implies continuous chains with every lithium site occupied. Not only do we know that, of course, the chains are discrete but we also know that ion hopping could not occur without local defects. The most likely source of such defects is the chain ends. The number of occurrences of a chain end along a tunnel in the polydispersed 6 : 1 structure will be greater than for a monodispersed structure in which the chain ends on each strand are coincident. This is clear from Fig. 7f where, in the case of the polydispersed material, there are four locations at which a chain meets a chain whereas the coincidence of chain ends in the monodispersed material results in only two such locations. As a result, the polydispersed material will exhibit more defects per unit length of the tunnel and hence a higher concentration of charge carriers, and a higher pre-exponential factor.

The model presented above for monodispersed $PEO_6:LiXF_6$ assumed the presence of only one chain length. However, 100% monodispersity is rarely, if ever, achieved in practice. In the present case the material consists almost exclusively of chains containing 22 EO units with only a minor contribution from chains that differ by only one EO unit. This constitutes only a ~ 2 Å difference in the 40 Å chain length and it is likely, in any case, to be observed within the chain end regions. There will be many more coincidences of chain ends in tunnels formed from monodispersed $PEO_6:LiXF_6$. There is clearly a difference between poly- and monodispersed $PEO_6:LiXF_6$, as reflected in the difference in unit cell parameters, and the interpretation given above remains valid.

Influence of chain ends

Investigating monodispersed PEO_6:$LiPF_6$ illustrated the important role that chain ends have in influencing conductivity, especially the number of charge carriers. To further investigate the role of chain ends, polydispersed PEO of average molar mass 1000 Da, but in which the chains are terminated by $-OC_2H_5$ groups instead of $-OCH_3$, was prepared as described in the experimental section, and this material was used to form a PEO_6:$LiPF_6$ complex.

The temperature dependent conductivities of the 1000 molecular weight PEO_6:$LiPF_6$ terminated by $-OCH_3$ and $-OC_2H_5$ are shown in Fig. 8. The activation energies are similar but the conductivity of the $-OC_2H_5$ terminated material is one order of magnitude higher than the $-OCH_3$ complex. The diffraction data for the two complexes are shown in Fig. 10. These data demonstrate that replacing $-OCH_3$ with $-OC_2H_5$ preserves the same crystal structure but with a shift in the peak positions corresponding to a change in the unit cell parameters from, for the $-OCH_3$ terminated material, $a = 11.924(4)$, $b = 17.335(6)$, $c = 9.202(3)$ Å, $\beta = 108.85(3)°$ to, for the $-OC_2H_5$ material, $a = 11.778(4)$, $b = 17.492(2)$, $c = 9.381(2)$ Å, $\beta = 111.33(5)°$. Both materials have the same polydispersity and therefore the same average chain length and number of chain ends. However, the expansion in the b and c directions with a contraction along a suggest that $-OC_2H_5$ groups rotate further out from the tunnels than $-OCH_3$. It is indeed remarkable that the introduction of an additional $-CH_2-$ group at the ends of the chains, which results in a sequence of 4 carbons between neighbouring ether oxygens along one strand of a tunnel (rather than the 2 carbons in the case of $-CH_3$ termination), still preserves the same crystal structure and translational symmetry of the $-CH_2-CH_2-O-$ units from one chain to a neighbouring chain along the tunnels. In fact, the preservation of the crystal structure, despite the bulkier $-C_2H_5$ end groups, must be achieved at the expense of greater local structural disorder, compared with $-CH_3$ groups. The conductivity data for the 6 : 1 complexes in Fig. 8 show that the activation energies are similar, implying similar barriers to ion hopping, but that the pre-exponential factor is greater than in the case of the $-OC_2H_5$ terminated compounds, consistent with greater local structural disorder and a resulting increase in the number of charge carriers.

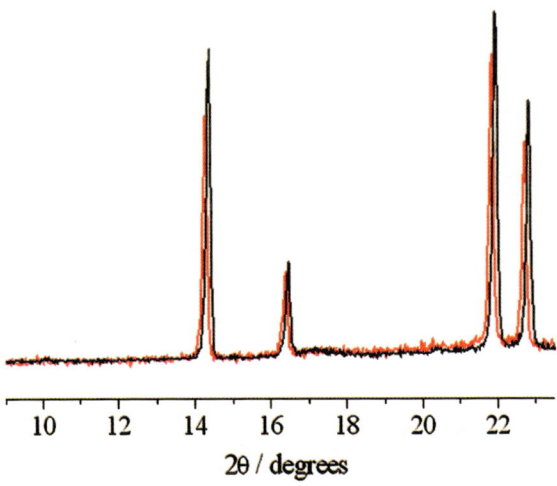

Fig. 10 Powder X-ray diffraction patterns of PEO_6:$LiPF_6$ prepared with $-OC_2H_5$ terminated PEO (red) and $-OCH_3$ terminated PEO (black).

Influence of doping

Given that the PEO_6:$LiXF_6$ materials are crystalline ionic conductors, in which ion transport occurs by hopping between neighbouring sites, it is reasonable to consider what effect doping might have on the conductivity. This is especially so since the 6 : 1 complexes are stoichiometric materials relying on the serendipitous generation of charge carriers through local defects, such as chain ends, as discussed above. Deliberate and judicious doping should enhance conductivity. Two doping strategies have been explored, isovalent doping, in which the XF_6^- anions are replaced, in part, by other anions of the same charge and aliovalent doping, in which the substituting anion has a different charge.

Considering isovalent doping first, up to 5 mol% of the AsF_6^- anions in PEO_6:$LiAsF_6$ have been replaced by the imide anion, $N(SO_2CF_3)_2^-$. The latter has the same charge but a different size and shape from the AsF_6^- anion, nevertheless space exists within the crystal structure for its accommodation. The rapid rise in conductivity by 1.5 orders of magnitude with just 5 mol% substitution is typical of doping, Fig. 11. The difference in size and shape of the anion is believed to disrupt the potential around the Li^+ ions, thus introducing more local disorder and hence greater conductivity.[19] The mechanism of enhanced conduction is considered to be

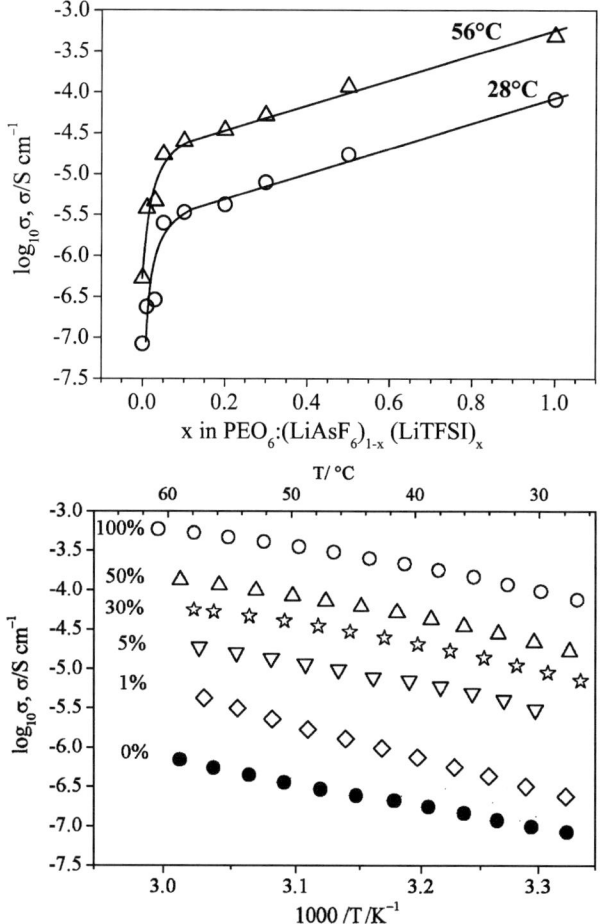

Fig. 11 PEO_6:$(LiAsF_6)_{1-x}(Li\ N(CF_3SO_2)_2)_x$. (Top) Conductivity isotherms as a function of x. (Bottom) Ionic conductivity as a function of temperature, the numbers are values of x in mol%.

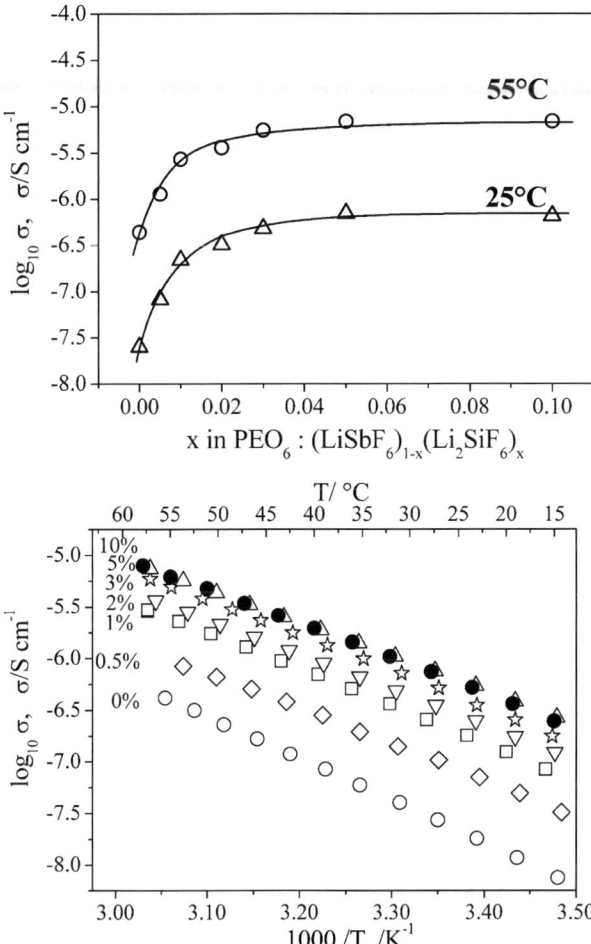

Fig. 12 PEO$_6$:(LiSbF$_6$)$_{1-x}$(Li$_2$SiF$_6$)$_x$. (Top) Conductivity isotherms as a function of x. (Bottom) Ionic conductivity as a function of temperature, the numbers are values of x in mol%.

similar to that observed in the silver ion conductor Ag(Br$_{1-x}$I$_x$), where such isovalent doping increases the conductivity by 3 orders of magnitude.[20] Above 5 mol%, a new phase, PEO$_6$:LiN(SO$_2$CF$_3$)$_2$, is formed but due to the low molar mass PEO (1000 Da) used, this is a liquid and it is conduction in this liquid phase that gives rise to the further gentle rise in conductivity above the 5 mol% solubility limit for incorporation of the imide anion in the 6 : 1 crystal structure.

The SbF$_6^-$ anion in the PEO$_6$:LiSbF$_6$ may be replaced by a divalent anion with the same shape and somewhat smaller size, namely SiF$_6^{2-}$. Despite the solubility limit being low, approximately 1 mol%, the conductivity again rises very rapidly on doping, Fig. 12. Above this solubility limit Li$_2$SiF$_6$ appears in the powder X-ray diffraction data. Since this phase is itself an insulator the conductivity no longer rises with increasing Li$_2$SiF$_6$ content in the system.[21]

Conclusions

Crystalline polymer electrolytes conduct, contrary to the traditional view of ion transport in polymers. The crystal structure of the PEO$_6$:LiXF$_6$, X = P, As, Sb polymer electrolytes supports conduction and consists of PEO tunnels formed from pairs of PEO chains within which Li$^+$ ions are located and coordinated by the ether oxygens. The anions are located between the tunnels.

In studying crystalline polymer electrolytes attention has been focussed on low molecular weights, predominantly 1000 Da, and hence the role of chain ends is significant. The crystals (typically of $\sim$2000 Å) are composed of many methoxy terminated PEO chains [CH$_3$O(CH$_2$CH$_2$O)$_n$CH$_3$] of average length $\sim$40 Å and hence many chain ends. A higher density of chain ends leads to higher conductivity. The chain end regions constitute defects within the crystal structure and lead to an increase in the number of charge carriers. Materials composed of polydispersed chain lengths will have more defects per unit length of a tunnel leading to higher conductivity than those composed of monodispersed chains, in which the chain ends on one PEO strand of the tunnel coincide with those on the other (see Fig. 7). Polydispersed PEO$_6$:LiPF$_6$ of average molecular weight 1000 Da has a conductivity 1 order of magnitude higher than the equivalent monodispersed material. Replacing the –OCH$_3$ chain termini in the polydispersed material by –OC$_2$H$_5$ leads to greater disruption at the chain ends and a further increase in conductivity by 1 order of magnitude (see Fig. 8). Doping by isovalent and aliovalent anions also increases the conductivity. Conductivities compatible to some of the best amorphous phases may now be obtained.

Acknowledgements

PGB is indebted to The Royal Society, the EPSRC and the EU for financial support.

References

1   M. Faraday, *Experimental Researches in Electricity*, Taylor and Francis, London, 1839, p. 1849.
2   D. E. Fenton, J. M. Parker and P. V. Wright, *Polymer*, 1973, **14**, 589.
3   *Solid State Electrochemistry*, ed. P. G. Bruce, Cambridge University Press, Cambridge, 1995.
4   *Applications of Electroactive Polymers*, ed. B. Scrosati, Chapman & Hall, London, 1993.
5   P. G. Bruce, *Philos. Trans. R. Soc. London, Ser. A*, 1996, **354**, 415.
6   M. B. Armand, J. M. Chabango and M. J. Duclot, in *Fast Ion Transport in Solids*, ed. P. Vashishta, J. N. Mundy and G. K. Shenoy, North-Holland, Amsterdam, 1979, p. 131.
7   C. A. Angell, C. Liu and E. Sanchez, *Nature*, 1993, **362**, 137.
8   F. Croce, S. Passerini and B. Scrosati, *J. Electrochem. Soc.*, 1994, **141**, 1405.
9   (*a*) Y. G. Andreev, P. Lightfoot and P. G. Bruce, *Chem. Commun.*, 1996, 2169; (*b*) Y. G. Andreev, P. Lightfoot and P. G. Bruce, *J. Appl. Crystallogr.*, 1997, **30**, 294.
10 (*a*) G. S. MacGlashan, Y. G. Andreev and P. G. Bruce, *Nature*, 1999, **398**, 792; (*b*) Z. Gadjourova, D. Martin, K. H. Andersen, Y. G. Andreev and P. G. Bruce, *Chem. Mater.*, 2001, **13**, 1282.
11 Z. Gadjourova, Y. G. Andreev, D. P. Tunstall and P. G. Bruce, *Nature*, 2001, **412**, 520.
12 Z. Stoeva, I. Martin-Litas, E. Staunton, Y. G. Andreev and P. G. Bruce, *J. Am. Chem. Soc.*, 2003, **125**, 4619.
13 (*a*) P. V. Wright, Y. Zheng, D. Bhatt, T. Richardson and G. Ungar, *Polym. Int.*, 1998, **47**, 34; (*b*) S. H. Chung, Y. Wang, S. G. Greenbaum, D. Golodnitsky and E. Peled, *Electrochem. Solid-State Lett.*, 1999, **2**, 553.
14 D. Brandell, A. Liivat, H. Kasemagi, A. Aabloo and J. O. Thomas, *J. Mater. Chem.*, 2005, **15**, 1422.
15 D. Brandell, PhD Thesis, University of Uppsala, Uppsala, Sweden, 2005.
16 E. Staunton, Y. G. Andreev and P. G. Bruce, *J. Am. Chem. Soc.*, 2005, **127**, 12176.
17 P. G. Bruce and A. R. West, *J. Electrochem. Soc.*, 1983, **130**, 662.
18 D. Brandell, A. Liivat, A. Aabloo and J. O. Thomas, *J. Mater. Chem.*, 2005, **15**, 4338.
19 A. M. Christie, S. J. Lilley, E. Staunton, Y. G. Andreev and P. G. Bruce, *Nature*, 2005, **433**, 50.
20 K. Shahi and J. B. Wagner, Jr, *Appl. Phys. Lett.*, 1980, **37**, 757.
21 C. Zhang, E. Staunton, Y. G. Andreev and P. G. Bruce, *J. Am. Chem. Soc.*, 2005, **127**, 18305.

Mass and charge transport in the PEO–NaI polymer electrolyte system: effects of temperature and salt concentration

N. A. Stolwijk, M. Wiencierz and Sh. Obeidi

Received 13th February 2006, Accepted 26th April 2006
First published as an Advance Article on the web 26th July 2006
DOI: 10.1039/b602143n

Ionic transport in amorphous complexes of poly(ethylene oxide) (PEO) and sodium iodide (NaI) was investigated by means of radiotracer diffusion and electrical conductivity measurements for three different compositions characterised by O-to-Na ratios of 20, 30, and 60. The diffusivity of ^{22}Na and ^{125}I as well as the charge diffusivity each show a systematic dependence on both temperature and salt concentration. The experimental data can be described within a transport model based on the occurrence of neutral ion pairs in addition to charged single cations and anions. It is found that both the enthalpy and entropy of ion pair formation decrease with increasing salt concentration. This indicates that the state of coordination between cations and polymer chain segments correlates with the number of available oxygen atoms per cation.

1 Introduction

Polymer electrolytes are used in batteries, fuel cells, and chemical sensors as ionically conducting materials with suitable thermo-mechanical properties and a high chemical stability.[1,2] However, even in simple polymer electrolyte systems consisting of, *e.g.*, poly(ethylene oxide) (PEO) complexed with an alkali metal salt, the ion conduction mechanisms are not well understood.[3] This is also true for the fully amorphous phase which exists at temperatures above the melting point of pure PEO, being 66 °C. Previous work by our group on PEO$_{30}$NaI[4] and a crosslinked polyether–NaI complex[5] has shown that the ion transport properties are greatly influenced by the formation of neutral cation–anion pairs having a high mobility. In particular, it was found that these ion pairs make the most significant contribution to the overall diffusivity of the cation species whereas the contribution of charged single cations turned out be almost negligible, at least at high temperatures.

This report focuses on the concentration dependence of ionic transport in the PEO–NaI system by investigating the ionic conductivity and tracer diffusivity at different compositions. Specifically, we performed Na and I tracer diffusion experiments along with frequency-dependent conductivity measurements over wide temperature ranges in PEO$_{20}$NaI and PEO$_{60}$NaI having O-to-Na ratios (y) of 20 and 60, respectively. Evaluating either system within the ionic transport model proposed earlier[4,6] yields best estimates for the seven parameters characterising the model. Combining these new results with the data previously obtained on a similar complex with y = 30 (PEO$_{30}$NaI) reveals distinct trends in most of the parameter values.

Universität Münster, Institut für Materialphysik und Sonderforschungs- bereich 458, Wilhelm-Klemm-Str. 10, 48149, Münster, Germany. E-mail: stolwij@uni-muenster.de

Most surprisingly, it is found that the (positive) enthalpy and (positive) entropy of pair formation decrease with increasing salt concentration. We interpret this in terms of changing conditions for cation binding to the polymer chains. At low salt concentration all cations can be optimally coordinated by the ether oxygen atoms leading to a large binding enthalpy and a corresponding great loss of configurational entropy of the polymer matrix. In contrast, at higher salt concentrations a competition among the cations for the restricted number of oxygen atoms on free chain segments gives rise to less strong energetic and entropic effects per cation or salt molecule upon solvation.

2 Experimental

Here, only a short description of the experimental procedures is given. More details can be found in previous publications.[5,8]

PEO with a molecular weight of 8×10^6 (98.4%,[9] Aldrich) and NaI (99%, Grüssing) were dried under dynamic vacuum of 10^{-3} Pa at temperatures of 50 and 100 °C, respectively. Proper amounts of both components, adjusted to O-to-Na ratios $y = 20$ or 60, were dissolved in dry acetonitrile. After vacuum evaporation of the solvent, differential scanning calorimetry showed a single peak near 66 °C which reflects the melting of PEO. Samples for diffusion and conductivity measurements were prepared by hot pressing in cylindrical Teflon molds under dry nitrogen atmosphere in a glovebox. Mass densities were determined by the Archimedes method and amounted to 1.33 g cm^{-3} for PEO$_{20}$NaI and 1.23 g cm^{-3} for PEO$_{60}$NaI.

The radiotracers ^{22}Na and ^{125}I, commercially supplied as aqueous solutions, were dried and subsequently dissolved in acetonitrile. After addition of gel-like PEO$_y$NaI (containing acetonitrile) the solution was cast onto an inert support and then subjected to prolonged solvent evaporation. The radioactive films produced in this way were of closely similar composition as the samples and thus suitable as diffusion-source material. In fact, the ratio of radioactive-to-natural isotopes in these films was less than $\sim 10^{-4}$ for either ionic species.

Samples with one or more source-film pieces on top of one face were diffusion annealed between about 70 and 190 °C in an oil-bath thermostat using suitable encapsulants, and subsequently quenched in water. Following microtome sectioning in a cold chamber, counting of the sections' radioactivity was performed by standard techniques. The dc conductivity σ_{dc} of PEO$_{20}$NaI and PEO$_{60}$NaI was determined by means of frequency-dependent impedance analysis using stainless steel electrodes in a temperature-controlled measuring cell operated under flowing nitrogen. Individual σ_{dc} data originate from multiple heating and cooling cycles covering temperature ranges which are slightly wider than those chosen for the tracer diffusion experiments, *i.e.*, 54–198 °C for PEO$_{20}$NaI and 57–205 °C for PEO$_{60}$NaI.

3 Results

3.1 Radiotracer data

3.1.1 Penetration profiles. Typical diffusion profiles are shown in Fig. 1 for PEO$_{20}$NaI and Fig. 2 for PEO$_{60}$NaI. Either figure displays a ^{22}Na and an ^{125}I profile arising from separate experiments at similar temperatures. These profiles are well fitted by the Gaussian function $C(x,t) = C_0 \exp(-x^2/4D^*t)$ (solid lines), where x denotes penetration depth, t diffusion time, D^* tracer diffusion coefficient, and C concentration (in arbitrary radioactivity units). The pre-factor C_0 is the concentration at $x = 0$. In several cases, the diffusion profile was much better described by the complementary error function $C(x,t) = C_0 \operatorname{erfc}(x/2\sqrt{D^*t})$. Both Gaussian and erfc-type profiles imply that the diffusivity D^* is independent of x, providing evidence for the homogeneity of our samples.

A number of diffusion profiles were found to be intermediate between Gaussian- and erfc-type. In such cases the diffusion coefficient was estimated by interpolating

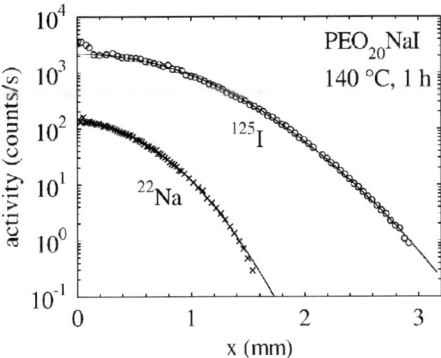

Fig. 1 Penetration profile of ^{22}Na and ^{125}I in PEO$_{20}$NaI resulting from separate diffusion treatments as indicated by temperature and duration. Solid lines are fits of the Gaussian function.

between the best-fit values arising from the two mathematical functions. The concomitant uncertainty in D^* is at most 10%. The observed variation in profile shape points to differences in the boundary conditions during diffusion, particularly those related to the strength of the diffusion source in proportion to the length of the diffusion anneal.[7]

3.1.2 Tracer diffusion coefficients. Diffusion coefficients D^*_{Na} obtained from fitting of the ^{22}Na penetration profiles in both PEO$_{20}$NaI and PEO$_{60}$NaI are displayed in Fig. 3 as a function of inverse temperature. For comparison, previous data on PEO$_{30}$NaI are also included.[4] It is seen that the D^*_{Na} data from the different compositions are rather close together. For any temperature within the range investigated the differences in magnitude are less than a factor of two. Moreover, a distinct trend is observed, with PEO$_{20}$NaI being the complex with the slowest Na diffusivity and PEO$_{60}$NaI having slightly higher values than PEO$_{30}$NaI on average. For each composition the data have been fitted by the Vogel–Tamann–Fulcher (VTF) equation (solid lines)

$$D_Z = D_{0Z}\exp[-B_Z/(T - T_{0Z})] \tag{1}$$

where D_Z is the diffusivity of interest (here Z = Na), D_{0Z} a pre-exponential factor, B_Z a pseudo activation energy, and T_{0Z} the zero mobility temperature. The VTF

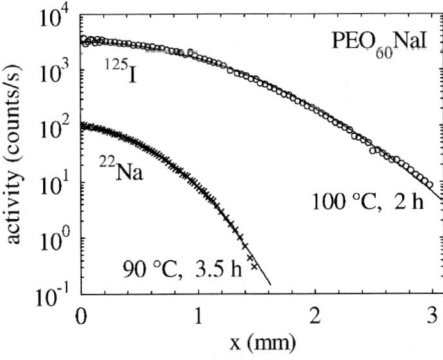

Fig. 2 Penetration profile of ^{22}Na and ^{125}I in PEO$_{60}$NaI resulting from separate diffusion treatments as indicated by temperature and duration. Solid lines are fits of the Gaussian function.

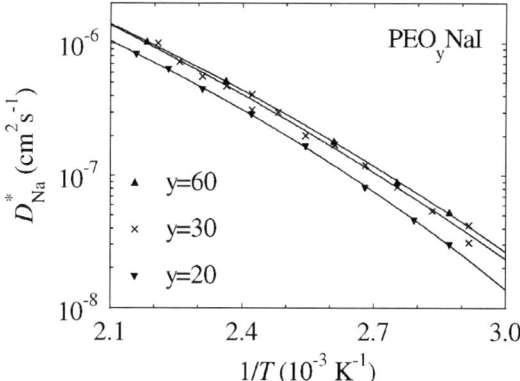

Fig. 3 Comparison of the sodium tracer diffusivity D^*_{Na} among three different compositions of the PEO–NaI electrolyte system. Solid lines are fits of the VTF function as characterised by the parameters given in Table 1.

parameter values resulting from the fitting procedures are listed in Table 1. At this stage of analysis no deeper meaning should be attributed to these parameter values; they just allow for interpolation and comparison with literature data.

Fig. 4 shows a survey of the D^*_I data arising from ^{125}I diffusion experiments on the PEO$_y$NaI system. We see a similar picture to that for Na in Fig. 3: the previously obtained data for the composition with $y = 30$[4] are 'sandwiched' by the present data for $y = 20$ at the low side and those for $y = 60$ at the high side. The differences in D^*_I between the latter compositions do not exceed a factor of two. The solid lines represent best fits based on eqn (1). The corresponding parameter values are compiled in Table 1.

3.2 Conductivity data

For each measuring temperature, the dc conductivity σ_{dc} was determined from the plateau region in a plot of the modulus of the complex impedance *versus* the ac frequency. Specifically, σ_{dc} was taken at the frequency value for which the phase angle was closest to $0°$.[8] The dc conductivity was converted into the charge diffusivity D_σ with the aid of the Nernst–Einstein equation, *i.e.*,

$$D_\sigma \equiv \frac{\sigma_{dc}k_B T}{C_s e^2},\qquad(2)$$

where k_B denotes the Boltzmann constant, T temperature, and e elementary charge. It should be noted that it is just the salt concentration C_s (*i.e.*, number density) that enters eqn (2), and not the total concentration of ionic species being equal to $2C_s$.

Table 1 VTF parameters from separate fits to D^*_{Na}, D^*_I and D_σ for each composition

y	Z	$D_{0Z}/cm^2\ s^{-1}$	B_Z/K	T_{0Z}/K
20	Na	$1\ 6 \times 10^{-4}$	1559	167
	I	2.7×10^{-5}	680	218
	σ	2.7×10^{-5}	852	203
30	Na	5.3×10^{-4}	2092	125
	I	6.1×10^{-5}	1009	176
	σ	7.1×10^{-6}	410	241
60	Na	3.1×10^{-4}	1829	138
	I	1.3×10^{-5}	388	251
	σ	2.9×10^{-6}	222	262

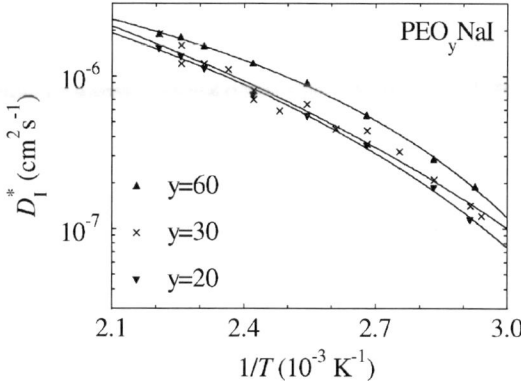

Fig. 4 Comparison of the iodine tracer diffusivity D_I^* among three different compositions of the PEO–NaI electrolyte system. Solid lines are fits of the VTF function as characterised by the parameters given in Table 1.

Due to this choice, D_σ stands for the charge diffusivity per salt molecule and thus comprises the joint effect of one cation and one anion.

Fig. 5 shows the D_σ data for the two PEO$_y$NaI complexes investigated in this work together with the earlier data for the intermediate composition with $y = 30$.[4] Again, we observe that the magnitude of D_σ increases with increasing value of y, that is, with decreasing salt concentration. However, in contrast to the Na and I tracer diffusivities, the differences in D_σ among the three different compositions distinctly depend on temperature. At the low end of the temperature range investigated, near 60 °C, the smallest and largest D_σ values are almost an order of magnitude apart. On the contrary, at the high end of the temperature interval under investigation, near 200 °C, the three individual D_σ plots seem to converge to a joint conductivity level at about 1×10^{-6} cm^2 s^{-1}. These findings are reproduced by the VTF-fits to the data which are based on eqn (1) and represented by the solid lines in Fig. 5. The associated VTF parameter values can be found in Table 1.

4 Modelling of ionic transport

The theoretical framework for the joint evaluation of diffusion and conductivity data was initially developed for PEO$_{30}$NaI[4,8] and also successfully applied to a cross-

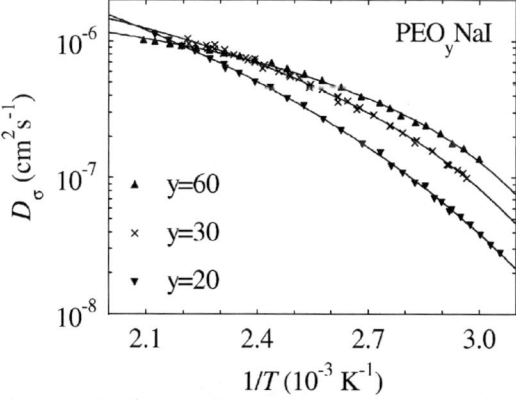

Fig. 5 Comparison of the charge diffusivity D_σ among three different compositions of the PEO–NaI electrolyte system. Solid lines are fits of the VTF function as characterised by the parameters given in Table 1.

linked polyether-based electrolyte.[5] A full account of this model has been given elsewhere.[6] Here we report only its major features. An important modification, however, concerns the formulation of the ion association/dissociation reaction. As outlined below, now allowance is made for the fact that the charged single cations are not free but coordinated by the oxygen atoms of the polymer chains.

4.1 Ion association

Association of the single ions Na^+ and I^- and their subsequent dissociation may be represented by the reaction

$$(\text{poly-O})_z Na^+ + I^- \rightleftharpoons NaI^0 + (\text{poly-O})_z. \tag{3}$$

Here, $(\text{poly-O})_z Na^+$ denotes a complex consisting of a cation coordinated by z polyether oxygen atoms, whereas $(\text{poly-O})_z$ is the corresponding 'free' polymer segment with a length of z monomer units which is reestablished after release of the cation. The coordination number z has been reported to adopt values of about 5 in PEO–NaI electrolytes.[10] In principle, several $(\text{poly-O})_z Na^+$ complexes may involve two or more polymer chains (or different segments of the same long chain) but they only tend to form a small fraction of the total number.[11,12] Therefore, this complication is considered to be of negligible significance in the present context.

Reaction 3 implicitly states that the neutral ion pair NaI^0 is not coordinated to the ether oxygens of the polymer chains. This admittedly strong simplification is supported by our finding, to be presented and discussed in sections 5 and 6, that the mobility of the ion pair is very high (*cf.* ref. 4). Also, the positive values obtained for the pair formation enthalpy and entropy (see below) comply with an appreciable loss of cation binding to the polymer upon pair formation.

A basic assumption of the model is that larger ion aggregates do not play a significant role. Using the index Na^+ for the $(\text{poly-O})_z Na^+$ complex, I^- for the anion, and p for the NaI^0 pair, the corresponding concentrations (number densities) are designated as C_{Na^+}, C_{I^-}, and C_p, respectively. These concentrations are subject to the constraints of charge neutrality, $C_{Na^+} = C_{I^-}$, and mass conservation, $C_{Na^+} + C_p = C_{I^-} + C_p = C_s$.

Application of the mass action law to eqn (3) yields the (true) reaction constant $\tilde{k}_p$ as

$$\tilde{k}_p = \frac{C_p(C_O - z C_{Na^+})}{C_{Na^+} C_{I^-}} \tag{4}$$

where C_O is the number density of oxygen atoms, or, equivalently, of monomer units. This expression takes into account that charged cations 'occupy' on average z consecutive oxygen atoms, and moreover, that any 'vacant' polyether oxygen may act as central docking site for Na^+.

Introducing the pair fraction $r_p \equiv C_p/C_s$ and the reduced salt concentration $\tilde{c}_s \equiv C_s/C_O = 1/y$, eqn (4) can be transformed into

$$\frac{r_p}{(1 - r_p)^2} = k_p[1 - z\tilde{c}_s(1 - r_p)]. \tag{5}$$

Here, the (reduced) reaction constant k_p must be identified with $\tilde{k}_p \tilde{c}_s$. We adopt the conventional form

$$k_p = k_{p0} \exp(-\Delta H_p/k_B T). \tag{6}$$

with the pair formation enthalpy ΔH_p and the (reduced) pre-factor k_{p0}. Hence the pre-factor $\tilde{k}_{p0}$ of the true reaction constant $\tilde{k}_p$, which contains the pair formation entropy ΔS_p, is obtained from the fitting parameter k_{p0} (see below) as

$$\tilde{k}_{p0} = \exp(\Delta S_p/k_B) = k_{p0} \tilde{c}_s^{-1} = k_{p0} y. \tag{7}$$

This journal is © The Royal Society of Chemistry 2006

In earlier treatments of the ion association kinetics,[4,5] the factor in square brackets on the right hand side of eqn (5) was ignored. This corresponds to the case of vanishing small salt concentration, in which no competition among the anions for the limited number of oxygen atoms exists. In this first analysis of the concentration dependence in the PEO–NaI system, the same approximate form of the mass action relationship is assumed, i.e., for reasons of simplicity and comparability. With this approximation, eqn (5) reduces to a quadratic equation which is solved as

$$r_p = 1 + \left(1 - \sqrt{1 + 4k_p}\right)/2k_p. \tag{8}$$

Hence r_p approaches k_p for small values ($k_p \ll 1$), as may be seen from the Taylor expansion $r_p = k_p - 2k_p^2 + \ldots$.

4.2 Ion migration

Each ionic species $X = Na^+$, I^-, or $p = NaI^0$ is characterised by its 'true' diffusivity D_X. A preliminary analysis reported for $PEO_{30}NaI$[6] indicated that these ionic diffusivities may be very similar concerning their (VTF-type) temperature dependence but dissimilar in magnitude. Therefore, we adopt the general expression

$$D_X = D_X^0 \exp[-B/(T - T_0)] \tag{9}$$

with the initially unknown VTF parameters (B, T_0) and pre-factors ($D_{NA^+}^0$, $D_{I^-}^0$, D_p^0). The underlying idea is that all transport processes in the polyether–salt complexes are driven by the segmental motion of the polymer chains as corroborated by previous work.[12–14]

Considering the overall migration of, e.g., Na, we have to distinguish between a contribution of the single ions Na^+ and a contribution due to the NaI^0 pairs. A more elaborate treatment[6] shows that this can be formulated in terms of 'effective' diffusivities $\hat{D}_X$ defined as

$$\hat{D}_X = \frac{C_X D_X}{C_s}. \tag{10}$$

In this expression the true diffusivity D_X is multiplied by the probability C_X/C_s that the configuration $X = Na^+$, I^-, or $p = NaI^0$ occurs. Eventually, the three different sets of experimental data can be described by the equations

$$D_{Na}^* = \hat{D}_{Na^+} + \hat{D}_p \tag{11}$$

$$D_I^* = \hat{D}_{I^-} + \hat{D}_p \tag{12}$$

$$D_\sigma = \hat{D}_{Na^+} + \hat{D}_{I^-} \tag{13}$$

with

$$\hat{D}_{Na^+} = (1 - r_p)D_{Na^+}^0 \exp[-B/(T - T_0)] \tag{14}$$

$$\hat{D}_{I^-} = (1 - r_p)D_{I^-}^0 \exp[-B/(T - T_0)] \tag{15}$$

$$\hat{D}_p = r_p D_p^0 \exp[-B/(T - T_0)], \tag{16}$$

Here $C_{Na^+}/C_s = C_{I^-}/C_s = 1 - r_p$ was used.

5 Evaluation of model parameters

5.1 Fitting procedure and results

For either composition, D_{Na}^*, D_I^*, and D_σ were simultaneously fitted by eqn (11)–(13) using the specific formulations of the effective diffusivities and ion-pair-related parameters given by eqn (14)–(16) and eqn (6)–(8), respectively. The fits closely follow the experimental data as shown by the solid lines in Fig. 6 for $PEO_{20}NaI$ and

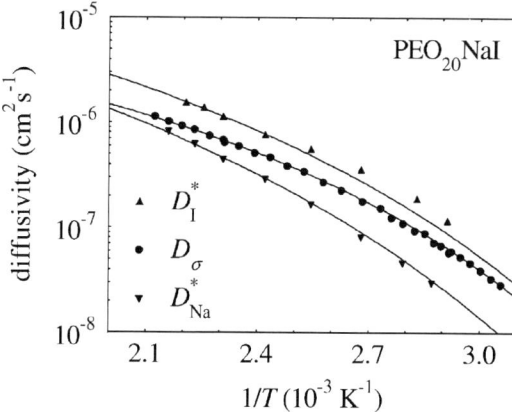

Fig. 6 Tracer diffusion coefficients D_{Na}^* and D_I^* compared to the charge diffusivity D_σ in PEO$_{20}$NaI. The solid lines result from a simultaneous fit to all experimental data based on the ion transport model described in the text.

Fig. 7 for PEO$_{60}$NaI. In particular, the intermediate position of D_σ with respect to the two tracer diffusivities is well reproduced by the model. A comparison between Fig. 6 and 7 also reveals the most significant difference between the two compositions. In PEO$_{20}$NaI, D_σ keeps a similar distance over the entire temperature range— in units of the depicted logarithmic diffusivity scale—to D_{Na}^* at the low side and to D_I^* at the high side. This relates to the fact that the curvature of D_σ in the Arrhenius plot of Fig. 6 is not too much different from either of the tracer diffusivities. In PEO$_{60}$NaI, however, D_σ approaches D_I^* at the low end of the temperature range, whereas it intersects the D_{Na}^* curve near the highest temperature investigated for this quantity. The previously examined PEO$_{30}$NaI composition[4] visually takes an intermediate position between PEO$_{20}$NaI and PEO$_{60}$NaI concerning the above features of the Arrhenius survey plots.

In the fitting procedure, the initially unknown model parameters were adjusted, i.e., either directly or in logarithmic form[15]: B, T_0, ln $D_{Na^+}^0$, ln $D_{I^-}^0$, ln D_p^0, ln k_{p0}, and ΔH_p. The resulting best estimates are compiled in Table 2 together with similar data for PEO$_{30}$NaI.[4] Although the number of free parameters, seven, may seem quite high, it favourably compares to the total number of nine required for each composition to obtain three individual VTF-fits (cf. Table 1). The validity of the model is further supported by the generally modest uncertainties in the resulting parameter values, which are also given in Table 2.[15]

It should be noted that the parameter sets obtained for PEO$_{20}$NaI and PEO$_{60}$NaI in this work and the earlier data on PEO$_{30}$NaI are compatible with each other concerning the magnitude levels of the individual parameters. In particular, it is seen that for all compositions the relationship $D_{Na^+}^0 \ll D_{I^-}^0 < D_p^0$ holds. For the composition $y = 20$ it was not possible to determine $D_{Na^+}^0$ within a finite confidence interval. Instead, an upper limit of 1×10^{-7} cm^2 s^{-1} was estimated (see Table 2), which relies on the plausible assumption that the contribution of the effective Na$^+$ diffusivity $\hat{D}_{Na^+}$ to both D_{Na}^* and D_σ (see eqn (11) and (13)) is below a detection threshold of 1%. Therefore, the PEO$_{20}$NaI data given in Table 2 essentially result from six-parameter fits (see Fig. 6). The large error in $D_{Na^+}^0$ for the PEO$_{60}$NaI composition (100%[15]) is consistent with the fact that the best estimate of 8.0×10^{-8} cm^2 s^{-1} is just above the detection threshold.

The absolute and relative values obtained for the model parameters have implications for the interpretation of the ionic migration mechanisms. In addition, several parameters in Table 2 vary monotonically with salt concentration whereas others show an extreme value at the intermediate composition. These issues are discussed in detail in the following sections.

Table 2 Model parameters and their statistical error from a simultaneous fit to D_{Na}^*, D_I^*, and D_σ, for each composition

y	B/K	T_0/K	$D_{Na^+}^0/$ cm^2 s^{-1}	$D_{I^-}^0/$ cm^2 s^{-1}	$D_p^0/$ cm^2 s^{-1}	$\ln \tilde{k}_{p0} =$ $\Delta S_p/k_B$	$\Delta H_p/eV$
20	772	212	$\leq 10^{-7}$	2.4×10^{-5}	1.8×10^{-4}	3.1 ($\pm$5%)	0.088
	($\pm$8%)	($\pm$4%)		($\pm$25%)	($\pm$35%)		($\pm$8%)
30	521	225	1.2×10^{-6}	9.7×10^{-6}	8.9×10^{-5}	6.0 ($\pm$7%)	0.19
	($\pm$15%)	($\pm$6%)	($\pm$30%)	($\pm$30%)	($\pm$40%)		($\pm$8%)
60	493	225	8.0×10^{-8}	1.4×10^{-5}	2.5×10^{-5}	10.9 ($\pm$4%)	0.26
	($\pm$8%)	($\pm$4%)	($\pm$100%)	($\pm$25%)	($\pm$12%)		($\pm$5%)

5.2 Parameter values and their trends with concentration

5.2.1 Vogel–Tamann–Fulcher parameters. The pseudo activation energy B takes values which are in the common range for PEO–salt systems.[16] Specifically, it is found that B decreases with increasing O-to-Na ratio y (or decreasing salt concentration $\tilde{c}_s$). Within the free volume theory[17,18] this may be interpreted as a gain of free volume upon reduction of the salt concentration. This plausible effect may be mainly due to an increase in the thermal expansion coefficient, since the zero mobility temperature T_0 does not change significantly with composition (Table 2).

The observed weak correlation of T_0 with $\tilde{c}_s$ may surprise since empirically T_0 is linked to the glass transition temperature T_g[18] which is known to increase upon salt addition.[19] In fact, the T_0 value obtained for the PEO$_{30}$NaI complex coincides with the pertaining T_g value of 225 K,[4,19] which is not uncommon for PEO-based electrolytes.[16] However, the physical meaning of T_0 is not unambiguously clear and a reliable data base for PEO–salt systems is missing in the literature. Rather, most of the reported T_0 results stem from the straightforward VTF-fitting of σ_{dc} data, that is, without taking into account the effects of ion aggregation. Furthermore, according to the Nernst–Einstein equation (eqn (2)) σ_{dc} and D_σ differ by a factor containing the temperature T. The impact of these circumstances may be recognised from a comparison of the T_0 values in Table 2 with those for σ_{dc} in Table 1. Therefore, we refrain from further discussion of the T_0 results.

5.2.2 Diffusivity pre-factors. Table 2 shows that the pre-factor of the true cation diffusivity $D_{Na^+}^0$ obtains its highest value at the intermediate composition $y = 30$.

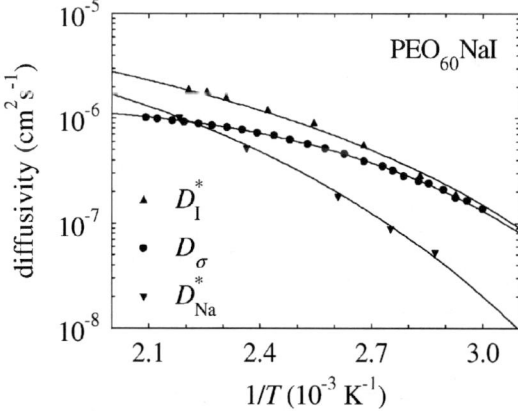

Fig. 7 Tracer diffusion coefficients D_{Na}^* and D_I^* compared to the charge diffusivity D_σ in PEO$_{60}$NaI. The solid lines result from a simultaneous fit to all experimental data based on the ion transport model described in the text.

For the complexes with $y = 20$ and $y = 60$, $D_{\mathrm{Na}^+}^0$ is at least one order of magnitude smaller. This may be tentatively explained as follows. At low salt concentration, $y = 60$, there are many 'free' oxygen atoms available to take part in site-exchanges of Na^+ ions. However, as will be outlined below, in these dilute complexes the binding of the cations to the polymer chains is strong. Conversely, at high salt concentration, $y = 20$, Na^+ migration suffers from site-blocking effects but profits from a smaller cation–polymer binding energy. As a consequence, the maximum mobility may occur at some intermediate composition.

The pre-factor of the true anion diffusivity $D_{\mathrm{I}^-}^0$ only varies weakly with composition. The $D_{\mathrm{I}^-}^0$ values listed in Table 2 do not differ by more than a factor of 2.5, which may be considered as moderate for pre-exponential factors. Moreover, these differences are almost within the indicated error ranges. The finding that I^- migration is less dependent on $\tilde{c}_s$ than Na^+ migration complies with the absence of any direct anion binding to the polymer matrix. This notion also explains that $D_{\mathrm{I}^-}^0$ is at least one order of magnitude larger than $D_{\mathrm{Na}^+}^0$.

The NaI^0 pairs diffuse even faster than the charged anions, as may be seen from the values of the pair diffusivity pre-factor D_p^0 listed in Table 2. However, the ratio $D_p^0/D_{\mathrm{I}^-}^0$, being 7.5 and 9 for $y = 20$ and 30, respectively, drops to less than 2 for $y = 60$. The high mobility of the neutral ion pairs indicates that freely moving contact pairs may be involved. Moreover, at high salt concentration the number of 'vacant' oxygen sites is limited. Hence, the probability for the pair, i.e., its Na^+ member, to 'dock' onto the polymer chain and thus to form—at least temporarily—a virtually immobile *solvent-shared*[21] pair seems small. This explains the very high D_p^0 data for $\mathrm{PEO}_{20}\mathrm{NaI}$ and $\mathrm{PEO}_{30}\mathrm{NaI}$ which are characteristic of small neutral molecules in polymer melts.[20] In contrast, at low $\tilde{c}_s$ values the fraction of contact pairs bound to the polymer will increase and concomitantly their diffusivity will decrease. This is formally expressed as

$$D_p^0 \approx \frac{C_{p,f} D_{p,f}^0}{C_p} = \frac{C_{p,f} D_{p,f}^0}{C_{p,f} + C_{p,b}}, \tag{17}$$

where $C_{p,f}$ and $D_{p,f}^0$ refer to free pairs and $C_{p,b}$ denotes the concentration of bound (solvent-shared) pairs. According to eqn (17) the maximum D_p^0 value, $D_{p,f}^0$, will be reduced by the factor $C_{p,f}/C_{p,b}$ in case of a high fraction of solvent-shared NaI^0 pairs.

5.2.3 Enthalpy and entropy of ion pairing. The most striking result of the present work is the distinct dependence on concentration found for ΔH_p and ΔS_p. As seen in Table 2, ΔH_p increases from 0.088 to 0.26 eV with increasing composition parameter y. Concomitantly, $\Delta S_p/k_B = \ln \tilde{k}_{p0}$ increases from 3.1 to 10.9. The data are plotted in Fig. 8 as a function of the reduced salt concentration $\tilde{c}_s$. Fig. 9 shows the resulting pair fraction r_p as a function of inverse temperature.

According to eqn (3), the pair formation enthalpy ΔH_p is a composite quantity given by

$$\Delta H_p = H_p + H_{\mathrm{poly-O}} - (H_{\mathrm{poly-O-Na}^+} + H_{\mathrm{I}^-}) \tag{18}$$

where the individual terms stand for the enthalpy of the species or complexes involved in the ion pairing reaction. The finding that ΔH_p is positive means that the pair state is not the preferred one from the viewpoint of enthalpy (cf. eqn (6)). This relates to the strong Na^+–polymer binding (low value of $\Delta H_{\mathrm{poly-O-Na}^+}$) and to the fact that the salt, with its high lattice energy, dissolves in the polymer at all.

We tentatively assume that three of the terms in eqn (18) are relatively independent of composition, i.e., the enthalpy of NaI^0 pairs H_p, the enthalpy of cation-free polymer segments consisting of z monomers $\Delta H_{\mathrm{poly-O}}$, and the enthalpy of I^- ions H_{I^-}. In this approximation, ΔH_p reflects the concentration dependence of $H_{\mathrm{poly-O-Na}^+}$, but with opposite sign. Thus the present results indicate that

This journal is © The Royal Society of Chemistry 2006

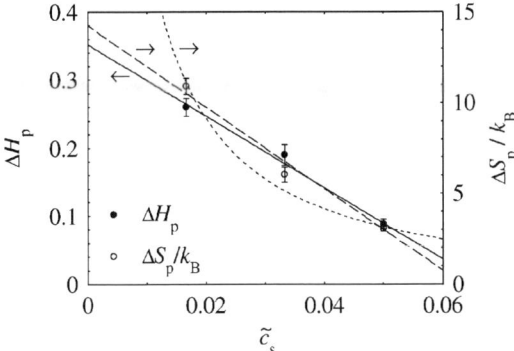

Fig. 8 Pair formation enthalpy ΔH_p (left axis) and pair formation entropy ΔS_p (in units of k_B; right axis) as a function of the reduced salt concentration $\tilde{c}_s$. The solid and coarsely-dashed line are fits based on a linear $\tilde{c}_s$-dependence of ΔH_p and ΔS_p, respectively. The finely-dashed line represents a fit to the ΔS_p data based on a linear function in $\tilde{c}_s^{-1}$.

$H_{poly-O-Na^+}$ increases with increasing $\tilde{c}_s$, which means that the binding of Na^+ to the polymer becomes weaker for higher salt concentrations. Intuitively, this may be understood from the notion that at very low salt concentrations all cations are optimally coordinated by the polymer chains, whereas morphology constraints begin to arise at higher $\tilde{c}_s$ values. Fig. 8 suggests a linear ΔH_p vs. $\tilde{c}_s$ dependence (solid line) predicting a maximum pair formation enthalpy of 0.35 eV in the low-concentration limit and a decrease of 0.05 eV per 0.01 unit of reduced salt concentration.

With a similar argument, $S_{poly-O} - S_{poly-O-Na^+}$ forms the most significant contribution to ΔS_p. Hence, the entropy gain due to releasing the polymer from its local conformations imposed by cation coordination appears to be a crucial factor in pair formation.[22] In dilute electrolytes this entropy gain is high, since for each cation long polymer segments are available for optimal coordination. In concentrated systems, however, the small number of monomers per cation hampers its perfect solvation and the entropy effects involved are accordingly weaker. The linear fit in Fig. 8 (coarsely-dashed line) predicts $\Delta S_p \approx 14.3\ k_B$ at $\tilde{c}_s \approx 0$, and moreover, that the decrease of ΔS_p per 0.01 (in units of $\tilde{c}_s$) amounts to $2.2k_B$.

For each composition, the combined effect of ΔH_p and ΔS_p gives rise to a specific temperature dependence of the pair fraction r_p which is depicted in Fig. 9. For the higher salt concentrations, $y = 20$ or 30, r_p remains below 0.1 over the entire temperature range. However, for the lowest concentration, $y = 60$, NaI^0 pairs may become abundant as indicated by r_p values approaching 0.5 at high temperatures.

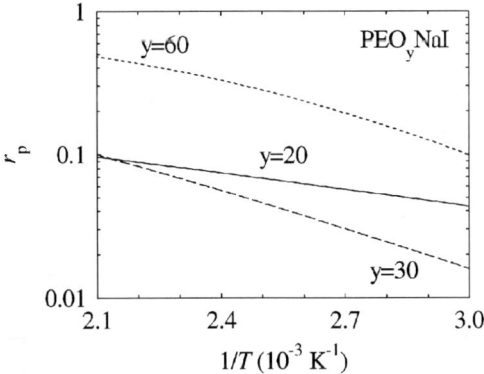

Fig. 9 The neutral ion pair fraction as a function of inverse temperature for three compositions of the PEO$_y$NaI system.

The increase of r_p with increasing temperature reflects the growing importance of the entropy term in the free enthalpy of pair formation $\Delta G_p = \Delta H_p - T\Delta S_p$. Furthermore, the high pair fractions in $PEO_{60}NaI$ are consistent with the finding that ΔS_p increases more strongly with decreasing $\tilde{c}_s$ than ΔH_p, as indicated by Fig. 8.

In Fig. 9, the r_p curves for $PEO_{30}NaI$ and $PEO_{60}NaI$ have a similar slope. However, this is an accidental effect which arises from a combination of two counteracting factors. First, the increase of k_p for $PEO_{60}NaI$, i.e., towards values greater than unity at high temperatures, is steeper than for $PEO_{20}NaI$. Second, according to the definition of the pair fraction, $r_p \leq 1$ must hold, which is warranted by the use of eqn (8). Thus for high k_p values, the difference with r_p becomes quite large.

For $PEO_{20}NaI$, the temperature dependence of r_p exhibits a smaller slope than for $PEO_{30}NaI$, as seen in Fig. 9. This is a consequence of the smaller ΔH_p value for $PEO_{20}NaI$ in conjunction with the fact that $k_p \ll 1$, since then $r_p \approx k_p$ holds true. Another feature is the relatively high pair fraction of $PEO_{20}NaI$ at low temperatures. According to the mass action law, this may be due, at least partly, to the high cation and anion concentrations for this composition.

6 Discussion

In elucidating the influence of salt concentration on ΔH_p and ΔS_p our reasoning was based on the polymer chain length per cation. This quantity is intimately connected with $y = 1/\tilde{c}_s = C_O/C_s$. In this expression, C_s is representative of the total cation concentration which is the sum of C_{Na^+} and C_p. However, in this context it would be more appropriate to allow only for the cations bound to the polymer, i.e., to replace C_s by $C_{Na^+} = C_s - C_p = C_s(1 - r_p)$. Consequently, the crucial parameter becomes $\tilde{c}_s(1 - r_p)$ instead of $\tilde{c}_s$. Since r_p changes with temperature, it is also expected that ΔH_p and ΔS_p become temperature-dependent through their functional dependence on $\tilde{c}_s(1 - r_p)$. Therefore, the present analysis, in which ΔH_p and ΔS_p were assumed to be constant, should be considered as preliminary only.

A more thorough evaluation of the data has to take into account the temperature-dependence of ΔH_p and ΔS_p by including, e.g., a linear dependence on $\tilde{c}_s(1 - r_p)$. However, ΔS_p may alternatively be a linear function of $1/\tilde{c}_s(1 - r_p)$, as suggested by the finely-dashed curve in Fig. 9 and supported by basic physical considerations which relate polymer entropy to the mean free chain length.[23] The two alternative relationships give rise to large differences in ΔS_p at low salt concentrations. This makes it promising to extend the present experiments on PEO_yNaI to compositions with $y > 60$.

In the present analysis, the factor in square brackets in eqn (6) was neglected. It can be estimated that this approximation has affected our results only to a minor extent. Future evaluations, however, should involve the full solution of the mass action equation. It is remarkable that the factor neglected so far also depends on $\tilde{c}_s(1 - r_p)$.

A striking result of our work is the high diffusivity found for the ion pairs, which led us to the concept of freely moving contact pairs. However, the question may be raised as to what extent the Na^+ partner of the pair interacts with the ether oxygens and whether a sharp distinction between free pairs and, e.g., solvent-shared pairs is justified. Therefore, it remains a task for future studies to clarify the nature of the ion pairs and their migration mechanism.

7 Conclusions

In this study on the PEO–NaI electrolyte system, the self-diffusivity of either ionic component and the overall ionic conductivity have been investigated for the first time over wide ranges of both concentration and temperature. As a general trend, the charge diffusivity is found to be intermediate between a lower [22]Na diffusivity and a higher ^{125}I diffusivity. This observation can be explained by the occurrence of neutral ion pairs which contribute to mass transport but not to charge transport. An extension of the ion-pairing reaction makes allowance for the coordination of the cations by polymer chain segments while long-range migration of all ionic species is conceived to be driven by the local motion of the polymer matrix. A mathematical formulation of these concepts yields good fits of the experimental data for different compositions, and, moreover, reveals trends in the model parameters with composition. The ion pair is found to be the species with the highest mobility at any temperature and composition whereas Na^+ ions are relatively slow. The pair fraction increases with increasing temperature and approaches values close to 50% for the composition with the lowest ion density. The most surprising result, however, is that the enthalpy and entropy of pair formation show a clear decrease with increasing salt concentration. This has been interpreted in terms of cation coordination being dependent on the mean polymer segment length per cation. An extension of the experiments to compositions with very low salt concentrations seems promising.

References

1 D. R. Sadoway and A. M. Mayes, *MRS Bull.*, 2002, **27**, 590.
2 F. M. Gray, *Solid Polymer Electrolytes*, VCH, New York, 1991.
3 P. V. Wright, *MRS Bull.*, 2002, **27**, 597.
4 N. A. Stolwijk and Sh. Obeidi, *Phys. Rev. Lett.*, 2004, **93**, 125901.
5 Sh. Obeidi, N. A. Stolwijk and S. J. Pas, *Macromolecules*, 2005, **38**, 10750.
6 N. A. Stolwijk and Sh. Obeidi, *Defect Diffusion Forum*, 2005, **237**, 1004.
7 J. Crank, *The Mathematics of Diffusion*, Clarendon Press, Oxford, 1995.
8 S. Obeidi, B. Zazoum and N. A. Stolwijk, *Solid State Ionics*, 2004, **173**, 77.
9 According to the pertaining lot specification, the base PEO material contained 1.4 wt% SiO_2 and 0.2 wt% CaO. These contaminations, which are inherent to the manufacturing process, are present in the form of small particles; see also: S. Suarez, S. Abbrent, S. G. Greenbaum, J. H. Shin and S. Passerini, *Solid State Ionics*, 2004, **166**, 407.
10 A. van Zon, G.-J. Bel, B. Mos, P. Verkerk and S. W. de Leeuw, *Comput. Mater. Sci.*, 2000, **17**, 265.
11 O. Borodin and G. D. Smith, *Macromolecules*, 1998, **31**, 8396.
12 O. Borodin and G. D. Smith, *Macromolecules*, 2000, **33**, 2273.
13 J. J. Fontanella, M. C. Wintersgill, M. K. Smith, J. Semancik and C. G. Andeen, *J. Appl. Phys.*, 1986, **60**, 2665.
14 O. Dürr, W. Dieterich and A. J. Nitzan, *Chem. Phys.*, 2004, **121**, 12732.
15 Pre-exponential factors, such as D_0, enter the fitting procedure in the form of $exp(\ln D_0)$ which yield better error statistics. The resulting standard deviations s_D in $\ln D_0$ (absolute values) lead to uncertainty factors of $exp(\pm s_D)$. The specification $D_0 \pm 100\%$ corresponds to uncertainty factors 2 (high bound) and 1/2 (low bound).
16 D. Bamford, A. Reiche, G. Dlubek, F. Alloin, J.-Y. Sanchez and M. A. Alam, *J. Chem. Phys.*, 2003, **118**, 9420.
17 M. H. Cohen and D. Turnbull, *J. Chem. Phys.*, 1959, **31**, 1164.
18 M. A. Ratner, in *Polymer Electrolyte Reviews*, ed. J. R. MacCallum and C. A. Vincent, Elsevier Applied Science, London, 1987, p. 173.
19 M. Minier, C. Berthier and W. Gorecki, *J. Phys. (Paris)*, 1984, **45**, 739.
20 R. H. Gee and R. H. Boyd, *Polymer*, 1995, **36**, 1435.
21 *Solid State Electrochemistry*, ed. P. G. Bruce, Cambridge University Press, Cambridge, UK, 1997.
22 M. A. Ratner and A. Nitzan, *Faraday Discuss. Chem. Soc.*, 1989, **88**, 19.
23 R. J. Young and P. A. Lovell, *Introduction to Polymers*, Chapman and Hall, London, 1991.

Calorimetric measurements of energetics of defect interactions in fluorite oxides

Alexandra Navrotsky,*[a] Petra Simoncic,[a] Harumi Yokokawa,[b] Weiqun Chen[a] and Theresa Lee[c]

Received 17th March 2006, Accepted 24th April 2006
First published as an Advance Article on the web 4th August 2006
DOI: 10.1039/b604014b

Direct measurement by oxide melt solution calorimetry of energetics of mixing in rare earth and yttrium doped zirconia, hafnia, and ceria systems provides support for spectroscopic and computational studies of the location and clustering of vacancies in these systems. Strongly negative heats of mixing are seen when the vacancy is transferred from being nearest neighbor to Y or RE in the sesquioxide to being nearest neighbor to Zr or Hf in the cubic solid solution. In the absence of such redistribution, small positive enthalpies of mixing are seen in CeO_2–$YO_{1.5}$ and CeO_2–$REO_{1.5}$ systems. Strongly positive enthalpies of mixing are seen in CeO_2–ZrO_2, which has a large difference in cation sizes and no vacancy formation. The system $Ce_{0.8}Y_{0.2}O_{1.9}$–$Zr_{0.8}Y_{0.2}O_{1.9}$ shows small positive heats of formation with less destabilization in the Ce-rich region, suggestive of "scavenging" of oxygen vacancies by Zr. The calorimetric data obtained in these studies offer direct comparison with the results of computations on defect clusters and their binding energies.

Introduction

Oxides having the fluorite structure offer a richness of defect chemistry, closely linked to their applications as solid electrolytes, gas separation membranes, and catalysts. Substituting the tetravalent ion, typically Zr, Hf, or Ce, by trivalent ions, typically yttrium and the rare earths, preserves the cubic fluorite structure to ambient conditions and produces materials with high oxygen mobility. Despite extensive studies of ionic conductivity, diffusion, and defect association by experimental and computational techniques, there has been little work till recently on the direct experimental determination of enthalpies of formation. The extreme refractory nature of these oxides has been a stumbling block to calorimetry, but recent advances in oxide melt solution calorimetry at 700–800 °C makes it possible to make such measurements.[1–3] The purpose of the new work reported here is to further quantify the energetic effects of cation and dopant size by measuring heats of formation in a number of rare earth doped systems. Specifically, oxide melt solution calorimetric data are presented for cubic fluorite phases $Hf_xRE_{1-x}O_{2-x/2}$ (RE =

[a] NEAT ORU & Thermochemistry Facility, University of California at Davis, 4440 Chemistry Annex, One Shields Ave., Davis, California, 95616, USA. E-mail: anavrotsky@ucdavis.edu; Fax: 1 (530) 752 9307; Tel: 1 (530) 752 3292
[b] National Institute of Advanced Industrial Science and Technology (AIST), Energy Electronics Institute, Central 5, Higashi 1-1-1, Tsukuba 305-8565, Japan
[c] Los Alamos National Laboratory, NMT-16, MS G721, New Mexico 87545, USA

Sm, Gd, Dy, Yb), $Ce_xRE_{1-x}O_{2-x/2}$ (RE = Gd, La), $Zr_xCe_{1-x}O_2$, and the $Ce_{0.8}Y_{0.2}O_{1.9}$–$Zr_{0.8}Y_{0.2}O_{1.9}$ system.

Recent prior calorimetric studies and their interpretation

Earlier published studies from this laboratory reported enthalpies of formation from the binary oxides stable at room temperature (monoclinic zirconia and hafnia, cubic fluorite-type ceria, cubic C-type yttria) for the three systems $(1 - x)ZrO_2$–$xYO_{1.5}$,[4] $(1 - x)HfO_2$–$xYO_{1.5}$,[5] and $(1 - x)CeO_2$–$xYO_{1.5}$.[6] The energetic behavior of CeO_2–$YO_{1.5}$ is strikingly different from that of ZrO_2–$YO_{1.5}$ and HfO_2–$YO_{1.5}$ (see Fig. 1).[7] In contrast to the small positive heat of formation from the end-member oxides, $\Delta H_{f,ox}$, of CeO_2–$YO_{1.5}$, both ZrO_2–$YO_{1.5}$ and HfO_2–$YO_{1.5}$ show strongly negative and curved heats of formation (from m–ZrO_2 or m–HfO_2 and C–$YO_{1.5}$) as a function of composition, with substantial stabilization (by 20–30 kJ mol^{-1}) at compositions near mole fractions of $YO_{1.5}$ of 0.4–0.5. Fit by quadratic equations, these give interaction parameters of -93.7 ± 12.0 kJ mol^{-1} for ZrO_2–$YO_{1.5}$ and -155.2 ± 10.2 kJ mol^{-1} for HfO_2–$YO_{1.5}$. These very negative interaction parameters indicate that regular solution behavior (with Raoultian entropies) almost certainly does not apply,[3] and the calorimetric data are consistent with extensive short-range order or defect association.

Since a maximum in ionic conductivity is seen in all three systems,[8–11] implying some degree of defect association, why is the enthalpy of mixing so different for the ceria system? The reason may lie in the nature of the dominant clusters. On the yttria–ceria system, computations suggest that the vacancies are mainly located nearest neighbor to the yttrium ions at all concentrations.[12,13] Thus, the primary cation–vacancy interactions do not change much with concentration, although the size of defect associates do change, and CeO_2 is acting, in a sense, as an inert diluent to the defect clusters containing yttrium, oxygen, and oxygen vacancies. The heat of mixing is thus slightly positive, and no significant energetic stabilization is seen.

ZrO_2–$YO_{1.5}$ and HfO_2–$YO_{1.5}$ with strongly negative heats of mixing are in marked contrast to CeO_2–$YO_{1.5}$. There is substantial evidence, from both experiment[14–16] and computation,[17,18] that the oxygen vacancy in the Zr and Hf systems, rather than being nearest neighbor to Y, has a strong tendency to associate with Zr (or Hf), rendering the tetravalent cation 7-coordinate, as in the monoclinic phase of ZrO_2 and HfO_2. Thus, rather than acting as a diluent to the Y-vacancy associates in $YO_{1.5}$, the ZrO_2 or HfO_2 causes a change in the dominant defect association, with a resulting large energetic stabilization. The large negative heats of mixing reflect this change. There is a relationship between the radii of the host and dopant cations and the favored location of the oxygen vacancy.[18] The Coulomb interaction between the charged defects makes a nearest neighbor oxygen vacancy energetically favorable. However, the size mismatch between the host and the dopant cations makes a second nearest neighbor favorable. In the case of significantly oversized dopant cations (like in ZrO_2–$YO_{1.5}$ and HfO_2–$YO_{1.5}$), the size effect is greater than the effect of Coulomb interaction, so the oxygen vacancies locate at the second nearest neighbor sites of Y, which are the nearest neighbors of Zr or Hf. In CeO_2–$YO_{1.5}$, since the radii of Y^{3+} and Ce^{4+} are similar, the oxygen vacancies locate at the nearest neighbor sites of Y. The calorimetric data support this interpretation and provide quantitative energetic values to compare with computational results.

New calorimetric results

Sample preparation and characterization and oxide melt drop solution calorimetry followed protocols very similar to our previous work[3–6] and will not be described here. Measured enthalpies of drop solution (ΔH_{DS}) for the newly studied systems are given in Table 1. The enthalpy of drop solution is the observed enthalpy when a small pellet (typically 5 mg) of the desired composition is dropped from room

This journal is © The Royal Society of Chemistry 2006

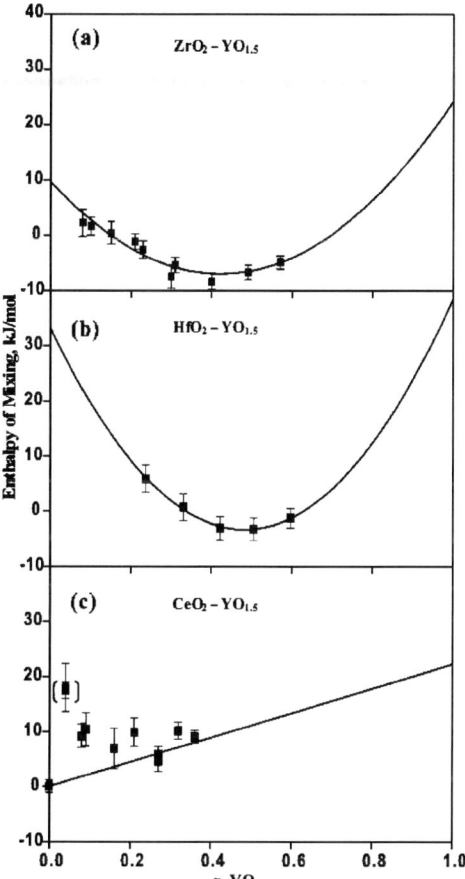

Fig. 1 Enthalpies of formation from the oxides stable at room temperature of cubic fluorite solid solutions in the zirconia–yttria, hafnia–yttria, and ceria–yttria systems.[7] Note that the point at lowest $YO_{1.5}$ content in ceria–yttria may be anomalous, as discussed in the original ref. 6 and the current text.

temperature into a molten salt solvent ($3Na_2O \cdot 4MoO_3$) in the hot calorimeter (700 °C). The enthalpies of formation with respect to the oxide end-members were calculated from the enthalpies of drop solution of the solid solutions (this work) and the binary oxides[2,4–6] using appropriate thermodynamic cycles. Enthalpies of formation of all samples are shown in Table 1.

The rare earth doped hafnia systems all show behavior similar to that in yttria doped zirconia and yttria doped hafnia, namely a strongly curved enthalpy of drop solution which implies strongly negative heats of mixing (see Fig. 2). Formation enthalpies become less exothermic with decreasing dopant radius. $Hf_{1-x}Sm_xO_{2-x/2}$ shows the most exothermic enthalpy of formation, while $Hf_{1-x}Yb_xO_{2-x/2}$ shows the most endothermic.

Enthalpies of formation from the component oxides of the $(1 - x)CeO_2–xREO_{1.5}$ solid solutions (RE = Y, Gd, La) are shown in Fig. 3. They are all slightly positive. Enthalpies of formation of the system $(1 - x)CeO_2–xZrO_2$ from fluorite–ceria and monoclinic zirconia are shown in Fig. 4. Large positive enthalpies of formation are seen. Enthalpies of formation of the $xCe_{0.8}Y_{0.2}O_{1.9}–(1 - x)Zr_{0.8}Y_{0.2}O_{1.9}$ solid solutions relative to the solid solution end-members $Ce_{0.8}Y_{0.2}O_{1.9}$ and $Ce_{0.8}Y_{0.2}O_{1.9}$

and relative to the oxide end-members, CeO_2, $m-ZrO_2$, and $C-YO_{1.5}$ are shown in Fig. 5. Both show small positive values.

Discussion

Hafnia-based systems

The enthalpies of formation of $Hf_{1-x}RE_xO_{2-x/2}$ (RE = Sm, Gd, Dy, Yb) can be described by a quadratic function, expressed as

$$\Delta H_{f,ox} = (1-x)\Delta H_{tr}(HfO_2) + x\Delta H_{tr}(REO_{1.5}) + \Delta H_{mix}$$

where ΔH_{mix} is the enthalpy of mixing relative to fluorite standard states and is approximated by

$$\Delta H_{mix} = \Omega x(1 - x).$$

with Ω a constant analogous to a regular solution parameter, $\Delta H_{tr}(HfO_2)$ is the enthalpy of transition of HfO_2 from the monoclinic to the cubic fluorite structure and $\Delta H_{tr}(REO_{1.5})$ is the enthalpy of transition of $REO_{1.5}$ from its structure at room temperature to the cubic fluorite structure. This formalism has been applied to the $ZrO_2-YO_{1.5}$[4] and $HfO_2-YO_{1.5}$[5] systems.

The interaction parameter Ω for all investigated $HfO_2-REO_{1.5}$ solid solutions is strongly negative which indicates a strong tendency to order in the solid solution. $HfO_2-SmO_{1.5}$ has the most negative interaction parameter (-269 ± 44 kJ mol^{-1}). The interaction parameter increases to less negative values for $HfO_2-GdO_{1.5}$ (-237 ± 10 kJ mol^{-1}) and $HfO_2-DyO_{1.5}$ (-196 ± 16 kJ mol^{-1}), while $HfO_2-YbO_{1.5}$ has the least negative Ω (-101 ± 84 kJ mol^{-1}). The interaction parameter $\Omega = -155 \pm 10$ kJ mol^{-1} for $HfO_2-YO_{1.5}$ by Lee et al.[4] is comparable to the values determined in this study and follows a trend of increasing Ω with decreasing dopant radius or increasing ionic potential (charge/radius) (see Fig. 6).

In all these cases, the trivalent dopant is larger than the tetravalent host, Hf. Thus the size difference between RE and Hf diminishes as the dopant radius decreases, implying less strain from size mismatch of the cations, and a smaller endothermic contribution to the enthalpy of mixing. Yet the interaction parameter becomes more endothermic; thus the above effect does not appear to dominate the energetics. Rather, the dominant effect may be the competition of Hf and RE for vacancy location, that is the difference in binding energy for a nearest neighbor versus a next nearest neighbor cluster. The trend in interaction parameters suggests that clusters with the vacancy located next nearest neighbor to the dopant (and nearest neighbor to Hf) are more stable with increasing dopant radius (decreasing ionic potential). In other words, the transfer of the vacancy from a REO_7 to an HfO_7 configuration becomes more energetically favorable as the RE radius increases. This is in accord with the calculations of Khan et al.,[18] who show that the difference in binding energies between nearest neighbor and next neatest neighbor dopant–vacancy clusters increases (favoring the latter) in the order Y (0.44 eV) , Gd (0.50 eV), Nd (0.56 eV), La (0.60 eV) in stabilized zirconia phases.

Further calorimetric studies are planned. We are analysing the data in these and other systems to obtain consistent values of the enthalpies of transformation of HfO_2 (m → c) and $REO_{1.5}$ (room temperature structure → fluorite). The effect of different sized trivalent dopants in zirconia will be compared to the results for hafnia presented here. The exact meaning of the defect energies from computations[12,13,17,18] must be defined by incorporating the computational results into chemically balanced reactions for doping and defect formation and clustering. Such reactions involve the differences in binding energies and/or lattice energies of different species and relating measured and computed energetic quantities is not always straightforward. The important point to realize is that the calorimetric measurements provide experimental values of quantities which can be related to computed values. This synergy

Table 1 Enthalpies of drop solution of oxides in molten $3Na_2O \cdot MoO_3$ at 700 °C

System	x	ΔH_{DS}/kJ mol^{-1}	$\Delta H_{f,ox}$/kJ mol^{-1}
$Sm_xHf_{1-x}O_{2-x/2}$ [a]	0.3	−4.9 ± 0.6	−4.1 ± 1.5
	0.33	−3.9 ± 0.9	−8.2 ± 1.2
	0.4	−6.0 ± 0.6	−12.6 ± 1.3
	0.62	−27.5 ± 1.3	−12.5 ± 0.5
$Gd_xHf_{1-x}O_{2-x/2}$ [a]	0.33	−7.2 ± 1.2	−4.1 ± 0.9
	0.4	−10.7 ± 1.0	−7.0 ± 1.0
	0.5	−19.6 ± 0.7	−7.6 ± 1.2
	0.62	−36.8 ± 1.7	−1.7 ± 0.9
$Dy_xHf_{1-x}O_{2-x/2}$ [a]	0.3	−4.0 ± 0.8	0.7 ± 1.1
	0.4	−5.6 ± 0.7	−5.3 ± 1.2
	0.5	−10.7 ± 1.0	−8.1 ± 0.6
	0.6	−20.2 ± 1.4	−6.2 ± 0.9
$Yb_xHf_{1-x}O_{2-x/2}$ [a]	0.25	−1.4 ± 0.6	4.1 ± 0.6
	0.3	−2.1 ± 0.3	1.3 ± 1.4
	0.4	−9.7 ± 0.6	2.0 ± 1.3
	0.5	−14.4 ± 0.7	−0.3 ± 1.3
	0.55	−21.0 ± 1.2	2.9 ± 0.5
$Gd_xCe_{1-x}O_{2-x/2}$ [b]	0.05	65.2 ± 1.6	2.8 ± 1.9
	0.1	50.1 ± 1.3	10.4 ± 1.6
	0.15	50.0 ± 1.1	3.0 ± 1.5
	0.2	44.3 ± 0.5	1.2 ± 1.0
	0.25	25.7 ± 1.3	2.3 ± 1.5
	0.3	25.5 ± 1.4	5.1 ± 1.6
	0.35	21.8 ± 0.7	1.3 ± 1.1
	0.39	12.8 ± 0.8	4.3 ± 1.1
$La_xCe_{1-x}O_{2-x/2}$ [b]	0.02	67.2 ± 1.5	2.5 ± 1.8
	0.05	58.5 ± 2.1	7.6 ± 2.3
	0.1	53.8 ± 0.9	2.9 ± 1.3
	0.15	42.9 ± 0.9	4.4 ± 1.3
	0.2	33.3 ± 0.8	4.5 ± 1.2
	0.25	25.4 ± 0.6	3.1 ± 1.1
	0.3	17.2 ± 0.4	2.6 ± 1.0
	0.35	7.1 ± 0.5	2.6 ± 1.0
$Ce_{0.8x}Zr_{0.8x}Y_{0.2}O_{1.9}$ [b]	0	3.5 ± 1.0	0.3 ± 1.2
	0.1	5.4 ± 1.0	2.8 ± 1.2
	0.2	9.6 ± 1.0	3.1 ± 1.2
	0.3	13.1 ± 1.0	4.1 ± 1.1
	0.4	16.1 ± 0.9	5.5 ± 1.1
	0.5	21.5 ± 1.1	4.6 ± 1.2
	0.6	27.6 ± 0.7	3.0 ± 0.9
	0.8	36.1 ± 0.9	3.3 ± 1.2
	0.9	40.2 ± 0.6	3.7 ± 1.0
	1	44.9 ± 1.0	3.5 ± 1.4
$Zr_xCe_{1-x}O_2$ [c]	0.05	64.5 ± 2.0	8.4 ± 2.3
	0.15	59.1 ± 2.0	8.1 ± 2.3
	0.25	48.8 ± 1.7	12.9 ± 1.9
	0.35	41.6 ± 1.8	14.6 ± 1.9
	0.45	35.5 ± 0.8	15.2 ± 1.1

[a] P. Simoncic and A. Navrotsky, *J. Mater. Res.*, 2006, submitted. [b] W. Chen and A. Navrotsky, in preparation. [c] T. Lee and A. Navrotsky, in preparation.

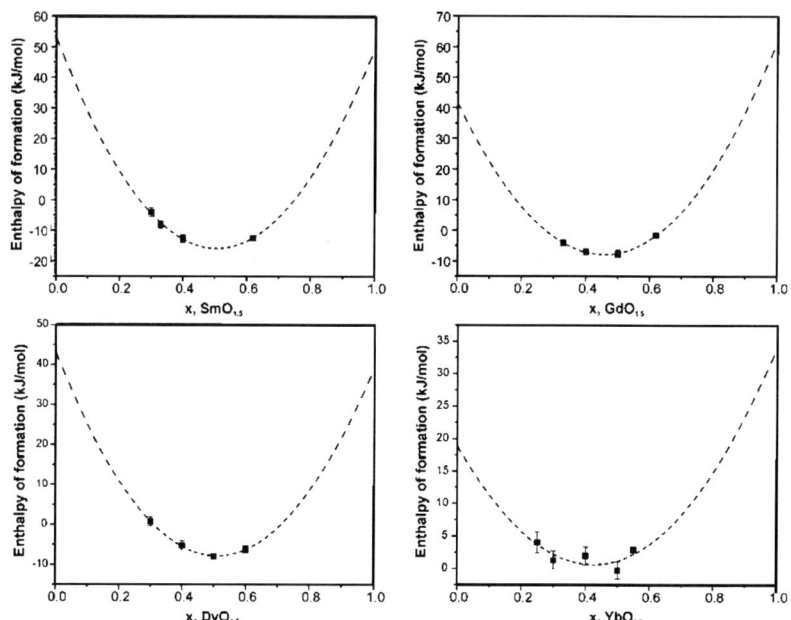

Fig. 2 Enthalpies of formation from binary oxides of cubic solid solutions in the HfO_2–$REO_{1.5}$ systems (Simoncic and Navrotsky, *J. Mater. Res.*, 2006, submitted).

between experiment and computation is worth exploring further by designing both experiment and theory to obtain the same parameters on the same systems.

The combination of calorimetry and structural studies suggests that the differences between "defect association", "short-range order", and "long-range order" are largely a matter of scale. As a defect cluster grows in extent, it becomes a short-range-ordered microdomain. As these domains grow, they become detectable as an ordered phase. The calorimetric data on these and many other systems suggest that the major decrease in energy and entropy occur at early stages of the process, with most of the thermodynamic stabilization occurring at the nanoscale.

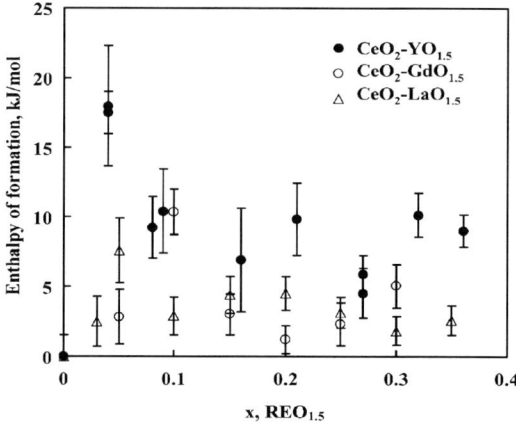

Fig. 3 Enthalpies of formation of the $(1-x)CeO_2$–$xREO_{1.5}$ solid solutions (RE = Y, Gd, La) from the oxide end-members, fluorite CeO_2, C-type $YO_{1.5}$, C-type $GdO_{1.5}$ and A-type $LaO_{1.5}$. From Chen and Navrotsky, in preparation.

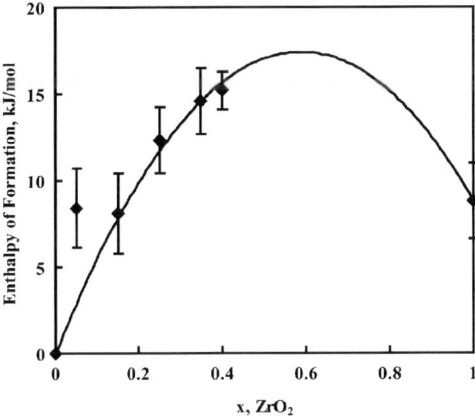

Fig. 4 Enthalpy of formation of ceria–zirconia cubic solid solution from fluorite ceria and monoclinic zirconia. The curve, a quadratic, is constrained to go through zero at $x = 0$ and 8.8 kJ mol^{-1} at $x = 1$, see text. Lee and Navrotsky, in preparation.

Ceria-based systems

The two additional RE-doped ceria systems studied (with Gd and La) behave very similarly to yttria–ceria. Enthalpies of formation are shown in Fig. 2. In all cases, small positive heats of formation are seen, with no evidence for strongly stabilizing interactions. This suggests that the ceria is acting as a diluent in all cases, and the vacancies remain predominantly nearest neighbor to the trivalent dopant. There is no tendency for Ce^{4+} to associate with a vacancy and become 7-coordinate. In all three cases, the most positive heats of formation occur near the Ce–rich end of the series.

The CeO_2–ZrO_2 system shows endothermic enthalpy of formation from oxides for composition in the range studied (up to 45% Zr substitution) (see Fig. 4). However, the roughly linear trend should not be extrapolated to $x = 1$, pure zirconia. Since no significant vacancy concentrations are created, one would expect no stabilization from ordering but a destabilization from size mismatch and a term arising from the enthalpy difference between cubic and monoclinic zirconia. Since CeO_2 possesses the fluorite structure but ZrO_2 is monoclinic, an appropriate expression to fit the enthalpy of formation in ceria–zirconia is

$$\Delta H_{f,ox} = x\Delta H_{tr}(ZrO_2) + \Omega x(1 - x)$$

where $\Delta H_{tr}(ZrO_2)$ is the enthalpy of transition of zirconia from monoclinic to cubic, 8.8 ± 2.2 kJ mol^{-1} (the sum of the enthalpies of the monoclinic–tetragonal and tetragonal cubic transitions), taken from a recent calorimetric study.[19] Constraining the enthalpies at both ends (zero at $x = 0$ and 8.8 at $x = 1$), and fitting the heat of formation data weighted by their errors gives an interaction parameter, $\Omega = 50.6$ kJ mol^{-1}. This large positive value is consistent with the large size difference between Zr and Ce.

Mixed Y-doped ceria zirconia system

The enthalpies of formation of the $xCe_{0.8}Y_{0.2}O_{1.9}$–$(1-x)Zr_{0.8}Y_{0.2}O_{1.9}$ samples with respect to the end-member solid solutions, $Ce_{0.8}Y_{0.2}O_{1.9}$ and $Zr_{0.8}Y_{0.2}O_{1.9}$, $(\Delta H_{f,ss})$ were calculated from the heat of drop solution data. As illustrated in Fig. 4a, $\Delta H_{f,ss}$ varies asymmetrically with the substitution level x. In the zirconia-dominated region $(0 \leq x < 0.4)$, $\Delta H_{f,ss}$ increases from 0 to 3.9 kJ mol^{-1}. In the ceria-dominated region

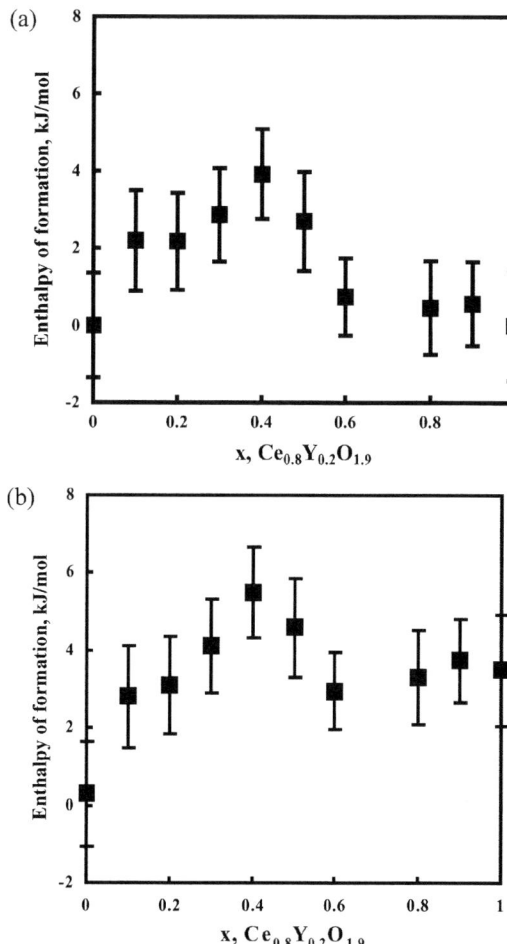

Fig. 5 Enthalpies of formation of the $x\mathrm{Ce_{0.8}Y_{0.2}O_{1.9}}$–$(1-x)\mathrm{Zr_{0.8}Y_{0.2}O_{1.9}}$ solid solutions at 25 °C: (a) relative to the solid solution end-members $\mathrm{Ce_{0.8}Y_{0.2}O_{1.9}}$ and $\mathrm{Ce_{0.8}Y_{0.2}O_{1.9}}$; (b) relative to the oxide end-members, $\mathrm{CeO_2}$, $\mathrm{m\text{-}ZrO_2}$, and $\mathrm{C\text{-}YO_{1.5}}$. Chen, Yokokawa, and Navrotsky, in preparation.

$(0.6 < x \leq 1)$, $\Delta H_{\mathrm{f,ss}}$ is almost constant at 0–0.7 kJ mol^{-1}. In the transitional region $(0.4 \leq x \leq 0.6)$, $\Delta H_{\mathrm{f,ss}}$ decreases from 3.9 to 0.7 kJ mol^{-1}.

The different energetic behavior of the $x\mathrm{Ce_{0.8}Y_{0.2}O_{1.9}}$–$(1-x)\mathrm{Zr_{0.8}Y_{0.2}O_{1.9}}$ system in the two regions (zirconia-dominated and ceria-dominated) recalls the different energetics of the $\mathrm{CeO_2}$–$\mathrm{YO_{1.5}}$[6] and $\mathrm{ZrO_2}$–$\mathrm{YO_{1.5}}$[4] systems. In the $\mathrm{CeO_2}$–$\mathrm{YO_{1.5}}$ system, the enthalpy of formation relative to the end-member oxides, fluorite $\mathrm{CeO_2}$ and C-type $\mathrm{YO_{1.5}}$ is slightly positive. Oxygen vacancies locate at the nearest neighbor sites of $\mathrm{Y^{3+}}$. Thus, $\mathrm{CeO_2}$ acts as a diluent to the Y-vacancy associates. However, in the $\mathrm{ZrO_2}$–$\mathrm{YO_{1.5}}$ system, the enthalpy of formation relative to the end-member oxides m-$\mathrm{ZrO_2}$ and C-type $\mathrm{YO_{1.5}}$ shows a strongly negative heat of mixing, related to the transfer of vacancies from Y to Zr environments.

In $x\mathrm{Ce_{0.8}Y_{0.2}O_{1.9}}$–$(1-x)\mathrm{Zr_{0.8}Y_{0.2}O_{1.9}}$ solid solutions, $\mathrm{Zr^{4+}}$ tends to trap oxygen vacancies created by $\mathrm{Y^{3+}}$ in both end-members to its nearest neighbor sites. This is the so-called scavenging effect, which was first proposed by Nowick and co-workers.[20] Similar phenomena have been reported in ceria systems doubly doped with $\mathrm{ScO_{1.5}}$–$\mathrm{YO_{1.5}}$ and with $\mathrm{ScO_{1.5}}$–$\mathrm{GdO_{1.5}}$.[21] The maximum energetic stabilization seen

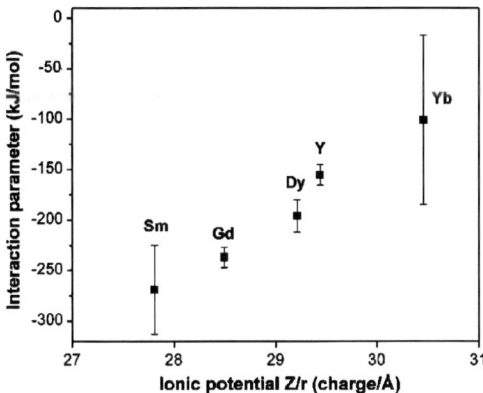

Fig. 6 Interaction parameter *versus* ionic potential for rare earth doped hafnates (Simoncic and Navrotsky, *J. Mater. Res.*, 2006, submitted).

near $x = 0.6$ may reflect this scavenging effect. At higher ceria concentrations, there may not be enough zirconium present to accommodate all the vacancies generated by the yttrium.

Based on the scavenging phenomena described above, a rough estimate on the composition range stabilized by this effect in the 20YDC–20YSZ system can be derived. Lee *et al.*[4] reported a deep minimum at about 40% $YO_{1.5}$ in the enthalpy of formation of the YSZ solid solution, suggesting extensive short-range order at this composition. We assume that such short-range order is the most stable configuration in both the YSZ and the 20YDC–20YSZ systems. The Zr^{4+} to oxygen vacancy ratio in 40YSZ is 3 : 1. This ratio is obtained in the 20YDC–20YSZ solid solution with 37.5% of 20YSZ. When less than 37.5% of 20YSZ is doped into the 20YDC solid solution, part of the oxygen vacancies in 20YDC may move from the nearest neighbor sites of Y^{3+} to the nearest neighbor sites of Zr^{4+}, changing the local defect configuration surrounding Zr^{4+} from that in 20YSZ to that in 40YSZ. The stabilization introduced by this process is comparable to the destabilization caused by size mismatch. Upon further addition of 20YSZ, the available oxygen vacancies are not sufficient for the most stable 40YSZ configuration. Therefore, the stabilizing effect decreases when the 20YSZ concentration is larger than 37.5%. This estimated stabilizing region is very close to that observed in the present study, 0–40% 20YSZ.

Conclusions

The measured energetics of the additional rare earth doped hafnia and ceria systems support the ideas developed in conjunction with yttria doped systems, namely that strong energetic stabilization occurs only when the oxygen vacancies become associated primarily with the tetravalent ions, Zr and Hf. The measured values of interaction parameters for several systems can provide direct comparison with defect models and calculations.

Acknowledgements

This work was supported by the U.S. Department of Energy (Grant DE-FG03ER46053).

References

1 A. Navrotsky, *Phys. Chem. Miner.*, 1997, **24**, 222.
2 K. B. Helean and A. Navrotsky, *J. Therm. Anal. Calorim.*, 2002, **69**, 751.

3　A. Navrotsky, *J. Chem. Thermodyn.*, 2001, **33**, 859.
4　T. A. Lee, A. Navrotsky and I. Molodetsky, *J. Mater. Res.*, 2003, **18**, 908.
5　T. A. Lee and A. Navrotsky, *J. Mater. Res.*, 2004, **19**, 1855.
6　W. Chen, T. A. Lee and A. Navrotsky, *J. Mater. Res.*, 2005, **20**, 144.
7　A. Navrotsky, *J. Mater. Chem.*, 2005, **15**, 1883.
8　T. H. Etsell and S. N. Flengas, *Chem. Rev.*, 1970, **70**, 339.
9　M. P. Trubelja and V. S. Stubican, *J. Am. Ceram. Soc.*, 1991, **74**, 2489.
10　J. D. Schieltz, J. W. Patterson and D. R. Wilder, *J. Electrochem. Soc.*, 1971, **118**, 1257.
11　H. Inaba and H. Tagawa, *Solid State Ionics*, 1996, **83**, 1.
12　L. Minervini, M. O. Zacate and R. W. Grimes, *Solid State Ionics*, 1999, **116**, 339.
13　M. S. Islam and G. Balducci, Computer Simulation Studies of Ceria-based Oxides, in *Catalysis by Ceria and Related Materials*, ed. A. Trovarelli, Imperial College Press, London, UK, 2002, p. 281.
14　J. Dexpert-Ghys, M. Feucher and P. Caro, *J. Solid State Chem.*, 1984, **54**, 179.
15　P. Li, J. W. Chen and J. E. Penner-Hahn, *Phys. Rev. B*, 1993, **48**, 10074.
16　R. M. Cillet, C. H. Deportes, G. Robert and G. Yittter, *Rev. Int. Hautes Temp. Refract.*, 1967, **4**, 269.
17　A. Bogicevic, C. Wolverton, G. M. Crosbie and E. B Stechel, *Phys. Rev. B*, 2001, **64**, 14106.
18　M. S. Khan, M. S. Islam and D. R. Bates, *J. Mater. Chem.*, 1998, **8**, 2299.
19　A. Navrotsky, L. Benoist and H. Lefebvre, *J. Am. Ceram. Soc.*, 2005, **88**, 2942.
20　A. S. Nowick, in *Diffusion in Crystilline Solids*, ed. G. E. Murch and A. S. Nowick, Academic Press, Orlando, FL, 1984.
21　R. Gerhardtanderson, F. Zamaninoor, A. S. Nowick, C. R. A. Catlow and A. N. Cormack, *Solid State Ionics*, 1983, **9**, 31.

Neutron diffraction and atomistic simulation studies of Mg doped apatite-type oxide ion conductors

E. Kendrick,[a] J. E. H. Sansom,[a] J. R. Tolchard,[b] M. S. Islam[c] and P. R. Slater*[a]

Received 15th February 2006, Accepted 30th March 2006
First published as an Advance Article on the web 19th July 2006
DOI: 10.1039/b602258h

In this paper, detailed studies of the effect of Mg doping in the apatite-type oxide ion conductor $La_{9.33}Si_6O_{26}$ are reported. Mg is confirmed as an ambi-site dopant, capable of substituting for both La and Si, depending on the starting composition. A large enhancement in the conductivity is observed for Si site substitution, with a reduction for substitution on the La site. Neutron powder diffraction studies show that in agreement with cation size expectations, an enlargement of the unit cell is observed on Mg substitution for Si, with a corresponding increase in the size of the tetrahedral sites. For Mg substitution on the La site, a contraction of the unit cell is observed, and the neutron diffraction results indicate that there is preferential occupancy of Mg on the La2 $(1/3, 2/3, \approx 0.5)$ site. Atomistic simulation studies show significant local structural changes affecting the oxide ion channels in both cases. Mg doping on the Si site leads to a local expansion of the channels, while doping on the La site results in a large displacement of the silicate O4 site, such that it encroaches the oxide ion channels. The observed differences in conductivities are discussed with respect to these observations.

Introduction

The identification and optimisation of oxide ion conductivity in solids has achieved considerable worldwide interest due to the technological applications of oxide ion conducting materials, e.g. as electrolytes for solid oxide fuel cells, O_2 separation membranes. Traditionally work has focused on materials with the fluorite (e.g. doped ZrO_2, CeO_2) or perovskite (e.g. doped $LaGaO_3$) structures.[1] Such materials are typically doped with aliovalent cations with the aim of introducing oxygen vacancies, and oxide ion conduction can then proceed via these vacant sites. The detailed research on fluorite and perovskite systems has essentially reached the state of the art that can be achieved in such structures. Therefore the search for new electrolytes has now begun to target alternative structures, and in this respect apatite-type materials have been attracting significant interest.

[a] Chemistry, UniS Materials Institute University of Surrey, Guildford, Surrey, UK GU2 7XH. E-mail: p.slater@surrey.ac.uk; Fax: +44 1483 686851; Tel: +44 1483 686847
[b] Department of Materials Technology, Norwegian University of Science and Technology, Sem Sælands vei 14, N-7491, Trondheim, Norway
[c] Department of Chemistry, University of Bath, Bath, UK BA2 7AY

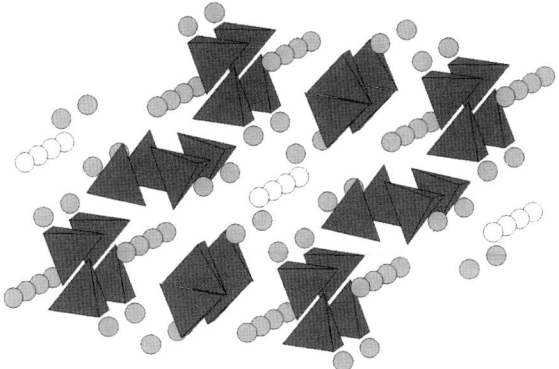

Fig. 1 The apatite structure, $Ln_{10-x}(SiO_4)_6O_{2+z}$ (dark spheres = Ln, tetrahedra = SiO_4, light spheres = O).

Research interest in this area has grown following the initial reports by Nakayama *et al.* of high oxide conductivities ($> 10^{-3}$ S cm^{-1} at 500 °C) in rare earth silicates, $Ln_{10-x}Si_6O_{26+y}$, with the apatite structure.[2–18] The structure of these materials is shown in Fig. 1. It consists of isolated SiO_4 units with the rare earth cations located in nine coordinate and seven coordinate channel sites. The remaining oxide ions occupy one-dimensional channels running through the structure, which are considered vital for the high oxide ion conduction in these materials.

Initial work in this area assumed that the conduction was mediated by oxide ion vacancies as for the fluorite, and perovskite systems. However, the weight of experimental and theoretical evidence indicates that these apatite-type materials conduct *via* an interstitial mechanism.[19–23] The presence of oxygen interstitials has been linked to the high observed conductivity in both oxygen excess samples, *e.g.* $La_9BaSi_6O_{26.5}$, and compositions showing vacancies on the La sublattice, *e.g.* $La_{9.33}Si_6O_{26}$. In contrast, samples which are fully stoichiometric, *e.g.* $La_8Ba_2Si_6O_{26}$, show poor conductivity due to a lack of interstitial oxygen site occupancy.[9,16–18] Computer modelling and neutron powder diffraction studies have contributed significantly to the further understanding of these materials. The modelling studies identified a new energetically favourable interstitial oxide ion position at the periphery of the oxide ion channels, in the vicinity of the SiO_4 groups.[19,20] Experimental studies have provided confirmation of this position, with Leon-Reina *et al.* showing occupation of a similar site in neutron diffraction studies of the $La_{9.33+x}(Si/Ge)_6O_{26+3x/2}$ systems, and Kharton *et al.* reporting penta-coordinated Fe^{3+} in Mössbauer studies of $La_{10}Si_5FeO_{26.5}$.[21–24] Moreover, the modelling results further suggested a complex sinusoidal conduction pathway down the channels. This pathway involves migration across the face of a silicate tetrahedral unit, coupled with considerable local relaxation of the silicate substructure (Fig. 2). Experimental observations of relevance to this proposed importance of the silicate substructure include the differing effects of rare earth *vs.* Si site doping outlined below, and recent solid state ^{29}Si NMR studies, which have shown a correlation between the silicon environment and the observed conductivity.[25]

In terms of doping studies, a wide range of dopants have been reported:[17,23–39]

Rare earth site: Mg, Ca, Sr, Ba, Bi, Mn, Co, Ni, Cu

Si site: Ge, Ti, P, B, Al, Ga, Mn, Co, Fe, Mg, Zn, Ni, Cu

Initial studies showed that doping with lower valent ions, *e.g.* B^{3+}, Al^{3+}, Ga^{3+}, on the Si site, while maintaining oxygen stoichiometry, *i.e.* $La_{9.33+x/3}Si_{6-x}M_xO_{26}$ (M = B, Al, Ga), leads to an enhancement in the conductivity. In contrast, similar doping with lower valent ions on the La site, *i.e.* $La_{9.33-x/3}M_{2x/3}Si_6O_{26}$ (M = Mg, Ca, Sr, Ba) produces no such enhancement but rather tends to lower the

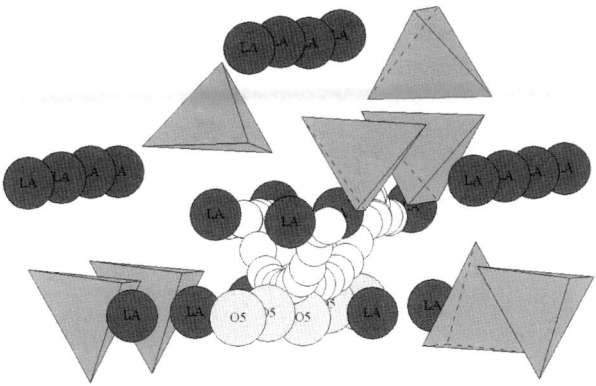

Fig. 2 Oxygen interstitial (white spheres) migration pathway through the apatite-type system, $La_{9.33}Si_6O_{26}$ (tetrahedra = SiO_4), looking down the oxide ion channels.[19]

conductivity.[31] Aside from noting the importance of the Si site in the conduction process, the origin of this different behaviour has still not been fully explained, and so requires further study. Most recently, it has been reported that Mg, Zn doping on the Si site, *i.e.* $La_{9.33 + 2x/3}Si_{6-x}M_xO_{26}$ (M = Mg, Zn) also produces a similar large enhancement in conductivity.[34,35,38] Transport measurements have indicated that the conductivity is indeed due to oxide ions.[34] Mg doping is particularly interesting as this dopant can also be substituted onto the La site, *i.e.* $La_{9.33-x/3}Mg_{2x/3}Si_6O_{26}$, in this case leading to a reduction in conductivity. Thus Mg is an ambi-site dopant in $La_{9.33}Si_6O_{26}$, substituting on either site depending on the starting composition. It therefore provides an ideal probe for the systematic investigation of the differing effects of La *versus* Si site substitution. With this in mind we report here a neutron powder diffraction structural study for 4 Mg doped samples, all having the same nominal oxygen content (O_{26}), to include 2 samples doped only on the Si site, $La_{9.5}Si_{5.75}Mg_{0.25}O_{26}$, $La_{9.67}Si_{5.5}Mg_{0.5}O_{26}$, one sample doped only on the La site, $La_9Mg_{0.5}Si_6O_{26}$, and one sample doped on both the La and Si site, $La_{9.33}Mg_{0.5}Si_{5.5}Mg_{0.5}O_{26}$. The neutron diffraction work is supported by atomistic simulation studies of the local perturbations caused by a Mg dopant on the La and Si sites, and the results are correlated with conductivity measurements.

Experimental

Four Mg doped samples were synthesised, $La_{9.5}Si_{5.75}Mg_{0.25}O_{26}$, $La_{9.67}Si_{5.5}Mg_{0.5}O_{26}$, $La_9Mg_{0.5}Si_6O_{26}$ and $La_{9.33}Mg_{0.5}Si_{5.5}Mg_{0.5}O_{26}$. The samples were prepared from the dried starting materials La_2O_3, SiO_2 and MgO. Stoichiometric mixtures were intimately ground and heated for 16 h at 1350 °C with a second firing at 1400–1500 °C for a further 16 h. Between firings the samples were reground to ensure homogeneous reaction, with phase purity being examined using a Seifert 3003TT powder X-ray diffractometer.

Conductivity measurements were performed using AC impedance spectroscopy (Solartron 1260 Impedance Analyser). Dense samples for measurement were prepared by pressing into 13 mm diameter pellets and firing for 2 h at 1550–1650 °C. Pt electrodes were then affixed to the pellets using Pt paste and the pellet fired again, at 850 °C for 30 min, to give good electrical contact between the sample and electrode.

Time of flight neutron diffraction data were recorded on diffractometer HRPD at the ISIS facility, Rutherford Appleton Laboratory. All structural refinements employed the GSAS suite of Rietveld refinement software.[40]

The computational studies of the local distortions introduced by a Mg dopant on the La and Si sites were performed using well established atomistic techniques, as embodied in the GULP code.[41] These methodologies are extensively documented elsewhere,[42] so only a brief summary is given here. The calculations are based on the Born model in which interactions between ions are evaluated in terms of long range coulombic terms and short-range interactions which take account of Pauli repulsion and van der Waals effects. To model these short-range interactions, the Buckingham potential was used:

$$V_{ij}(r_{ij}) = A \exp(-r_{ij}/\rho) - C/r_{ij}^6 \qquad (1)$$

where r is the interatomic separation and A, ρ and C are empirically derived parameters. A well established three-body term for the angle-dependant [SiO$_4$] tetrahedral unit was used.[19,20] Because charged defects will polarise other ions in the lattice, it is important to include a good description of ionic polarisability in the model. The shell model was used in this work, which describes each ion in terms of a separate core connected to a mass-less shell via a harmonic spring.

An important feature of these simulations is the treatment of lattice relaxation around the dopant Mg^{2+} ion. The Mott–Littleton approach was employed, in which the crystal lattice is partitioned into two regions: the immediate environment surrounding the defect is relaxed explicitly, whilst the remainder of the crystal, where the defect forces are relatively weak, is treated by more approximate quasi-continuum methods. In this way the local relaxation is effectively modelled and the crystal is not considered simply as a rigid lattice.

Interatomic potentials were transferred directly from our previous study of $La_{9.33}Si_6O_{26}$, which accurately reproduce the observed complex structure.[19,20] For this work, calculations were based upon a $1 \times 1 \times 3$ supercell (124 atoms) of the $La_{9.33}Si_6O_{26}$ crystallographic unit cell, with periodic boundary conditions applied to generate the full crystal. The La vacancies were placed on the La1/La2 sites in accordance with previous experimental and simulation work, and the lattice was then allowed to relax to find the lowest energy configuration. This structural model was then used for the local structure analysis on incorporation of Mg^{2+} dopant ions onto the Si and La sites.

Results

X-Ray powder diffraction showed that the four samples, $La_{9.5}Si_{5.75}Mg_{0.25}O_{26}$, $La_{9.67}Si_{5.5}Mg_{0.5}O_{26}$, $La_9Mg_{0.5}Si_6O_{26}$ and $La_{9.33}Mg_{0.5}Si_{5.5}Mg_{0.5}O_{26}$, were successfully prepared, confirming that Mg is an ambi-site dopant, which can be substituted onto both the La and the Si site. The bulk conductivity data (Table 1) shows that doping Mg onto the Si site results in an enhancement in conductivity of an order of magnitude while doping onto the La site reduces the conductivity by a similar amount, a result which accords well with previous reports.[34,35,38] The sample, $La_{9.33}Mg_{0.5}Si_{5.5}Mg_{0.5}O_{26}$, doped on both sites displays a conductivity similar to undoped $La_{9.33}Si_6O_{26}$ as the two effects counterbalance each other.

Table 1 Conductivity data for Mg doped $La_{9.33}Si_6O_{26}$ samples

Sample	σ/S cm^{-1} at 500 °C	E_a/eV
$La_{9.33}Si_6O_{26}$	1.1×10^{-4}	0.74
$La_{9.5}Si_{5.75}Mg_{0.25}O_{26}$	1.8×10^{-3}	0.68
$La_{9.67}Si_{5.5}Mg_{0.5}O_{26}$	3.0×10^{-3}	0.67
$La_9Mg_{0.5}Si_6O_{26}$	2.1×10^{-5}	0.98
$La_{9.33}Mg_{0.5}Si_{5.5}Mg_{0.5}O_{26}$	1.6×10^{-4}	0.80

Structure determination

Initial Rietveld analysis of the diffraction data focused on the determination of the appropriate symmetry for the Mg doped systems. In accordance with previous studies on apatite-type materials, two space groups were examined: $P6_3/m$, and $P6_3$, with the assignment being based on the residual R-factors. For all four samples an improved fit was obtained by lowering the symmetry to $P6_3$. There was no evidence for any anisotropic peak broadening observed previously for Co doping on the Si site.[32]

The Mg occupancies of the La, Si site were fixed at the value expected from the starting composition. In accordance with previous experimental observations, and predictions from modelling, the 6c La3 site was found to be fully occupied by La, with La non-stoichiometry preferring the [1/3,2/3,z] La1/La2 positions. The occupancies of these two sites were refined with the constraint that the final stoichiometry equalled the starting composition of the material. For the samples with Mg on the La site, $La_9Mg_{0.5}Si_6O_{26}$ and $La_{9.33}Mg_{0.5}Si_{5.5}Mg_{0.5}O_{26}$, the presence of both cation vacancies and dopant complicates the location of the dopant. Therefore, in the initial stages of the refinement, the Mg was distributed statistically over the La1/La2 sites. An inspection of the bond distances showed a significant difference in the La2–O distances compared to samples without Mg doping on the La site. The bond length changes were consistent with modelling predictions for Mg substituting on this site (see later) and therefore the results suggested that Mg preferentially occupies the La2 site in $La_9Mg_{0.5}Si_6O_{26}$ and $La_{9.33}Mg_{0.5}Si_{5.5}Mg_{0.5}O_{26}$. Therefore, in the final refinement of these two samples the Mg was assumed to all be located on the La2 site. The number of cation vacancies on the La1/La2 sites was then refined.

In terms of the oxygen positions, the occupancy of the channel oxygen (O5) site was initially varied. In all cases, the refined site occupancy dropped below 1.0 with a value lying between 0.84–0.90. Along with this reduced site occupancy, the anisotropic thermal displacement parameter U_{33} for O5 in all samples were very high. This suggested that there were significant displacements of the oxygens in the O5 site into interstitial positions. Therefore a number of refinements were performed to try to find the location of the remaining oxygens. Difference Fourier maps suggested the presence of small levels of unfitted scattering along z close to the centre of the channels, as well as at the periphery, close to the silicate groups, in positions similar to those reported by previous modelling and neutron diffraction studies. Attempts to refine occupancy in these positions were, however, unsuccessful, which may be related to the resultant small occupancies of these sites. Therefore in the final refinements, no interstitial sites were included and the occupancies of the O5 site were fixed at 1.0 in order to achieve electroneutrality.

Final refined structural data are given in Tables 2 and 3, with selected bond distances in Table 4. The neutron diffraction profiles are shown in Fig. 3–6.

To supplement the neutron diffraction studies, the local bond length changes accompanying the Mg doping were determined from atomistic simulations. These local bond length changes are shown in Table 5.

Discussion

The calculated cell parameter data (Tables 2 and 3) were all in agreement with those expected from the dopant size. For example, doping Mg onto the La site ($La_9Mg_{0.5}Si_6O_{26}$) results in a decrease in cell size compared to $La_{9.33}Si_6O_{26}$ due to the smaller size of Mg^{2+} compared to La^{3+}. Similarly substitution on the Si site ($La_{9.5}Si_{5.75}Mg_{0.25}O_{26}$, $La_{9.67}Si_{5.5}Mg_{0.5}O_{26}$) results in a significant increase in the cell size as Mg^{2+} is larger than Si^{4+}. The sample doped with Mg on both sites ($La_{9.33}Mg_{0.5}Si_{5.5}Mg_{0.5}O_{26}$) has cell parameters in between those of the single site doped samples $La_9Mg_{0.5}Si_6O_{26}$ and $La_{9.67}Si_{5.5}Mg_{0.5}O_{26}$.

Table 2 Atomic parameters for $La_{9.33}Mg_{0.5}Si_{5.5}Mg_{0.5}O_{26}$, $La_{9.5}Si_{5.75}Mg_{0.25}O_{26}$, $La_{9.67}Si_{5.5}Mg_{0.5}O_{26}$, and $La_9Mg_{0.5}Si_6O_{26}{}^a$

	$La_{9.33}Mg_{0.5}Si_{5.5}Mg_{0.5}O_{26}$	$La_{9.5}Si_{5.75}Mg_{0.25}O_{26}$	$La_{9.67}Si_{5.5}Mg_{0.5}O_{26}$	$La_9Mg_{0.5}Si_6O_{26}$
	La/Si-0.5Mg	Si-0.25Mg	Si-0.5Mg	La-0.5Mg
a, b	9.709 29(2)	9.723 32(2)	9.734 15(1)	9.690 54(2)
c	7.160 94(4)	7.193 29(2)	7.216 05(2)	7.136 92(2)
Rwp	0.0835	0.0731	0.065	0.0857
χ^2	1.662	1.145	1.07	1.244
La1, 2b, $(1/3, 2/3, z)$				
SOF	1.0	0.92(3)	0.93(4)	0.94(3)
Z	−0.007(2)	−0.003(1)	−0.003(1)	−0.005(2)
La2/Mg, 2b, $(1/3, 2/3, z)$				
SOF	0.6667/0.25	0.83(3)	0.91(4)	0.56(3)/0.25
Z	0.492(2)	0.496(1)	0.496(1)	0.494(2)
La3, 6c, (x, y, z)				
X	0.2412(1)	0.2410(1)	0.24159(9)	0.2438(1)
Y	0.0113(2)	0.0123(1)	0.0120(1)	0.0130(2)
Z	0.241(1)	0.243(1)	0.2428(9)	0.245(1)
Si /Mg, 6c, $(x, y, 0.25)$				
SOF	0.9167/0.0833	0.9583/0.0417	0.9167/0.0833	1
X	0.3746(3)	0.3720(2)	0.3731(2)	0.3733(2)
Y	0.3995(3)	0.3994(2)	0.4011(2)	0.4014(3)
O1, 6c, (x, y, z)				
X	0.4910(2)	0.4858(2)	0.4876(2)	0.4881(3)
Y	0.3296(3)	0.3249(2)	0.32570(2)	0.3262(1)
z	0.240(2)	0.245(1)	0.246(1)	0.243(2)
O2, 6c, (x, y, z)				
x	0.4701(2)	0.4721(2)	0.4716(1)	0.4703(2)
y	0.5987(3)	0.5964(2)	0.5977(2)	0.5948(2)
z	0.243(2)	0.247(1)	0.247(1)	0.249(2)
O3, 6c, (x, y, z)				
x	0.2579(7)	0.2581(6)	0.2569(6)	0.2551(8)
y	0.3534(6)	0.3516(6)	0.3515(6)	0.3507(8)
z	0.054(1)	0.058(1)	0.058(1)	0.060(2)
O4, 6c, (x, y, z)				
x	0.2543(6)	0.2527(5)	0.2540(5)	0.2539(7)
y	0.3312(9)	0.3369(7)	0.3392(7)	0.3309(10)
z	0.414(1)	0.419(1)	0.421(1)	0.418(2)
O5, 2a, $(0, 0, z)$				
z	0.248(5)	0.256(4)	0.250(3)	0.249(4)

^{a} All oxygen (O1–O5) and La3 sites are fully occupied.

All four Mg doped samples were refined with space group $P6_3$. The lowering of symmetry from $P6_3/m$ to $P6_3$ leads to a splitting of some of the crystallographic sites, and there was evidence for ordering of La vacancies/Mg in the split La sites. In particular Mg was shown to occupy the La2 (1/3, 2/3, ≈ 0.5) position in preference

Table 3 Anisotropic thermal displacement paramters for $La_{9.33}Mg_{0.5}Si_{5.5}Mg_{0.5}O_{26}$, $La_{9.5}Si_{5.75}Mg_{0.25}O_2$, $La_{9.67}Si_{5.5}Mg_{0.5}O_{26}$, and $La_9Mg_{0.5}Si_6O_{26}$. ($La_{9.33+2x/3-2y/3}Mg_ySi_{6-x}Mg_xO_{26}$)

	x, y	$x, y = 0.5$	$x = 0.25$	$x = 0.5$ $y = 0.5$		$x, y = 0.5$	$x = 0.25$	$x = 0.5$	$y = 0.5$
La1					**O1**				
U_{11}	2.7(3)	1.9(3)	1.7(3)	2.0(3)	U_{11}	2.4(1)	3.0(1)	2.62(8)	3.1(1)
U_{22}	2.7(3)	1.9(3)	1.7(3)	2.0(3)	U_{22}	5.4(2)	4.4(1)	4.50(8)	4.2(1)
U_{33}	2.9(6)	3.6(4)	2.5(4)	4.3(5)	U_{33}	4.0(2)	2.5(1)	2.07(9)	4.1(1)
U_{12}	1.4(1)	0.9(2)	0.8(2)	1.0(2)	U_{12}	2.6(1)	2.75(9)	2.48(7)	2.9(1)
U_{13}	0	0	0	0	U_{13}	0.7(4)	0.2(3)	0.2(3)	1.6(6)
U_{23}	0	0	0	0	U_{23}	−1.4(4)	−1.5(3)	−1.2(3)	−0.3(4)
La2					**O2**				
U_{11}	0.7(2)	1.0(3)	1.5(3)	0.6(3)	U_{11}	1.4(1)	1.48(8)	1.27(7)	1.60(9)
U_{22}	0.7(2)	1.0(3)	1.5(3)	0.6(3)	U_{22}	2.0(1)	1.75(8)	1.66(7)	1.56(9)
U_{33}	1.6(5)	1.3(4)	1.0(3)	1.0(4)	U_{33}	6.0(2)	3.6(1)	3.37(9)	6.3(2)
U_{12}	0.4(1)	0.5(2)	0.8(2)	0.3(2)	U_{12}	0.30(9)	0.31(7)	0.36(6)	0.35(8)
U_{13}	0	0	0	0	U_{13}	−0.5(4)	1.4(2)	1.0(3)	−0.2(5)
U_{23}	0	0	0	0	U_{23}	1.8(4)	−0.4(2)	−0.4(2)	−0.4(5)
La3					**O3**				
U_{11}	1.79(7)	1.77(6)	1.66(5)	1.65(6)	U_{11}	3.9(3)	4.1(2)	4.0(2)	3.7(3)
U_{22}	1.64(6)	1.56(5)	1.54(4)	1.52(6)	U_{22}	3.9(2)	3.4(2)	3.0(2)	5.2(3)
U_{33}	1.60(6)	1.55(5)	1.44(4)	1.49(5)	U_{33}	1.9(3)	2.0(2)	2.3(2)	2.1(3)
U_{12}	0.69(6)	0.72(5)	0.82(4)	0.82(6)	U_{12}	2.8(2)	2.8(2)	2.6(2)	2.9(3)
U_{13}	−0.3(2)	0.5(2)	0.4(2)	0.5(2)	U_{13}	−0.4(2)	−0.4(2)	−0.1(2)	−0.6(2)
U_{23}	−0.3(3)	0.1(2)	0.22(2)	0.3(3)	U_{23}	−0.9(2)	−1.2(1)	−0.6(1)	−1.0(2)
Si					**O4**				
U_{11}	2.0(1)	1.8(1)	1.54(8)	1.3(1)	U_{11}	0.7(2)	0.2(2)	0.7(2)	0.8(2)
U_{22}	2.3(1)	2.1(1)	1.57(8)	1.5(1)	U_{22}	12.1(5)	8.9(4)	7.6(3)	11.9(5)
U_{33}	3.2(2)	2.7(1)	1.7(1)	2.5(2)	U_{33}	2.2(3)	1.7(2)	1.4(2)	1.8(2)
U_{12}	0.9(1)	0.96(9)	0.53(7)	0.6(1)	U_{12}	2.1(3)	1.0(2)	1.3(2)	2.5(3)
U_{13}	2.0(3)	0.6(3)	0.9(2)	1.6(4)	U_{13}	0.5(2)	0.4(2)	0.3(2)	0.6(2)
U_{23}	−1.8(3)	−2.6(2)	−1.3(3)	0.4(4)	U_{23}	3.4(3)	1.9(2)	1.9(2)	3.5(3)
					O5				
					U_{11}	2.2(1)	2.09(9)	1.77(7)	1.97(10)
					U_{22}	2.2(1)	2.09(9)	1.77(7)	2.0(1)
					U_{33}	14.6(6)	15.2(6)	10.7(3)	11.3(4)
					U_{12}	1.08(6)	1.05(4)	0.89(3)	0.99(5)
					U_{13}	0	0	0	0
					U_{23}	0	0	0	0

to La1 (1/3, 2/3, ≈ 0.0). In addition, the former site generally contained more cation vacancies than the latter. For example, in the case of the sample $La_{9.33}Mg_{0.5}Si_{5.5}Mg_{0.5}O_{26}$, the occupancy of the La1 site refined to 1.0 indicating no vacancies on this site, with all the vacancies correspondingly on the La2 site.

A comparison of the bond distances for the Mg doped samples shows that on doping on the Si site an expansion of the tetrahedra is observed (Table 4) in agreement with the larger size of Mg^{2+} compared to Si^{4+}. Atomistic simulation studies indicate that Mg doping results in a large local distortion, with the predicted Mg–O bond distances ranging between 1.84 and 1.96 Å (Table 5). This corresponds to an average 17% expansion in the tetrahedra.

In the case of Mg doping on the La site, the observed data show that the La2–O bond distances change significantly, indicating that Mg substitutes preferentially on this site. From the data it can be seen that six of the bonds (La2–O1, O2) shorten while there is a lengthening of the remaining three (La2–O4). This has the effect of creating further distortion in the coordination of this site, such that the original distorted nine coordination environment approaches six coordination, with the remaining three bonds being very long. The neutron diffraction results are in

Table 4 Bond lengths for $La_{9.33}Mg_{0.5}Si_{5.5}Mg_{0.5}O_{26}$, $La_{9.5}Si_{5.75}Mg_{0.25}O_2$, $La_{9.67}Si_{5.5}Mg_{0.5}O_{26}$, and $La_9Mg_{0.5}Si_6O_{26}$

	$La_{9.33}Mg_{0.5}Si_{5.5}Mg_{0.5}O_{26}$	$La_{9.5}Si_{5.75}Mg_{0.25}O_{26}$	$La_{9.67}Si_{5.5}Mg_{0.5}O_{26}$	$La_9Mg_{0.5}Si_6O_{26}$
La1_O1 × 3	2.479(8)	2.498(7)	2.487(7)	2.479(10)
La1_O2 × 3	2.503(9)	2.534(7)	2.534(7)	2.538(11)
La1_O3 × 3	2.783(5)	2.805(5)	2.806(5)	2.801(6)
La2_O1 × 3	2.447(8)	2.486(6)	2.486(6)	2.460(9)
La2_O2 × 3	2.503(10)	2.534(6)	2.528(6)	2.496(11)
La2_O4 × 3	3.002(8)	2.948(6)	2.931(6)	2.995(9)
La3_O1	2.818(3)	2.769(2)	2.790(2)	2.761(2)
La3_O2	2.525(2)	2.522(2)	2.513(1)	2.519(2)
La3_O3	2.635(6)	2.641(5)	2.627(5)	2.622(7)
La3_O3	2.443(7)	2.461(5)	2.469(5)	2.436(7)
La3_O4	2.570(6)	2.576(5)	2.586(5)	2.584(6)
La3_O4	2.473(7)	2.477(5)	2.479(5)	2.453(7)
La3_O5	2.290(1)	2.288(1)	2.3007(8)	2.298(1)
Si1_O1	1.584(3)	1.598(3)	1.612(2)	1.601(3)
Si1_O2	1.677(4)	1.660(3)	1.658(2)	1.622(3)
Si1_O3	1.720(9)	1.687(7)	1.701(7)	1.680(10)
Si1_O4	1.551(8)	1.577(6)	1.588(6)	1.565(9)

complete agreement with atomistic simulation predictions, which show that the introduction of the small Mg^{2+} cation on the La site causes a large change in the bond distances (Table 5) resulting in an effective six coordination for the Mg.

Apart from the bond distance and cell parameter changes due to the dopant Mg, reported above, there are no other significant structural differences between the four samples. In all four samples, very high thermal displacement parameters are observed for the O5 channel oxygen site consistent with previous structural studies of apatite-type rare earth silicates/germinates.[9,21–23] These thermal displacement parameters are highly anisotropic, with U_{33} (corresponding to the direction of the oxygen channels) being particularly large. As noted in the structural determination section, the freely refined occupancy of this site drops below 1.0. This coupled with

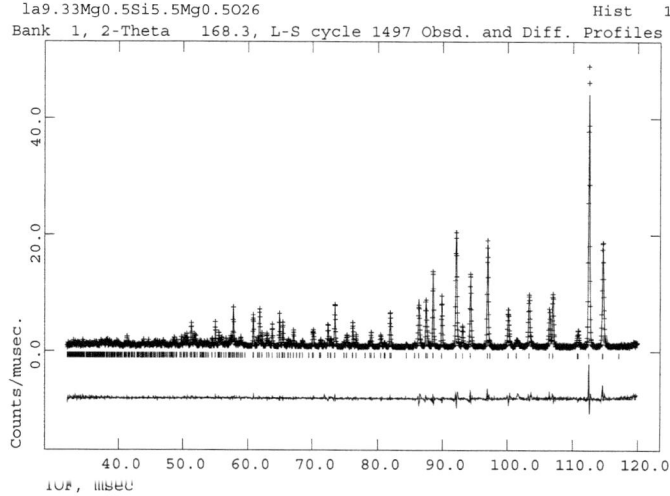

Fig. 3 Fitted neutron diffraction data for $La_{9.33}Mg_{0.5}Si_{5.5}Mg_{0.5}O_{26}$.

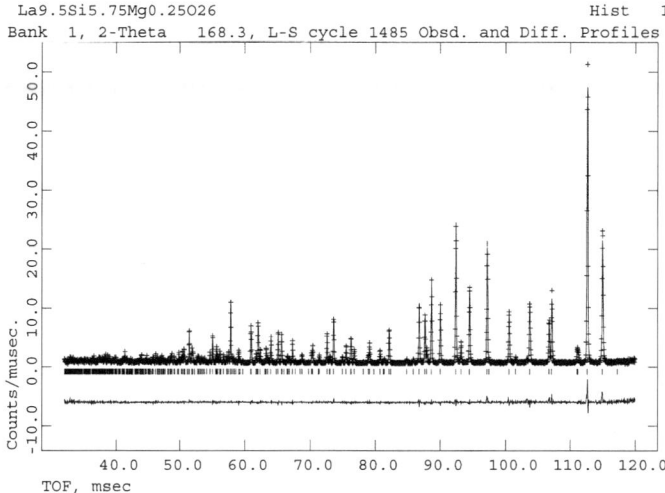

Fig. 4 Fitted neutron diffraction data for $La_{9.5}Si_{5.75}Mg_{0.25}O_{26}$.

the high thermal displacement parameters indicates significant displacement of oxygens into interstitial sites. Difference Fourier maps were investigated to locate the exact positions of the interstitial sites, and indicated small regions of unfitted scattering along the channels near to the channel centre, along with regions at the periphery of the channels in positions similar to the favourable interstitial site predicted by our modelling studies of $La_{9.33}Si_6O_{26}$,[19,20] and subsequently reported in neutron diffraction studies of $La_{9.33+x}(Si/Ge)_6O_{26+3x/2}$ by Leon-Reina *et al.*[21-23] However, attempts to refine occupancy in these sites were unsuccessful, which may be due to the displaced oxygens being distributed over not just one, but a range of interstitial sites, leading to low occupancies for each site.

In addition to the high thermal displacement parameters for the O5 site, the values for the silicate tetrahedra oxygens (O1–O4) are also high, which are also consistent with previous structural studies.[9,21-23] It should be noted that the presence of Mg on

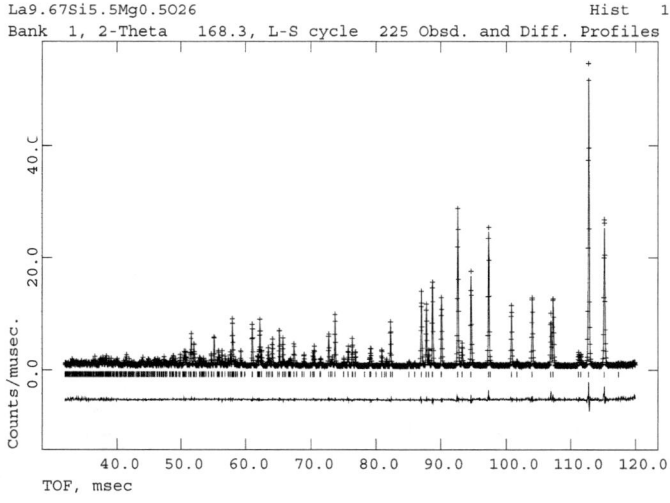

Fig. 5 Fitted neutron diffraction data for $La_{9.67}Si_{5.5}Mg_{0.5}O_{26}$.

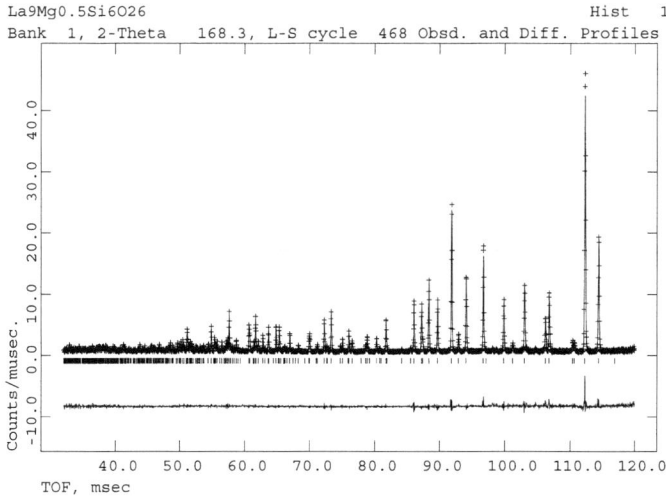

La9Mg0.5Si6O26   Hist   1
Bank   1, 2-Theta   168.3, L-S cycle   468 Obsd. and Diff. Profiles

Fig. 6   Fitted neutron diffraction data for La$_9$Mg$_{0.5}$Si$_6$O$_{26}$.

the Si site would be expected to produce an increase in these parameters due to the significantly longer Mg–O bond distances compared to Si–O, but nevertheless the dopant levels are relatively low and so the high thermal displacement parameters are unlikely to simply be correlated with the effect of the dopant. Moreover, high values are also observed for the sample without Si site doping. In terms of the other sites,

Table 5   Calculated local interatomic separations on incorporation of Mg on the La2 or Si site

	Undoped	Mg on La2	Mg on Si
La2–O1	2.44	2.30	2.48
La2–O1	2.44	2.30	2.48
La2–O1	2.44	2.30	2.63
La2–O2	2.54	2.27	2.34
La2–O2	2.54	2.27	2.46
La2–O–O2	2.54	2.27	2.63
La2–O4	2.92	3.37	2.45
La2–O4	2.92	3.37	3.31
La2–O4	2.92	3.38	3.35
La1–O1	2.43	2.51	2.46
La1–O1	2.43	2.51	2.48
La1–O1	2.43	2.51	2.58
La1–O2	2.65	2.70	2.42
La1–O2	2.65	2.70	2.66
La1–O2	2.65	2.70	2.78
La1–O3	2.67	2.55	2.44
La1–O3	2.67	2.55	2.77
La1–O3	2.67	2.55	2.88
Si–O1	1.59	1.59	1.84
Si–O2	1.64	1.63	1.92
Si–O3	1.64	1.67	1.96
Si–O4	1.63	1.62	1.91
O3–O5	3.47	3.56	3.86
O4–O5	3.26	2.95	3.72
O1–O5	3.95	4.13	3.86

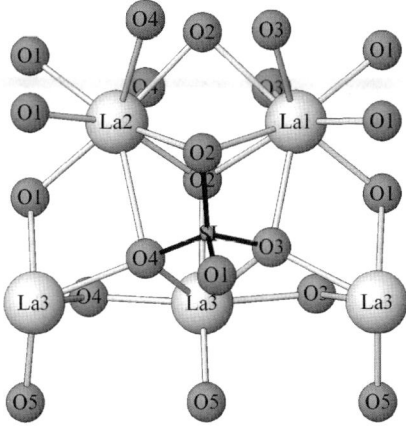

Fig. 7 Atomic configuration of La1 and La2 polyhedra above the O5 channel with no dopant from atomistic modelling calculations.

the thermal parameters for the La1 site are also relatively high, which may be related to the local distortions caused by neighbouring La vacancies and/or Mg in the adjacent La2 site.

Overall the results suggest significant structural distortions in these Mg doped apatite-type lanthanum silicates. However from the structural parameters derived from neutron diffraction, there is no clear indication why the samples doped with Mg on the La site have significantly lower conductivities than the samples doped on the Si site. This deficiency can be correlated with the fact that neutron diffraction only gives an average structure, and since the dopant levels are relatively low, the structural changes observed are masked within this average structure. In order to determine the origin of the conductivity differences, we therefore need to analyse the *local* structural changes, and in this respect the atomistic simulation studies provide detailed results. These atomistic simulation studies show the structural changes occurring in the vicinity of a dopant Mg^{2+} cation, whether it is on the La or the Si site.

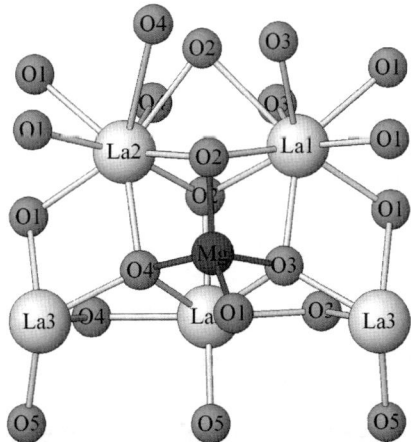

Fig. 8 Atomic configuration of La1 and La2 polyhedra above the O5 channel with Mg doping on the Si site from atomistic modelling calculations. Note the shortened distances between the La1/La2 sites and the O4, O3, O2 sites around the Mg dopant.

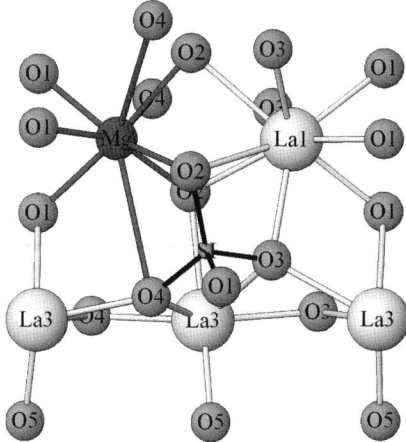

Fig. 9 Atomic configuration of La1 and La2 polyhedra above the O5 channel with Mg on the La2 site from atomistic modelling calculations. Note the large Mg–O4 distance causing the O4 site to encroach on the oxide ion channels.

The modelled local environments for the undoped, Mg on Si site, and Mg on La2 site systems are shown in Fig. 7–9. As outlined above, doping with Mg on the Si site leads to an expansion of the tetrahedra. In addition to affecting the tetrahedral site, such large changes also have a large affect on the surrounding sites (Table 5, Fig. 8). In particular, it is seen that the La1/La2–O distances to the oxygens of the MgO$_4$ tetrahedra all shorten to a similar length (≈ 2.4 Å). In addition, there is a change in the tilting of the tetrahedra (MgO$_4$) with respect to the O5 channels as well as an increase in the average distance between the O5 site and O1, O3, O4 sites (average distance = 3.82 Å (Mg on Si site) *versus* 3.56 Å (undoped)). The latter indicates that Mg doping for Si leads to an expansion of the oxide ion channels, while the former results in more of the O1–O3–O4 face showing to the channel. Since the predicted interstitial oxide ion sites are at the periphery of the channels near the silicate tetrahedra, and their conduction pathway involves migration across the face of a silicate tetrahedral unit, then both the above local structural changes might account for the observed increase in conductivity by enhancing the favourability of the interstitial oxide ion defects as well as lowering their migration energy. In addition, the ability of Mg to expand its coordination sphere might also enhance the conductivity for samples doped with Mg on the Si site.

In contrast to the beneficial effects of Mg doping on the Si site, the incorporation of a Mg^{2+} dopant cation on the La site leads to a significant reduction in conductivity. The simulation data shows that the incorporation of Mg on the La2 site results in a large change in the bond distances of this site (Table 5, Fig. 9). In particular six of the bonds (La2–O1, O2) shorten, while the remaining three (La2–O4) lengthen considerably. The overall effect is that the coordination of the site changes from a distorted nine coordinate environment to an effective six coordinate environment for Mg. The effect of the large (0.45 Å) expansion in the La2–O4 distance is to cause the O4 site to encroach on the oxide ion channels, as can be seen from the significant reduction in the O5–O4 distance (from 3.255 to 2.954 Å). In addition, this reduction in O5–O4 distance results in a significant change in the tilting of the adjacent SiO$_4$ tetrahedra, such that the oxide ion channels see less of the O1–O3–O4 face, as evidenced by the large differences in the O1–O5, O3–O5 and O4–O5 distances, which vary between 2.95 and 4.13 Å (*cf.* the much smaller variation for Mg doping on the Si site, 3.72–3.86 Å). Both these structural changes might be expected to adversely affect the generation of interstitial oxide ion defects as well as their migration, accounting for the reduction in conductivity.

Conclusions

Mg has been successfully doped into the apatite-type lanthanum silicate, $La_{9.33}$ Si_6O_{26}, substituting onto either the La or Si site depending on the starting stoichiometry. In agreement with cation size expectations, an enlargement of the unit cell is observed on Mg substitution for Si, with a contraction of the unit cell for Mg substitution on the La site. For the latter, the neutron diffraction results indicate that there is ordering of Mg on the La2 (1/3, 2/3, ≈ 0.5) site, with corresponding significant changes in the bond lengths for this site, consistent with predictions from atomistic simulation studies that the coordination effectively reduces from a distorted nine coordination to six coordination. These atomistic simulation studies show that both doping strategies influence the oxide ion channels, with Mg doping on the Si site leading to a local expansion of the channels, while doping on the La site results in a large displacement of the silicate O4 site, such that it encroaches the oxide ion channels. The observed local structural changes are therefore consistent with Mg doping on the Si site enhancing the conductivity, while La site doping reduces it.

Acknowledgements

The authors would like to thank EPSRC for funding this work. We would also like to express thanks to ISIS for neutron diffraction beam time and to Kevin Knight for his help in the neutron data collection.

References

1   J. B. Goodenough, *Annu. Rev. Mater. Res.*, 2003, **33**, 91.
2   S. Nakayama, H. Aono and Y. Sadaoka, *Chem. Lett.*, 1995, 431.
3   S. Nakayama and M. Sakamoto, *J. Eur. Ceram. Soc.*, 1998, **18**, 1413.
4   S. Nakayama, M. Sakamoto, M. Higuchi, K. Kodaira, M. Sato, S. Kakita, T. Suzuki and K. Itoh, *J. Eur. Ceram. Soc.*, 1999, **19**, 507.
5   M. Higuchi, H. Katase, K. Kodaira and S. Nakayama, *J. Cryst. Growth*, 2000, **218**, 218.
6   S. Nakayama, M. Sakamoto, M. Higuchi and K. Kodaira, *J. Mater. Sci. Lett.*, 2000, **19**, 91.
7   S. Nakayama and M. Niguchi, *J. Mater. Sci. Lett.*, 2001, **20**, 913.
8   S. Tao and J. T. S. Irvine, *Mater. Res. Bull.*, 2001, **36**, 1245.
9   J. E. H. Sansom, D. Richings and P. R. Slater, *Solid State Ionics*, 2001, **139**, 205.
10 H. Arikawa, H. Nishiguchi, T. Ishihara and Y. Takita, *Solid State Ionics*, 2000, **136–137**, 31.
11 J. E. H. Sansom, L. Hildebrandt and P. R. Slater, *Ionics*, 2002, **8**, 155.
12 S. Nakayama and M. Sakamoto, *J. Mater. Sci. Lett.*, 2001, **20**, 1627.
13 P. Berastegui, S. Hull, F. J. Garcia Garcia and J. Grins, *J. Solid State Chem.*, 2002, **168**, 294.
14 L. Leon-Reina, M. E. Martin-Sedeno, E. R. Losilla, A. Caberza, M. Martinez-Lara, S. Bruque, F. M. B. Marques, D. V. Sheptvakov and M. A. G. Aranda, *Chem. Mater.*, 2003, **15**, 2099.
15 E. J. Abram, C. A. Kirk, D. C. Sinclair and A. R. West, *Solid State Ionics*, 2005, **176**, 1941.
16 Y. Masubuchi, M. Higuchi, S. Kikkawa, K. Kodaira and S. Nakayama, *Solid State Ionics*, 2004, **175**, 357.
17 P. R. Slater and J. E. H. Sansom, *Solid State Phenomena*, 2003, **90–91**, 195.
18 Y. Masubuchi, M. Higuchi, H. Katase, T. Takeda, S. Kikkawa, K. Kodaira and S. Nakayama, *Solid State Ionics*, 2004, **166**, 213.
19 M. S. Islam, J. R. Tolchard and P. R. Slater, *Chem. Commun.*, 2003, 1486.
20 J. R. Tolchard, M. S. Islam and P. R. Slater, *J. Mater. Chem.*, 2003, **13**, 1956.
21 L. Leon-Reina, E. R. Losilla, M. Martinez-Lara, S. Bruque and M. A. G. Aranda, *J. Mater. Chem.*, 2004, **14**, 1142.
22 L. Leon-Reina, E. R. Losilla, M. Martinez-Lara, M. C. Martin-Sedeno, S. Bruque, P. Nunez, D. V. Sheptyakov and M. A. G. Aranda, *Chem. Mater.*, 2005, **17**, 596.
23 L. Leon-Reina, E. R. Losilla, M. Martinez-Lara, S. Bruque, A. Llobet, D. V. Sheptyakov and M. A. G. Aranda, *J. Mater. Chem.*, 2005, **15**, 2489.
24 V. V. Kharton, A. L. Shaula, M. V. Patrakeev, J. C. Waerenborgh, D. P. Rojas, N. P. Vyshatko, E. V. Tsipis, A. A. Yaremchenko and F. M. B. Marques, *J. Electrochem. Soc.*, 2004, **151**, A1236.

25   J. E. H. Sansom, J. R. Tolchard, D. Apperley, M. S. Islam and P. R. Slater, *J. Mater. Chem.*, 2006, **16**, 1410.
26   E. J. Abram, D. C. Sinclair and A. R. West, *J. Mater. Chem.*, 2001, **11**, 1978.
27   J. McFarlane, S. Barth, M. Swaffer, J. E. H. Sansom and P. R. Slater, *Ionics*, 2002, **8**, 149.
28   J. R. Tolchard, J. E. H. Sansom, P. R. Slater and M. S. Islam, *Solid State Ionics*, 2004, **167**, 17.
29   A. L. Shaula, V. V. Kharton, M. V. Patrakeev, J. C. Waerenborgh, D. P. Rojas, N. P. Vyshatko, E. V. Tsipis, A. A. Yaremchenko and F. M. B. Marques, *Mater. Res. Bull.*, 2004, **39**, 763.
30   A. A. Yaremchenko, A. L. Shaula, V. V. Kharton, J. C. Waerenborgh, D. P. Rojas, M. V. Patrakeev and F. M. B. Marques, *Solid State Ionics*, 2004, **171**, 51.
31   A. Najib, J. E. H. Sansom, J. R. Tolchard, M. S. Islam and P. R. Slater, *Dalton Trans.*, 2004, **19**, 3106.
32   J. R. Tolchard, J. E. H. Sansom, M. S. Islam and P. R. Slater, *Dalton Trans.*, 2005, **20**, 1273.
33   J. R. Tolchard, P. R. Slater and M. S. Islam, *J. Am. Chem. Soc.*, submitted.
34   H. Yoshioka and S. Tanase, *Solid State Ionics*, 2005, **176**, 2395.
35   H. Yoshioka, *Chem. Lett.*, 2004, **33**, 392.
36   J. E. H. Sansom and P. R. Slater, *Proc. 5th Euro SOFC forum*, 2002, **2**, 627.
37   J. E. H. Sansom, P. A. Sermon and P. R. Slater, *Solid State Ionics*, 2005, **176**, 1765.
38   J. E. H. Sansom, E. Kendrick, J. R. Tolchard, M. S. Islam and P. R. Slater, *J. Solid State Electrochem.*, 2006, **10**, 562.
39   P. R. Slater, J. E. H. Sansom, E. Kendrick, A. Scullard, C. Olsen, M. S. Islam and P. A. Sermon, *Proc. 7th Euro SOFC Forum*, in press.
40   A. C. Larson and R. B. Von Dreele, *Report. No LA-UR-86-748*, Los Alamos National Laboratory, 1987.
41   J. D. Gale, *J. Chem. Soc., Faraday Trans.*, 1997, **93**, 629.
42   C. R. A. Catlow, *Computer Modelling in Inorganic Crystallography*, Academic Press, San Diego, 1997.

A computational investigation of stoichiometric and calcium-deficient oxy- and hydroxy-apatites

Nora H. de Leeuw,[*ab] James R. Bowe[a] and Jeremy A. L. Rabone[a]

Received 13th February 2006, Accepted 4th May 2006
First published as an Advance Article on the web 15th August 2006
DOI: 10.1039/b602012g

Computer modelling techniques have been employed to qualitatively and quantitatively investigate the dehydration of hydroxyapatite to oxyapatite and the defect chemistry of calcium-deficient hydroxyapatite, where a number of vacancy formation reactions are considered. The dehydration of hydroxyapatite into oxyhydroxyapatite is calculated to be endothermic by $E = +83.2$ kJ mol^{-1} in agreement with experiment, where thermal treatment is necessary to drive this process. Calcium vacancies are preferentially charge-compensated by carbonate ions substituting for phosphate groups ($E = -5.3$ kJ mol^{-1}), whereas charge-compensating reactions involving PO$_4$ vacancies are highly endothermic ($E \geqslant 652$ kJ mol^{-1}). The exothermicity of the charge compensation of a Ca vacancy accompanied by a PO$_4$/CO$_3$ substitution agrees with their co-occurrence in natural bone tissue and tooth enamel. Our calculations of a range of defect structures predict (i) that calcium vacancies as well as substitutional sodium and potassium ions would occur together with carbonate impurities at phosphate sites, but that other charge compensations by replacement of the phosphate groups are unfavourable, and (ii) that the hydroxy ions in the channel are easily replaced by carbonate groups, but that the formation of water or oxygen defects in the channels is thermodynamically unfavourable. Calculated elastic constants are reported for the defect structures.

1. Introduction

Apatites Ca$_{10}$(PO$_4$)$_6$(F,Cl,OH)$_2$ are a complex and diverse class of materials, which are becoming increasingly important as candidates for use as bio-materials. In the geological environment, they are the most abundant phosphorus-bearing minerals, found extensively in igneous, metamorphic and sedimentary rocks.[1] Apatites are often used as geo- and thermo-chronometers, either by measuring fission tracks of thorium and uranium,[2–4] argon dating[5] or the retention of ^{4}He, the decay product of uranium and thorium.[6,7] In addition, the presence and isotopic composition of noble gases in apatites can give insight into mantle processes in the past.

More recently, apatites have gained additional prominence due to their biological rôle as one of the main constituents of mammalian bones and tooth enamel.[8] The presence of hydroxyapatite in bone and tooth enamel gives rise to its utility in a

[a] School of Crystallography, Birkbeck College, University of London, Malet Street, London, UK WC1E 7HX
[b] Department of Chemistry, University College London, 20 Gordon Street, London, UK WC1H 0AJ . E-mail: n.h.deleeuw@ucl.ac.uk

range of biomedical applications, for example, in the manufacture of artificial bone material and as a coating on surgical implants. Research has shown that polycrystalline calcium phosphate can directly bond to bone, which is regarded as the precursor to bone apatite formation *in vivo*.[9] The good biocompatibility of these calcium phosphates indicates their suitability in repair or replacement of damaged or diseased bone. To improve the strength of the relatively brittle hydroxyapatite, a metallic or ceramic implant can be coated with hydroxyapatite, which will encourage bonding with the living bone, aiding the acceptance of the implant material by the body.[10,11]

Apart from the regular F–Cl–OH solid solutions of the different end-members, both the natural bone mineral and geological apatites also contain a variety of other defects, including carbon-based impurities,[8] which affect the physical and chemical properties of the material.[11–15] Often carbon is taken up in the material as substitutional carbonate defects, either replacing OH groups, labelled A-type defects, or PO_4 groups in the structure, B-type defects.[16–19] However, in addition to these carbonate impurities, recent carbon isotope research has suggested that in high-temperature igneous apatites, carbon may also occur in the structure in its elemental state (of the order of 100 ppm carbon).[20] Another high-temperature effect in hydroxyapatite is the loss of volatile water from the structure, leading to dehydration of the hydroxy groups into oxygen ions, in the ultimate situation leading to the pure oxyapatite material,[21–24] sometimes also known as the mineral voelckerite, although its existence has long been disputed.[21] In addition, apatites are often found to be deficient in calcium compared to the stoichiometric material,[25–27] and this calcium deficiency is frequently accompanied by carbonate defects.[28,29]

The present study reports a detailed, atomic level computational study of the incorporation of various defects which may contribute to the calcium deficiency in the apatitic phase of bone, including OH vacancies, PO_4 vacancies and substitutional carbonate defects. Computational methods are well placed to calculate at the atomic level the defect formation energies and structures of the calcium-deficient apatite materials, which can be difficult to obtain experimentally, especially for natural bone material.[12,13] Our approach for these defect calculations is to employ primarily interatomic potential methods, as these methods combine the accuracy, required to investigate the defect structures at the atomic level and to calculate the energetics of the exchange reactions, with the computational efficiency required to sample the necessarily large numbers of different defect configurations to obtain the thermodynamically preferred structures for direct comparison. However, we have also employed extended plane wave methods based on the density functional theory to calculate the dissocation of hydroxy groups in the apatite structure into water and oxygen, as modelling bond dissociation necessitates the use of electronic structure methods. In addition, the comparison of the electronic structure methods with the interatomic potential methods gives an indication of the agreement between the two techniques.

2. Theoretical methods

The perfect and some of the defective apatite lattices were modelled using both density functional theory (DFT) methods and interatomic potential-based simulation techniques. The latter techniques are based on the Born model of solids,[30] which assumes that the ions in the crystal interact *via* long-range electrostatic forces and short-range forces, including both the repulsion and van der Waals attraction between neighbouring electron charge clouds, which are described by simple parameterised analytical functions. The electronic polarisability is included *via* the shell model of Dick and Overhauser,[31] where each polarisable ion, in our case the oxygen ion, is represented by a core and a massless shell, connected by a spring. The polarisability of the model ion is then determined by the spring constant and the charges of the core and shell. When necessary, angle-dependent forces are included

This journal is © The Royal Society of Chemistry 2006

to allow directionality of bonding as, for example, in the covalent phosphate and carbonate anions. We have employed the energy minimisation code METADISE[32] to calculate the various defects and impurities in the apatite lattice in a three-dimensional periodic boundary approach, at all times ensuring that sufficiently large supercells are employed to avoid finite size effects and interactions between the repeating images. METADISE has been used successfully for a range of simulations of complex oxide materials, including surface adsorption[33] and reconstruction simulations[34] and, more relevant to the present work, bulk defect calculations.[35–37]

When calculating the energies of the different locations of the various defects, we had to investigate a range of different configurations to ensure that as far as possible the lowest-energy configuration was obtained, where the system was minimised by a Newton Raphson energy minimisation method.

2.1 Potential model

In recent publications, we have empirically derived interatomic potential parameters for the simulation of a comprehensive range of apatite materials,[38–40] including fluor- and hydroxyapatite, which were also tested against electronic structure calculations, and shown to give excellent agreement with the experimental structures and properties.[38,39] As carbonate impurities in the apatite material were always an important consideration, even at the stage of first deriving the fluor- and hydroxy-apatite interatomic potential parameters, the apatite potential models were fitted to be compatible with existing calcium carbonate potential models,[41] which have been used successfully in a range of simulations of bulk and surface properties,[41,42] defect and sorption calculations.[43,44] In this study we have employed this combined calcium carbonate/hydroxyapatite potential model, which was first used to calculate the structures and energies of A-type and B-type carbonate defects in the hydroxyapatite lattice.[37] The potential parameters for the lattice oxygen ions, which were used instead of OH groups to charge compensate certain of the defect structures, were taken from the SiO_2 potential of Sanders et al.[45] These parameters are compatible with the potential parameters of the oxygen and hydrogen atoms in the hydroxy groups, as used together in previous simulations, e.g.,[46,47] and the parameters to describe its interactions with the rest of the apatite crystal have been derived and employed in a series of simulations of apatite thin films grown on silicate substrates.[48–50] The complete potential model used in this work is listed in Table 1.

2.2 DFT Calculations

For the electronic structure calculations of the water defects, we have employed the Vienna ab initio simulation program (VASP).[51–54] The spin-polarised calculations were performed within the generalized-gradient approximation (GGA), using the exchange–correlation potential developed by Perdew et al.[55] and the spin interpolation formula of Vosko, Wilk and Nusair.[56] Although spin-polarised calculations are not necessary for the closed-shell systems described in this work, we have used them here to be able to make direct comparison with calculations of open-shell defects, such as iron, which we intend to carry out in the future.

The interaction between the valence electrons and the core was described with the projected augmented wave method (PAW)[57] in the implementation of Kresse and Joubert.[58] The number of plane waves in VASP is controlled by the cut-off energy, which in our calculations was 500 eV with a convergence criterion of 10^{-5} eV. The k-point density used was a Monkhorst–Pack mesh[59] of $3 \times 3 \times 3$ points, where an odd number of divisions is used so that the mesh is centred on the Γ-point, as is required in VASP for structures with hexagonal symmetries. The optimisation of the atomic coordinates (and unit cell size/shape for the bulk materials) was performed via a conjugate gradients technique which utilises the total energy and the Hellmann–Feynman forces on the atoms (and stresses on the unit cell).

Table 1 Potential parameters (short-range cutoff 20 Å, short-range parameters between shells unless stated otherwise)

Ion	Charges/e		Core-shell interaction/eV Å^{-2}
	Core	Shell	
Ca	+2.0000		
P	+1.1800		
C	+1.1350		
Hydroxy hydrogen (Hh)	+0.4000		
Water hydrogen (Hw)	+0.4000		
Phosphate/carbonate oxygen (O)	+0.5870	−1.6320	507.4000
Hydroxy oxygen (Oh)	+0.9000	−2.3000	74.920 38
Interstitial oxygen (Od)	+0.848 19	−2.848 19	74.920 38
Water oxygen (Ow)	+1.2500	−2.0500	209.449 602

	Buckingham Potential parameters (cutoff 20 Å)		
Ion pair	A/eV	ρ/Å	C/eV Å^6
Ca–O	1550.0	0.297	0.0
Ca–Oh	1250.0	0.344	0.0
Ca–Od	2966.5	0.297	0.0
Ca–Ow	1186.6	0.297	0.0
C–Oh	709.4	0.344	0.0
H–O	312.0	0.250	0.0
H–Oh	312.0	0.250	0.0
H–Od	396.27	0.250	0.0
H–Ow	396.27	0.250	0.0
Hw–O	396.27	0.230	0.0
Hw–Oh	311.97	0.250	0.0
Hw–Od	396.27	0.250	0.0
Hw–Ow	396.27	0.250	10.0
O–O	16 372.0	0.213	3.47
Oh–Oh	22 764.0	0.149	6.97
Od–Od	22 764.3	0.149	27.88
O–Oh	22 764.0	0.149	4.92
O–Od	16 372.0	0.213	3.47
Oh–Od	22 764.0	0.149	13.94
O–Ow	12 533.6	0.213	12.09
Oh–Ow	22 764.0	0.149	6.97
Od–Ow	22 764.0	0.149	28.92

	Lennard-Jones Potential (cutoff 20 Å)	
Ion pair	A/eV Å^{12}	B/eV Å^6
Ow–Ow	39 344.98	42.15

	Morse Potential		
	DeV	α/Å^{-1}	r_0/Å
P–O$_{core}$	3.47	1.900	1.600
C–O$_{core}$	4.7100	3.8000	1.1800
Hh–Oh	7.0525	3.1749	0.9485
Hw–Ow	6.203 713	2.220 03	0.923 76

Table 1 (*continued*)

	Three-body potential	
	$k/eV\ rad^{-2}$	Θ_0
O_{core}–P–O_{core}	1.322 626	109.47
O_{core}–C–O_{core}	1.690 00	120.00
Hw–O_{shell}–Hw	4.199 78	108.69

	Four-body potential	
	$k/eV\ rad^{-2}$	Θ_0
C–O_{core}–O_{core}–O_{core}	0.1129	180.0

	Intramolecular Coulombic interaction (%)
H–Oh	0
H–Ow	50
Hw–Hw	50

3. Results and discussion

Biological apatite material has a hexagonal crystal structure with space group $P6_3/m$, $a = b = 9.36–9.64$ Å, $c = 6.78–6.90$ Å, depending on the halide present in the material, and $\alpha = \beta = 90°, \gamma = 120°.$[1,8] The halide ions in the apatite structure are all stacked above each other in hexagonal channels in the c-direction, where each F^- is coordinated to three surrounding Ca ions which lie in the same a/b plane, but each OH^- position is slightly offset from this symmetry position, either above or below it. Alternate rotation of the surrounding Ca positions (the Ca(II) positions[8]) in the a/b plane gives rise to the hexagonally shaped channels (Fig. 1). In the ideal structure, the OH groups are stacked in a regular head-to-tail column within the channels, although the direction of the OH groups in the columns may differ randomly between neighbouring channels.[60,61] However, as synthetic hydroxy apatite crystallizes in a fully-ordered monoclinic structure with space group $P2_1/b$, which is in effect a doubling of the hexagonal unit cell, but with all possible OH positions defined, we have used the monoclinic cell for our simulations, when necessary using (2 × 1) and

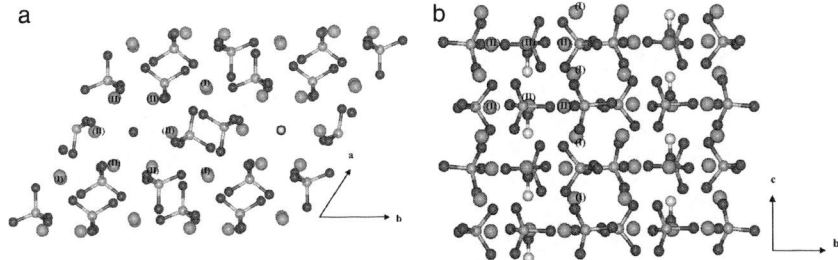

Fig. 1 Plan and side views of the hydroxy apatite structure, showing the OH^- groups in hexagonal channels surrounded by Ca ions in the Ca(II) positions, where the Ca(I) positions are within the rest of the lattice (O = black, Ca = dark gray, P = pale gray, H = white, PO_4 groups shown bonded, Ca(I) and Ca(II) indicated on some positions).

Table 2 Calculated defect formation energies from interatomic potential-based simulations (unless otherwise indicated)

Defect formation energies Defect	Reaction equation	E_{def}/kJ mol^{-1}
1	$2OH^x_{OH} \rightarrow O'_{OH} + (H_2O)^{\bullet}_{OH}$	$+6.3$ $+19.0$ (DFT)
2	$2OH^x_{OH} \rightarrow O'_{OH} + V^{\bullet}_{OH} + H_2O_{(g)}$	$+83.2$
3	$Ca^x_{Ca} + 2OH^x_{OH} \rightarrow V''_{Ca} + 2V^{\bullet}_{OH} + Ca^{2+}_{(aq)} + 2OH^-_{(aq)}$	$+239.9$
4	$Ca^x_{Ca} + 2OH^x_{OH} + H_2O \rightarrow V''_{Ca} + 2V^{\bullet}_{OH} + (H_2O)^x_i + Ca^{2+}_{(aq)} + 2OH^-_{(aq)}$	$+570.7$
5	$Ca^x_{Ca} + (PO_4)^x_{PO_4} + OH^x_{OH} + OH^-_{(aq)} \rightarrow V''_{Ca} + V^{\bullet}_{OH} + (O)^{\bullet}_{PO_4} + Ca^{2+}_{(aq)} + (PO_4)^{3-}_{4(aq)} + H_2O_{(aq)}$	$+706.0$
6	$Ca^x_{Ca} + (PO_4)^x_{PO_4} + OH^x_{OH} + OH^-_{(aq)} \rightarrow V''_{Ca} + (H_2O)^{\bullet}_{OH} + (O)^{\bullet}_{PO_4} + Ca^{2+}_{(aq)} + (PO_4)^{3-}_{4(aq)}$	$+651.9$
7	$Ca^x_{Ca} + (PO_4)^x_{PO_4} + OH^x_{OH} + OH^-_{(aq)} \rightarrow V''_{Ca} + V^{\bullet\bullet\bullet}_{PO_4} + (O)'_{OH} + Ca^{2+}_{(aq)} + (PO_4)^{3-}_{4(aq)} + H_2O_{(aq)}$	$+824.1$
8	$Ca^x_{Ca} + 2(PO_4)^x_{PO_4} + 2CO^{2-}_{3(aq)} \rightarrow V''_{Ca} + 2(Co_3)^{\bullet}_{PO_4} + Ca^{2+}_{(aq)} + 2PO^{3-}_{4(aq)}$	-5.3
9	$(PO_4)^x_{PO_4} + Ca^x_{Ca} + CO^{2-}_{3(aq)} + K^+_{(aq)} \rightarrow (CO_3)^{\bullet}_{PO_4} + K'_{Ca} + PO^{3-}_{4(aq)} + Ca^{2+}_{(aq)}$	-5.6
10	$(PO_4)^x_{PO_4} + Ca^x_{Ca} + CO^{2-}_{3(aq)} + Na^+_{(aq)} \rightarrow (CO_3)^{\bullet}_{PO_4} + Na'_{Ca} + PO^{3-}_{4(aq)} + Ca^{2+}_{(aq)}$	-71.1
11	$(PO_4)^x_{PO_4} + CO^{2-}_{3(aq)} + OH^-_{(aq)} \rightarrow (CO_3)^{\bullet}_{PO_4} + (OH)'_i + PO^{3-}_{4(aq)}$	-0.7
12	$2OH^-_{(aq)} + CO^{2-}_{3(aq)} + \rightarrow 2V^{\bullet}_{OH} + (CO_3)''_i + 2OH^-_{(aq)}$	-404.0
13	$(PO_4)^x_{PO_4} + (OH)^x_{OH} + 2CO^{2-}_{3(aq)} \rightarrow (CO_3)^{\bullet}_{PO_4} + (CO_3)'_{OH} + PO^{3-}_{4(aq)} + OH^-_{(aq)}$	-518.7

(2×2) supercells. Once the simulation cell is created, it is essentially a $P1$ structure, as no symmetry constraints are used in the simulations and all species in the simulation cell are free to move independently from one another. In addition, the simulation cell itself is allowed to contract/expand and deform anisotropically.

We first discuss the formation of oxyapatite from hydroxyapatite, followed by various calcium-deficient hydroxyapatite structures, where the calcium vacancies are charge-compensated by a range of further defects, namely hydroxy group vacancies, phosphate vacancies and substitutional carbonate groups. These defect calculations were carried out using interatomic potential simulation using the potential model in Table 1, with DFT calculations carried out for comparison. The defect formation energies for all defect structures are listed in Table 2.

3.1 Oxy(hydroxy)apatites

Pure oxyapatite $Ca_{10}(PO_4)_6O$ can be produced from synthetic hydroxyapatite by careful sintering at high temperature,[22] although usually it is insufficiently dehydrated and more often found as oxy(hydroxy)apatite,[23,24] for example in plasma sprayed hydroxyapatite coatings for use as biomaterials.[62,63] Although not widespread in calcium phosphate apatites, oxyapatite structures are more common in substituted compositions, for example lanthanum and vanadium-containing compounds[64] and others, where they may be important as fast oxide ion conductors.[65,66]

3.1.1 Replacement of $2OH^-$ by O^{2-} and water.
We first investigated the replacement of OH groups in the hydroxyapatite structure by an oxygen ion and water

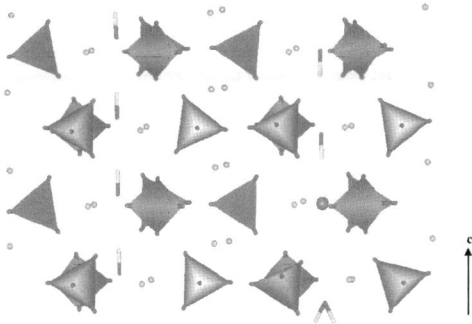

Fig. 2 Lowest-energy structure of the $(1 \times 1 \times 2)$ simulation cell of hydroxyapatite with two hydroxy groups replaced by one oxide ion and one water molecule (O = black, Ca = gray, H = white, PO_4 groups shown as tetrahedra).

molecule, in effect the products of the dissociation of one hydroxy ion and the addition of the resulting proton to the second hydroxy ion. This reaction is a necessary intermediate step in the dehydration process, before removal of the volatile water from the channel.[23,24]

In order to obtain the energetically preferred configuration of the oxygen ion/water molecule in the channel, we have simulated a number of different starting configurations, varying both the positions of the replaced hydroxy groups and the orientation of the substitutional species. As could perhaps be expected, the lowest energy structure for this defect, as shown in Fig. 2, was obtained when two adjacent hydroxy groups were replaced by the oxygen ion plus water molecule. Both the single oxygen ion and the oxygen of the water molecule are approximately located at oxygen lattice positions of the replaced hydroxy groups. The defect oxygen ion in the channel is strongly coordinated to calcium ions at a closest Ca–O distance of 2.18 Å (*cf.* Ca–O_{OH} of 2.47 Å in the perfect crystal). Conversely, the Ca–O_{water} distance for the defect water molecule has increased to 2.62 Å. There is a strong hydrogen-bonded interaction between the hydrogen atom of one of the remaining hydroxy groups to the defect oxygen ion at $O \cdots H_{OH}$ of 2.12 Å, whereas the O–O_{water} distance between the two defect species has lengthened to 4.41 Å (*cf.* O_{OH}–O_{OH} distance of 3.41 Å in the perfect crystal), partly as a result of the mutual attraction between the defect oxygen and the hydrogen of its neighbouring hydroxy group and partly as a result of the attraction of the water molecule's hydrogen atoms to the oxygen of its neighbouring hydroxy group (O_{OH}–H_{water} = 1.80 Å). Finally, all these shifts along the channel to increase O–H and Ca–O interactions lead to an increase in the O_{OH}–O_{OH} distance of 0.26 Å with respect to the OH groups in the defect-free channel.

The defect formation energy is calculated according to the following reaction equation, in Kröger–Vink notation:[67,68]

$$2OH^x_{OH} \rightarrow O'_{OH} + (H_2O)^\cdot_{OH} \qquad (1)$$

where OH^x_{OH} is an OH$^-$ group at a lattice hydroxy position, O'_{OH} is an oxygen atom at an OH position with a charge of -1 with respect to the lattice and $(H_2O)^\cdot_{OH}$ is a water molecule at an OH position with a charge of $+1$ with respect to the lattice. However, calculation of this defect involves the dissociation of an OH group into its constituent ions and the subsequent recombination of the second OH group with the proton, which is not straightforward to calculate using potential-based methods. In order to calculate the defect formation energy, we also need to know the energy of dissociation of the water molecule, but this reaction requires the second electron affinity of oxygen, which is material-dependent. However, this energy can be obtained from an energy cycle using experimental enthalpies for some of the reactions, as shown below and fully described in ref. 69 and 70.

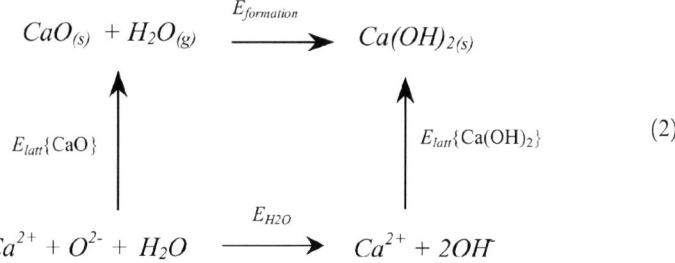

$$(2)$$

where $E_{\text{formation}}$ is the experimental heat of formation of calcium hydroxide from CaO and water and $E_{\text{latt}}\{Ca(OH)_2\}$ and $E_{\text{latt}}\{CaO\}$ are the calculated lattice energies of $Ca(OH)_2$ and CaO. From the experimental enthalpies and calculated lattice energies, listed in Table 3, we can then obtain the energy of reacting two OH groups to form one water molecule and one oxygen ion, where we considered that it was most appropriate to use a calcium (hydr)oxide system for our calculation of this material-dependent property, as previously employed successfully in a study of the dissociative adsorption of water at calcium oxide surfaces.[69] E_{H2O} was then calculated at -661.4 kJ mol^{-1}, which gave a defect energy for the process in eqn (1) of $E_{\text{def}} = +6.3$ kJ mol^{-1}.

However, because we suspected that the interstitial water molecule might under certain conditions dissociate to recombine into the two hydroxy groups of the ideal structure, we have also modelled this defect by electronic structure calculations, having used the low energy structures identified by the interatomic potential simulations as input for the DFT calculations. In addition, the agreement or otherwise between the two methods would give us some measure of the accuracy of our methods. The lowest-energy structure of this defect in a $(1 \times 1 \times 1)$ simulation cell had the water molecule located in the channel with both hydrogen atoms pointing towards the oxygen atom and the inter-hydrogen mirror plane along the hexagonal c-axis, while the other structure we considered had one of the hydrogen atoms pointing towards the oxygen atom with the OH bond along the hexagonal c-axis, but with the other hydrogen pointing away from the c-axis at an angle of 75°. A schematic representation of both starting configurations is shown in Fig. 3. Upon DFT geometry optimisation, the water and oxygen atom in the first structure remain as they are, whereas the second structure eventually relaxes back to hydroxyapatite, with one of the hydrogen atoms rejoining the oxygen atom to reform into a hydroxy group. These calculations indicate that there is no significant energy barrier for the dissociation of the water molecule and recombination into OH groups in the second structure, but that the first structure is either a stable structure or that there is a significant activation energy. Fig. 4 shows a charge density difference plot of the oxyapatite with water in the channel compared to the same structure without the water molecule. As we can see, there is an increase in charge density along the H···O axis between the water molecule and the oxygen atom (with a sharp decrease in the immediate surrounding of the oxygen atom), and to a lesser extent between the

Table 3 Lattice energies and enthalpies of formation, used for the calculation of the material-dependent water dissociation reaction

	Enthalpy of formation/kJ mol^{-1}	Lattice Energy/kJ mol^{-1}
$CaO_{(s)}$	-635.1	-3468.5
$Ca(OH)_{2(s)}$	-986.1	-2916.3
$H_2O_{(g)}$	-241.8	—

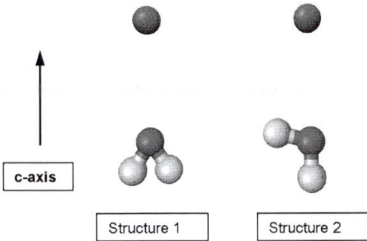

Fig. 3 Schematic representation of the two defect starting structures for the DFT calculations, showing substitutional oxygen atoms and water molecules only (O = black, H = white).

second hydrogen atom and an oxygen of the lattice (shown on the right hand side of the figure).

The DFT defect formation energy for the first structure was calculated to be $E_{def} = +19.0$ kJ mol^{-1} in quite acceptable agreement with the calculated defect energy from the interatomic potential simulations above ($E_{def} = +6.3$ kJ mol^{-1}), especially taking into account the approximations that are required in both methods and which were decided upon independently. The second step in the dehydration process is the removal of the water molecule from the channel, which process we calculated according to eqn (3) at $E = +76.9$ kJ mol^{-1}, and the complete two-step process of 25% dehydration of the hydroxyapatite lattice (into Ca$_{20}$(PO$_4$)$_{12}$O(OH)$_3$) is thus endothermic by $E = +83.2$ kJ mol^{-1}.

$$O'_{OH} + (H_2O)\overset{..}{_{OH}} \rightarrow O'_{OH} + V\overset{..}{_{OH}} + H_2O_{(g)} \qquad (3)$$

However, although this process is endothermic, it is not an insurmountable energy barrier to overcome, in agreement with experimental studies which find that this process does occur in hydroxyapatite upon heating up to a 78% dehydration.[22–24,63]

3.1.2 The structure of oxyapatite Ca$_{10}$(PO$_4$)$_6$O. Having established that the interatomic potential gives acceptable agreement with DFT calculations for the replacement of two channel hydroxy groups by an oxide ion and water molecule, we next investigated a series of solid solutions between hydroxyapatite and oxyapatite, where each time two hydroxy groups in a channel were replaced by one oxygen ion. We have used a simulation supercell, incorporating two hydroxy channels and containing a total of eight hydroxy groups, leading to 25, 50 and 75% substitutions

Fig. 4 Charge density difference plot of the hydroxyapatite channel with an oxygen atom at one OH position and a water molecule in a neighbouring OH position, where blue is increased and red is decreased charge density in the defect structure compared to the oxyhydroxyapatite without the water molecule (O = red, H = silver).

of the hydroxy groups. For each solid solution, we have again calculated a number of possible locations for the oxygen atom, namely at a hydroxy position, at the symmetry position or midway between two symmetry positions, whereas for the 50% substituted structure, we also had to calculate a number of configurational combinations of the two substitutional oxygen ions within the lattice. For the pure oxyapatite, we found that the channel oxygen ion was preferentially located at a symmetry position (*i.e.* where the fluoride ion would reside in fluorapatite, in the same a/b plane as one of the calcium(II) triangles), which maximises the $Ca–O_{channel}$ interactions, as already observed in section 3.1.1. This position agrees well with a recently published structure of oxyapatite determined by Henning *et al.*[71]

As the replacement of two hydroxy groups for one channel oxygen ion leads to one hydroxy vacancy in the channel for each substituted oxygen, there are a number of ways that these vacancies can be distributed in the channels, for example randomly distributed within the channel or in strict O–vacancy–O–vacancy ordering. In addition, there may be longer-range effects between the channels, leading to ordering in the a/b planes. We first used a hexagonal ($1 \times 1 \times 2$) oxyapatite supercell, containing two oxygen ions and four symmetry positions per channel, to investigate the ordering of the vacancies within the channel. To this end, we kept one oxygen ion fixed at one of the symmetry positions while systematically varying the position within the channel of the second oxygen atom, calculating the lattice energy for each configuration, where in this instance the lattice was kept fixed. The different positions of the second oxygen ion within the lattice are shown in Fig. 5(a), where the fixed oxygen is shown in dark blue at positions 0 and 16, and the second oxygen ion is located sequentially at positions 1–15. The lattice energy for the system with the second oxygen at each position is plotted in Fig. 5(b), from which it is clear that position 8 is a minimum energy position. Position 8 is also one of the four available symmetry positions, namely the one exactly midway between the fixed oxygen ion at positions 0 and 16. The other two symmetry positions are 4 and 12, which are equivalent positions with respect to the fixed oxygen. However, although the lattice energy surface between positions 4 and 12 is much flatter than the sites closer to the fixed oxygen (sites 1–3 and 13–15 in Fig. 5), their lattice energies are still about 320 kJ mol^{-1} per $Ca_{10}(PO_4)O$ unit higher than for position 8, *i.e.* much less favourable sites for the second oxygen ion. It thus appears that again the symmetry positions are

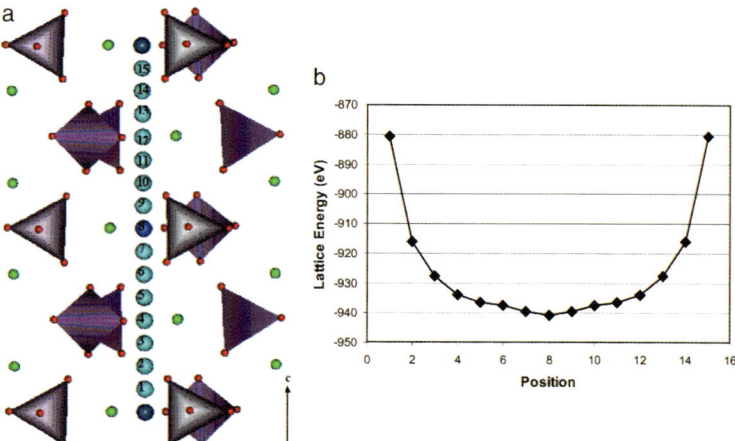

Fig. 5 Variation of position of second channel oxygen ion in a ($1 \times 1 \times 2$) unit cell of the oxyapatite structure, (a) numbering of the different positions investigated, and (b) the variation of energy with position (O_{PO_4} = red, Ca = green, H = white, $O_{channel}$ = dark blue (fixed position and lowest energy position 8) $O_{channel}$ = pale blue (unstable positions 1–7 and 9–15), PO_4 groups shown as tetrahedra).

the favoured locations for the oxygen ion, but that in addition the channel oxygen ions prefer to remain as far apart from each other as possible, probably due to charge repulsion, leading to an ordered array of alternating oxygen ions and vacancies in the channels. Once the lowest energy positions for the oxygen atoms were determined, the lattice was relaxed to give the optimised geometry of the oxyapatite material.

With regard to ordering of the channel oxygen atoms in the a/b plane, $i.e.$ between neighbouring hydroxy channels, again we found that an ordered structure was energetically preferred over a random distribution of oxygen ions. We found that the lowest-energy configuration had all oxygen ions situated in one a/b plane, located at a halide symmetry position, with the next a/b plane of symmetry positions vacant. This structure somewhat resembles the results we obtained in earlier work on solid solutions of fluor- and hydroxyapatite, where we found that the structure with equal amounts of fluoride and hydroxy ions also had alternating a/b planes of hydroxy groups and fluoride ions.[36] The reason behind this layering of the solid solution was the absence of distortion in the structure, which retained its hexagonal angles of $\alpha = \beta = 90°$, $\gamma = 120°$. This symmetry was lost when the anions were distributed in different planes, when the structure became distorted. The situation here is analogous, as the fully ordered oxyapatite structure, with oxygen ions at halide symmetry positions and alternating a/b planes of channel oxygen ions and vacancies keeps the hexagonal symmetry of the material with a small uniform expansion in the c-direction. However, the difference in energy between this strict alternation of O^{2-}/ vacancies in the a/b planes and the next most stable configuration, where the a/b planes contain both channel oxygen ions and vacancies, is only 7.2 kJ mol^{-1} per $Ca_{10}(PO_4)_6O$ unit, and we can therefore expect other distributions in the a/b directions to occur as well.

3.1.3 Solid solutions of $Ca_{40}(PO_4)_{24}(OH)_8$ and $Ca_{40}(PO_4)_{24}O_4$.

The stability of an oxyhydroxyapatite solid solution $(OH_{2-2x}O_x)$ is considered with respect to the two perfect end members, $i.e.$ hydroxyapatite (OH) and oxyapatite (O), and calculated as an excess energy of solid solution per unit cell, as follows:

$$\Delta E_x (OH,O) = E(OH_{2-2x}O_x) - \{[2 - 2x]E(OH) + xE(O)\} \qquad (3a)$$

where x is the fraction of hydroxy groups replaced by oxide ions. The excess energies of solid solution, plotted in Fig. 6, are not excessively high for a low oxyapatite content, for example $\Delta E = \sim 8$ kJ mol^{-1} at 50% substitution, depending on the structure of the 50% solid solution. However over 50% substitution, the excess energy of solid solution is well over 10 kJ mol^{-1}, and we therefore would not expect much higher levels of substitution of OH groups by oxygen ions under ambient conditions, which agrees with the rarity of the natural pure oxyapatite mineral

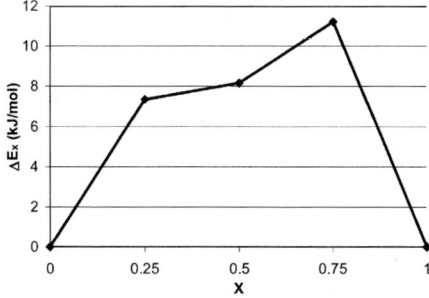

Fig. 6 Graph of the excess energies of solid solutions of oxy- and hydroxyapatite, where x is the fraction of oxyapatite.

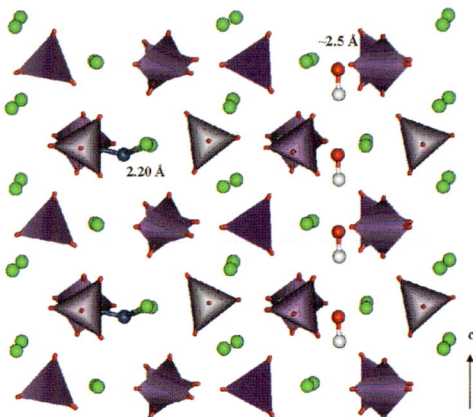

Fig. 7 Lowest-energy configuration of the 50% oxyhydroxyapatite solid solution with half the hydroxy channels replaced by complete channels of oxygen ions (O_{PO_4} = red, Ca = green, P = pink, H = white, $O_{channel}$ = blue, PO_4 groups shown as tetrahedra).

voelckerite, whose existence is disputed,[21] and the finding that synthetically produced oxyapatite decomposes into a number of other calcium phosphate phases.[22]

For the 50% oxyhydroxyapatite solid solution, we considered different configurations of the substitutional oxygen ions in the channels, where the lowest energy structure, shown in Fig. 7, has all alternating channels fully filled with hydroxy groups and fully substituted with oxygen ions, rather than a distribution over the channels. This structure is clearly different from the fluor-hydroxyapatite solid solutions described above, where we obtained alternating a/b planes of fluoride and hydroxy ions. However, a similar structure is found in solid solutions of fluor- and chlorapatite, where the 50% solid solution contains an equal number of channels completely filled with either fluoride or chloride ions,[72] and the reasons may be the same as in this case. When both oxygen ions and hydroxy groups are present within one channel, significant rearrangement of these species takes place within the channel to optimise interactions between them and with the atoms in the lattice, which involves a degree of competition. When only one species is present in the channel, a much more regular structure is obtained, more closely approximating the pure end-member. The preference of fluor-hydroxyapatite solid solutions for mixed columns is due to the particular affinity of the fluoride ion to the hydroxy group, which form OH–F pairs in the structure, hence especially stabilising the 50% solid solution. Although OH–O pairing also occurs in the oxyhydroxyapatite solid solution (as observed in the 25 and 75% solid solutions), there is no significant advantage over OH–OH pairing and the distortion of the structure due to the vacancies accompanying each substitutional oxygen ion on balance stabilises the structure with an array of regular hydroxy- or oxy-channels over mixed channels.

Although pure oxyapatite retains an undistorted hexagonal structure, our solid solutions all show structural distortions. Alberius-Henning *et al.* attained a maximum level of 78% dehydration of hydroxyapatite before the onset of decomposition of the oxyhydroxyapatite. His structure refinement showed a decrease in the *a*- and *b*-axes (resulting *b*-axis < *a*-axis) and an increase in the *c*-axis of the hydroxyapatite upon dehydration, which was accompanied by loss of symmetry of the structure to form a triclinic structure, which he attributed to a possible tilting of the hydroxy groups away from the channel axis.[23] Our calculated structure for the 75% substituted hydroxyapatite does indeed show some tilting of the hydroxy groups, although the effect is small. In agreement with experiment, our calculated structure also shows a triclinic distortion, where the calculated *a*- and *b*-axes decrease by 0.6

This journal is © The Royal Society of Chemistry 2006

and 0.2%, respectively (*cf.* exp. at 78% of 0.2% and 0.25%), the *c*-axis increases by 0.8% (exp. 0.3%) and $\alpha = 89.68°$, $\beta - 91.69°$ and $\gamma = 120.36°$ (exp. $\alpha = 90.06°$, $\beta = 89.75°$ and $\gamma = 119.997°$).

They also investigated a lower level of dehydration of 39% of the structure, which gave an experimental distortion of *a*-axis = −0.12%, *b*-axis = −0.09%, *c*-axis = +0.02%, $\alpha = 90.04°$, $\beta = 89.85°$ and $\gamma = 120.001°$,[23] whereas our calculated structure for the lowest energy 50% substituted oxyhydroxyapatite gave the following structural changes: *a*-axis = −0.8%, *b*-axis = −0.8%, *c*-axis = +0.3%, $\alpha = 90.26°$, $\beta = 89.73°$ and $\gamma = 120.04°$. The other 50% solid solutions, gave slightly different triclinic distortions, with more tilting of the hydroxy groups than in the 75% solid solution, and a smaller γ angle of approximately $\gamma = 119.7°$, in better agreement with the 78% dehydrated structure, although now $\alpha < 90°$ and $\beta > 90°$. Clearly, our calculated triclinic distortions are more extreme than the experimentally determined structures for oxyhydroxyapatite,[23] but we noticed that the distortions were very dependent on the configuration of the substituted hydroxy groups in the structure, which could be expected to be more random in an experimental structure. We would therefore suggest that the experimental distortion may be an average for a range of local structural distortions.

3.2   Calcium-deficient hydroxyapatites

Having considered the dehydration of hydroxyapatite into oxyhydroxyapatite solid solutions, we next investigated a series of calcium-deficient hydroxyapatite materials, where calcium vacancies introduced in the lattice were charge-compensated in a number of different schemes: by OH vacancies, through PO$_4$ substitution and by the introduction of substitutional carbonate groups.

3.2.1 Ca^{2+} vacancies charge-compensated by OH^- vacancies.

We first investigated a calcium-deficient apatite, where each Ca^{2+} vacancy was accompanied by two OH^- vacancies in the halide channel. In order to find the lowest energy combination of these vacancies, we have investigated a series of Ca–(OH)$_2$ vacancy pairs, systematically varying the calcium and hydroxy vacancies to find the minimum energy defect structure. Perhaps not surprisingly, in the energetically preferred configuration, shown in Fig. 8, the calcium vacancy and the charge-compensating hydroxy vacancies are situated closely together in the hydroxyapatite lattice (Fig. 8(a)), where the Ca vacancy occurs preferentially at a Ca(II) position, *i.e.* one of the Ca ions surrounding the OH channel between the two OH vacancies. After removal of the calcium and hydroxy ions, rotation of the surrounding PO$_4$ groups occurs, especially the PO$_4$ group near the calcium vacancy.

The defect formation energy (Table 2) is calculated for the following process:

$$Ca^x_{Ca} + 2OH^x_{OH} \rightarrow V''_{Ca} + 2V^{\bullet}_{OH} + Ca^{2+}_{(aq)} + 2OH^-_{(aq)} \tag{4}$$

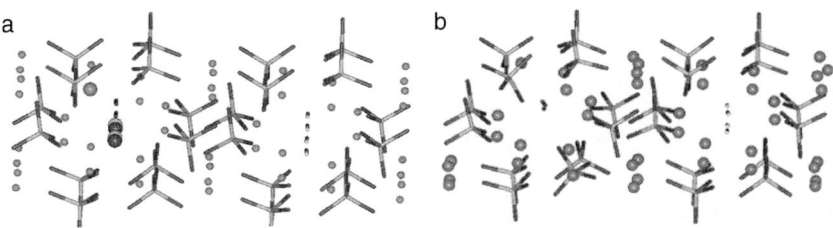

Fig. 8 $Ca^{2+}/(OH^-)_2$ vacancy pair formation in hydroxy apatite, (a) the perfect lattice viewed down the *c*-axis with the ions to be removed shown large and (b) the geometry optimised structure after removal of the ions (O = black, Ca = dark gray, P = pale gray, H = white).

Table 4 Calculated lattice energies of hydrated species from molecular dynamics simulations

Ion	Calculated energy of hydrated ion/kJ mol^{-1}
$PO_4^{3-}{}_{(aq)}$	−4698.8
$CO_3^{2-}{}_{(aq)}$	−4081.3
$OH^-{}_{(aq)}$	−723.6
$Na^+{}_{(aq)}$	−279.8
$K^+{}_{(aq)}$	−183.3
$Ca^{2+}{}_{(aq)}$	−1321.8
$H_2O_{(aq)}$	−891.4

where Ca_{Ca}^x and OH_{OH}^x are a calcium ion and a hydroxy group at a calcium and hydroxy lattice position, respectively (with neutral charge with respect to the lattice), V''_{Ca} is a vacancy at a calcium lattice position with a charge of −2 with respect to the lattice and $V_{OH}^{\cdot}$ is a vacancy at a hydroxy lattice position, with a charge of +1 with respect to the lattice. The energies of the various solvated ions, which we need to calculate the defect formation energies, were obtained from MD simulations of each ion in a simulation cell filled with 255 water molecules, and are listed in Table 4. Using these calculated energies for the solvated ions together with the lattice energies of the perfect and defective apatite materials, gives us a defect formation energy for the process in eqn (4) of +239.9 kJ mol^{-1}, indicating that this defect is unlikely to occur on thermodynamic grounds. We also calculated whether this defect could be stabilised by an interstitial water molecule in the channel, essentially replacing the two hydroxy groups by a water molecule, according to eqn (5). However, the defect formation energy then became +570.7 kJ mol^{-1}, indicating that this defect is much less stable even than the simple calcium and hydroxy vacancies in the dehydrated structure.

$$Ca_{Ca}^x + 2OH_{OH}^x + H_2O_{(aq)} \rightarrow V''_{Ca} + 2V_{OH}^{\cdot} + (H_2O)_i^x + Ca_{(aq)}^{2+} + 2OH_{(aq)}^- \quad (5)$$

3.2.2 Ca^{2+} vacancies charge-compensated by PO$_4^{3-}$ vacancies.

In the next range of calcium-deficient apatite structures, we have charge-compensated the calcium vacancies by accompanying phosphate vacancies. However, as the negative charge of the PO$_4$ group does not exactly cancel the positive charge of the calcium ion, we either had to introduce three calcium vacancies for two phosphate vacancies or introduce extra defects in the structure, which is the course we have taken here. In the first instance, we have considered the replacement of a PO$_4$ group and an OH ion by an oxygen ion in the lattice to charge-compensate for calcium vacancies in the apatite structure, according to eqn (6):

$$Ca_{ca}^x + (PO_4)_{PO_4}{}^x + OH_{OH}^x + OH_{(aq)}^- \rightarrow V''_{ca} + V_{OH}^{\cdot} + (O)_{PO_4}^{\cdot} + Ca_{(aq)}^{2+}$$
$$+ PO_{4(aq)}^{3-} + H_2O_{(aq)} \quad (6)$$

where we considered the interstitial oxygen ion both in the hydroxy channel and at the PO$_4$ location in the lattice. As before, the most stable configurations have all vacancies and interstitial oxygen clustered together, and the lowest energy configuration with the oxygen ion in the PO$_4$ position is shown in Fig. 9(b), where we see that the OH vacancy leads to a tilting of the neighbouring OH away from the vacancy and towards a phosphate oxygen ion. The interstitial oxygen ion interacts with a neighbouring Ca ion at a normal Ca–O$_{interstitial}$ bond distance of 2.39 Å.

As we were once more considering the dissociation of an OH group into a water molecule, we again needed the energy E_{H2O} as calculated in section 3.1 above. The calculated defect formation energy for this defect then became $E_{def} = +706.0$ kJ

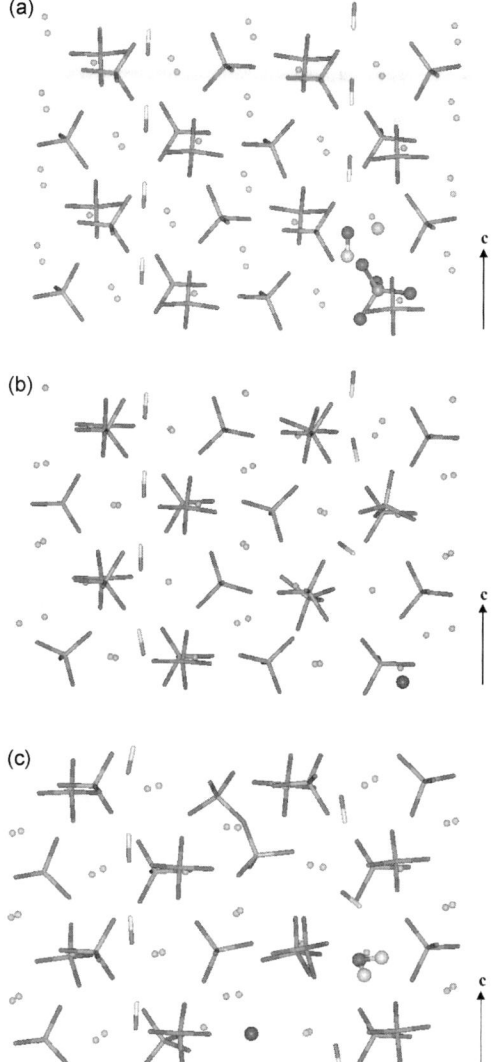

Fig. 9 $Ca^{2+}/PO_4^{3-}/OH^-$ vacancy formation in the hydroxy apatite, (a) the perfect lattice with the ions to be removed shown large, (b) the geometry optimised structure after removal of the ions and replacement by an O^{2-} ion (dark gray) at the PO_4 position, and (c) the geometry optimised defect structure with an extra water molecule in the channel (O = black, Ca = dark gray, P = pale gray, H = white).

mol^{-1}, *i.e.* highly exothermic and we therefore would not expect this defect to occur. Once again, we considered whether an interstitial water molecule at the OH position would stabilise this defect, according to the process in eqn (7), but although the formation of this defect is energetically slightly less expensive than the process in eqn (6), it remains energetically highly unfavourable with a calculated defect formation energy of $E_{def} = +651.9$ kJ mol^{-1}.

$$Ca^x_{Ca} + (PO_4)_{PO_4}{}^x + OH^x_{OH} + OH^-_{(aq)} \rightarrow V''_{Ca} + (H_2O)^{\cdot}_{OH} + (O)^{\cdot}_{PO_4}$$
$$+ Ca^{2+}_{(aq)} + PO_4{}^{3-}_{(aq)} \qquad (7)$$

As can be seen in Fig. 9(c) the water molecule in the channel is more or less aligned with the neighbouring OH groups, although one is still rather tilted in the channel, and it forms hydrogen-bonded interactions to its neighbours at HOH–OH and H₂O–HO distances of 2.71 and 1.90 Å respectively, hence stabilising the channel species. The Ca–O$_\text{interstitial}$ bond distance has shortened somewhat to 2.32 Å.

Finally, we investigated the location of the interstitial oxygen ion in the channel position, *i.e.* replacing the OH⁻ vacancy with an O²⁻ ion and charge-compensating this defect by accompanying Ca and PO₄ vacancies. The defect formation energy was calculated according to eqn (8)

$$Ca_{Ca}^{x} + (PO_4)_{PO_4}^{x} + OH_{OH}^{x} + OH_{(aq)}^{-} \rightarrow V''_{Ca} + V^{\cdots}_{PO_4} + (O)'_{OH} + Ca_{(aq)}^{2+} + PO_{4(aq)}^{3-} H_2O_{(aq)} \tag{8}$$

at $E_{\text{def}} = +824.1$ kJ mol⁻¹, *i.e.* even more endothermic than the two defects with the interstitial oxygen ion in the PO₄ position, described above. The higher exothermicity is probably due to the remaining PO₄ vacancy, which leads to an excess of positive charge in that location, which is no longer offset by the interstitial oxygen ion. The oxygen ion in the channel does form very strong interactions with the three surrounding ions at Ca–O distances of 2.19–2.48 Å.

3.2.3 Ca²⁺ vacancies charge-compensated through substitutional CO₃²⁻ groups.
Finally, we investigated the substitution of phosphate groups by carbonate groups to charge-compensate for the introduction of calcium vacancies. In a recent paper, the same methods were used to study a range of carbonate defects in the hydroxyapatite lattice, both the A-type defect, where the carbonate is found within the hydroxy channel, and the B-type defect, where carbonate groups are found at PO₄ positions.[37] In this previous study we did not investigate charge-compensation of the carbonate substitution at the PO₄ position by the introduction of calcium vacancies in the lattice. However, as biological apatite is often both calcium-deficient and rich in carbonate, we clearly need to include this combination of defects in our calculations to consider its possible occurrence in the apatite mineral. As such, we have calculated the process described in eqn (9), where two carbonate groups are substituted for two phosphate groups with the coincident creation of a calcium vacancy:

$$Ca_{Ca}^{x} + 2(PO_4)_{PO_4}^{x} + 2CO_{3(aq)}^{2-} \rightarrow V''_{ca} + 2(CO_3)^{\cdot}_{PO_4} + Ca_{(aq)}^{2+} + 2PO_{4(aq)}^{3-}. \tag{9}$$

We have again calculated this process for all possible combinations of vacancy locations and substitution positions within the simulation supercell, not only removing calcium ions from all positions, but also investigating which three of the four phosphate oxygen positions remain occupied by oxygens of the carbonate group and which becomes an oxygen vacancy. The lowest-energy defect structure is shown in Fig. 10, where the Ca ion that was preferentially removed was located in

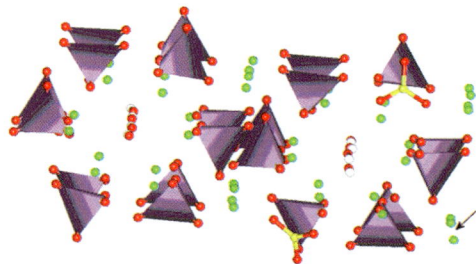

Fig. 10 Lowest energy geometry-optimised structure of a Ca²⁺ vacancy charge-compensation by two carbonate groups substituting for PO₄ groups, where the calcium vacancy is indicated by an arrow (O = red, Ca = green, P = pink, C = yellow, H = white).

the Ca(I) position, *i.e.* a lattice calcium ion. There is little distortion of the lattice caused by this defect, as the carbonate groups fit well into the PO_4 positions and are close to the calcium vacancy, hence avoiding an unequal distribution of charge within the lattice.

From our calculated energies for the different configurations of the substitutional carbonate groups, we found that the variations in energy were not particularly large, less than 30 kJ mol^{-1}, which is only about 0.2% of the total energy of the defect structure. As such we may expect to see a number of rotations of the carbonate groups within the apatite lattice. The defect formation energy for the process shown in eqn (9) was calculated to be E_{def} = −5.3 kJ mol^{-1}, *i.e.* the process is just exothermic, which is in agreement with the widespread occurrence of carbonate-rich, calcium-deficient apatite minerals. However, charge-compensation of the B-type carbonate defect in the apatite lattice by half a calcium vacancy is not the energetically most favourable process, as was shown in our previous work on carbonate defects in apatite. For example, if the carbonate defect is accompanied by the substitution of a monovalent Na$^+$ or K$^+$ ion for the calcium ion, the defect formation energies are calculated at −71.1 and −5.6 kJ mol^{-1}, respectively. However, whichever of these three charge-compensations is used, whether a vacancy or a substitutional monovalent ion, the resulting material is still calcium-deficient, unlike other carbonate defects in apatite, also calculated in our previous work,[37] namely the substitution of both a PO_4 group (B-type defect) and an OH$^-$ ion (A-type defect) by two carbonate groups, according to eqn (10):

$$(PO_4)_{PO_4}^x + (OH)_{OH}^x + 2CO_{3(aq)}^{2-} \rightarrow (CO_3)_{PO_4}^\cdot + (CO_3)_{OH}' + PO_{4(aq)}^{3-} + OH_{(aq)}^-. \quad (10)$$

This process was calculated to release 518.7 kJ mol^{-1} and hence can be expected to be the major process in the incorporation of carbonate into the apatite lattice, in agreement with experimental syntheses of carbonated apatite, which finds a combination of A-type and B-type carbonate defects in the products.[17,73,74] However, despite the preference for A-type defects in synthetic apatites (calculated as shown in eqn 12 in Table 2),[17,73–75] B-type carbonate defects are prevalent in natural bone,[16,73,76,77] possibly due to their prior incorporation in the lattice of calcium phosphate precursor phases to the biological hydroxyapatite.[37] As we have shown here, when only B-type carbonate defects are present in the apatite lattice, charge compensation through calcium deficiency is an energetically favourable process.

3.2.4 Effect of defects on elastic constants. In Table 5, we have listed the calculated elastic constants of the various defective apatite structures, where the numbering of the structures is the same as in Table 2. The original fluorapatite potential model was fitted by Mkhonto and de Leeuw[33] to experimentally obtained elastic constants by Sha *et al.*[78] However, defects in the lattice can have a large effect on the elastic constants of a material and it is important for biomaterial applications of apatite minerals that the elastic constants for defective structures are known.[79] For example, although it is generally accepted that carbonate defects are widespread in biologically occurring apatites, a recent calculation of the stress-distribution in bone between hydroxyapatite and its matrix by measuring X-ray reflexes of hydroxyapatite had to assume that the elastic modulus of the bone apatite mineral was the same as that of pure synthetic hydroxyapatite.[80] We trust that our calculated elastic constants reported in Table 5 may thus be of use in future experimental studies, where elastic moduli of substituted apatite minerals are required.

4. Conclusions

In this study we have used computer simulation techniques to investigate a number of defects in the hydroxyapatite structure. We have first shown that our potential model adequately reproduces DFT calculations of the same defect, namely the

Table 5 Calculated elastic constants of perfect and defective structures (defect numbering as in Table 2)

Calculated elastic constants of pure and defective apatite structures/10^{10} pa

Material	C_{11}	C_{33}	C_{44}	C_{12}	C_{13}
Experiment[78]	152.0	185.7	42.8	50.0	63.1
Fluorapatite	150.6	176.6	53.2	62.8	73.6
Hydroxyapatite	134.4	184.7	51.4	48.9	68.5
Oxyapatite	142.2	142.5	44.7	51.5	55.2
1	130.5	161.1	45.9	55.0	58.2
2	137.1	169.9	48.6	52.6	61.6
3	135.0	178.3	50.1	53.6	65.3
4	124.4	148.1	41.1	59.5	58.0
5	127.1	165.3	44.2	55.9	56.6
6	129.5	151.8	42.5	58.9	52.5
7	111.5	146.3	43.2	44.9	51.2
8	132.8	157.4	44.9	53.9	55.2
9	130.2	172.1	45.1	46.4	60.6
10	130.9	175.0	45.4	48.7	61.4
11	117.1	167.8	48.6	46.4	62.4
12	130.0	181.3	49.9	53.2	70.0
13	144.0	178.1	42.4	52.8	68.1

dissociation of one hydroxy group into its constituent ions with the association of the proton with a second hydroxy group to form a water molecule in the channel. The stepwise dehydration of hydroxyapatite into oxyhydroxyapatite, whilst endothermic, is energetically not prohibitively expensive, which agrees with experimental evidence of partial dehydration of hydroxyapatite by thermal methods. In addition, the excess energies of oxy- and hydroxyapatite solid solutions are relatively small and we thus may expect considerable mixing of the two materials, especially with lower oxyapatite content, again in agreement with experiment, where oxyhydroxyapatite is found to be relatively stable, but pure oxyapatite decomposes into other calcium phosphate materials.

Our calculations of calcium-deficient hydroxyapatites show that charge-compensation by carbonate groups is energetically favourable, where the calcium-deficiency can either occur as vacancies or as monovalent ions substituting at the calcium positions. Charge compensation for calcium vacancies by the introduction of phosphate vacancies is calculated to be very endothermic for all structures considered here, especially if the phosphate vacancy remains in the lattice without any charge compensation in the form of, for example, an interstitial oxygen ion. Stoichiometric hydroxyapatite is relatively easy to synthesise in the laboratory, whereas calcium-deficient hydroxyapatite is primarily found in bone and other biological tissue, which also has high carbonate content. Our simulations strongly suggest that the calcium vacancies are preferentially compensated by carbonate defects, which agrees well with the simultaneous occurrence of both defects in the same material.

Acknowledgements

NHdL thanks the Engineering and Physical Sciences Research Council, UK, for an Advanced Research Fellowship and for grant no. GR/S67142/01.

References

1   W. A. Deer, R. A. Howie and J. Zussman, *An introduction to the rock-forming minerals*, Longman, Harlow, UK, 1992.

This journal is © The Royal Society of Chemistry 2006

2   R. L. Romer, *Geochim. Cosmochim. Acta*, 1996, **60**, 1951.
3   M. Menzies, K. Gallagher, A. Yelland and A. J. Hurford, *Geochim. Cosmochim. Acta*, 1997, **61**, 2511.
4   F. Corfu and D. Stone, *Geochim. Cosmochim. Acta*, 1998, **62**, 2979.
5   P. Pellas, C. Fieni, M. Trieloff and E. K. Jessberger, *Geochim. Cosmochim. Acta*, 1997, **61**, 3477.
6   R. A. Wolf, K. A. Farley and L. T. Silver, *Geochim. Cosmochim. Acta*, 1996, **60**, 4231.
7   K. A. Farley, R. A. Wolf and L. T. Silver, *Geochim. Cosmochim. Acta*, 1996, **60**, 4223.
8   T. S. B. Narasaraju and D. E. Phebe, *J. Mater. Sci.*, 1996, **31**, 1.
9   A. M. A. Ambrosio, J. S. Sahota, Y. Khan and C. T. Laurencin, *J. Biomed. Mater. Res. (Appl. Biomater.)*, 2001, **58**, 295.
10 S. B. Cho, F. Miyaji, T. Kokubo, K. Nakanishi, N. Soga and T. Nakamura, *J. Mat. Sci.: Mat. Medicine*, 1998, **9**, 279.
11 A. A. Baig, J. L. Fox, R. A. Young, Z. Wang, J. Hsu, W. I. Higuchi, A. Chettry, H. Zhuang and M. Otsuka, *Calcif. Tissue Int.*, 1999, **64**, 437.
12 A. Yasukawa, K. Kandori and T. Ishikawa, *Calcif. Tissue Int.*, 2003, **72**, 243.
13 W. Kolodziejski, *Topics in Current Chemistry*, 2004, **246**, 235.
14 D. G. A. Nelson, J. D. B. Featherstone, J. F. Duncan and T. W. Cutress, *J. Dent. Res.*, 1982, **61**, 1274.
15 R. Z. LeGeros, R. Kijkowska, C. Bautista and J. P. LeGeros, *Connect. Tissue Res.*, 1995, **32**, 525.
16 E. F. Morgan, D. N. Yetkinler, B. R. Constantz and R. H. Dauskardt, *J. Mater. Sci.: Mat. Medicine*, 1997, **8**, 559.
17 Y. Suetsugu, Y. Takahashi, F. P. Okamura and J. Tanaka, *J. Solid State Chem.*, 2000, **155**, 292.
18 M. E. Fleet, X. Liu and P. L. King, *Amer. Mineral.*, 2004, **89**, 1422.
19 A. Peeters, E. A. P. De Maeyer, C. Van Alsenoy and R. M. H. Verbeeck, *J. Phys. Chem. B*, 1997, **101**, 3995.
20 W. H. Peck, private communication, 2005.
21 T. A. Stolyarova, *Experiment in Geosciences*, 1999, **7**, 80.
22 J. M. Zhou, X. D. Zhang, J. Y. Chen, S. X. Zeng and K. De Groot, *J. Mat. Sci.: Mat. In Medicine*, 1993, **4**, 83.
23 P. Alberius-Henning, E. Adolfsson and J. Grins, *J. Mater. Sci.*, 2001, **36**, 663.
24 T. Wang and A. Dorner-Reisel, *Mater. Lett.*, 2004, 3025.
25 J. C. Elliott, *Structure and chemistry of the apatites and other calcium orthophosphates*, Studies in inorganic chemistry 18, Elsevier, 1994.
26 R. M. Wilson, J. C. Elliott, S. E. P. Dowker and L. M. Rodriguez-Lorenzo, *Biomaterials*, 2005, **26**, 1317.
27 L. M. Rodriguez-Lorenzo, *J. Mat. Sci. Mat. in Medicine*, 2005, **16**, 393.
28 S. J. Joris and C. H. Amberg, *J. Phys. Chem.*, 1971, **75**, 3172.
29 J. L. Meyer and B. O. Fowler, *Inorg. Chem.*, 1982, **21**, 3029.
30 M. Born and K. Huang, *Dynamical Theory of Crystal Lattices*, Oxford University Press, Oxford, 1954.
31 B. G. Dick and A. W. Overhauser, *Phys. Rev.*, 1958, **112**, 90.
32 G. W. Watson, E. T. Kelsey, N. H. deLeeuw, D. J. Harris and S. C. Parker, *J. Chem. Soc., Faraday Trans.*, 1996, **92**, 433.
33 D. Mkhonto and N. H. de Leeuw, *J. Mater. Chem.*, 2002, **12**, 2633.
34 Z. Du and N. H. de Leeuw, *Surf. Sci.*, 2004, **554**, 193.
35 N. H. de Leeuw NH and T. G. Cooper, *Phys. Chem. Chem Phys*, 2003, **5**, 433.
36 N. H. de Leeuw, *Phys. Chem. Chem. Phys.*, 2004, **6**, 1860.
37 S. Peroos, Z. Du and N. H. de Leeuw, *Biomaterials*, 2006, **27**, 2150.
38 D. Mkhonto and N. H. de Leeuw, *J. Mater. Chem.*, 2002, **12**, 2633.
39 N. H. de Leeuw, *Phys. Chem. Chem. Phys.*, 2004, **6**, 1860.
40 J. A. L. Rabone and N. H. de Leeuw, *J. Comput. Chem.*, 2006, **27**, 253.
41 A. Pavese, M. Catti, S. C. Parker and A. Wall, *Phys. Chem. Miner.*, 1996, **23**, 89.
42 N. H. de Leeuw and S. C. Parker, *J. Phys. Chem. B*, 1998, **102**, 2914.
43 N. H. de Leeuw, *J. Phys. Chem. B*, 2002, **106**, 5241.
44 N. H. de Leeuw and T. G. Cooper, *Cryst. Growth Des.*, 2004, **4**, 123.
45 M. J. Sanders, M. Leslie and C. R. A. Catlow, *J. Chem. Soc., Chem. Commun.*, 1984, 1271.
46 N. H. de Leeuw, G. W. Watson and S. C. Parker, *J. Phys. Chem.*, 1995, **99**, 17219.
47 N. H. de Leeuw, F. M. Higgins and S. C. Parker, *J. Phys. Chem. B*, 1999, **103**, 1270.
48 N. H. de Leeuw, D. Mkhonto and C. R. A. Catlow, *J. Phys. Chem. B*, 2003, **107**, 1.
49 N. H. de Leeuw and D. Mkhonto, *Chem. Mater.*, 2003, **15**, 1567.
50 N. H. de Leeuw and D. Mkhonto, *J. Mater. Chem.*, 2005, **15**, 3272.
51 G. Kresse and J. Hafner, *Phys. Rev. B*, 1993, **47**, 558.

52   G. Kresse and J. Hafner, *Phys. Rev. B*, 1994, **49**, 14251.
53   G. Kresse and J. Furthmüller, *Phys. Rev. B*, 1996, **54**, 11169.
54   G. Kresse and J. Furthmüller, *Comput. Mater. Sci.*, 1996, **6**, 15.
55   J. P. Perdew, J. A. Chevary, S. H. Vosko, K. A. Jackson, M. R. Pederson, D. J. Singh and
      C. Fiolhais, *Phys. Rev. B*, 1992, **46**, 6671.
56   S. H. Vosko, L. Wilk and M. Nusair, *Can. J. Phys.*, 1980, **58**, 1200.
57   P. E. Blochl, *Phys. Rev. B*, 1994, **50**, 17953.
58   G. Kresse and D. Joubert, *Phys. Rev. B*, 1999, **59**, 1758.
59   H. J. Monkhorst and J. D. Pack, *Phys. Rev. B*, 1976, **13**, 5188.
60   N. H. de Leeuw, *Chem. Commun.*, 2001, **17**, 1646.
61   N. H. de Leeuw, *Phys. Chem. Chem. Phys.*, 2002, **4**, 3865.
62   K. A. Gross, C. C. Berndt, P. Stephens and R. Dinnebier, *J. Mater. Sci.*, 1998, **33**, 3985.
63   P. Hartmann, C. Jäger, St. Barth, J. Vogel and K. Meyer, *J. Sol. State Chem.*, 2001, **160**,
      460.
64   A. Bouhaouss, A. Laghzizil, A. Bensaoud, M. Ferhat, G. Lorent and J. Livage, *Int. J.
      Inorg. Mater.*, 2001, **3**, 743.
65   M. S. Islam, J. R. Tolchard and P. R. Slater, *Chem. Commun.*, 2003, **13**, 1486.
66   J. R. Tolchard, J. E. H. Sansom, M. S. Islam and P. R. Slater, *Dalton Trans.*, 2005, **7**, 1273.
67   F. A. Kröger, *The chemistry of imperfect crystals*, North Holland Press, Amsterdam, 1964.
68   W. van Gool, *Principles of defect chemistry of crystalline solids*, Academic Press, New
      York, 1966.
69   N. H. de Leeuw, G. W. Watson and S. C. Parker, *J. Phys. Chem.*, 1995, **99**, 17219.
70   Z. Du and N. H. de Leeuw, *Surf. Sci.*, 2004, **554**, 193.
71   P. A. Henning, A. R. Landa-Canovas, A.-K. Larsson and S. Lidin, *Acta Crystallogr., Sect.
      B*, 1999, **55**, 170.
72   N. H. de Leeuw, *Chem. Mater.*, 2002, **14**, 435.
73   M. Vignoles, G. Bonel, D. W. Holcomb and R. A. Young, *Calcif. Tissue Int.*, 1988, **43**, 33.
74   P. D. Moens, F. J. Callens, P. F. Matthys and R. M. Verbeeck, *J. Chem. Soc., Faraday
      Trans.*, 1994, **90**, 2653.
75   D. U. Schramm, J. Terra, A. M. Rossi and D. E. Ellis, *Phys. Rev. B*, 2000, **63**, 024107–1.
76   W. E. Brown and L. C. Chow LC, *Annu. Rev. Mater. Sci.*, 1976, **6**, 213.
77   J. Sadlo, P. Matthys, G. Vanhaelewyn, F. Callens, J. Michalik and W. Stachowicz, *J.
      Chem. Soc., Faraday Trans.*, 1998, **94**, 3275.
78   M. C. Sha, Z. Li and R. C. Brandt, *J. Appl. Phys.*, 1994, **75**, 7784.
79   I. L. Jäger, 2005, private communication.
80   I. L. Jäger, *J. Biomechanics*, 2005, **38**, 1451.

General Discussion

Professor Islam opened the discussion of Professor Nazar's paper: You mentioned the cation "doping" results of Chung et al.[1] What are your current views on their enhancement of conductivity in $LiFePO_4$ and whether there is true lattice doping (of Nb, Zr etc.)? Our recent simulation results suggest that the energetics of "supervalent" dopant incorporation are unfavourable (on both Li and Fe sites).

1   S. Y. Chung, J. T. Bloking and Y. M. Chiang, *Nat. Mater.*, 2002, **1**, 123.
2   M. S. Islam, D. Driscoll, C. A. J. Fisher and P. R. Slater, *Chem. Mater.*, 2005, **17**, 5085.

Professor Nazar replied: I think your simulation results are very important, given that it's difficult to rule out the possibility of very limited (<2%) "supervalent" doping by experiment since the change in lattice parameters would be extremely small. However, I think the issue of supervalent doping is something of a red herring. Even if it occurs even to a limited degree, it would have to be accompanied by lithium substoichiometry to account for any changes in electronic conductivity. The fact that $LiFePO_4$ can exhibit conductivities as high as 10^{-2} S cm^{-1} in the absence of any dopant[1] demonstrates that the dopant is not the major contributor. Our current views have not changed since then. We find no evidence for true, homogeneous lattice doping using TEM-elemental mapping, for example. Conversely, the presence of metallic phosphides that contribute to the increase in conductivity is unequivocal.

1   Nazar et al., *Nat. Mater.*, 2004.

Professor Maier asked: With regard to the explanation by Chung et al., it is hard to understand how Nb-doping, which should result in a donor effect, leads to an extremely high p-type conductivity. Do you agree?

Professor Nazar agreed: Absolutely. High p-type conductivity could only arise from lithium substoichometry that would cause the formation of hole carriers (Fe^{3+}) in the lattice. Inclusion of a supervalent dopant ion such as Nb^{5+} would formally decrease the iron valence state (below 2+) in the absence of this.

Professor Heitjans asked: Is there direct experimental evidence that Li diffusion is one-dimensional apart from obvious structural considerations?

Professor Nazar responded: None published in the literature to date, but Professor Maier has some new results on Li conductivity in $LiFePO_4$ single crystals that were presented at the recent IMLB meeting in June 2006.

Professor Maier added: We recently measured ionic and electronic conductivities in single crystalline $LiFePO_4$:

$$\sigma_{eon}(b) \approx \sigma_{eon}(c) \gg \sigma_{eon}(a) \gg \sigma_{ion}(b) \approx \sigma_{ion}(c) \gg \sigma_{ion}(a)$$

Also, $D_{Li}{}^{chem}(b) \approx D_{Li}{}^{chem}(c) \gg D_{Li}{}^{chem}(a)$, partly in contradiction to theoretical predictions. How much this is specific to the T-range considered or influenced by the preparation temperature (frozen-in anti-site disorder) is yet to be found out.

Professor Heitjans said: I would like to mention that *via* NMR it is possible in principle by measuring on a polycrystalline material to determine the dimensionality (1-D, 2-D, 3-D) of the diffusion process, without the need for a single crystal. Confirmation that it is a 1-D ion conductor can be obtained from the functional dependence of the diffusion induced spin–lattice relaxation rate T_1^{-1} on the NMR frequency at a temperature well above that of the characteristic $T_1^{-1}(T)$-maximum (see *e.g.* ref. 1). The question, of course, is whether in this particular case the relevant T_1^{-1} data for ^{7}Li can be measured.

1   P. Heitjans, A. Schirmer and S. Indris, in *Diffusion in Condensed Matter - Methods, Materials, Models*, ed. P. Heitjans and J. Kärger, Springer, Berlin/Heidelberg, 2005, p. 367.

Professor Maier replied: That is an important point.

Professor Nazar said: The study of Li ion transport using field gradient ^{7}Li NMR is an excellent idea, and one we have considered ourselves *via* collaboration. The problem lies with the magnitude of the diffusion coefficient, and the presence of the paramagnetic $Fe^{2+/3+}$ centres.

Professor Islam addressed Professor Maier: From your recent studies of $LiFePO_4$ single crystals you find identical ionic conductivities and activation energies for two axes. Could this coincidence be due to the same conduction mechanism? For this orthorhombic olivine structure (space group *Pbna*) I would expect preferential Li ion conduction down the *b*-axis channel.

Professor Maier replied: What is clear from the experiments is that the two axes with the higher ion conductivities have quite similar conductivities, indeed. So far we could only measure this in a small temperature range (upper limit: melting point of Li, lower limit: waiting time). I know that this contradicts theoretical predictions, yet it does not appear certain to me that low activation energies in a given one-dimensional path indeed lead to a statistically relevant long-range transport. In fact, the measured activation enthalpy is significantly higher than the calculated one.

Professor Nazar asked: The activation energies you show for electronic conduction along the three crystal axes are effectively identical, which is contrary to expectation. Electronic conduction should be extremely hindered perpendicular to the FeO_6 planes. What exactly are you measuring in these experiments? Is it not the activation energy for formation of the carrier—not the activation energy for transport?

Professor Maier responded: The similar activation energies indeed point towards a negligible migration energy as compared with the formation value, the latter not being affected by anisotropies.

Professor Navrotsky asked a general question: There was some early work on diffusion in Fe_2SiO_4 olivine that showed considerable anisotropy. This may be relevant, being a property of the olivine structure. Does anyone remember the details?

Professor Nazar answered: I'm afraid I don't.

Professor Dieckmann responded: Our group has grown fayalite (Fe_2SiO_4) single crystals with different orientation and measured the electrical conductivity and the iron tracer diffusion at 1130 °C as a function of oxygen activity a_{O_2}, and orientation. It was observed that the tracer diffusion of Fe was fastest in the $\langle 001 \rangle$ direction (*Pbnm*) and slowest in the $\langle 100 \rangle$ direction. The ratio $D_{Fe}^*\langle 001 \rangle / D_{Fe}^*\langle 100 \rangle$ was about 10, almost independent of the oxygen activity a_{O_2}. The electrical conduction was fastest in the $\langle 001 \rangle$ direction and slowest in the $\langle 010 \rangle$ direction. The ratio $\sigma\langle 001 \rangle / \sigma\langle 010 \rangle$ is about 1.5 to 2 and decreases with increasing oxygen activity. The sequence $D_{Fe}^*\langle 001 \rangle > D_{Fe}^*\langle 010 \rangle > D_{Fe}^*\langle 100 \rangle$ is different from that reported in the literature for interdiffusion experiments in the system $(Fe_xMg_{1-x})_2SiO_4$ with x of about 0.1 for which $\tilde{D}\langle 001 \rangle > \tilde{D}\langle 100 \rangle > \tilde{D}\langle 001 \rangle^1$ but identical with that observed for the chemical diffusion related to point defect relaxation.[2]

1   D. K. Buening and P. R. Busek, *J. Geophys. Res.*, 1973, **78**, 6852–6862
2   K. Ullrich, K. D. Becker, *Solid State Ionics*, 2001, **141–142**, 307–312

Professor Irvine addressed Professor Nazar: How well do we understand the ion and electron transport properties of intermediate olivine composition that might exist under actual cycling conditions?

Professor Nazar replied: We are still learning much about these systems. At room temperature, it seems clear that the miscibility gap between $LiFePO_4$ and $FePO_4$ is quite large, although the gap may be narrowed to some degree in nanocrystallites of very small dimensions. Therefore the ion and electron transport properties of the intermediate compositions formed during cycling reflect those of the parent phases, except at very low and high Li stoichiometry where the limits of solid solubility become an interesting question.

Professor Bruce asked a general question: Can anyone comment on the recent TEM work of Tom Richardson suggesting that ion/electron transport occurs along the interface between the two phases rather than through the phases?[1] Surely ion/electron transport could be very different in this region than in the bulk phases?

1   G. Y. Chen, X. Y. Song and T. Richardson, *Electrochem. Solid State Lett.*, 2006, **9**, A295.

Professor Maier replied: Only a principal remark: The contact $FePO_4/LiFePO_4$ should necessarily result in electron and Li^+ redistribution, leading to varied conductivities and then also to varied conductances parallel to the interface.

Professor Bruce then asked: Do you think that the interface between the two phases consists of a solid solution?

Professor Maier replied: It is difficult to say; but even when phase diagrams suggest a sharp boundary, elastic and electric effects as well as gradient effects can smear it out compositionally.

Professor Nazar answered: Yes, I do; a very limited solid solution, however.

Dr Gray-Weale opened the discussion of Professor Bruce's paper: Roughly what fraction of lithium ions are outside the polymeric tubes? If polydispersity and larger end-groups both increase the conductivity, and the alignment of ends reduces it, is it not possible that the lithium ions are wandering between tubes, and the imperfections in the arrangement of the tubes produced by polydispersity *etc.* make it easier for the lithium ions to get in and out of the tubes? I suppose there may not be room for the lithiums outside the tubes, but that is where the anions are and a lithium ion is small and highly polarising, so perhaps it might find a way.

Professor Bruce replied: To re-cap: the crystal structure of the conducting $PEO_6:LiX_6$ (X = P, As, Sb) electrolytes is composed of pairs of poly(ethylene oxide) chains that together form tunnels, within which the Li^+ ions reside, with the anions located between the tunnels. The average molar mass is 1000 [$CH_3O(CH_2CH_2O)_{22}CH_3$], the average chain length is ~ 40 Å and the crystallite size is in excess of 2000 Å. There are many chains inside each crystallite (see Fig. 1). This is not the usual semicrystalline polymer where the chains are longer than the ordered regions. The crystallography does not of course identify the chain ends, which are

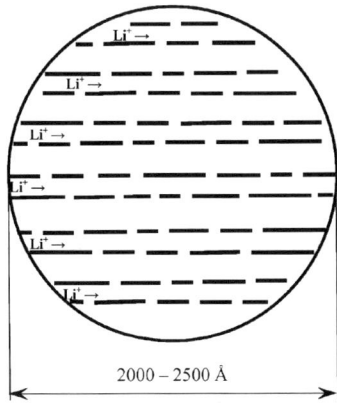

2000 – 2500 Å

Fig. 1 Schematic representation of the arrangement of polymer chains within a crystallite of PEO$_6$:LiXF$_6$ (X = P, As, Sb) prepared with polydisperse PEO of an average molecular weight 1000–2000 Da. Solid lines represent polymer chains forming the tunnels containing the Li$^+$ ions.

randomly distributed in the polydispersed material, instead seeing the tunnels as continuous.

Thermodynamically there must be departure from a perfect crystal structure. Energetically one might anticipate that defects would form most easily at the chain ends. The model I offered during the discussion was meant purely to illustrate how defects *could* arise. The model (see Fig. 2) proposed that Li$^+$ ions and their charge compensating anions are missing near the chain ends (Schottky defects), thus generating Li$^+$ vacancies within the tunnels. Other mechanisms are also possible, *e.g.*, displacement of Li$^+$ from the chain ends to form LiX ion pairs just outside the chains (Frenkel pair). In any case, charge carriers must exist for ion transport. Li$^+$ vacancies at or near the chain ends could then migrate by cooperative displacement of neighbouring Li$^+$ ions to form a vacancy at the next chain end along the tunnel, which was previously occupied by a Li$^+$ ion (see upper region of Fig. 2).

Such ion transport processes mirror those in ceramic ionic conductors, especially 1-D conductors such as the hollandites.

Now addressing specifically your first question. Given that the model is a proposition without experimental or computational verification, it is not possible to estimate the proportion of defects and hence the fraction of Li$^+$ ions outside the tubes.

Considering that replacement of the polydisperse chains by chains of equal length increases the possibility of chain end alignment perpendicular to the tube axis, yet reduces the conductivity, we might argue that Li$^+$ transport by ions moving between tunnels is less likely than along the tunnels.

Professor Chadwick asked: You are proposing a vacancy mechanism. Is there any way two lithium ions could get past each other inside the PEO helix?

Professor Bruce answered: Remember that the specific mechanism offered during the discussion is only, as you say, a proposal, however it is difficult to envisage one Li$^+$ ion passing another along the tunnel.

Professor Chadwick then asked: Do you have any transport number data for these systems?

Professor Bruce replied: Yes, we have NMR data for PEO$_6$:LiPF$_6$ that demonstrates Li$^+$ ion transport, with little evidence of significant anion transport.

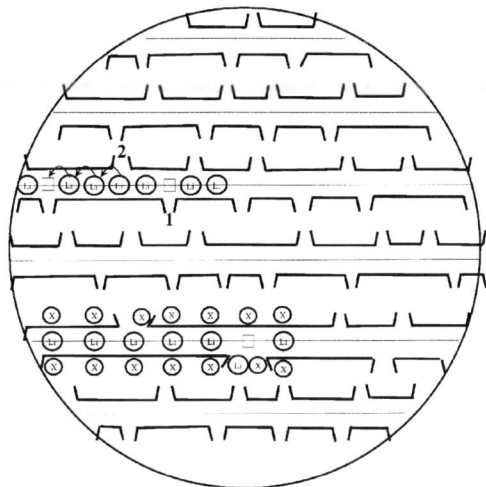

Fig. 2 Possible mechanisms of defect formation and conduction in the crystalline $PEO_6:LiXF_6$ (X = P, As, Sb) polymer electrolytes. Black lines represent PEO chains, dotted lines show axes of tunnels; encircled Li and X represent the cations and anions respectively; squares represent Li^+ vacancies. The conduction mechanism shown in the upper region of the crystallite involves a Li^+ vacancy at a chain end (1) migrating to the next chain end (2) by cooperative displacement of Li^+ ions.

Dr Stolwijk asked: You indicated in your answer to the previous question that vacancy formation is accompanied by the creation of a cation–anion pair. Do you have any experimental evidence for the occurrence of these pairs?

Professor Bruce responded: No. The problem with defects, the concentration of which is by definition small (in this case probably <5 mol%), is that it is difficult to study their structure directly. Indeed the history of defects in ceramics is that conductivity was used as a probe to study defects. Vibrational spectroscopies have so far failed to identify the nature of the defects (*e.g.* ion pairs) in the crystalline polymers, consistent with their concentration being small. Defects of some type must be present thermodynamically and given the observed conductivities.

Mr Martin asked: Do you envisage building cross-links of tunnels between the main conducting tunnels that add at least another dimension to the system?

Professor Bruce replied: Of course 1-D conductors, such as $PEO_6:LiXF_6$ (X = P, As, Sb), are not the best. Unwanted defects can block conduction channels leading to suppression of ionic conductivity. 2-D or 3-D transport would be better. It is difficult to simply introduce tunnels perpendicular to those present in the PEO_6:-$LiXF_6$ structure. However, Peter Wright, Sheffield, has taken a different approach to ordered polymeric ionic conductors and has prepared 2-D systems that have high ionic conductivity.[1]

1 Y. G. Zheng, A. Gibaud, N. Cowlam. T. H. Richardson, G. Ungar and P. V. Wright, *J. Mater. Chem.*, 2000, **10**, 69.

Professor Yashima asked: I want to ask about the details of the crystal structure. Did you observe the positional disorder for the mobile lithium cations?

I have observed large positional disorder and diffusion paths of Li^+ cations in $Li_{0.16}La_{0.64}TiO_3$–Li conductor.[1] I want to know the difference between the polymer and ceramic systems.

1 M. Yashima, M. Itoh, Y. Inaguma and Y. Morii, *J. Am. Chem. Soc.*, 2005, **127**, 3491.

Professor Bruce answered: I am aware of your very nice work determining diffusion pathways from crystallographic data from the scattering density (X-ray or neutron) along the diffusion pathways. However, this relies on the potential energy profile along the diffusion paths being relatively flat (shallow and broad potential wells) so that the mobile ions spend a significant proportion of time away from the centres of their sites. This is associated with small activation energies for ion hopping and high conductivity (as for $Li_{0.16}La_{0.64}TiO_3$, 10^{-3} S cm^{-1}). Our materials, in the absence of doping, have lower conductivities (10^{-7} S cm^{-1} at room temperature). Although this increases substantially on doping, the increase is due to an increase in the number of defects, not necessary their mobility. As a result we do not see abnormally high thermal parameters in the X-ray or neutron diffraction data.

Dr Yashima addressed Professor Bruce and Dr Andreev: How reliable was the structure determination of your electrolytes with respect to determining the position of Li?

Dr Andreev answered: The bulk of the structure was determined from X-ray diffraction, however, the position of Li was verified by neutron diffraction which is more sensitive when it comes to lighter atoms.

Professor Yashima said: It was very difficult to determine the positions of Li$^+$ cations due to the flat surface of R_{wp} values for various Li$^+$ positions and large disorder. Did you observe large amplitudes of thermal vibration for Li?

Dr Andreev replied: No, the thermal parameters for Li were compatible with those for other atoms comprising the structure.

Mr O'Dell addressed Professor Bruce: Could the expansion/contraction of the polymer tunnel diameter as the Li ions move through it have any effect on the anions?

Professor Bruce answered: The distortion of the polymer chain was not large enough to dramatically affect the anions.

Professor Islam opened the discussion of Dr Stolwijk's paper: You find that the enthalpy (and entropy) of pair formation shows a decrease with increasing salt concentration. What is your rationalisation for this?

Dr Stolwijk replied: Our tentative interpretation is that both the enthalpy and entropy of pair formation depend on the mean polymer length (number of oxygen atoms) per cation. The binding of the cation to the ether oxygens of the polymer is the main factor in the solvation of the salt. Thus it may be assumed that the associated enthalpy (entropy) constitutes a most significant contribution to the pair formation enthalpy (entropy). Apparently, for low salt concentrations the polymer chains can optimally coordinate to the cations and therefore the binding enthalpy is high. By contrast, for more concentrated systems the topological constraints increase, which leads to a lower mean binding enthalpy per cation and consequently to a smaller pair formation enthalpy. For the entropy the argument is similar.

Professor Islam then asked: Do you have any evidence for such an interpretation?

Dr Stolwijk responded: As already indicated, our interpretation is tentative and should be verified by further experiments including more dilute electrolytes and, if possible, other methods of analysis.

Professor Bruce asked: Are you surprised that the highest aggregates are ion pairs?

Dr Stolwijk replied: Our experiments do not yield direct information about the nature of the charge carriers, that is, whether we are mainly dealing with single ions, singly charged triple ions, or higher-order aggregates. Our approach was to take the simplest model that works, that is, one based on the single-ion/ion-pair hypothesis. Allowing for triple ions in addition to single ions would increase the number of model parameters but due to the limited experimental data, which remain as they are, there is no chance to determine these additional parameters by fitting. It may be interesting to check if a migration model that is exclusively based on triple ions (and ion pairs) could lead to a satisfactory description of the data. Anyway, if future studies provide evidence for the involvement of higher-order aggregates, we shall have to adapt our model correspondingly.

Professor Maier asked: Did you measure the degree of ion pairing at extremely small concentration, where mass action demands it must vanish?

Dr Stolwijk answered: The lowest salt concentration investigated in our work occurs in the composition $PEO_{60}NaI$, which has a monomer-to-salt ratio of 60. For this composition we find a rather high degree of ion pairing, as indicated by pair fractions of up to 50% at the upper end of the temperature range. Measurements for more dilute compositions are currently in progress.

Professor Bruce commented: Some years ago we measured the conductivity of polymer electrolytes down to very low ion concentrations and we saw the increase then decrease of ion pairs on reduction of ion concentration. The maximum in the concentration of ion pairs occurs at very low concentrations because of the low dielectric constant of the polymer.

Professor Maier responded: I could not comment on the first point, but the latter one then follows from mass action (configurational entropy).

Dr Stolwijk responded: Our results may fit into this general scheme since we expect a further increase in the ion-pairing tendency for the next composition to be examined, that is, $PEO_{120}NaI$. It will be interesting to find out at which low salt concentration the ion-pairing tendency will be reversed.

Professor Maier addressed Dr Stolwijk: Your technique seems to be an excellent tool to study the filler effect in polymers. Did you perform experiments in this direction?

Dr Stolwijk replied: We do plan to perform experiments on PEO-NaI complexes containing SiO_2 or Al_2O_3 nanoparticles and to compare the results with the present data on similar complexes without filler materials. We expect to see shifts in the tracer and charge diffusivities which can be analyzed within our ion migration model. Very likely, some of the model parameters will significantly change and thus provide clues for the chemical and physical processes affected by the nanoparticles.

Professor Bruce asked: Are you surprised that the effects you see, including the difficulty of PEO chains coordinating the cations appearing at such high PEO : salt ratios?

Dr Stolwijk answered: Yes, I am surprised that the cation coordination seems so affected at relatively high monomer-to-salt ratios in the range from 60 to 20. However, it should be noted that $PEO_{20}NaI$ represents a 'borderline' composition in the PEO-NaI phase diagram. For complexes with a higher salt concentration the phase diagram predicts the precipitation of a stoichiometric crystalline phase (PEO_3NaI) from the amorphous matrix within the temperature range under investigation. This implies that a monomer-to-salt ratio of, say, 15 is no longer compatible with complete salt solvation, at least at the low end of the temperature range. Interestingly, our Fig. 8 seems to indicate that the pair formation enthalpy and entropy become negative close to this composition ($c_s = 0.07 = 1/15$).

Professor Irvine opened a general discussion, addressing Professor Bruce: You stated that you could improve conductivity by controlling crystallinity. Would you agree that there is an optimum degree of crystallinity for good ionic conductivity?

Professor Bruce replied: Considering the degree of crystallinity there are different aspects.

(1) First, there is no evidence of an amorphous phase. It is important to remember that the average chain length is 40 Å, whereas the crystallite size is 2000 Å. There are many chains, and hence random chain ends, within each crystallite. The picture is reversed from the usual semicrystalline polymer.

(2) Considering crystallite size. Each grain is composed of a mosaic of misaligned crystallites. The larger the crystallites then the fewer the number of interfaces which ions have to cross between misaligned tunnels and hence the higher the conductivity.

(3) Considering the degree of order within the crystallites, then yes, there is an optimum degree for such order. Sufficient disorder is required at the randomly distributed chain ends to ensure a relatively large number of defects but not such that disruption of the pathways for ion transport occurs. In other words, local disorder favours a high concentration of charge carriers but an excess of it can impede ion transport (lower ion mobility).

Professor Navrotsky asked a general question: There are many ceramics (*e.g.* zeolites, manganese oxides) with tunnels in 1, 2 or 3 directions. Why are these seemingly less popular as Li ion conductors and are there systems which should be studied further?

Professor Bruce answered: High ionic conductivity requires a large number of mobile ions and a high mobility. The latter requires a close match between the ion size and the bottleneck through which it must pass along the diffusion path. In zeolites and other large framework ceramics the channels are too wide for fast Li^+ transport.

Another issue is the polarisability of the anion sublattice. By moving from oxides to sulfides, ionic conductivity can be enhanced. More work examining crystalline sulfides should be carried out.

Professor Maier said: It may be doubted if an ideal 1-D transport within a chain is really efficient in view of the statistics involved.

Professor Bruce responded: Yes, 1-D transport is not the best because any unwanted defects blocking the tunnels suppresses conductivity, although the alkali metal hollandites do exhibit very respectable levels of conductivity. We have achieved $> 10^{-6}$ S cm^{-1} at 25 °C/$> 10^{-4}$ at 50 °C in our 1-D system, so clearly they do conduct. 2-D and 3-D conduction pathways are of course preferred and, as I stated in my response to Mr Martin above, Peter Wright has prepared 2-D conductors that exhibit high levels of ion transport.

Professor Nazar commented: Li ion conductivity in most zeolites would likely be driven by hydration effects, which present a problem for most practical electrolytes.

Professor Navrotsky replied: This would obviously be a problem for high temperature applications but may not be a problem in a temperature range overlapping or slightly higher than that for polymer electrolytes.

Professor Bruce said: The reason for low Li^+ conductivity in zeolites was addressed in my previous answer to Professor Navrotsky's questioning of why ceramics are seemingly less popular as Li ion conductors.

It should be added, however, that by hydrating framework solids with zeolite-like architectures, it is possible to induce high ionic conductivity. A conductivity of 1.8×10^{-2} Ω^{-1} cm^{-1} at 25 °C and >85% relative humidity is possible[1] but the conductivity falls off significantly on dehydration.

1   N. F. Zheng, X. H. Bu and P. X. Feng, *J. Am. Chem. Soc.*, 2003, **426**, 428–432.

Dr Gray-Weale addressed Professor Bruce: You mentioned a problem with conduction in one dimension: if a tube is blocked anywhere along its length, then that whole tube is out of action. Might it be possible to use this to identify the mechanism of conduction more clearly? If some concentration of blockages were introduced, perhaps by attaching some group to the polymeric tubes, or substituting some part of them, then for one dimensional conduction the conductivity would fall drastically. If lithium ions can wander between the tubes then the effect of blockage would be less significant.

Professor Bruce answered: In principle, this could be used to confirm 1-D conduction along the tunnels. One would however have to be sure that the blockages were located in the tunnels and not preferentially at the chain ends, otherwise interpretation would be ambiguous because such chain end segregation could hinder any ion transport between tunnels.

Professor Nazar commented: In LiFePO$_4$, one can prepare the material at "low" temperatures ($\sim$130 °C), and sustain a fraction of Li/Fe site disorder, on the order of 3–5%. This indeed appears to block transport along the tunnels as shown by the poor electrochemical behaviour these materials exhibit.

Professor Bruce then said: Although we should not push the analogy too far one can think of the ion transport in the tunnel (1-D) structure of PEO$_6$:LiXF$_6$ (X = P, As, Sb) as being similar to ion transport in the hollandites. Mobile defects are introduced into the hollandite tunnels by substitution of ions in the static lattice and conductivity can be high in such systems. Such substitution is similar to the presence of chain ends along the tunnels in the polymer structure.

Professor Forsyth asked: If the lithium ion is moving through the tunnels, how does doping with different anions affect the defect structure? How are the anions involved?

Can you use positron annihilation lifetime spectroscopy (PALS) to determine the number and size of defects?

Professor Bruce answered: If the anion is of a higher charge, such as substitution of SbF$_6^-$ by SiF$_6^{2-}$ (aliovalent substitution), then additional Li^+ ions must accompany the SiF$_6^{2-}$. Sites exist in the tunnels that could accommodate the interstitial ions, leading to an interstitialcy mechanism of ion transport.[1] Where isovalent substitution occurs, *e.g.*, AsF$_6^-$ by N(SO$_2$CF$_3$)$^-$, then the larger size and different shape of the anion can distort the neighbouring tunnels and hence the potential of the Li^+ in the tunnels, leading to more and/or more mobile defects. An analogy may be drawn with AgBr$_{1-x}$I$_x$, where a three orders of magnitude increase in conduction occurs for $x = 0.2$ due to lattice strain effects.

Yes, position annihilation spectroscopy is a potentially useful tool and we have been exploring its use in these systems.

1   C. Zhang, E. Staunton, Y. G. Andreev and P. G. Bruce, *J. Am. Chem. Soc.*, 2005, **127**, 18305.

Professor Islam addressed Professor Nazar: As we know, size is important. For the $LiFePO_4$ system, is there the possibility of reducing the need for carbon coatings by smaller (nano)crystallite sizes and hence shorter diffusion path lengths?

Professor Nazar replied: Definitely, the shorter diffusion length sustained in nanocrystallites is of tremendous benefit, and will reduce (but probably not eliminate) the need for conductive coatings. A combination of the two effects is necessary to optimize the electrochemical properties.

Professor Yashima asked: Can you show the detailed picture of a polaron? Can you show the coupling figure between the polaron and mobile Li^+ cations?

Professor Nazar responded: A small polaron is a space-limited lattice distortion caused by the transient localization of either an electron or a hole on the iron site. Thus, there is an electrostatic interaction between the mobile Li^+ cation and that localized charge density.

Professor Maier asked: What is the experimental evidence for coupling of Li^+ and e^- transport?

Professor Nazar replied: It is indirect, but clear. The temperature at which the Li^+ ions disorder in the olivine lattice exactly coincides with the onset temperature for electron (small polaron) hopping, as seen in the Mössbauer experiments described in our paper. The activation energy for "free" small polaron transport determined from first principles calculations[1] is also much lower than determined from experiment—either from conductivity or from Mössbauer data—unless the binding energy of the Li^+ to the small polaron is included.

1   Ceder *et al.*, *Phys. Rev. B*, 2006.

Professor Bruce commented: If there is strong ion/electron coupling of the transport then it should be possible to measure the cross coefficients in the transport equations from irreversible thermodynamics (L_{12}), these should be high.

Professor Maier responded: That is correct, but these are not easy to measure as the driving forces are not easy to vary independently in a given experiment.

Dr Stolwijk addressed Professor Bruce: If one compares your crystalline polymer electrolyte system with our amorphous system, there is a significant difference concerning charge transport. In the crystalline system, as you indicated, the cation transference number is close to one. By contrast, in our amorphous complexes the cation transference number was determined to be as low as 0.1 or even smaller. This may imply that the crystalline systems perform better in terms of say Li^+ transport than comparable amorphous systems although the dc conductivity of the latter is higher.

Professor Bruce replied: Crystallinity imposes selectivity in crystalline polymer electrolytes, selecting for cation transport in contrast to amorphous polymers where anions dominate. When comparing the conduction of amorphous and crystalline polymers it is important to remember that a crystalline polymer electrolyte with a cation transport number of 1 will in a cell give the same transport of Li^+ as an amorphous polymer with a ten times higher conductivity but a cation transport number of 0.1. Also there is no polarisation for an ionic conductor of $T+ = 1$ in a cell.

Professor Maier opened the discussion of Professor Navrotsky's paper: I would be interested in your opinion on the following: how far is it correct to consider a mixed oxide phase as a mixture of oxide constituents even though they are dissociated, *i.e.* to circumvent the defect thermodynamical treatment by a pseudo-molecular approach?

This point should be essentially critical when referring to the entropy.

Example: $NaCl/H_2O$; μ = chemical potential,

$\mu_{NaCl(H_2O)} \neq$ const. + $RT\ln c_{NaCl}$ even if dilute, rather μ_{NaCl} = const. + $RT\ln c_{NaCl}^2$ (!) owing to dissociation.

Even though this does not affect your evaluation, I am interested in your opinion.

Professor Navrotsky replied: I think a pseudomolecular approach is fundamentally wrong because complex oxides contain several sublattices and charge balance is a strong constraint. There are already too many thermodynamic formalisms in the literature in which large deviations from ideality are the result of choosing unrealistic components.

Professor Yashima asked: What is the origin of the difference of the ΔH mix between melted and coprecipitated samples?

I feel that the coprecipitated sample is inhomogeneous due to the difference in solubility product in aqueous solution between Ce and Zr ions.

The sintering of oxide mixtures is the easiest way to obtain homogeneous t'-solid solutions.

Professor Navrotsky responded:

(1) Probably the Ce^{3+} content is the major difference; the melted samples are black.

(2) The initial solutions are concentrated and precipitation occurs fast, so such effects are probably minimized. We see no evidence for second phases or inhomogeneity at the micron scale (electron microprobe analysis).

(3) We were unable to get complete reaction by sintering.

Professor Haile said: I find it somewhat surprising that you can draw conclusions about atomic scale defects from macroscopic measurements of heats of formation. Specifically, can you comment on whether it is possible to distinguish between anion and cation ordering? Also, the notion of clusters strikes me as a phenomenon that accompanies phase separation/segregation rather than a negative heat of mixing and ordering.

Professor Navrotsky replied: One cannot directly distinguish the type of order from calorimetry but one can draw inferences from systematic variations in thermodynamic properties with composition and heat treatment, especially when there are structural and spectroscopic and computational studies to draw on. Thermodynamics is yet another source of evidence, to be combined with the others. The clustering issue is one of nomenclature. You are thinking of clusters of like atoms which leads to phase separation. I mean the formation of clusters or associates of unlike atoms and/or vacancies, leading to ordered phases. The former generally is associated with positive heat of mixing, the latter with negative.

Professor Haile asked: Can you relate your observations to those of Professor Irvine and his discussion of defect association and clustering? He has assigned these to two different energy terms but it appears in your case these are all one phenomenon only differentiated by length scale.

Professor Navrotsky said: If the extent of clustering changes significantly with temperature (dissociation with increasing temperature), one might expect to see an excess heat capacity. Transposed temperature drop calorimetric experiments in our laboratory did not detect any excess enthalpy (to about ± 1 kJ mol^{-1}) for several YSZ samples compared to yttria and zirconia between room temperature and about 1200 °C. This suggests that the energetic effects of any cluster dissociation are small.

Professor Irvine commented: The question raised about cluster size is rather important. We would suggest that cluster size decreases with temperature and so there may be some disparity between calorimetry and structural studies.

The anomaly at low dopant contents seems rather interesting. This would be at the point where cluster–cluster regions no longer dominate, with intercluster, cluster and disordered regions co-existing.

Professor Kilner commented: One way of looking at the interactions between the dopant cations and vacancies and the clusters that may form is to look at the variation of the lattice parameter with the addition of dopant. We have recently looked at the CeO_2–Nd_2O_3 system and observed deviations from Vegard's Law which would indicate stony clustering of defects. This is consistent with early data of Bevan and co-workers who showed that the lattice parameter has a quadratic variation with dopant concentration.[1] The quadratic factors extracted from the data scale with ion size and would indicate a different propensity for clustering in the different CeO_2–R_2O_3 systems.

1 J. M. Bevan, W. W. Barker and T. C. Parks, in *Proceedings of the Fourth Conference on Rare Earth Research*, ed. L. Eyring, Gordon and Breach, New York, 1965, p. 441.

Professor Ishihara opened the discussion of Dr Slater's paper: According to the data, the activation energy for oxide ion conductivity in this system is very small. Could you comment on such a small activation energy? Activation energy is usually large in the case of interstitial conductivity because of the large size of O^{2-}. In this apatite structure, what is the difference in free volume between apatite and fluorite oxide?

Dr Slater answered: The lowest activation energy observed was 0.67 eV, which compares quite favourably with that calculated from our modelling studies[1] on $La_{9.33}Si_6O_{26}$ for an interstitial oxide ion conduction mechanism (0.56 eV). These modelling studies show that the flexibility of the silicate framework aids the conduction process, hence accounting for the low activation energies for interstitial oxide ion conduction.

Regarding the free volume difference between the apatite and fluorite oxides, this is difficult to assess, since a standard calculation would only give you the overall free volume, whereas a more accurate comparison would be with the free volume of the apatite conduction channels.

1 J. R. Tolchard, M. S. Islam and P. R. Slater; *J. Mater. Chem.*, 2003, **13**, 1956.

Professor Vannier asked: Are the O(5) oxygen ions involved in the conduction process? On one hand, oxygen vacancies are evidenced on this site, moreover, it exhibits high thermal parameters along the x axis. On the other hand, it is quite strongly bonded to the La(3) site with a small bond length.

Dr Slater replied: The mechanism deduced from the modelling studies involves the interstitial oxide ions moving past the regular (O(5)) oxygen ions within the channels. Whether a further process where an interstitial oxygen displaces an O(5) oxygen in a type of knock-on mechanism requires further investigation, and in this respect we are investigating MD simulations. The high thermal parameters for O(5) can be somewhat misleading, as high values are also observed for poorly conducting samples, e.g., $La_8Sr_2Si_6O_{26}$. It is clear from the structural and modelling work on these apatite materials that there is likely to be significant localised disorder, which will cause displacement of the O(5) oxygens from their ideal site. Thus the high thermal parameters are likely to be related to such static displacements.

Professor Haile asked: Can you elaborate on the conduction mechanism? Are the interstitial oxygen ions possibly moving past the regular oxygen ions within the channels and not necessarily involving the regular sites in the transport process? What is the distance between interstitial sites?

Dr Slater answered: The mechanism deduced from the modelling studies involves the interstitial oxide ions moving past the regular oxygen ions within the channels. Whether a further process involving a kind of knock-on mechanism is also involved needs further study, and in this respect we are investigating MD simulations. On an average crystallographic scale the interstitial sites are separated by approx. 3.6 Å. However, the presence of interstitial oxide ions causes considerable local distortions (e.g., a displacement of 1.3 Å of the O(3) position of the neighbouring SiO_4 unit[1]), and so locally there could be shorter distances, or indeed a series of interstitial positions.

1   J. R. Tolchard, M. S. Islam and P. R. Slater, *J. Mater. Chem.*, 2003, **13**, 1956.

Professor Vannier asked: In your paper, you assume that the actual structural composition fit the nominal composition. In the case of the $La_{9.67} Si_{5.5} Mg_{0.5} O_{26}$ nominal composition, could you also have some Mg on the La site? What evidence was there for the Mg ions being mainly localised on the Si site? Is there any difference in the atomistic calculation of reaction energies, indicating which site is the more favourable?

Dr Slater replied: The structural data showed doping on the Si site led to an increase in cell volume in agreement with the larger size of Mg^{2+} *versus* Si^{4+}, while doping on the La site led to a decrease, again in agreement with ion sizes. Moreover the atomistic simulation studies suggested that Mg^{2+} could substitute on both La and Si sites supporting the "ambi-site" behaviour observed in the experimental work. As with any "ambi-site" dopant, there is always the possibility of some incorporation of Mg on both sites. Refinements of site occupancies, however, suggested that any such interexchange was small, and so the occupancies were then fixed.
  In terms of the atomistic simulation data, both site energies were similar.

Professor Aranda said:
  (1) The Mg substitution at the La and Si sites has been proved by neutron powder diffraction and atomistic simulation. However, it would be very important to know if the oxygen interstitial stability is different or the same in these two compositions.
  (2) The statement of better results in $P6_3$ over $P6_3/m$ space groups should be supported by R_F, R_I and R_{wp} values.

Dr Slater responded:
  (1) That is an interesting point, and a key feature may be whether the incorporation of Mg on the Si site may lead to more favourable interstitial sites.

(2) As an example, the R factors for the different space groups for $La_{9.67}Si_{5.5}$ $Mg_{0.5}O_{26}$ are given below, showing the improved fit for $P6_3$.

$P6_3/m$ $R_{wp} = 0.0665$, $R_P = 0.0586$, $\chi^2 = 1.115$
$P6_3$ $R_{wp} = 0.0647$, $R_P = 0.0570$, $\chi^2 = 1.058$

Professor Haile asked: Can you make any comments relating the atomistic simulations to more macroscopic properties for which measurements can be made?

Dr Slater answered: Measurements of single crystals of a Nd containing analogue gave an activation energy for conduction along c of 0.62 eV[1,2] which accords well with the activation energy from atomistic simulation of 0.56 eV for $La_{9.33}Si_6O_{26}$.

1 S. Nakayama, M. Sakamoto, M. Higuchi and K. Kodaira, *J. Mater. Sci. Lett.*, 2000, **19**, 91.
2 S. Nakayama and M. Niguchi, *J. Mater. Sci. Lett.*, 2001, **20**, 913.

Professor Islam commented: The atomistic simulation studies provide useful trends in the energetics of dopant substitution at different sites. The calculations on apatite silicates suggest that Mg^{2+} could substitute on both La and Si sites, showing possible "ambi-site" (or "amphoteric") behaviour. I also want to add that there is now significant experimental data that strongly point to interstitial oxygen in these apatite silicates from neutron diffraction,[1] Mössbauer[2] and ^{29}Si NMR studies.[3]

1 L. Leon-Reina, E. R. Losilla, M. Martinez-Lara, S. Bruque and M. A. G. Aranda, *J. Mater. Chem.*, 2004, **14**, 1142.
2 V. V. Kharton, A. L. Shaula, M. V. Patrakeev, J. C. Waerenborgh, D. P. Rojas, N. P. Vyshatko, E. V. Tsipis, A. A. Yaremchenko and F. M. B. Marques, *J. Electrochem. Soc.*, 2004, **151**, A1236.
3 J. E. H. Sansom, J. R. Tolchard, D. Apperley, M. S. Islam and P. R. Slater, *J. Mater. Chem.*, 2006, **16**, 1410.

Dr De Souza asked: Do you see a grain boundary resistance in your impedance spectra? If so, is it due to an insulating grain boundary phase or to an intrinsic phenomenon, *e.g.*, space charge?

Dr Slater replied: At low–intermediate temperatures, a grain boundary resistance is observed. The grain boundary resistance is highest for poorly sintered samples, and so is probably an intrinsic phenomenon.

Professor Beck asked: What is really meant by "interstitial"? It is known as a general feature that the size of the anions in the column is much too large. This necessitates the formation of vacancies to accommodate the rest of the anions. There could also be the possibility of "buckling" of the chains. I wonder whether your interstitial sites are just a reflection of this kind of buckling?

Dr Slater responded: In the phosphate systems to which you are referring, you are correct that anion vacancies are more likely. However, in the oxide ion conducting apatite silicates and germanates, oxygen excess is more favourable. Experimental studies have shown difficulties in incorporating oxygen vacancies, whereas a number of groups have shown that the structure can accommodate oxygen excess. Thus oxygen contents up to $O_{26.5-27.0}$ have been reported. In the case of such oxygen excess samples, the extra oxygens must occupy interstitial sites, and this will cause displacements of neighbouring channel oxygens, explaining the high thermal parameters observed for these. In terms of oxygen stoichiometic O_{26} systems, the interstitial oxygens are oxygens that have been displaced from their ideal site in the centre of the channels to sites at the periphery near the silicate groups. The presence of these interstitial sites at the periphery of the channels is supported by neutron diffraction,[1] Mössbauer[2] and ^{29}Si NMR.[3]

1   L. Leon-Reina, E. R. Losilla, M. Martinez-Lara, S. Bruque and M. A. G. Aranda, *J. Mater. Chem.*, 2004, **14**, 1142.
2   V. V. Kharton, A. L. Shaula, M. V. Patrakeev, J. C. Waerenborgh, D. P. Rojas, N. P. Vyshatko, E. V. Tsipis, A. A. Yaremchenko and F. M. B. Marques, *J. Electrochem. Soc.*, 2004, **151**, A1236.
3   J. E. H. Sansom, J. R. Tolchard, D. Apperley, M. S. Islam and P. R. Slater, *J. Mater. Chem.*, 2006, **16**, 1410.

Professor Yashima asked: You have refined the anisotropic thermal parameters. What do you think of the relationship between the anisotropy and conductivity?

Are your results consistent with single-crystal work by Dr Okudera? If not, what is the reason?

Dr Slater replied: The paper by Dr Okudera *et al.*[1] showed high anisotropic thermal parameters for the channel oxygens and therefore proposed that conduction was directly down the channels with the silicate groups not involved. However, the anisotropic thermal parameters include contributions from both static displacements as well as thermal vibrations. Therefore it is difficult to make a direct correlation between this anisotropy and conductivity. For example, the anisotropic thermal parameters for $La_{9.67}Si_{5.5}Mg_{0.5}O_{26}$ and $La_9Mg_{0.5}Si_6O_{26}$ are similar and yet the conductivity of the former at 500 °C is two orders of magnitude higher than the latter.

There is considerable experimental support for the interstitial oxide ion conduction mechanism proposed by our earlier atomistic work[2] from neutron diffraction,[3] Mössbauer,[4] and more recently ^{29}Si NMR.[5]

1   H. Okudera, Y. Masubuschi, S. Kikkawa and A. Yoshiasa, *Solid State Ionics*, 2005, **176**, 1473.
2   J. R. Tolchard, M. S. Islam and P. R. Slater, *J. Mater. Chem.*, 2003, **13**, 1956.
3   L. Leon-Reina, E. R. Losilla, M. Martinez-Lara, S. Bruque and M. A. G. Aranda, *J. Mater. Chem.*, 2004, **14**, 1142.
4   V. V. Kharton, A. L. Shaula, M. V. Patrakeev, J. C. Waerenborgh, D. P. Rojas, N. P. Vyshatko, E. V. Tsipis, A. A. Yaremchenko and F. M. B. Marques, *J. Electrochem. Soc.*, 2004, **151**, A1236.
5   J. E. H. Sansom, J. R. Tolchard, D. Apperley, M. S. Islam and P. R. Slater, *J. Mater. Chem.*, 2006, **16**, 1410.

Professor Beck opened the discussion of Dr de Leeuw's paper: I question whether there is a complete ordering within the *a,b* plane. It is known that the chain containing vacancies may be one-dimensionally disordered. This is seen in the form of diffuse planes in the reciprocal lattice.

Dr de Leeuw responded: Complete ordering of oxygen ions and vacancies in the *a,b* plane is calculated to be energetically preferred. However, the difference between the ordered *a,b* plane and a more random structure is only about 7 kJ mol^{-1}. This small energetic advantage is unlikely to preclude disorder in the *a,b* plane at the elevated temperatures, where oxyapatite is a stable structure.

Professor Yashima asked: Is your sample monoclinic or hexagonal? Stoichiometric HAp is monoclinic when the OH$^-$ is ordered. What effect will the ordering/ symmetry difference have on the mobility/conductivity?

Dr de Leeuw replied: We have used the monoclinic structure for our calculations, which is fully ordered as to the OH positions. However, we have checked with the hexagonal structure and found that a comparable fully ordered structure is also the lowest-energy structure in the hexagonal unit cell. As such, the calculations give the same results for the monoclinic and the hexagonal structures. In any case, we have used supercells in our calculations to make the simulated system sufficiently large, so that defects do not interact across the periodic boundary conditions. The discussion in the paper uses the hexagonal unit cell parameters as the use of these is more widespread in the experimental literature, seeing that natural apatites occur as hexagonal systems.

Professor Haile asked: Can you speculate at all on the mobilities of various ionic species along the channels, particularly under non-stoichiometric conditions?

Dr de Leeuw answered: Previous molecular dynamics simulations of hydroxy-apatite surfaces in solution have shown that the OH species are mobile in the channel and easily dissolve into solution. However, MD simulations of carbonate defects in the channels have shown that at room temperature the carbonate species is free to rotate and spin within the channel, but that it does not move along the channel due to the interactions of two of its oxygen ions with calcium ions in the lattice. I would expect that mobility of the channel species will be fairly high when the system is non-stoichiometric or at elevated temperatures, for example in the oxyapatite structure, where the channel oxygen ions could move into neighbouring vacant sites.

Professor Maier asked: How important are oxygen vacancies in the apatites? A local condensation $[(P_2O_7)_{(PO_4)_2}]^{\bullet\bullet}$ would be equivalent to $V_O^{\bullet\bullet}$ (Norby mentioned this at E-MRS, Nice, 2006).

Dr de Leeuw replied: I would not expect a significant amount of P_2O_7 to occur in the hydroxyapatite lattice as its phosphate groups are all isolated *ortho*-phosphates in a fairly dense structure. In addition, in recent simulations of phosphate-containing bio-glasses, we found that even there the phosphates tend to remain isolated, neither forming chains with silicate groups nor with other phosphate groups. However, it would be interesting to investigate this defect, as HPO_4 groups do occur in the lattice and P_2O_7 could thus be a condensation product, maybe at elevated temperatures.

Dr Kohanoff asked: Why did you use shell models to describe apatites? Are electronic polarisability effects that important to require this approach, rather than rigid ion models? How important are polarisation effects on the energetics and structure of defects?

Dr de Leeuw responded: The shell model is a simple but effective model for the electronic polarisability, which we consider essential, when carrying out calculations of defects. I have never used rigid ion potentials myself, but the relaxation of the structures is better described when polarisability in the form of the shell model or otherwise is included in the simulations. In particular, in surface simulations and calculations of glasses or liquids we have found that only the use of a polarisable potential will give good agreement with experiment.

Dr Wilson commented: Rigid-ion models (RIM), in which the ion electron density is effectively frozen, can be very effective, particularly in (highly ordered) crystalline environments. However, once the ions readily encounter more asymmetric environ-ments (in defects, surfaces, amorphous states, whilst moving along activated diffusion pathways . . .) then the inclusion of the response of the ion electron density to the resulting electric fields is vital. Polarization effects will tend to lower defect energies and activation energies for diffusion processes (with respect to the RIM). Such effects may, in some sense, be entangled into the RIM parameters (depending upon the details of the parameterisation procedure) but such potentials will tend to be less reliable at state-points distinct from those around which the model was parameterised (*i.e.*,the models will lack transferability). The shell model[1] represents a particular traditional mechanical representation of the dipolar ionic response to both electric fields and short-range effects. The historic origin of these models lay in the requirement to minimise the parameter set as, generally speaking, only a small amount of experimental information was available to parameterise the models. More modern models, such as the polarizable-ion model (PIM),[2] describe the ion polarization in a more physically-transparent manner and, as a result, may be more flexible both in terms of the retention of the physical meaning of the parameters and their transferability between chemically-related systems. Further-more, higher order moments may be readily included. The greater number of parameters required to parameterise these models are not problematic as they may be obtained by reference to high level electronic structure calculations.[3]

1   B. G. Dick and A. W. Overhauser, *Phys. Rev.*, 1958, **112**, 90.
2   P. A. Madden and M. Wilson, *Chem. Soc. Rev.*, 1996, **25**, 339.
3   M. Wilson, S. Jahn and P. A. Madden, *J. Phys.: Condens. Matter*, 2004, **16**, S2795–S2810.

Dr de Leeuw said: In reply to Dr Kohanoff's question and Dr Wilson's comment, we find that the polarisation afforded by the shell model in interatomic potential-based simulations is essential to calculate the geometries, relaxations and energies of defects and surfaces in ionic materials.

Professor Haile opened a general discussion by asking: I'd like to return to the question that Professor Navrotsky posed to the group. Can we design computations that connect more directly to the quantities (for example, enthalpies of formation) that we can measure experimentally?

Dr de Leeuw replied: Computer simulations of solid solutions of two materials have been shown to be comparable to experimental measurements. However, averaging of the solid solution structures has been shown not to give good comparison with experiment and, therefore, in order to calculate an accurate representation of an experimental solid solution, it is necessary to simulate a large number of different geometries, which can be prohibitively expensive computation-ally. It is also necessary to include full relaxation of the structures to obtain the correct energies, which again adds to the computational cost.

Professor Navrotsky asked: What one measures by solution calorimetry is the enthalpy of some chemically balanced and charge balanced reaction in the solid state. What one calculates often is the energy of creation of an isolated defect, sometimes with the removal of an atom or ion to infinity. The two processes are different. My question to all is how can we relate them so that the computations can be compared to experiment?

Dr de Leeuw answered: There are a number of experimental measurements, which are directly comparable with calculated processes. One very good example is measurements from the temperature programmed desorption (TPD) technique, which can be directly compared with adsorption energies as calculated by computer simulations.

Professor Islam responded: I agree that the links and interactions between computational results and experimental data are important. For the specific case of the energy of creation of an isolated defect, this calculated term is often combined with other isolated defect energies to derive total energies for charge-balanced defect reactions, for example, Schottky or Frenkel formation energies. These calculated energies can then be compared directly with available experimental data, and indeed have generally found good agreement for a range of binary halides and oxides.[1] Of course, for more complex ternary oxides such experimental data are sometimes not readily available. Another area worth noting has been the calculation of migration activation energies for ion transport in solid state ionics,[1,2] which again are compatible with observed values from conductivity and diffusion experiments. However, our studies on dopants in perovskite-oxides[2,3] calculate energies of dopant incorporation (termed "solution" energies), which have provided useful "trends" in dopant-ion solubility (in accord with observation), but the calculated solution energies are not linked directly with experimental enthalpies. Therefore, further debate about relating such computed energies to data from calorimetric measurements would be very worthwhile.

1   C. R. A. Catlow, R. G. Bell and J. G. Gale, *J. Mater. Chem.*, 1994, **4**, 781; C. R. A. Catlow, *Ann. Rev. Mater. Sci.*, 1986, **16**, 517.
2   M. S. Islam, *J. Mater. Chem.*, 2000, **10**, 1027.
3   M. S. Islam and R. A. Davies, *J. Mater. Chem.*, 2004, **14**, 86; R. A. Davies, M. S. Islam, A. V. Chadwick and G. E. Rush, *Solid State Ionics*, 2000, **130**, 115.

Professor Islam then asked Professor Navrotsky: A strong theme of our simulation work is the close interaction with experimental studies such as conductivity, diffraction and EXAFS. It's good to see that your calorimetric measurements of energetics of mixing link up well with our trends on dopant–vacancy association/clustering in ZrO_2 from atomistic simulation.[1] Have you looked at dopant–vacancy clustering in perovskites (such as Sr/Mg-doped $LaGaO_3$), which is much less studied in comparison to fluorite oxides?

1   M. S. Khan, M. S. Islam and D. R. Bates, *J. Mater. Chem.*, 1998, **8**, 2299.

Professor Navrotsky answered: Yes, and LSGM with substitution on both Ga and La sites is the most stable. But for all these perovskites the variation of heat of formation is linear with composition, with no minimum like that seen in the fluorite systems.

Dr Sokol commented: We should recognise two different situations when we are concerned with defect energies. In one case, we have "noninteracting" defects, which are reasonably well described in the limit of infinite dilution. Here, we can compare the calculated values with experimental data, available for example from temperature programmed diffusion and chemical kinetics studies. We do then often obtain agreement at the level of chemical accuracy. On the other hand, when defects are present in high concentrations, possibly forming complexes, or solid solutions are of interest, the energy of formation of single defects should not of course be compared with experiment directly, but rather suitable models of defective solids should be studied, and the defect energies then obtained using statistical averaging over studied configurations.

Dr de Leeuw said: I would like to add to Dr Sokol's comment, that indeed calculated heats of mixing of solid solutions have been shown to be in good agreement with experimentally obtained energies when full lattice relaxation is taken into account in the calculations.[1]

1   See, for example: C. R. A. Catlow, B. E. F. Fender and P. J. Hampson, Thermodynamics of MnO + CoO and MnO + NiO solid solutions, *J. Chem. Soc., Faraday Trans. 2*, 1977, **73**, 911–925.

Professor Haile addressed Dr Sokol: Is the challenge with solid solutions that the supercell sizes become exorbitantly large in order to represent the desired stoichiometries?

Dr Sokol agreed: Yes, you are right. Indeed, in these simulations, we need to evaluate very high numbers of configurations, even when using symmetry constraints. The cost of the simulations also grows rapidly for the systems where dopants or defects form associates/clusters, as large size supercells should be used to accommodate such correlated structures. For example, when increasing the linear size of a cubic simulation box by a factor of 2, typically we increase the length of our single point calculation by a factor of 8. Furthermore, this increase also leads to a rapid increase in the number of configurations, which should be sampled.

Capacitance of single crystal and low-angle tilt bicrystals of Fe-doped $SrTiO_3$

R. A. De Souza*[ab] and J. Maier[a]

Received 27th February 2006, Accepted 5th May 2006
First published as an Advance Article on the web 21st July 2006
DOI: 10.1039/b602914k

We used a.c. impedance spectroscopy to study the capacitance of single crystal and bicrystal Fe-doped $SrTiO_3$. Measurements performed on a single crystal sample indicate unequivocally that the bulk dielectric permittivity is dependent on defect concentration. Three symmetrical [001] tilt bicrystals with misorientation angles θ = 2.3, 5.4 and 7.8° were examined. The area specific capacitances obtained for the 5.4 and 7.8° boundaries are consistent with values predicted from a one-dimensional double-Schottky-barrier model. For the 2.3° boundary, more complex behaviour was observed. This is attributed to the electrical non-uniformity of the interface becoming significant at large dislocation separation. The effects of a d.c. bias on the impedance of the bicrystals was also investigated.

1. Introduction

The dielectric response of the perovskite oxide $SrTiO_3$ is, from a variety of standpoints, a subject of great interest. Much work has focused on the behaviour of this incipient ferroelectric at cryogenic temperatures, and in particular on the reduced dielectric permittivity of thin films in comparison with single crystals[1–4] and on the large dielectric nonlinearity in an applied electric field.[5–7] In the first part of this contribution we demonstrate a new unusual aspect of strontium titanate's dielectric response. At temperatures high enough to allow equilibration of the oxygen stoichiometry, that is, at temperatures well above ambient, we show that the capacitance of single crystal Fe-doped $SrTiO_3$ depends on oxygen partial pressure and thus on defect concentration.

In polycrystalline $SrTiO_3$ space-charge layers at grain boundaries act as tiny parallel plate capacitors and as a result allow artificially high dielectric permittivities to be achieved.[8] Our interest lies in how the crystallographic structure of an interface determines the strength and extent of the associated space-charge zones and thus the dielectric response of the interface. Elucidation of such structure–property relationships is hampered, when studying polycrystalline ceramics, by the difficulty in varying the interfacial structure of grain boundaries in a controlled manner: there is a multitude of random grain boundary orientations and most boundaries exhibit considerable curvature. Our approach to this problem has been to study symmetrical low-angle [001] tilt bicrystals of acceptor-doped $SrTiO_3$.[9–11] Besides containing single, planar interfaces, such samples offer the advantage of the misorientation

[a] Max-Planck-Institut für Festkörperforschung, Stuttgart, Germany
[b] Institut für Physikalische Chemie, RWTH Aachen, Germany. E-mail: desouza@pc.rwth-aachen.de; Fax: +49 241 80 92128; Tel: +49 241 80 94739

being accommodated by periodic arrays of edge dislocations whose separation d_{dis} is given by Frank's formula,[12]

$$d_{dis} = |\boldsymbol{b}|/\{2 \sin (\theta/2)\} \qquad (1)$$

where $\boldsymbol{b}$ is the Burgers vector and θ the misorientation angle. In other words, one can vary the areal density of identical, discrete interfacial structural units simply by varying the misorientation angle.

In a series of publications[9–11] we developed an electrical model of low-angle tilt grain boundaries in acceptor-doped $SrTiO_3$. The basis of the model is that oxygen vacancies prefer to reside in the dislocation cores.[9,13] Segregation of oxygen vacancies from the bulk to a dislocation core results in the formation of a space-charge tube, depleted of vacancies, surrounding each dislocation. Since a low-angle grain boundary consists, then, of an array of space-charge tubes, one parameter governing the boundary's electrical properties is the ratio of the dislocation separation d_{dis} to the radius of the space-charge tube λ^*_{dis}. For grain boundaries composed of closely spaced dislocations with very large depletion radii ($d_{dis} \ll \lambda^*_{dis}$), one may regard the charge distribution at the interface core as being virtually uniform, and one may therefore treat the interface in a one-dimensional model, in which the local properties vary only perpendicularly to the interface. As d_{dis} increases relative to λ^*_{dis}, the non-uniformity of the charge distribution will become increasingly apparent and deviations from the one-dimensional model are expected, i.e., the local properties now also vary along the interface. By means of detailed simulations, we ascertained that the electrical resistance of such boundaries is, surprisingly, more or less unaffected by confining the interface charge to the dislocations as opposed to distributing it homogeneously.[10] In the second part of this contribution we turn our attention to the capacitance of these low-angle tilt grain boundaries.

2. Theory

Fe substitutes for Ti in $SrTiO_3$, and for the oxidising conditions and moderate temperatures of this study is present as Fe'_{Ti} and $Fe^\times_{Ti}$. The trivalent iron dopants are predominantly compensated by oxygen vacancies but also some electron holes (which greatly reduces the concentration of electrons). The defect concentrations in the bulk, $[X]_b$, can therefore be summarised as

$$[Fe'_{Ti}]_b \approx 2[V_{\ddot{O}}]_b > [h^\bullet]_b \gg [e']_b \qquad (2)$$

Our treatment of the capacitance of low-angle grain boundaries is based on a simple one-dimensional double-Schottky-barrier model of grain boundaries in acceptor-doped $SrTiO_3$.[14–17] The validity of assuming a one-dimensional model is examined later in the light of the experimental results. In this simple model, the grain boundary core possesses a uniform positive charge per unit area, Q_{core}, which leads to the depletion of positive charge carriers and accumulation of negative charge carriers in the surrounding material. A schematic illustration of the model is shown in Fig. 1.

Our aim is an analytical expression for the capacitance as a function of misorientation angle, and to this end we assume that the dopant ions are immobile and completely ionised and that Q_{core} is independent of applied voltage. The latter allows us to treat the grain boundary as two space-charge regions, each with capacitance C_{sc}. Considering, therefore, one of the space-charge regions, we can write the relevant Poisson–Boltzmann equation as

$$\begin{aligned}
\frac{\partial^2 \phi(x)}{\partial x^2} = \frac{e}{\varepsilon_0 \varepsilon_r} &([Fe'_{Ti}]_b - 2[V_{\ddot{O}}]_b e^{-2\beta\{\phi(x)-\phi(\infty)\}} - [h^\bullet]_b e^{-\beta\{\phi(x)-\phi(\infty)\}} \\
&+ [e']_b e^{\beta\{\phi(x)-\phi(\infty)\}})
\end{aligned} \qquad (3)$$

where $\phi(x)$ is the electrical potential and $\phi(\infty)$ is the potential in the bulk far from the space-charge zone; $\varepsilon_r\varepsilon_0$ is the dielectric permittivity and taken to be that of the

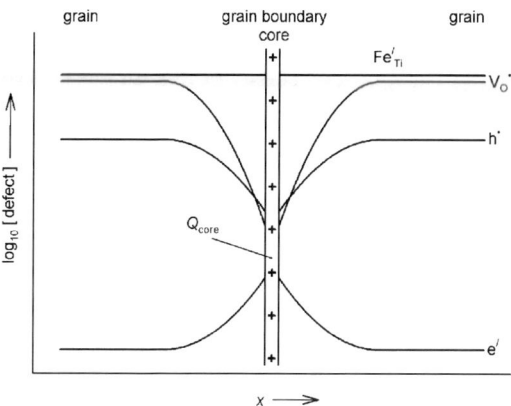

Fig. 1 Schematic illustration of the simple one-dimensional double-Schottky-barrier model of grain boundaries in acceptor-doped $SrTiO_3$.

bulk; and $\beta = e/k_B T$ with elementary charge e, Boltzmann constant k_B and absolute temperature T. As long as oxygen vacancies are the dominant defects in the bulk and as long as the negative space charge is due to a lack of vacancies, then the concentrations of electrons and electron holes in eqn (3) can be neglected. The Poisson–Boltzmann equation thus becomes

$$\frac{\partial^2 \phi(x)}{\partial x^2} = \frac{e}{\varepsilon_0 \varepsilon_r} ([\mathrm{Fe}'_{\mathrm{Ti}}]_b - 2[\mathrm{V_O^{\bullet\bullet}}]_b e^{-2\beta\{\phi(x)-\phi(\infty)\}}). \tag{4}$$

In this case it can be shown by means of an elegant integration[18] that the electrical field at $x = 0$ is given, to a good approximation, by

$$\left.\frac{\partial \phi(x)}{\partial x}\right|_{x=0} = -\left\{\frac{2e[\mathrm{Fe}'_{\mathrm{Ti}}]_b}{\varepsilon_0 \varepsilon_r}\left(\Phi_0 - \frac{1}{2\beta}\right)\right\}^{1/2}. \tag{5}$$

with space-charge potential $\Phi_0 = \phi(0) - \phi(\infty)$. Subsequently, one can make use of Gauss' law to express the charge due to the uncompensated acceptors in one of the space-charge layers in terms of the space-charge potential

$$Q_{\mathrm{sc}} = -\left\{2e[\mathrm{Fe}'_{\mathrm{Ti}}]_b \varepsilon_0 \varepsilon_r\left(\Phi_0 - \frac{1}{2\beta}\right)\right\}^{1/2}. \tag{6}$$

Defining the capacitance per unit area of one layer as

$$C_{\mathrm{sc}} = \frac{\partial |Q_{\mathrm{sc}}|}{\partial \Phi_0} \tag{7}$$

we thus find the grain boundary capacitance per unit area to be given by

$$C_{\mathrm{gb}} = \frac{C_{\mathrm{sc}}}{2} = \sqrt{\frac{e\varepsilon_0\varepsilon_r[\mathrm{Fe}'_{\mathrm{Ti}}]_b}{8(\Phi_0 - \frac{1}{2\beta})}}. \tag{8}$$

It is noted that eqn (8) is not valid for values of Φ_0 less than $1.5\beta^{-1}$, for which eqn (5) is no longer a valid approximation,[18] and for values of Φ_0 greater than $\beta^{-1} \ln\{[\mathrm{Fe}'_{\mathrm{Ti}}]_b/[e']_b\}$, for which the concentration of accumulated electrons exceeds that of the dopant and therefore can no longer be neglected from eqn (3).

From previous work on the resistance of the 5.4° boundary[9] we obtain the variation in Q_{core} with θ, assuming that the charge per unit length of dislocation

Q_{dis} is independent of dislocation spacing and subsequently assuming for our one-dimensional treatment that this charge is uniformly distributed at the interface, $Q_{core} = Q_{dis}/d_{dis}$. Combining $Q_{core} = 2Q_{sc}$ with eqn (6) and (8) yields an expression for C_{gb} in terms of Q_{core}, from which we can then predict C_{gb} as a function of θ

$$C_{gb} = \frac{e\varepsilon_0\varepsilon_r[\text{Fe}'_{Ti}]_b|\boldsymbol{b}|}{2Q_{dis}\sin(\theta/2)}.$$ (9)

3. Experimental

Single crystal and bicrystal samples of Fe-doped SrTiO$_3$ were obtained from CrysTec, Berlin, Germany and Mateck, Jülich, Germany. Samples were of the order of 5 mm × 5 mm × 1 mm. In each of the three bicrystal samples the grain boundary was parallel to the large sample faces and situated approximately in the middle of the slab. Small slivers of each sample were analysed with ICP-OES, in order to determine the dopant concentration [Fe$_{Ti}$]$_b$ (see Table 1).

Thin films of YBa$_2$Cu$_3$O$_{6+x}$ (200 nm thick), which were deposited by pulsed laser deposition onto the large sample faces, were employed as reversible electrodes. The use of YBa$_2$Cu$_3$O$_{6+x}$ electrodes minimises the possibility of the grain boundary arc being obscured by the electrode response.[9–11,17] Impedance spectra were acquired in the frequency range from 1 MHz down to 20 Hz (and in some cases down to 100 mHz) with a signal amplitude of 20 mV. Various oxygen partial pressures were set by diluting O$_2$ with N$_2$; the flow rate was kept constant at 40 mL min^{-1}. For each measurement point, the sample was first equilibrated at the temperature and oxygen partial pressure of interest, after which at least ten spectra were acquired. The ZView program (Ver. 2.1b, Scribner Associates, Inc.) was used to extract the bulk and grain boundary contributions from each spectrum.

4. Results and discussion

4.1 Single crystal

In Fig. 2(a) and (b) the resistance and capacitance of an Fe-doped single crystal sample are shown as a function of oxygen partial pressure at constant temperature. The resistance is found to obey a power law, $R_b \propto pO_2^{-m}$, with $m = 0.161 \pm 0.002$. These data are in good agreement with values calculated, as suggested by Rodewald et al.,[19] with the defect chemical model parameters given by Denk et al.,[20] but with the hole mobility given by Moos and Härdtl.[21] The small difference between experiment and prediction can be attributed to the fact that the defect model parameters were derived for a higher temperature range than that considered here (see also ref. 9). The deviation of m from 0.25 arises from [Fe$'_{Ti}$]$_b$ varying with oxygen partial pressure.

The results presented in Fig. 2(b) suggest that the bulk capacitance, and hence the relative dielectric permittivity ($\varepsilon_r = C_b d_b \varepsilon_0^{-1} A_b^{-1}$), is also dependent on oxygen partial pressure. The dependence is far weaker than that observed for R_b: assuming a power law, $C_b \propto pO_2^{-m}$, we find $m = (1.27 \pm 0.07) \times 10^{-2}$. In order to exclude the

Table 1 Selected characteristics of the bicrystals examined in this study. d_{dis}/λ^*_{dis}, n_{CPE} and C_{gb} refer to $T = 623$ K and $pO_2 = 0.1$ bar

$\theta/^\circ$	$d_{dis}/$nm	[Fe$_{Ti}$]$_b/$m^{-3}	d_{dis}/λ^*_{dis}	n_{CPE}	$C_{gb}/$F m^{-2}
2.3	9.73	7.0×10^{24}	0.94	0.18 ± 0.01	—
5.4	4.14	9.0×10^{24}	0.45	0.854 ± 0.001	1.69×10^{-2}
7.8	2.87	4.2×10^{24}	0.22	0.959 ± 0.001	7.72×10^{-3}

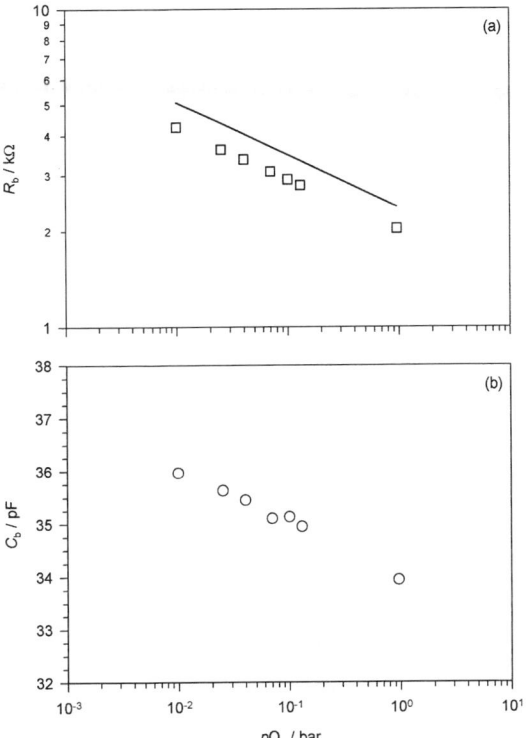

Fig. 2 (a) Bulk resistance and (b) bulk capacitance of an Fe-doped $SrTiO_3$ single crystal ($[Fe'_{Ti}]_b = 2.2 \times 10^{25}$ m^{-3}) as a function of oxygen partial pressure at $T = (672.5 \pm 0.5)$ K. Solid line in (a) refers to values predicted by defect chemical modelling (see text). NB: the ordinate scale is logarithmic in (a) and linear in (b). Values determined by impedance spectroscopy, with $YBa_2Cu_3O_{6+x}$ thin films as electrodes.

possibility that experimental artefacts are responsible for this effect (in particular, a variation in T with pO_2[22]), we performed a chemical diffusion experiment on the single crystal and monitored the incorporation of oxygen with impedance spectroscopy. From the resultant set of spectra we obtained as a function of time R_b and C_b, and subsequently, the normalised conductivity σ and the normalised relative dielectric permittivity ε_r plotted in Fig. 3. Since both quantities exhibited identical responses, since σ is dependent on defect concentration, as shown in Fig. 2(a), and since the responses occurred over a timescale (hours) that is consistent with the chemical incorporation and diffusion of oxygen into the sample,[23,24] this experiment unambiguously indicates that ε_r is dependent on defect concentration.

If we assume that the defects responsible for this effect are those present in the largest amounts, then the explanation for this behaviour must involve one or more of $Fe^{\times}_{Ti}$, Fe'_{Ti} and $V^{\cdot\cdot}_{O}$. The reorientation of $\{Fe'_{Ti} - V^{\cdot\cdot}_{O}\}$ associates in the applied a.c. field is unlikely to be the cause, however, since the concentration of such defect dipoles is negligible above $T \approx 573$ K,[25] and the results shown in Fig. 2(b) and Fig. 3 refer to $T = 673$ K. In addition, the reorientation of $\{Fe^{\times}_{Ti} - V^{\cdot\cdot}_{O}\}$ dipoles can be ruled out, for, if they exist, their binding energy is presumably lower than that of $\{Fe'_{Ti} - V^{\cdot\cdot}_{O}\}$ and one can therefore expect such dipoles to be fully dissociated at temperatures even below $T \approx 573$ K. The effect could be tentatively ascribed to oxygen vacancies alone. If one considers the dielectric response of $SrTiO_3$ in terms of the displacement of the Ti^{4+} ions relative to the other ions in the unit cell, one could imagine that, by introducing oxygen vacancies into $SrTiO_3$, some Ti^{4+} ions may be

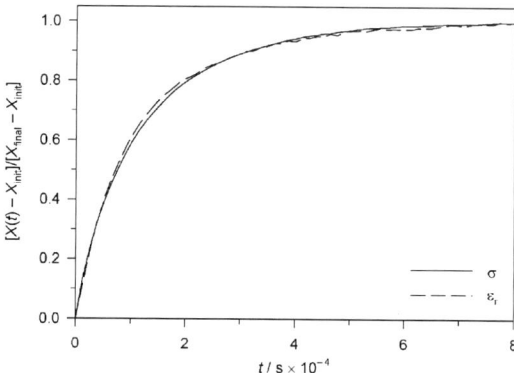

Fig. 3 Relaxation curves of an Fe-doped $SrTiO_3$ single crystal ($[Fe'_{Ti}]_b = 2.2 \times 10^{25}$ m^{-3}) following a change in oxygen partial pressure from 10^{-2}–1 bar at $T = (672.5 \pm 0.5)$ K: conductivity σ (solid line) and relative dielectric permittivity ε_r (broken line). The curve for ε_r is noisier, because of the weaker dependence of ε_r on pO_2. Curves determined by impedance spectroscopy, with $YBa_2Cu_3O_{6+x}$ thin films as electrodes.

allowed to make larger displacements. This would lead to an increase in dielectric permittivity with decreasing pO_2, in accordance with the results of Fig. 2(b). One important implication of oxygen vacancies being solely responsible for this effect is that grain boundaries that exhibit strong depletion of oxygen vacancies will be characterised by $\varepsilon_{gb} < \varepsilon_b$; without confirmation, however, we ignore this complication when evaluating the grain boundary data.

4.2 Bicrystals: at equilibrium

Impedance spectra obtained for the three bicrystals at $T = 623$ K and $pO_2 = 0.1$ bar are shown in Fig. 4, together with the fitted responses from the equivalent circuit $(R_bC_b)(R_{gb}Q_{gb})$, where Q_{gb} is a constant phase element whose impedance is $Z = Q^{-1}(i\omega)^{-n_{CPE}}$. It is emphasised that, although there is apparently only a single (bulk) arc in Fig. 4(a), $(R_bC_b)(R_{gb}Q_{gb})$ provided a significantly better fit to the data (χ^2 was more than an order of magnitude lower) than either (R_bC_b) or (R_bQ_b).

From ten spectra of each bicrystal at $T = 623$ K and $pO_2 = 0.1$ bar we determined mean values of the ideality parameter n_{CPE}. The results are listed in Table 1. In comparison with values determined for the other two boundaries, n_{CPE} for the 2.3° boundary is much lower and also exhibits a larger standard deviation. Such details, combined with the absence of a clear boundary arc in Fig. 4(a), may cast some doubt on the reliability of this value, or indeed, as to whether $(R_{gb}Q_{gb})$ provides the correct physical description. Attempts to effect the appearance of a clear grain boundary arc in the impedance response of the 2.3° boundary by varying the temperature or the oxygen partial pressure were unsuccessful. Nevertheless, values of n_{CPE} extracted from such measurements remained consistently below 0.4. We therefore contend, and results in the following section provide further support, that, if one can describe the response of this grain boundary in terms of a resistance and constant phase element connected in parallel, then low values of n_{CPE} are characteristic of the boundary.

In Table 1 we compare values of n_{CPE} for the three bicrystals with their respective d_{dis}/λ^*_{dis} ratios, λ^*_{dis} being given approximately by $\sqrt{Q_{dis}/\pi[Fe'_{Ti}]_b e}$. One sees that there is a clear trend towards smaller n_{CPE} with increasing d_{dis}/λ^*_{dis}. Since deviations of n_{CPE} from unity indicate increased dispersion of time constants from a single value, this trend reflects the electrical non-uniformity of the boundary becoming more pronounced as d_{dis}/λ^*_{dis} increases.

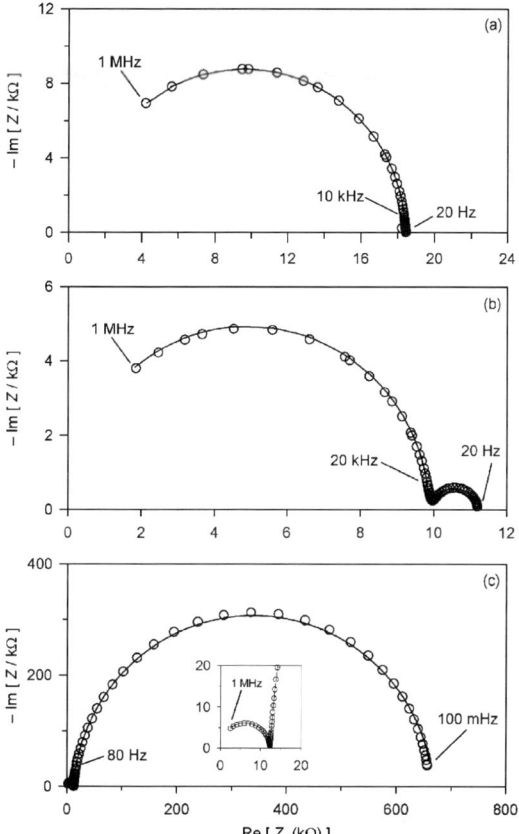

Fig. 4 Impedance spectra acquired for Fe-doped $SrTiO_3$ bicrystals with $YBa_2Cu_3O_{6+x}$ electrodes at $T = (623 \pm 0.5)$ K and $pO_2 = 0.1$ bar: (a) $\theta = 2.3°$, (b) $\theta = 5.4°$, (c) $\theta = 7.8°$ (inset shows region around origin). Solid lines correspond to fitted equivalent circuit $(R_bC_b)(R_{gb}Q_{gb})$.

For n_{CPE} close to unity there are sound reasons[26] for calculating an area specific capacitance according to

$$C_{gb} = (R_{gb}^{1-n_{CPE}}Q_{gb})^{1/n_{CPE}}A_{gb}^{-1},\qquad(10)$$

and values for the 5.4 and 7.8° boundaries thus calculated are listed in Table 1. No value is given for the 2.3° boundary, because there is no expression available for determining the capacitance associated with such low values of n_{CPE}. In Fig. 5 we plot the experimental C_{gb} values as a function of misorientation angle together with the behaviour predicted from eqn (9), that is, from a one-dimensional double-Schottky-barrier model. The value given for the 5.4° boundary in Table 1 was multiplied by a factor of $(4.2 \times 10^{24}/9.0 \times 10^{24})$ {see eqn (9)} to correct for the different dopant concentrations of the two bicrystals and thus permit a consistent comparison. The agreement between the experimental data and predicted behaviour is satisfactory given that no adjustable parameters were used in predicting C_{gb}.

The predicted curve in Fig. 5 is shown as a dashed line at low θ, since the results obtained for the 2.3° boundary indicate that for d_{dis}/λ^*_{dis} close to unity a two-dimensional space-charge model is required. We have employed an arbitrary limit of $d_{dis}/\lambda^*_{dis} = 0.5$ in Fig. 5 (which is equivalent to $\theta = 3.3°$), although one can imagine that at some lower value of d_{dis}/λ^*_{dis} deviations from the one-dimensional model will

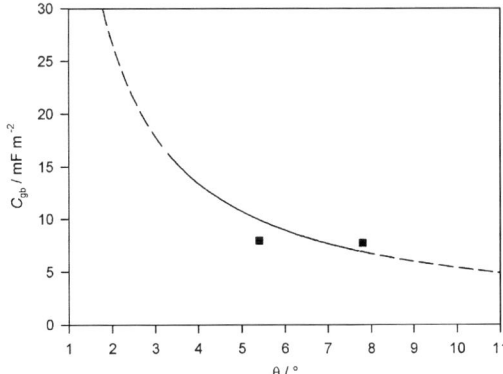

Fig. 5 Grain boundary capacitance per unit area as a function of the misorientation angle at $T = 673$ K and $pO_2 = 0.1$ bar for $[Fe'_{Ti}]_b = 4.2 \times 10^{24}$ m^{-3}: predicted variation (solid/dashed line), experimental data (■).

already be apparent and will increase with increasing d_{dis}/λ^*_{dis}. As a consequence it is impossible at present to judge whether the difference between prediction and experiment seen for the 5.4° boundary, being slightly larger than experimental error, is in part due to deviations from the one-dimensional model.

At high θ the predicted curve in Fig. 5 is also shown dashed on account of the predicted space-charge potential being large enough to cause an electron accumulation layer, $\Phi_0 > \beta^{-1} \ln\{[Fe'_{Ti}]_b/[e']_b\}$, and thus invalidate eqn (9). High-temperature defect models[20,21] give the electron concentration in the bulk as being of the order of 10^{10} m^{-3} for the conditions of interest, but this is certainly an overestimate.[10] Taking a more reasonable value of 10^8 m^{-3} for this lower temperature[10] still puts the upper limit of eqn (9) around $\theta = 7.8°$. Thus the acceptable agreement between prediction and experiment obtained for the 7.8° boundary suggests that either $[e']$ is 10^8 m^{-3} or lower, or that C_{gb} does not exhibit large departures from the predicted behaviour for $\Phi_0 > \beta^{-1} \ln\{[Fe'_{Ti}]_b/[e']_b\}$.

4.3 Bicrystals: under d.c. bias

In Fig. 6 we compare impedance spectra that were obtained for the three bicrystals with and without a d.c. bias. The conditions under which the spectra were acquired were not identical, but this is not of primary importance, as our interest here is in the qualitative behaviour.

We begin by examining the spectra obtained for the 5.4° bicrystal, which are shown in Fig. 6(c). The d.c. bias produced essentially two effects: a slightly larger, slightly depressed bulk arc ($R_b \uparrow$, $n^b_{CPE} \downarrow$); and a much larger, less depressed grain boundary arc ($R_{gb} \uparrow$, $n^{gb}_{CPE} \uparrow$). Such behaviour can be explained in terms of the stoichiometry polarisation of the bulk,[9] as in a Hebb–Wagner experiment, the boundary being more blocking for doubly charged, and thus more strongly depleted, oxygen vacancies than for singly charged electron holes (see Fig. 1).

Turning now to the 2.3° bicrystal (Fig. 6(b)), we see that the d.c. bias caused the appearance of a small grain boundary contribution, which is characterised by a low value of n_{CPE}. More importantly, we recognise that the d.c. bias produced the same effects as those observed for the 5.4° bicrystal: $R_b \uparrow$, $n^b_{CPE} \downarrow$; $R_{gb} \uparrow$, $n^{gb}_{CPE} \uparrow$. Since no such effects were observed for the single crystal, as one can see in Fig. 6(a), this result therefore supports our view that, if it can be described by a CPE, this boundary is characterised by a low n_{CPE}.

The behaviour of the 7.8° bicrystal, shown in Fig. 6(d), was different. This is because R_{gb}/R_b was much larger than unity and hence the applied voltage fell principally over the grain boundary. Thus, in comparison with the two other

This journal is © The Royal Society of Chemistry 2006

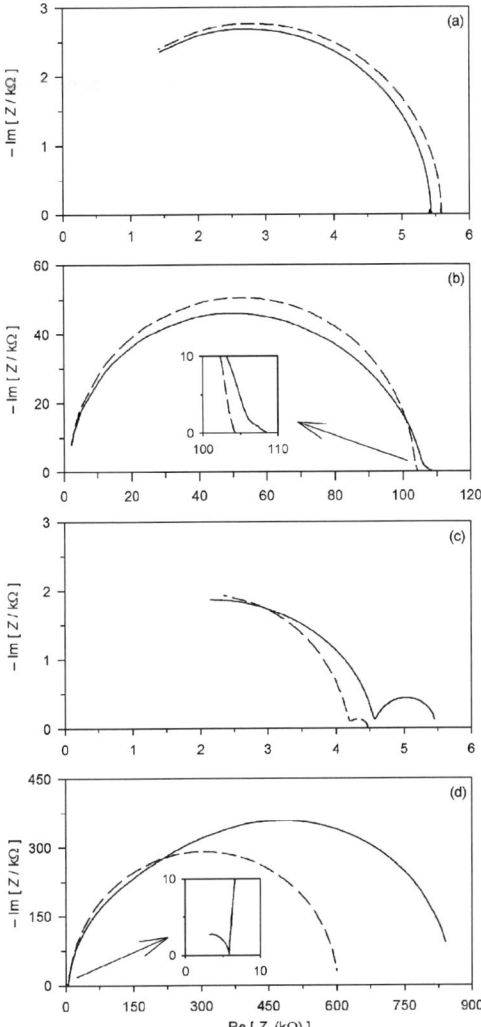

Fig. 6 Effect of d.c. bias on the impedance response of the Fe-doped $SrTiO_3$ bicrystals. Dashed lines refer to response without bias, solid lines to response with bias; the same frequency range was used in both cases. (a) Single crystal, $U_{dc} = 0.25$ V, $T = 623$ K; (b) $\theta = 2.3°$, $U_{dc} = 0.5$ V, $T = 570$ K; (c) $\theta = 5.4°$, $U_{dc} = 0.5$ V, $T = 673$ K; (d) $\theta = 7.8°$, $U_{dc} = 0.4$ V, $T = 673$ K.

bicrystals, for which $R_{gb}/R_b < 1$, there was negligible stoichiometry polarisation of the bulk, as shown in the inset of Fig. 6(d), and only changes in the grain boundary contribution were observed. Under an applied bias, the impedance response is in fact better described by the equivalent circuit $(R_bC_b)(R^A_{gb}Q^A_{gb})(R^B_{gb}Q^B_{gb})$, rather than $(R_bC_b)(R_{gb}Q_{gb})$; that is, the grain boundary contribution seems to split into two arcs, A and B. Further measurements were undertaken to determine the variation of the A and B grain boundary parameters as a function of applied bias. The results are plotted in Fig. 7, with the data points for zero applied bias being taken from the spectrum without bias and being given by $R_{gb}/2$ and $2C_{gb}$.

We suggest that the splitting of the grain boundary response into two arcs results from one of the space-charge layers at the boundary being forward-biased and the other being reverse-biased. Given that this boundary is characterised by $p-n$

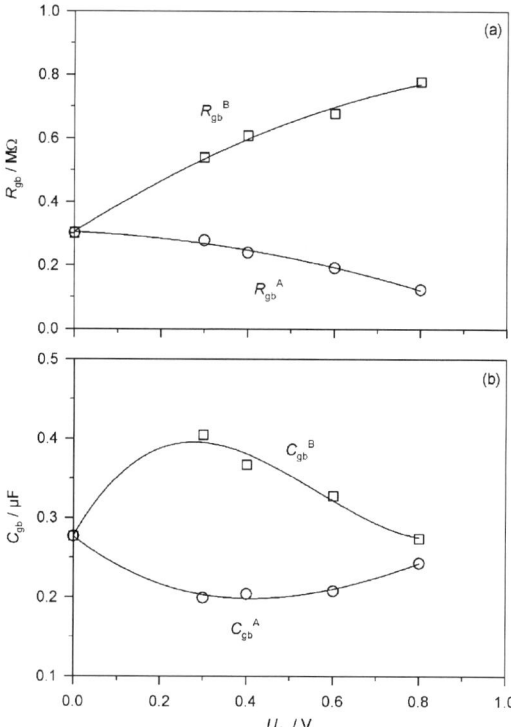

Fig. 7 Resolved grain boundary resistances and capacitances as a function of applied d.c. voltage bias for the 7.8° boundary at $T = 673$ K and $pO_2 = 0.1$ bar. Lines are guides to the eye.

inversion layers,[10] one can therefore explain the changes in the resistances in terms of a forward biased p–n layer leading to reduced resistance (R_{gb}^A) and a reverse biased p–n layer leading to increased resistance (R_{gb}^B). One would consequently expect the capacitance of the forward-biased region, C_{gb}^A, to increase and that of the reverse-biased, C_{gb}^B, to decrease. In Fig. 7(b), however, C_{gb}^B is larger than C_{gb}^A, and both exhibit complex behaviour. At present a simple explanation for this anomalous behaviour is unavailable.

5. Concluding remarks

Three main points emerge from this study:

• Equilibrium measurements as a function of oxygen partial pressure and a chemical relaxation experiment indicate that the relative dielectric permittivity of single crystal Fe-doped $SrTiO_3$ depends on defect concentration.

• We found a correlation between the grain boundary's degree of electrical non-uniformity, defined in terms of the ratio of the dislocation separation d_{dis} to the radius of the space-charge tube surrounding each dislocation λ^*_{dis}, and the ideality parameter of the grain boundary's constant phase element, n_{CPE}. This necessarily assumes that the impedance response of a low-angle grain boundary can be described by a parallel connection of a resistance and a constant phase element.

• Despite using a one-dimensional model with several simplifying assumptions, we found satisfactory agreement between the predicted variation in area specific capacitance with misorientation angle and the experimental data for boundaries with minimal electrical non-uniformity.

Future work will focus on numerical modelling of the drift-diffusion equations, in one dimension, to examine the behaviour of grain boundaries in mixed conductors under d.c. bias; and in two dimensions, to examine the impedance response of boundaries with significant electrical non-uniformity.

Acknowledgements

R. A. De S. gratefully acknowledges helpful discussions with R. Meyer, J. Fleig and J. Jamnik.

References

1 C. Zhou and D. M. Newns, *J. Appl. Phys.*, 1997, **82**, 3081.
2 S. K. Streiffer, C. Basceri, C. B. Parker, S. E. Lash and A. I. Kingon, *J. Appl. Phys.*, 1999, **86**, 4564.
3 A. A. Sirenko, C. Bernhard, A. Golnik, A. M. Clark, J. Hao, W. Si and X. X. Xi, *Nature*, 2000, **404**, 373.
4 L. J. Sinnamon, M. M. Saad, R. M. Bowman and J. M. Gregg, *Appl. Phys. Lett.*, 2002, **81**, 703.
5 A. B. Kozyrev, T. B Samoilova, A. A. Golovkov, E. K. Hollmann, D. A. Kalinikos, V. E. Loginov, A. M. Prudan, O. I. Soldatenkov, D. Galt, C. H. Mueller, T. V. Rivkin and G. A. Koepf, *J. Appl. Phys.*, 1998, **84**, 3326.
6 D. Fuchs, C. W. Schneider, R. Schneider and H. Rietschel, *J. Appl. Phys.*, 1999, **85**, 7362.
7 A. K. Tagantsev, V. O. Sherman, K. F. Astafiev, J. Venkatesh and N. Setter, *J. Electroceram.*, 2003, **11**, 5.
8 A. J. Moulson and J. M. Herbert, *Electroceramics*, Chapman and Hall, London, 1990.
9 R. A. De Souza, J. Fleig, J. Maier, O. Kienzle, Z. Zhang, W. Sigle and M. Rühle, *J. Am. Ceram. Soc.*, 2003, **86**, 922.
10 R. A. De Souza, J. Fleig, J. Maier, Z. Zhang, W. Sigle and M. Rühle, *J. Appl. Phys.*, 2005, **97**, 053502.
11 Z. Zhang, W. Sigle, R. A. De Souza, W. Kurtz, J. Maier and M. Rühle, *Acta Mater.*, 2005, **53**, 5007.
12 See, *e.g.*, *Interfaces in Crystalline Materials*, ed. A. P. Sutton and R. W. Balluffi, Clarendon Press, Oxford, 1995, p. 105.
13 Z. Zhang, W. Sigle and M. Rühle, *Phys. Rev. B*, 2002, **66**, 094108.
14 G. E. Pike, in *Materials Science and Technology*, ed. M. V. Swain, VCH, Weinheim, 1994, vol. 11, p. 731.
15 Y.-M. Chiang and T. Takagi, *J. Am. Ceram. Soc.*, 1990, **73**, 3278.
16 M. Vollmann and R. Waser, *J. Am. Ceram. Soc.*, 1994, **77**, 235.
17 I. Denk, J. Claus and J. Maier, *J. Electrochem. Soc.*, 1997, **144**, 3526.
18 E. H. Rhoderick and R. H. Williams, *Metal-Semiconductor Contacts*, Clarendon Press, Oxford, UK, 1988.
19 S. Rodewald, J. Fleig and J. Maier, *J. Am. Ceram. Soc.*, 2001, **84**, 521.
20 I. Denk, W. Münch and J. Maier, *J. Am. Ceram. Soc.*, 1995, **78**, 3265.
21 R. Moos and K.-H. Härdtl, *J. Am. Ceram. Soc.*, 1997, **80**, 2549.
22 Variation of temperature with oxygen partial pressure was minimised by using N_2 as an inactive diluent rather than, say, Ar.
23 J. Claus, M. Leonhardt and J. Maier, *J. Phys. Chem. Solids*, 2000, **61**, 1199.
24 M. Leonhardt, R. A. De Souza, J. Claus and J. Maier, *J. Electrochem. Soc.*, 2002, **149**, J19.
25 R. Merkle and J. Maier, *Phys. Chem. Chem. Phys.*, 2003, **5**, 2297.
26 J. Fleig, *Solid State Ionics*, 2002, **150**, 181.

Structure and thermodynamic stability of hydrogen interstitials in BaZrO$_3$ perovskite oxide from density functional calculations

Mårten E. Björketun, Per G. Sundell and Göran Wahnström

Received 13th February 2006, Accepted 18th April 2006
First published as an Advance Article on the web 20th July 2006
DOI: 10.1039/b602081j

Density functional calculations have been used to study the electronic structure, preferred sites in the lattice, formation energies and vibrational frequencies for hydrogen interstitials in different charge states in the cubic phase of perovskite-structured BaZrO$_3$. By combining *ab initio* results with thermodynamic modeling, defect formation at finite temperature and pressure has been investigated. We demonstrate how the site selectivity and spatial distribution of dopant atoms in the lattice can be affected by changes in the environmental conditions (atomic chemical potentials, oxygen partial pressure and temperature) used during processing of the material. In addition, we have calculated the thermodynamic parameters of the water uptake reaction for an acceptor-doped BaZrO$_3$ crystal in equilibrium with a humid atmosphere. The interaction energies between a protonic defect and the investigated Ga, Gd, In, Nd, Sc, and Y dopants were found to be attractive, and we show that a simple model of defect association may reproduce an experimentally observed trend in the hydration enthalpy.

1 Introduction

Many materials based on perovskite-structured oxides exhibit significant oxygen ion and proton conductivity, and have therefore attracted a lot of attention for their potential use as solid membranes in various electrochemical applications.[1] In particular, there has been recent interest in BaZrO$_3$, which, when doped with lower-valent cations on the Zr^{4+} sites and exposed to a humid atmosphere, is known to become a protonic conductor.[2–9] This material appears to combine a high proton conductivity with good chemical and mechanical stability,[2,3,10] which makes it a potential candidate for solid oxide fuel cell (SOFC) applications. The high proton mobility was however long overlooked because conductivity measurements were performed on ceramic samples and at high temperatures—conditions where the bulk conductivity is covered by large grain boundary resistances.[11] Finally, since this particular perovskite system has a near cubic structure over a wide range of temperatures and dopant concentrations,[2] it is also an important model system for the general phenomenon of fast protonic transport in oxides.

According to the well-established defect chemistry model of perovskite oxides,[2,12] the introduction of aliovalent dopants results in the formation of charge-compensating oxygen vacancies. Upon the subsequent hydration, oxygen vacancies are

Department of Applied Physics, Chalmers University of Technology, S-412 96, Göteborg, Sweden. E-mail: martebjo@fy.chalmers.se

replaced by protonic defects (in the form of hydroxyl ions residing on oxygen ion sites) *via* dissociative absorption of water from the gas phase according to

$$H_2O(g) + V_O^{\cdot\cdot} + O_O^{\times} \rightleftharpoons 2(OH_O^{\cdot}) \tag{1}$$

in Kröger–Vink notation. In the ideal situation, the oxygen vacancy concentration in the 'dry' state would therefore equal half the dopant concentration, whereas the concentration of protonic defects in a fully hydrated sample would exactly match the dopant concentration. However, the available thermodynamic data for reaction (1) show a considerable variation between different oxides.[12] Even for the $BaZrO_3$ system, the hydration enthalpy and entropy varies significantly with the dopant used.[3,7,13] In addition, the maximum water uptake is often considerably less than the theoretical limit, and it also seems to depend on the choice of dopant.[3] This deviation from the nominal value has been explained by a distribution of dopant atoms over both cation sites.[14] A detailed understanding of the microscopic mechanisms for this behavior is important in order to be able to predict and control the protonic concentration in new materials and material classes for future energy applications.

Several theoretical studies based on computer modeling techniques have previously addressed the formation and mobility of protonic defects in perovskite-type zirconates.[15–23] In particular, atomistic simulations using inter-atomic potentials have been used to calculate the formation energies of intrinsic defects, solution energies for impurity atoms and activation energies for ionic migration.[15,16] There is also a growing interest in quantum mechanical (or *ab initio*) approaches,[15,17,18,20–22] which are able to provide insights regarding the electronic structure of defects.[19,23] However, since these approaches are usually to be considered as zero-temperature, zero-pressure techniques, direct comparison with experiments carried out at ambient conditions is by no means straightforward.[24]

We have previously investigated dopant substitution and the formation of anion and cation vacancies in $BaZrO_3$ in equilibrium with an oxygen-containing atmosphere.[25] This will now be extended to also include the technologically important case of hydrogen incorporation. Using quantum mechanical simulations within the framework of the density functional theory (DFT), we have investigated the structural, electronic and vibrational properties of hydrogen interstitials in various charge states in the cubic phase of $BaZrO_3$. By combining these *ab initio* results with thermodynamic modeling, we have been able to study defect formation at finite temperature and pressure. In particular, we have calculated the change in free energy associated with the hydration reaction (1). This allows for a direct comparison with experiments conducted under typical fuel cell operating conditions. We have also investigated how the structure and stability of hydrogen defects can be influenced by the choice of dopants. To this end, we have considered the incorporation of Ga, Gd, In, Nd, Sc, and Y impurities on different sites in the lattice. We demonstrate how the environmental conditions used during synthesis (Ba and Zr chemical potentials, oxygen partial pressure and temperature) can seriously affect the dopant site-selectivity, and thus limit the water uptake during subsequent hydration. In addition, we have assessed the interaction energies of various defect clusters, and we show that a simple model of defect association (or 'trapping') may qualitatively explain experimentally observed trends in hydration enthalpies for different dopants.

2 Theoretical formalism

The necessary formalism to determine the formation energies of defects in various charge states in semiconductors and insulators from *ab initio* total-energy calculations has been well established during the last decade.[26–28] More recently, it has also been demonstrated how such results can be extended to take into account the effects of finite temperature and pressure.[29] In this section, we will apply the formalism to a $BaZrO_3$ crystal in contact with a humid oxygen-containing atmosphere.

2.1 Defect formation energies

Defect formation energies can be defined as the total energy of the system containing a defect minus the total energy of the perfect system. When investigating defects that change the composition of a material, the formation energies are usually evaluated with respect to a set of fixed external chemical potentials for the atomic species involved. The results will then depend on the actual values of these atomic chemical potentials and, in the case of charged defects, on the chemical potential of electrons (*i.e.* the Fermi energy).[27]

The total energy required to create a hydrogen interstitial (H_i) in charge state q can be calculated from the difference in total energy E^{tot} of two supercells containing n primitive cells as[26]

$$\Delta E_{H_i^q}^f = E^{tot}[n(BaZrO_3H_{1/n}); q] - E^{tot}[n(BaZrO_3)] - \mu_H + q\mu_e \quad (2)$$

where $1/n$ is the effective hydrogen concentration, μ_H is the chemical potential of hydrogen atoms, and μ_e is the chemical potential of electrons (*i.e.* the Fermi energy). By $E^{tot}[\ldots;q]$ we denote the total energy of a supercell where the number of electrons has been adjusted to create a defect in charge state q. Similarly, the solution energy of a dopant atom M on a barium site (M_{Ba}) or a zirconium site (M_{Zr}) can be calculated as the total energy difference[30]

$$\Delta E_{M_{Ba}^q}^f = E^{tot}[n(Ba_{1-1/n}M_{1/n}ZrO_3);q] - E^{tot}[n(BaZrO_3)] + \mu_{Ba} - \mu_M + q\mu_e \quad (3)$$

$$\Delta E_{M_{Zr}^q}^f = E^{tot}[n(BaM_{1/n}Zr_{1-1/n}O_3);q] - E^{tot}[n(BaZrO_3)] + \mu_{Zr} - \mu_M + q\mu_e \quad (4)$$

where $1/n$ is the effective dopant concentration and μ_{Ba}, μ_{Zr}, and μ_M denote the chemical potential of Ba, Zr and dopant atoms, respectively.

When investigating defect formation in the dilute limit, care must be taken to choose a sufficiently large supercell to avoid any interactions between defects in neighboring cells. However, defect interactions are interesting in their own right, since they may result in important phenomena such as 'trapping' of hydrogen interstitials near dopants and dopant–oxygen vacancy association. The interaction energy ΔE^{int} for a pair of point defects can be calculated as the difference in formation energy of the two defects occupying neighboring sites in the lattice and the sum of the formation energies of the individual defects in the dilute limit[15] as

$$\Delta E^{int} = \Delta E_{pair}^f - \Sigma \Delta E_{isolated\ defects}^f. \quad (5)$$

2.2 Environmental conditions. In order to calculate defect formation energies using eqn (2)–(4), we need to define the chemical potentials of atoms being removed from or added to the system. These quantities will in general depend on the experimental conditions. For instance, if the system is in equilibrium with an atmosphere containing hydrogen gas, the hydrogen chemical potential μ_H can be set equal to half the total energy of a H_2 dimer. In this work, however, we assume that an atmosphere containing oxygen and water vapor is present in all cases. At $T = 0$ we may therefore define μ_H as

$$\mu_H = \frac{1}{2}(E^{tot}[H_2O] - \mu_O), \quad (6)$$

where $E^{tot}[H_2O]$ is the total energy of an isolated water molecule, and the oxygen chemical potential

$$\mu_O = \frac{1}{2}E^{tot}[O_2] \quad (7)$$

is half the total energy of an isolated oxygen molecule. When calculating dopant solution energies (eqn (3) and (4)), we must also take into account that $BaZrO_3$ may exist within a range of chemical potentials for barium and zirconium. Two different limits will be considered: in the *Ba rich* limit the system is in equilibrium with BaO(s) and hence the barium chemical potential is given by $\mu_{Ba} = \mu_{Ba}^{BaO} \equiv E^{tot}[BaO] - \mu_O$. In the *Zr rich* limit the system is instead in equilibrium with ZrO$_2$(s) and the zirconium chemical potential is given by $\mu_{Zr} = \mu_{Zr}^{ZrO2} \equiv E^{tot}[ZrO_2] - 2\mu_O$. We have previously shown[25] that the allowed range for μ_{Ba} can be written as

$$\mu_{Ba}^{BaO} + \Delta E_{BaZrO_3}^f < \mu_{Ba} < \mu_{Ba}^{BaO}, \tag{8}$$

where $\Delta E_{BaZrO_3}^f$, the heat of formation for BaZrO$_3$, is negative. For any value of the chemical potential for Ba within this range, the chemical potential for Zr must assume the value

$$\mu_{Zr} = \mu_{Zr}^{ZrO2} + \Delta E_{BaZrO_3}^f + \mu_{Ba}^{BaO} - \mu_{Ba}. \tag{9}$$

Finally, the chemical potential μ_M of a dopant M can be determined from the total energy of the corresponding binary oxide M_2O_3. In the present work we will, however, only study differences in solution energies of a dopant on Ba and Zr sites (*cf.* section 4.1) and may therefore disregard μ_M.

In a semiconductor or an insulator, the electronic chemical potential μ_e can assume values ranging from the energy of the valence band maximum, ε_{VBM}, to the energy of the conduction band minimum, ε_{CBM}. These limiting values of μ_e can, in practical supercell calculations, be obtained as the ground state total-energy differences $\varepsilon_{VBM} = E^{tot}[0] - E^{tot}[+1]$ and $\varepsilon_{CBM} = E^{tot}[-1] - E^{tot}[0]$, where $E^{tot}[q]$ denotes the total energy of a perfect lattice supercell where q electrons have been removed.[20,29]

2.3 Finite temperatures and pressures

Formation energies calculated using eqn (2)–(4) are strictly valid only at $T = 0$, and they also exclude zero-point motion effects. In order to compare with experiments conducted at finite pressure p and temperature $T > 0$, the relevant thermodynamical potential becomes Gibb's free energy $G(p,T)$. At equilibrium, and in the dilute limit, the concentration c of a particular defect is given by

$$c(p,T) = e^{-\Delta G^f(p,T)/k_B T}, \tag{10}$$

where k_B is the Boltzmann's constant,

$$\Delta G^f(p,T) = \Delta H^f(p,T) - T\Delta S^f(p,T), \tag{11}$$

and ΔH^f and ΔS^f is the change in enthalpy and entropy, respectively, associated with creation of the defect. There will in general be several different contributions to the defect formation free energy, eqn (11). For the solid phases, in the harmonic approximation, Gibb's free energy can be written as

$$G(p,T) = E^{tot} + F^{vib}(T), \tag{12}$$

where E^{tot} is the total electronic energy of the solid, and $F^{vib}(T)$ is the vibrational part. The latter is given by

$$F^{vib}(T) = \sum_{s=1}^{3N} \left\{ \frac{h v_s}{2} + k_B T \ln[1 - \exp(-h v_s/k_B T)] \right\}, \tag{13}$$

where s runs over the $3N$ normal mode branches in the lattice of N atoms. When a defect is created, the vibrational free energy of the system will change in two distinct ways: firstly, if an atom is removed (vacancy formation) or added (interstitial formation) the defective lattice will have a different number of vibrational modes compared to the perfect crystal. Secondly, the frequencies of the remaining modes may change in the presence of the defect. Since we have assumed that our crystal is in equilibrium with a humid oxygen-containing atmosphere, atoms may also be exchanged between the solid and the gas phase when defects are formed. Assuming ideal gas behavior, the free energy per gas molecule is

$$g(p, T) = h^0(T) - Ts^0(T) + k_B T \ln \frac{p}{p^0}, \tag{14}$$

where $h^0(T)$ and $s^0(T)$ are the enthalpy and absolute entropy, respectively, per gas molecule at the reference partial pressure p^0. As the electronic total energy obtained from DFT calculations corresponds to zero temperature, we set the reference enthalpy equal to the zero-point vibrational energy of the molecule.

We will explicitly consider the hydration reaction, eqn (1), where two hydrogen interstitials are formed by dissociative absorption of a water molecule into an oxygen vacancy. The change in total energy ΔE_{hydr}^{tot} can then be calculated as

$$\Delta E_{hydr}^{tot} = 2\Delta E_{H_i^q}^f - \Delta E_{V_O^q}^f, \tag{15}$$

where $\Delta E_{H_i^q}^f$ is the formation energy of a hydrogen interstitial eqn (2) and $\Delta E_{V_O^q}^f$ is the formation energy of an oxygen vacancy.[25] For the vibrational part, we apply an Einstein model where it is assumed that all oxygens in the perfect lattice have the same vibrational frequencies $\{\nu_{O, s}\}_{s=1}^3$. Moreover, when an oxygen is either removed from the crystal or becomes coordinated to a hydrogen interstitial, we assume that only the eight nearest oxygens are affected so that their vibrational frequencies change to $\{\nu_{O,s}'\}_{s=1}^3$ or $\{\nu_{O,s}''\}_{s=1}^3$, respectively. The free energy of the nine extra lattice modes that appear when an oxygen vacancy is filled and two regular oxygen sites are hydrated then gives a contribution

$$\Delta \tilde{G}_{hydr}^{vib}(T) = 2\sum_{s=1}^3 \left\{ \frac{h\nu_{H,s}}{2} + k_B T \ln[1 - \exp(-h\nu_{H,s}/k_B T)] \right\} \\ + \sum_{s=1}^3 \left\{ \frac{h\nu_{O,s}}{2} + k_B T \ln[1 - \exp(-h\nu_{O,s}/k_B T)] \right\}, \tag{16}$$

where $\{\nu_{H, s}\}_{s=1}^3$ are the hydrogen vibrational frequencies. Similarly, the change in the vibrational frequencies of the surrounding oxygens is given by

$$\Delta G_{hydr}^{vib}(T) = 2 \times 8 \sum_{s=1}^3 \left\{ \frac{h(\nu_{O,s}'' - \nu_{O,s})}{2} + k_B T \ln \frac{1 - \exp(-h\nu_{O,s}''/k_B T)}{1 - \exp(-h\nu_{O,s}/k_B T)} \right\} \\ + 8 \sum_{s=1}^3 \left\{ \frac{h(\nu_{O,s} - \nu_{O,s}')}{2} + k_B T \ln \frac{1 - \exp(-h\nu_{O,s}/k_B T)}{1 - \exp(-h\nu_{O,s}'/k_B T)} \right\}, \tag{17}$$

The increased number of vibrational degrees of freedom of the crystal is however compensated by an increase in free energy due to the removal of a water molecule from the gas phase. The latter is given by

$$\Delta G_{hydr}^{gas}(p, T) = -g_{H_2O}(p, T), \tag{18}$$

where $g_{H_2O}(p, T)$ is the free energy per $H_2O(g)$ obtained from eqn (14). In total, the change in free energy upon absorption of one water molecule can hence be written as

$$\Delta G_{hydr}^f(p, T) = \Delta E_{hydr}^{tot} + \Delta \tilde{G}_{hydr}^{vib}(T) + \Delta G_{hydr}^{vib}(T) + \Delta G_{hydr}^{gas}(p, T), \tag{19}$$

where ΔE_{hydr}^{tot} is the change in electronic total energy, $\Delta \tilde{G}_{hydr}^{vib}$ and ΔG_{hydr}^{vib} are the vibrational contributions defined by eqn (16) and (17), respectively, and finally ΔG_{hydr}^{gas} is the gas phase contribution given by eqn (18).

3 Computational details

In the present work we carried out DFT calculations based on the plane-wave/ psuedopotential method as implemented in the Vienna *ab initio* simulation package

$(VASP)$[31,32] to determine the electronic structures and formation energies of defects that affect the hydration of $BaZrO_3$. For the exchange–correlation functional, we used a generalized gradient approximation (GGA) due to Perdew and Wang.[33] Electron–ion interaction was described by the projector augmented wave method.[34] A plane-waves basis set with a cutoff energy of 400 eV was used in all calculations. Brillouin zone sampling was performed using a $6 \times 6 \times 6$ k-point grid for the five-atom primitive cell. The equilibrium structure was determined by calculating the total energy for several different lattice parameters and fitting to Murnaghan's equation-of-state. That gave a cubic structure with a lattice constant $a_0 = 4.25$ Å, which is only 1.4% larger than the experimental value 4.19 Å.[35] The heat of formation of $BaZrO_3$ from the binary oxides BaO and ZrO_2,

$$\Delta E_{BaZrO_3}^f = E^{tot}[BaZrO_3] - E^{tot}[BaO] - E^{tot}[ZrO_2], \qquad (20)$$

was calculated to be -1.31 eV, in good agreement with the experimental value -1.33 eV.[36] However, the calculated band gap is around 3 eV, which is significantly smaller than the experimental results 5.3 eV.[37] That the electronic band gaps of insulators and semiconductors come out grossly underestimated is a well-known artifact of the traditional DFT.[38] It is likely to affect the calculated formation energies of defects that introduce electrons in donor states with conduction band character.[27] Such results may be improved by a better treatment of the non-local exchange part of the total-energy functional, or more heuristically by a simple correction scheme where the conduction band is assumed to be rigidly shifted upward to match the experimental band gap.[26,39] This will be discussed further in section 4.2. Point defects were modeled using periodically repeated supercells consisting of either $2 \times 2 \times 2$ or $3 \times 3 \times 3$ primitive cells with the number of k-points in each direction reduced accordingly. This corresponds to defect concentrations $c = 1/8$ and $c = 1/27$, respectively. In the simulations of charged defects, electrons were removed from or added to the supercell. In order to avoid divergence of the Coulomb energy of the (infinite) crystal, the resulting electronic charge was neutralized by the standard means of including a jellium background. Structural optimizations were performed at constant volume until all residual forces were smaller than 0.05 eV Å^{-1}. Vibrational frequencies were calculated within the harmonic approximation by evaluating and diagonalizing a dynamical matrix. Except for the O_2 molecule, spin-polarization was not taken into account. However, although the electronic ground state of pure $BaZrO_3$ is non-magnetic, this cannot be assumed *a priori* for the investigated point defects in the material.[40] Therefore, we have also performed a few spin-polarized test calculations to check the accuracy of our results.

4 Results and discussion

4.1 Dopant atoms

To start with, we will investigate the incorporation of Ga, Gd, In, Nd, Sc and Y dopants in $BaZrO_3$. Without specifying the actual value of the atomic chemical potentials μ_M, it is possible to address several interesting questions about these impurities.

Firstly, we will discuss site selectivity, *i.e.* whether a particular dopant M prefers to substitute for Ba or Zr. This is an important issue, since occupation of Ba sites will reduce the number of charge compensating oxygen vacancies created and hence result in a suppression of the water uptake at low temperatures. From eqn (3) and (4) it is seen that the difference in solution energy of a dopant atom on a Ba site and on a Zr site is independent of μ_M. However, it will depend on μ_{Ba} and μ_{Zr}, which are not fixed but may depend on the environment as discussed in section 2.2. For instance, the energy to incorporate a dopant on a Ba site is lower under Zr rich conditions (low μ_{Ba}, high μ_{Zr}) than under Ba rich conditions (high μ_{Ba}, low μ_{Zr}) and *vice versa*. In addition, the formation energy of a charged impurity atom will depend on the prevailing doping of the material, *i.e.* on the actual position of the Fermi energy μ_{e}. We have previously shown[25] that substituting a Zr atom in $BaZrO_3$ with any of the investigated dopants in the present study introduces acceptor levels close to (less than 0.06 eV above) the top of the valence band. When substituting a Ba atom instead, donor levels appear close to the bottom of the conduction band for the large dopants Gd, Nd, and Y, further down in the gap for Sc, and close to the top of the valence band for Ga and In. Here we will assume a $BaZrO_3$ crystal that initially has p-type character, $(\mu_{\mathrm{e}} \approx \varepsilon_{\mathrm{VBM}})$ so that dopants are present as M'_{Zr} on Zr sites, but as M_{Ba} on Ba sites (except for M = Ga and In which are neutral). The difference in solution energy for the preferred charge state on Ba and on Zr sites is shown in Fig. 1 as a function of the ionic radius[41] of the dopant. The calculations were performed using $3 \times 3 \times 3$ supercells containing one dopant, *i.e.* corresponding to dopant concentrations $c = 1/27$. In particular, results are presented for the two stability limits of $BaZrO_3$, the Ba rich and the Zr rich limit, *cf.* section 2.2. If changes in vibrational free energy between the solid phases are neglected it can be shown from eqn (3), (4), (8) and (9) that the differences in solution free energy will vary with temperature and pressure as the temperature and pressure dependent part of μ_{O}, $\frac{1}{2}g_{\mathrm{O}_2}(p,T)$, given by eqn (14). Accordingly, the lower horizontal line in Fig. 1 marks the crossover between Zr site and Ba site preference at $T = 0$ K for any oxygen partial pressure. Solution energy differences found over this line indicate a preference for the Zr site. Likewise, the upper horizontal line serves as reference when $T = 1500$ K and the oxygen partial pressure $p = 0.2$ atm, with the same interpretation concerning the site selectivity. Our results indicate that dopants with small ionic radii (Ga, Sc, and In) prefer to substitute for Zr regardless of the chemical potentials for Ba and Zr at all realistic temperatures and pressures. However, dopants with large ionic radii (such as Gd and Nd) instead have a lower solution energy when substituting for Ba if μ_{Zr} is sufficiently high. Moreover, this tendency increases with temperature: at 1500 K, for instance, Nd seems to prefer the Ba site even in the Ba rich limit. This prediction is in accordance with the finding by Kreuer *et al.* that Y doped $BaCeO_3$ single crystals prepared at 2000 °C take up much less protonic defects than the corresponding ceramics prepared at 1650 °C,[42] a result they have explained by a more even distribution of dopant atoms over the two cation sites in the single crystal. Irrespective of what cation site appears to be preferred, the formation of M_{Ba} defects can be expected to raise the Fermi energy of the crystal. As can be seen from eqn (3) and (4), this would result in a lowering of $\Delta E^{\mathrm{f}}_{M'_{\mathrm{Zr}}}$ and an increase of $\Delta E^{\mathrm{f}}_{M_{\mathrm{Ba}}}$, until an equilibrium distribution of dopant atoms over Ba and Zr sites is reached.

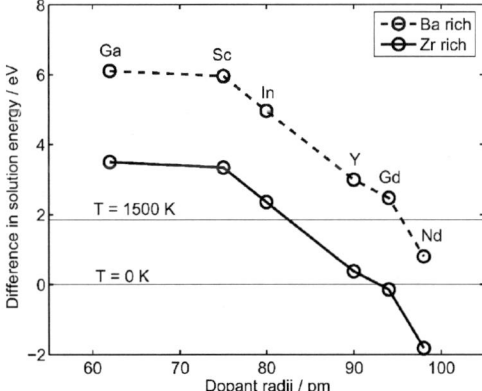

Fig. 1 Difference in solution energy for various dopants on Ba or Zr sites in $BaZrO_3$ calculated for a low defect concentration ($c = 1/27$). Only the most stable charge state of each dopant on either site is considered. Results are shown for two temperatures and for the two stability limits for $BaZrO_3$; the Ba rich (high μ_{Ba}) and the Zr rich limit (high μ_{Zr}).

Islam *et al.* found a similar correlation between dopant radii and site selectivity in $BaZrO_3$ using atomistic simulations.[15] That the solution of trivalent dopants is mainly driven by size effects has also been shown for other perovskite oxides.[16] Haile *et al.*[10] have discussed the possible influence of non-stoichiometry in doped ABO_3 type alkaline earth cerates and zirconates, and suggested that a divalent ion deficiency might drive the dopant incorporation onto the A site instead of the intended B site. This is also consistent with the finding in the present work.

Secondly, we will present data for and discuss some consequences of dopant–dopant interactions in $BaZrO_3$. The main reason we do that is because the distribution of dopants in the material may affect both the hydration process[13] and the diffusion of hydrogen interstitials,[43] something that will be briefly touched upon in sections 4.2 and 4.3. For simplicity, we will assume that dopants only occupy Zr sites, even though we have seen that does not necessarily have to be the case. According to eqn (5), the dopant–dopant interactions can then be obtained as the difference in solution energy of a pair of dopants occupying neighboring Zr sites (total effective charge $q = -2$), and twice the solution energy of an isolated dopant on a Zr site ($q = -1$) from eqn (4). Interaction energies ΔE^{int} thus calculated, employing $3 \times 3 \times 3$ supercells, have previously been presented[25] for the same set of dopants as that considered here. All ΔE^{int} were then found to be positive, which means that dopants tend to repel each other. It was also concluded that the strength of the repulsion was strongly correlated with the size of the dopant—ranging from weak (<0.05 eV) for dopants with approximately the same radius as Zr to strong (0.65 eV) for large dopants. To get a rough estimate of how these interactions will affect the distribution of dopants in the material, we will now turn our attention to the probability P of finding the dopants at neighboring Zr sites after equilibration in a supercell of general size, containing two dopants. This probability is a function of interaction energy and temperature as well as of dopant concentration c. Assuming a Boltzmann distribution over two different kinds of configurations (one where the two dopants occupy neighboring sites and one where they are further apart) it can be written as

$$P(\Delta E^{int}, T, c) = \frac{6\exp(-\Delta E^{int}/k_B T)}{2/c - 7 + 6\exp(-\Delta E^{int}/k_B T)}. \tag{21}$$

In Fig. 2 we have plotted the relative probability

$$P^{rel}(\Delta E^{int}, T, c = 2/27) \frac{P(\Delta E^{int}, T, c = 2/27)}{P(0, T, c = 2/27)} \tag{22}$$

as a function of dopant radius for three realistic sintering temperatures, where $c = 2/27$ was chosen to match the dopant concentration in the previous interaction energy calculations. The reference probability $P(0, T, c = 2/27)$ corresponds to a totally random distribution of non-interacting dopants. From inspection of Fig. 2 we note that the strong repulsion between large dopants effectively obstructs occupation of neighboring sites even at high temperatures, whereas small dopants are almost uniformly distributed over the Zr sites at these temperatures. Furthermore, we see that irrespective of dopant radius P^{rel} decreases with decreasing temperature, thus making neighboring site occupation less likely at low temperatures. Consequently, in samples containing a considerable amount of small dopants occupation of neighboring sites is expected to occur quite frequently at ordinary working temperatures given that the material has been quenched from high temperatures. To avoid such configurations the samples should instead be cooled slowly enough for equilibrium to be established. On the other hand, in samples with large dopants occupation of neighboring sites should never be very common as long as the dopant concentration is not extremely high.

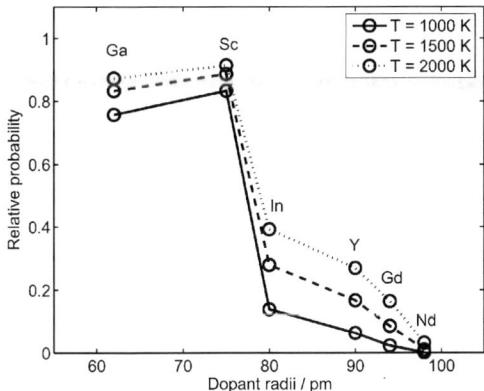

Fig. 2 The relative probability P^{rel} that two dopants occupy neighboring Zr sites in BaZrO$_3$, calculated for various dopants and temperatures assuming a dopant concentration $c = 2/27$. $P^{rel} = 1$ corresponds to a totally uniform distribution of dopant atoms over the Zr sites.

4.2 Hydrogen interstitials

We will now consider the incorporation of hydrogen atoms in an otherwise defect-free region of BaZrO$_3$. To find the preferred hydrogen sites, we first map out adiabatic potential energy surfaces (PESs) by calculating the total energy for a H atom—in various charge states—placed at several different positions in a fixed lattice. This will not only show the stable equilibrium sites for a hydrogen interstitial, but may also indicate the general features of its local vibrational motion and diffusion pathways.[43] The results are shown in Fig. 3. It is seen that the PESs for H$^+$ and H^0 are rather similar, with a potential minimum where the hydrogen is located close to one of the oxygen atoms. At this stable site, the O–H distance is approximately 1 Å, and the O–H axis is oriented along the bisector of two oxygen–oxygen connecting lines. Due to the cubic symmetry of BaZrO$_3$, there are thus four equivalent sites around each oxygen. This configuration of hydrogen defects has previously been suggested for various perovskite oxides, both experimentally from muon spin relaxation measurements,[44] and theoretically from atomistic modeling[15,16,45] and quantum mechanical[17–19,22,46,47] simulations. However, alternative positions have also been proposed. For instance, Sata et al. have performed neutron diffraction measurements on Sc doped SrTiO$_3$ and found a stable proton site 1.2 Å from the oxygen, slightly off the oxygen–oxygen connecting line toward Ti (Sc).[48] For H$^-$, on the other hand, the PES looks quite different. The most stable site is located halfway along a barium–barium connecting line, indicating that a negatively charged hydrogen atom may interact repulsively with the host lattice oxygens.

To understand the behavior of hydrogen in the different charge states in detail, we show the electronic density of states (DOS) for hydrated BaZrO$_3$ in Fig. 4. If a positively charged or neutral hydrogen atom is added to the crystal, it will occupy an interstitial site close to an oxygen atom so that a strong OH bond can be formed. This will modify the electronic structure of the host, and introduce two new sets of electronic states with bonding and antibonding character, respectively (cf. Fig. 4a and b). The bonding states at around −5 eV lie well below the edge of the valence band, and are thus fully occupied. The antibonding states, on the other hand, appear at +4 eV, which is above the edge of the calculated conduction band. For H$^+$ these states are empty, so the highest occupied state of the crystal is at the top of the valence band (similar to the situation in a perfect crystal). For H^0 an extra electron should be added to the system. However, this electron will not occupy any of the hydrogen-induced states, since it may relax to the bottom of the conduction band in a way similar to that for H/ZnO.[26] This means that locally H$^+$ and H^0 will have a

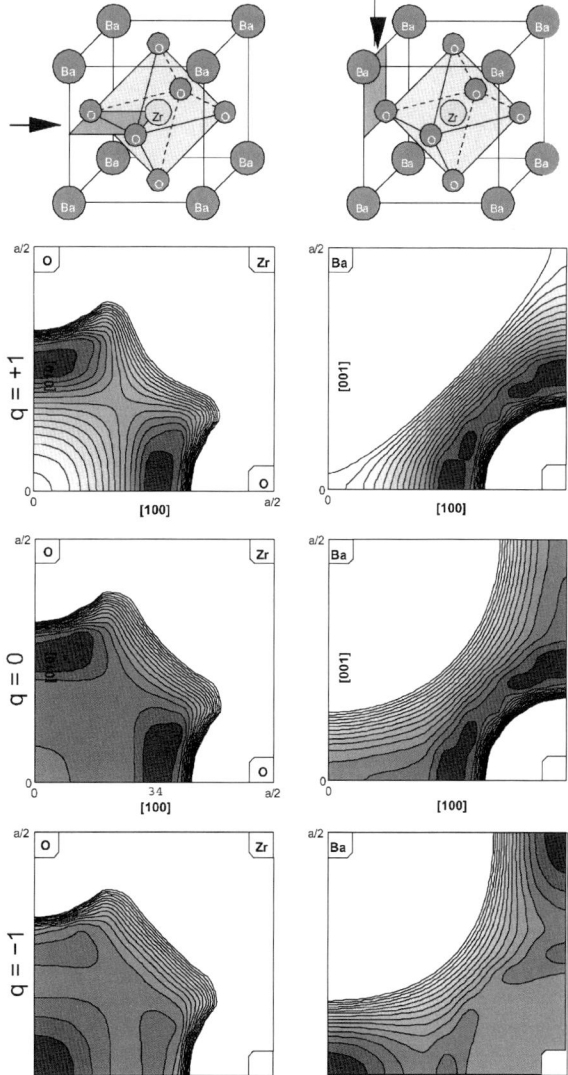

Fig. 3 Static potential energy surfaces (PESs) show the calculated variation in total energy with the position of a hydrogen interstitial in charge state $q = +1, 0,$ or -1 for two different crystal planes in $BaZrO_3$ ($a = 4.25$ Å).

very similar electronic structure, which suggests that these defects should have approximately the same O–H bond length and bond angles. For H^-, on the other hand, the system is no longer stable with the two additional electrons accommodated in states above the band gap. Instead, the O–H unit dissociates and the negatively charged hydrogen moves away from the lattice oxygen. For this configuration there is only a single set of occupied hydrogen-induced states that appear deep in the band gap (cf. Fig. 4c). In a previous theoretical study, Xiong and Robertson[19] investigated the electronic structure of hydrogen defects in several different perovskite oxides, including $BaZrO_3$. While our results for H^+ are in excellent agreement, they find a significant elongation of the O–H bond length for the neutral charge state so that the H^0 level also appears deep in the band gap. Although the reasons for this

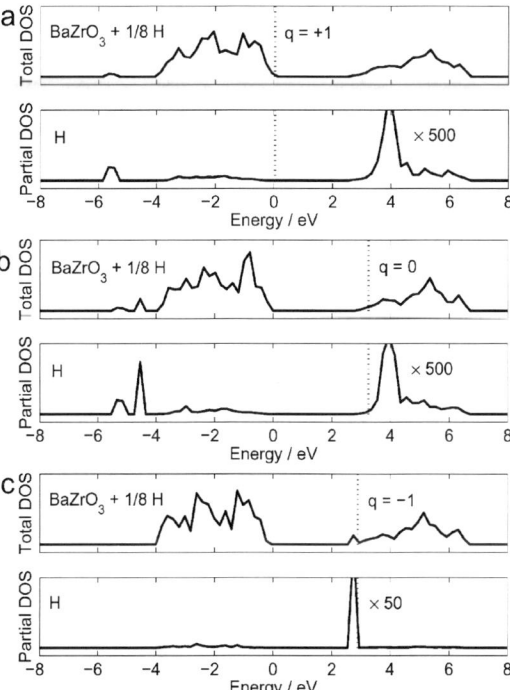

Fig. 4 Calculated electronic density of states (DOS) around the band gap region for hydrated BaZrO$_3$. Here, (a)–(c) show the total and hydrogen site-projected DOS for the system with a H$^+$, H^0, and H$^-$ defect, respectively, introduced on their stable sites in the lattice.

discrepancy are still not clear, our studies seem to show a similar variation of the position of the hydrogen-induced electronic states with the location of the interstitial in the lattice. The situation also resembles that of hydrated SrTiO$_3$, where a combined experimental and theoretical study suggests that apparent hydride ions can be present in the form of extended defect species (consisting of a neutral hydrogen with a tendency to associate with electrons, *e.g.* on neighboring Ti ions).[49]

The hydrogen-induced lattice distortion was investigated by placing a H atom in various charge states at the equilibrium positions found for the perfect lattice (*cf.* Fig. 3), and then allowing all ionic coordinates to relax. We find a large distortion of the positions of the surrounding atoms, suggesting that a hydrogen interstitial interacts strongly with the host lattice. The resulting structures obtained in the low-concentration limit are illustrated in Fig. 5. Near a H$^+$ or a H^0, there is a marked increase of the (OH)–Zr separation and a corresponding decrease of the (OH)–O separation. This is in good agreement with the findings of previous computational studies of hydrogen defects in BaZrO$_3$.[17,18,22] For a H$^-$, on the other hand, the neighboring oxygens are repelled by the defect, so that there is a significant increase of the O–O distance.

We have also calculated the harmonic frequencies of vibration for a hydrogen with the atoms of the host lattice held rigidly at their relaxed positions. The results are given in Table 1. For H$^+$ and H^0 there is a high frequency mode at around 3500 cm^{-1} (corresponding to O–H stretch vibrations) and two lower frequency modes at around 900 and 600 cm^{-1}, respectively (corresponding to frustrated reorientations of the O–H axis). These results are consistent with infrared (IR) spectroscopy data on acceptor-doped perovskite oxides, which typically show a broad absorption band around in the range 3000–3500 cm^{-1} after hydration.[2] For H$^-$ there is instead a

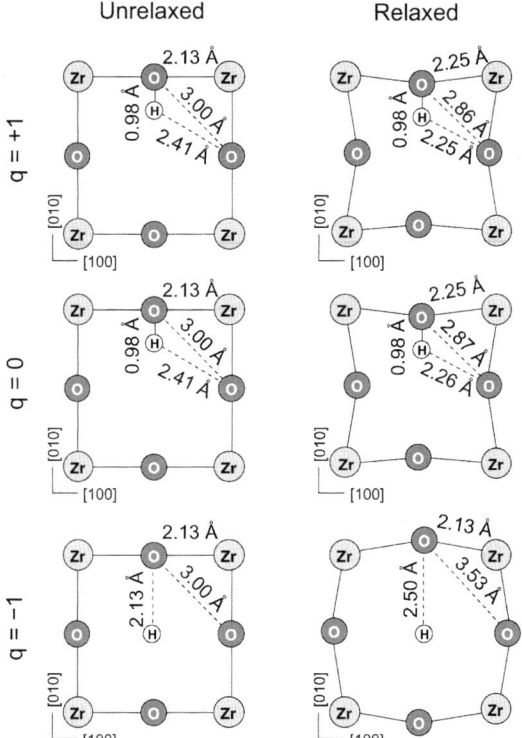

Fig. 5 Lattice distortion surrounding a hydrogen interstitial at the equilibrium position for different charge states q in BaZrO$_3$.

high-frequency mode at around 1400 cm^{-1}, and a two-fold degenerated mode at around 700 cm^{-1}.

In order to determine the preferred charge state for hydrogen atoms in BaZrO$_3$, we have calculated the formation energies $\Delta E^f_{H_i^q}$ for $q = +1$, 0 and -1 using eqn (2). The results for a predominantly p-type material ($\mu_e = \varepsilon_{VBM}$) are given in Table 2. It is seen that allowing for lattice relaxations lowers the formation energies considerably, which is consistent with the large structural distortions surrounding a hydrogen defect (*cf.* Fig. 5). Nevertheless, our results appear reasonably well converged with respect to the size of the supercell. Test calculations also show that allowing for spin polarization only changes the formation energies by a few meV. In a recent density functional study, Shi *et al.*[18] reports a formation energy of -1.14 eV per H$^+$ for In doped BaZrO$_3$, assuming hydrogen-rich conditions ($\mu_H = 1/2\ (E^{tot}[H_2])$). In order to compare this value with our results obtained for the water-rich limit $\mu_H = 1/2(E^{tot}[H_2O] - \mu_O)$, we should add half the formation energy of a H$_2$O molecule

Table 1 Calculated vibrational frequencies for hydrogen interstitials at the equilibrium position in a defect-free region of BaZrO$_3$

Mode	Frequency/cm^{-1}		
	ν_1	ν_2	ν_3
H$^+$	3502	900	601
H^0	3505	865	573
H$^-$	1418	726	726

This journal is © The Royal Society of Chemistry 2006

Table 2 Formation energies ΔE^f for hydrogen in different charge states in an (unrelaxed) relaxed $BaZrO_3$ lattice, calculated using either a small $2 \times 2 \times 2$ supercell corresponding to $c = 1/8$ or a larger $3 \times 3 \times 3$ supercell corresponding to $c = 1/27$. For H^+ and H^-, the Fermi energy has been set equal to the energy of the top of the valence band, ε_{VBM}, of the pure oxide

| | $\Delta E^f/eV$ | |
Defect	$c = 1/8$	$c = 1/27$
H^+	(0.95) 0.05	(1.35) 0.21
H^0	(4.22) 3.42	(4.27) 3.24
H^-	(7.48) 6.36	(7.50) 6.31

(which amounts to -1.24 eV at $T = 0^{50}$) to the formation energies in Table 2. This gives $\Delta E_{H_i^+}{}^f = -1.03$ eV in reasonable agreement with the results of Shi et al.[18]

For hydrogen interstitials carrying a charge, the formation energy will however depend on the actual position of the Fermi level in the material. Therefore, in Fig. 6, we show the results obtained in the dilute limit ($c = 1/27$) for $\varepsilon_{VBM} < \mu_e < \varepsilon_{CBM}$. It is readily seen that the $+1$ charge state has the lowest formation energies for all Fermi level positions in the calculated band gap, which means that hydrogen will act exclusively as a donor in $BaZrO_3$. This behavior was first reported for H in ZnO by Van de Walle,[26] and later also for H in a perovskite oxide by Yoshino et al.[20] We note, however, that the different charge states have similar energies of formation when $\mu_e \approx \varepsilon_{CBM}$, which gives some support to previous speculations about the possible existence of hydride ions in oxides under strongly reducing conditions.[12,49] Finally, our results might be sensitive to the band gap error inherent in the present DFT approach. In previous density functional studies of hydrogen in oxides it is argued that if a band gap correction is applied, the formation energies of H^0 and H^- should increase relative to that of H^+, so that the positive charge state is preferred even throughout the larger experimental band gap.[20,26]

So far, we have only considered the behavior of hydrogen interstitials in a region far from other defects. In particular, the presence of dopant atoms has only been taken into account implicitly, via the homogeneous background charge included in the supercell during the simulation of H^+ and H^-. Although conductivity data show no such effect,[42] several other experimental[44,51] as well as theoretical[15,45,52] works have clearly indicated the possibility of dopant–proton association, or 'trapping', in acceptor-doped perovskite oxides. Therefore, we have also investigated configurations M'_{Zr}–$OH^•_O$–$Zr^×_{Zr}$ where a proton is bound to an oxygen neighboring a single M'_{Zr} defect. In the relaxed geometries, the O–H axis becomes significantly tilted toward the adjacent dopant atom. As shown in a previous combined experimental and theoretical study of hydrated $BaIn_xZr_{1-x}O_{3-x/2}$, such configurations will result in an increased tendency of hydrogen bond formation accompanied by a spectral broadening of the O–H stretch band.[53] To check this, we have calculated the harmonic frequencies of vibration for a protonic defect in the vicinity of different

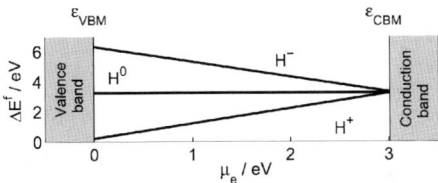

Fig. 6 Formation energies of hydrogen interstitials in different charge states in $BaZrO_3$ as a function of the Fermi level position μ_e within the calculated band gap.

Table 3 Calculated vibrational frequencies for hydrogen interstitials at the equilibrium position in the vicinity of various dopant atoms in $BaZrO_3$

Dopant	Frequency/cm^{-1}		
M	ν_1	ν_2	ν_3
Ga	3090	1025	962
Gd	3400	977	699
In	3455	922	719
Nd	3366	1013	713
Sc	3157	998	981
Y	3491	916	626

dopant atoms. The results are given in Table 3. In all cases there is a softening of the stretch mode and a corresponding hardening of the wag modes compared with the frequencies at a regular protonic site (cf. Table 1). This effect is particularly pronounced for small dopants, such as Ga or Sc. The interaction energies of the different dopant–proton clusters, calculated using eqn (5) and a large $3 \times 3 \times 3$ supercell, are given in the second column of Table 4. It is seen that all ΔE^{int} are negative, indicating that the protonic sites near a dopant atom are energetically lowered compared with the regular sites. If complete dopant–proton association is assumed, the formation energies given in Table 2 would therefore be reduced by as much as 0.13–0.38 eV, depending on the choice of dopant. This is consistent with the trapping energy of 0.2 eV deduced from muon spin relaxation measurements on Sc doped $SrZrO_3$ by Hempelmann et al.[44] Islam et al.[15] have previously investigated doping and defect association in several different perovskite-type oxides. However, while for $BaZrO_3$ we find a similar atomic configuration for the proton–dopant clusters, and the same general trend for the absolute value of the interaction energies (Y < In < Sc), their results seem to show a stronger tendency of defect association with binding energies in the range 0.26–0.74 eV. Yoshino et al.[23] calculated the local electronic structure around hydrogen and acceptor ions in different zirconates. For $BaZrO_3$, they predict that the strength of the trapping effect changes with dopant as Y < In ≈ Nd < Ga, which is in excellent agreement with the findings in the present study. To connect to the discussion about dopant distributions in section 4.1 we have also calculated interaction energies for M'_{Zr}–$OH^{\cdot}_O$–M'_{Zr} clusters, i.e. interactions between a protonic defect and two dopants occupying neighboring Zr sites. These are presented in the third column of Table 4. It is seen that these interactions are even stronger than those between protons and single dopants. For comparison, we also include in the fourth column in Table 4 the interaction energies for dopant–oxygen vacancy clusters similarly obtained in a previous study.[25]

Table 4 Calculated interaction energies ΔE^{int} for dopant–proton and dopant–oxygen vacancy[25] clusters in $BaZrO_3$

Dopant	ΔE_{int}/eV		
M	M'_{Zr}–$OH^{\cdot}_O$–$Zr^{\times}_{Zr}$	M'_{Zr}–$OH^{\cdot}_O$–M'_{Zr}	M'_{Zr}–$V^{\cdot\cdot}_O$
Ga	−0.38	−0.58	−0.68
Gd	−0.14	−0.41	−0.55
In	−0.20	−0.47	−0.62
Nd	−0.19	−0.32	−0.89
Sc	−0.23	−0.42	−0.37
Y	−0.13	−0.37	−0.45

Table 5 Calculated vibrational frequencies for oxygens in the perfect $BaZrO_3$ lattice, oxygens neighboring an oxygen vacancy, and oxygens neighboring a hydrogen interstitial

Mode	Frequency/cm^{-1}		
	ν_1	ν_2	ν_3
Regular	557	250	250
Near $V_O^{\bullet\bullet}$	532	274	234
Near $OH_O^{\bullet}$	540	258	234

4.3 Hydration at finite temperature and pressure

Now that we have shown that H will always be in charge state $+1$, we will study proton incorporation through dissociative absorption of water (*cf.* eqn (1)) at finite water partial pressure p and temperature T. This process is governed by the change in Gibb's free energy $\Delta G_{hydr}^f (p, T)$ as given by eqn (19). Combining the proton formation energy given in Table 2 with the oxygen vacancy formation energy $\Delta E_{V_O^{2+}}^f$ = 1.21 eV obtained previously with the same reference for μ_e,[25] the change in electronic total energy becomes $\Delta E_{hydr}^{tot} = 2\Delta E_{H^+}^f - \Delta E_{V_O^{2+}}^f = -0.79$ eV per absorbed water molecule. To evaluate the vibrational contributions to the free energy (eqn (16) and (17)) a set of hydrogen and oxygen frequencies are needed. In addition to the vibrational modes of a H^+ interstitial already given in Table 1, we have therefore performed harmonic analysis on individual oxygen atoms in a $2 \times 2 \times 2$ supercell containing either no defects, a hydrogen interstitial in the $+1$ charge state, or an oxygen vacancy in the $+2$ charge state. The results are presented in Table 5. Finally, for the free energy of the gas phase H_2O molecules, we use data from thermodynamic tables.[50]

The resulting change in free energy per absorbed water molecule is shown in Fig. 7 as a function of T at constant water pressure $p = 0.02$ atm. At low temperatures, $\Delta G^f < 0$ due to the negative contribution from the change in electronic total energy. This means that the oxide will take up water until the hydration limit is reached. However, at high temperatures, the remaining terms in eqn (19) become sufficiently large to bring $\Delta G^f > 0$. The dominating contribution to the free energy of formation

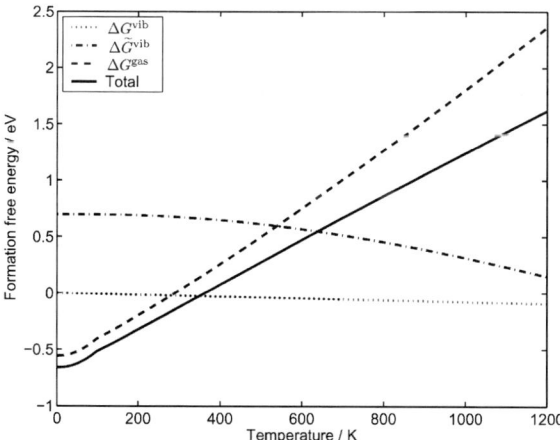

Fig. 7 The calculated electronic, vibrational and gas phase contributions to the change in free energy upon absorption of one water molecule in $BaZrO_3$ as a function of temperature at a constant water partial pressure $p = 0.02$ atm.

then comes from the entropy decrease associated with the absorption of gas molecules from the atmosphere. The change in vibrational free energy of the lattice is smaller and is dominated by the gain of the nine lattice modes (three oxygen modes plus six hydrogen modes). For instance, at $T = 1000$ K, $\Delta G^{\mathrm{vib}} + \Delta \tilde{G}^{\mathrm{vib}} = 0.19$ eV, $\Delta G^{\mathrm{gas}} = 1.81$ eV, and thus $\Delta G_{\mathrm{hydr}}^{\mathrm{f}} = 1.20$ eV. This means that protonic defects are unstable, and water is expected to leave the sample.

Experimentally, the enthalpy and entropy of hydration are determined through linearization of the hydration free energy:

$$\Delta G_{\mathrm{hydr}}^{\mathrm{f}} = \Delta H_{\mathrm{hydr}}^{0} - T \Delta S_{\mathrm{hydr}}^{0}. \tag{23}$$

Kreuer et al.[3] find enthalpies ranging from -1.24 to -0.69 eV and entropies ranging from -1.29 to -0.89 meV K^{-1} for doped BaZrO$_3$ and Schober and Bohn[7] report an enthalpy -0.77 ± 0.03 eV and an entropy -0.9 ± 0.1 meV K^{-1} for Y doped BaZrO$_3$. By fitting experimental data to a two-state model based on Fermi–Dirac statistics Groß et al.[13] obtain an enthalpy of -0.99 ± 0.05 eV for regular sites and an entropy of -0.86 ± 0.10 meV K^{-1} for Y doped BaZrO$_3$. Linearization of our calculated $\Delta G_{\mathrm{hydr}}^{\mathrm{f}}(p = 0.02, T)$ in the temperature interval $T = 800$–1200 K yields $\Delta H_{\mathrm{hydr}}^{0} = -0.65$ eV and $\Delta S_{\mathrm{hydr}}^{0} = -1.8$ meV K^{-1}. While the enthalpy is in fairly good agreement with experimental data,[3,7,13] the entropy is approximately twice as low as that obtained experimentally.

So far we have neglected the fact that dopant–proton and dopant–oxygen vacancy interactions might affect the hydration free energy. Hence, the calculated enthalpies and entropies will always assume the same values, regardless of which type of dopant the system contains. However, in the experiments carried out by Kreuer et al. on Gd, In, Sc, and Y doped BaZrO$_3$ samples,[3] the enthalpy and entropy varied significantly with dopant, suggesting that such interaction effects might be important. To distinguish between different dopant atoms, we have therefore investigated two simple trapping models. In the first model (henceforth referred to as model 1) all protons are assumed to be bound to single dopants, in the form of M'_{Zr}–OH$_{\mathrm{O}}^{\cdot}$–Zr clusters, under equilibrium conditions, while oxygen vacancies do not interact with the dopants at all. It has been argued[13] that M'_{Zr}–OH$_{\mathrm{O}}^{\cdot}$–M'_{Zr} clusters—called the trap site by Groß et al.—are also of importance for the hydration process. Indeed, according to Table 4, we find a large trapping energy for a proton in such configurations. However, due to the repulsive interaction between dopant pairs (cf. section 4.1) they should not occur very frequently at realistic dopant levels. Their main effect would therefore be to passify a small fraction of protons and prevent them from leaving the material at high temperatures, and they will not be considered further here. In the second model (model 2) both protons and oxygen vacancies are assumed to be associated with single dopants. The associative interactions will contribute an extra term to the net change in electronic total energy in the hydration process. More precisely, in model 1 twice the dopant–proton binding energy

Table 6 Change in electronic total energy upon hydration of doped BaZrO$_3$. Calculated values are presented for three different cases: no defect associations involved, only dopant–proton association (model 1), and both dopant–proton and dopant–oxygen vacancy association (model 2)

Dopant M	$\Delta E_{\mathrm{hydr}}^{\mathrm{tot}}/\mathrm{eV}$		
	No trapping	Model 1	Model 2
Ga	−0.79	−1.54	−0.86
Gd	−0.79	−1.06	−0.51
In	−0.79	−1.19	−0.57
Nd	−0.79	−1.18	−0.29
Sc	−0.79	−1.26	−0.89
Y	−0.79	−1.05	−0.60

(*cf*. Table 4, column 2) should be added to $\Delta E_{\text{hydr}}^{\text{tot}}$ to give the true change in electronic total energy, whereas model 2 also requires an additional subtraction of the dopant–oxygen vacancy binding energy (*cf*. Table 4, column 4) to give the corresponding total energy. The results for Ga, Gd, In, Nd, Sc, and Y doped samples, as given by the two trapping models, are presented in Table 6. Comparison with the data of Kreuer *et al.* shows that model 2 correctly reproduces the experimental trend in hydration enthalpy, Sc < Y < In < Gd, but that model 1, in which the dopant–oxygen vacancy interaction is ignored, does not. Even so, on an absolute scale model 1 renders better estimates of the hydration enthalpies ΔH_{hydr}^0 for Sc and Y doped samples. The experimental values for Gd and In doped samples, on the other hand, are found approximately halfway between the two theoretical predictions. The varying accuracy of the two models with dopant might be related to differences in the dopant–oxygen vacancy associations. For instance, this attractive interaction is much stronger in Gd and In doped samples than in samples with Sc and Y (*cf*. Table 4). At ordinary sintering temperatures, most oxygen vacancies in Gd and In doped samples should thus be trapped at dopants, whereas in Sc and Y samples they are probably more randomly distributed. Due to restricting kinetics, a substantial amount of the oxygen vacancies in the two latter systems is therefore expected to become frozen in at regular sites after subsequent quenching. That would lessen the importance of dopant–oxygen vacancy interactions in the hydration reaction (eqn (1)) and could be one of the reasons why model 1 gives quantitatively better predictions for those two systems. Finally, both models predict a low hydration enthalpy in Ga doped samples (comparable to Sc) whereas it should be significantly larger in an Nd doped sample. It would be interesting if this prediction could be tested experimentally.

Although the two trapping models seem to capture some of the observed differences in hydration enthalpies among the doped samples, they can in principle not explain variations in entropy, since in that respect all systems are treated on an equal footing. Protons are always assumed to be trapped at dopants, whereas oxygen vacancies are either completely trapped or totally randomly distributed. Still, entropic effects stemming from changes in protonic vibrational frequencies in the vicinity of dopants (*cf*. Table 3) may be incorporated. However, these effects turn out to be negligibly small. Their contribution to the total entropy ΔS_{hydr}^0 does not exceed 0.1 meV K^{-1} at 1000 K.

5 Conclusions

To summarize, we have combined density functional calculations and thermodynamic modeling to investigate the structure and thermal stability of hydrogen interstitials in a BaZrO$_3$ crystal in equilibrium with a humid oxygen containing atmosphere. To this end we first determined the preferred sites in the lattice, and calculated the electronic structures, formation energies, and vibrational frequencies for hydrogen interstitials in different charge states in BaZrO$_3$. We found that a hydrogen atom preferably assumes the $+1$ charge state, and that its stable positions are located approximately 1 Å from the oxygen atoms, with the O–H axis oriented along the bisector of two oxygen–oxygen connecting lines. Moreover, hydrogen interstitials were found to interact strongly with the host lattice, resulting in a large structural distortion of the positions of the surrounding atoms. Calculated vibrational frequencies for a H$^+$ in the relaxed lattice were found to be consistent with IR data on acceptor-doped perovskite oxides.

From the calculated solution energies of Ga, Gd, In, Nd, Sc, and Y dopants on Ba and Zr sites, and the dopant–dopant interaction energies, we have also investigated the arrangement of dopant atoms in the lattice in the dilute limit under equilibrium conditions. Firstly, we have presented results for the dopant occupation of Ba and Zr sites for two different limits of the Ba and Zr chemical potentials at different temperatures and oxygen pressures. We find that most of the investigated dopants preferably incorporate on Zr sites in the lattice, but that large dopants have a tendency to occupy Ba sites under Zr rich conditions. It is also concluded that the drive toward the Ba site is enhanced when the temperature is raised. This site selectivity will affect the hydration limit at low temperatures. Secondly, we have estimated the dopant distribution over the Zr sites at elevated temperatures. The interaction between Zr site dopants is always repulsive but the aversion to nearest Zr site occupation is clearly correlated to dopant radius and temperature. This type of dopant ordering will have implications both for the hydration process and the mobility of the hydrogen interstitials, since protonic defects were found to interact attractively with acceptor dopants. Proton–dopant cluster formation is predicted to result in lowered site energies, a tilting of the O–H axis toward the dopant, and a softening of the O–H stretch mode accompanied by a corresponding hardening of the O–H wag modes.

Finally, we have shown that protonic defects are thermodynamically stable only at low temperatures, in agreement with empirical wisdom. Calculated thermodynamic parameters for the water incorporation reaction were comparable to those found in thermogravimetric measurements. An experimental trend for the hydration enthalpies of $BaZrO_3$ doped with Gd, In, Sc, and Y could be correctly reproduced when dopant–proton and dopant–oxygen vacancy interactions were taken into account.

Acknowledgements

This work was supported by the National Graduate Schools in Scientific Computing and Materials Science, and by the Foundation for Strategic Research *via* the ATOMICS program, Sweden. Allocations of computer resources through the Swedish National Allocation Committee are gratefully acknowledged.

References

1 H. Iwahara, in *Proton conductors: Solids, membranes and gels—materials and devices*, ed. P. Colomban, Cambridge University Press, Cambridge, 1992, ch. 8.
2 K. D. Kreuer, *Solid State Ionics*, 1999, **125**, 285.
3 K. D. Kreuer, S. Adams, W. Münch, A. Fuchs, U. Klock and J. Maier, *Solid State Ionics*, 2001, **145**, 295.
4 W. Wang and A. V. Virkar, *J. Power Sources*, 2005, **142**, 1.
5 H. Iwahara, T. Yajima, T. Hibino, K. Ozaki and H. Suzuki, *Solid State Ionics*, 1993, **61**, 65.
6 R. C. T. Slade, S. D. Flint and N. Singh, *Solid State Ionics*, 1995, **82**, 135.
7 T. Schober and H. G. Bohn, *Solid State Ionics*, 2000, **127**, 351.
8 H. G. Bohn and T. Schober, *J. Am. Ceram. Soc.*, 2000, **83**(4), 768.
9 V. P. Gorelov, V. B. Balakireva, Y. N. Kleschev and V. P. Brusentsov, *Inorg. Mater.*, 2001, **37**(5), 535.
10 S. M. Haile, G. Staneff and K. H. Ryu, *J. Mater. Sci.*, 2001, **36**, 1149.
11 K. D. Kreuer, *Annu. Rev. Mater. Res.*, 2003, **33**, 333–359.
12 T. Norby, M. Widerøe, R. Glöckner and Y. Larring, *Dalton Trans.*, 2004, 3012.
13 B. Groß, J. Engeldinger, D. Grambole, F. Herrmann and R. Hempelmann, *Phys. Chem. Chem. Phys.*, 2000, **2**, 297–301.
14 K. D. Kreuer, E. Schönherr and J. Maier, *Solid State Ionics*, 1994, **70/71**, 278–284.
15 M. S. Islam, P. R. Slater, J. R. Tolchard and T. Dinges, *Dalton Trans.*, 200419), 3061–3066.
16 R. A. Davies, M. S. Islam and J. D. Gale, *Solid State Ionics*, 1999, **126**, 323–335.
17 M. A. Gomez, M. A. Griffin, S. Jindal, K. D. Rule and V. R. Cooper, *J. Chem. Phys.*, 2005, **123**, 094703.
18 C. Shi, M. Yoshino and M. Morinaga, *Solid State Ionics*, 2005, **176**, 1091–1096.
19 K. Xiong and J. Robertson, *Appl. Phys. Lett.*, 2004, **85**(13), 2577.
20 M. Yoshino, Y. Liu, K. Tatsumi, I. Tanaka, M. Morinaga and H. Adachi, *Mater. Trans.*, 2002, **43**, 1444–1450.
21 K. D. Kreuer, *Solid State Ionics*, 1997, **97**, 1.
22 W. Münch, K. D. Kreuer, G. Seifert and J. Maier, *Solid State Ionics*, 2000, **136**, 183–189.
23 M. Yoshino, K. Kato, E. Mutiara, H. Yukawa and M. Morinaga, *Mater. Trans.*, 2005, **46**(6), 1131–1139.

24   C. Stampfl, M. V. Ganduglia-Pirovano, K. Reuter and M. Scheffler, *Surf. Sci.*, 2002, **500**, 368–394.
25   P. G. Sundell, M. E. Björketun and G. Wahnström, *Phys. Rev. B*, 2006, **73**, 104112.
26   C. G. Van de Walle, *Phys. Rev. Lett.*, 2000, **85**(5), 1012.
27   A. F. Kohan, G. Ceder, D. Morgan and C. G. Van de Walle, *Phys. Rev. B*, 2000, **61**(22), 15019.
28   C. G. Van de Walle and J. Neugebauer, *Nature*, 2003, **423**, 626.
29   A. Kuwabara and I. Tanaka, *J. Phys. Chem. B*, 2004, **108**, 9168–9172.
30   H. Moriwake, I. Tanaka, K. Tatsumi, Y. Koyama, H. Adachi, H. Yakabe and I. Yasuda, *Mater. Trans.*, 2002, **43**, 1456–1459.
31   G. Kresse and J. Furthmüller, *Phys. Rev. B*, 1996, **54**(16), 11169.
32   G. Kresse and J. Hafner, *Phys. Rev. B*, 1993, **48**(17), 13115.
33   Y. Wang and J. P. Perdew, *Phys. Rev. B*, 1991, **44**(24), 13298.
34   P. E. Blöchl, *Phys. Rev. B*, 1994, **50**(24), 17953.
35   W. Pies and A. Weiss, *Landolt-Börnstein, New Series*, Springer-Verlag, Berlin, 1975, vol. 7b1.
36   K. T. Jacob and Y. Waseda, *Metall. Mater. Trans. B*, 1995, **26**, 775.
37   J. Robertson, *J. Vac. Sci. Technol., B*, 2000, **18**, 1785.
38   R. M. Martin, *Electronic Structure: Basic Theory and Practical Methods*, Cambridge University Press, Cambridge, 2004.
39   M. Yoshino, K. Nakatsuka, H. Yukawa and M. Morinaga, *Solid State Ionics*, 2000, **127**, 109–123.
40   D. Ricci, G. Bano, G. Pacchioni and F. Illas, *Phys. Rev. B*, 2003, **68**, 224105.
41   R. D. Shannon, *Acta Crystallogr., Sect. A*, 1976, **32**, 751.
42   K. D. Kreuer, W. Münch, M. Ise, T. He, A. Fuchs, U. Traub and J. Maier, *Ber. Bunsen-Ges. Phys. Chem.*, 1997, **101**, 1344–1350.
43   M. E. Björketun, P. G. Sundell, G. Wahnström and D. Engberg, *Solid State Ionics*, 2005, **176**, 3035–3040.
44   R. Hempelmann, M. Soetratmo, O. Hartmann and R. Wäppling, *Solid State Ionics*, 1998, **107**, 269–280.
45   M. S. Islam, R. A. Davies and J. D. Gale, *Chem. Mater.*, 2001, **13**, 2049–2055.
46   W. Münch, G. Seifert, K. D. Kreuer and J. Maier, *Solid State Ionics*, 1997, **97**, 39–44.
47   W. Münch, G. Seifert, K. D. Kreuer and J. Maier, *Solid State Ionics*, 1996, **86–88**, 647–652.
48   N. Sata, K. Hiramoto, M. Ishigame, S. Hosoya, N. Niimura and S. Shin, *Phys. Rev. B*, 1996, **54**, 15795.
49   M. Widerøe, W. Münch, Y. Larring and T. Norby, *Solid State Ionics*, 2002, **154–155**, 669–677.
50   M. W. Chase, *JANAF Thermochemical Tables*, American Chemical Society and the American Institute of Physics, New York, 3rd edn, 1986.
51   T. Matzke, U. Stimming, C. Kramonik, M. Soetratmo, R. Hempelmann and F. Güthoff, *Solid State Ionics*, 1996, **86–88**, 621–628.
52   G. C. Mather and M. S. Islam, *Chem. Mater.*, 2005, **17**, 1736.
53   M. Karlsson, M. E. Björketun, P. G. Sundell, A. Matic, G. Wahnström, D. Engberg, L. Börjesson, I. Ahmed, S. Eriksson and P. Berastegui, *Phys. Rev. B*, 2005, **72**, 094303.

Point defects in ZnO

Alexey A. Sokol,*[a] Samuel A. French,[ab] Stefan T. Bromley,[c] C. Richard A. Catlow,[ad] Huub J. J. van Dam[e] and Paul Sherwood[e]

Received 25th May 2006, Accepted 30th May 2006
First published as an Advance Article on the web 25th August 2006
DOI: 10.1039/b607406e

We have investigated intrinsic point defects in ZnO and extended this study to Li, Cu and Al impurity centres. Atomic and electronic structures as well as defect energies have been obtained for the main oxidation states of all defects using our embedded cluster hybrid quantum mechanical/molecular mechanical approach to the treatment of localised states in ionic solids. With these calculations we were able to explain the nature of a number of experimentally observed phenomena. We show that in zinc excess materials the energetics of zinc interstitial are very similar to those for oxygen vacancy formation. Our results also suggest assignments for a number of bands observed in photoluminescence and other spectroscopic studies of the material.

1 Introduction

Zinc oxide is a material of fundamental importance and great interest to both academic and industrial communities. A large number of experimental and theoretical studies have been carried out to date, aimed at the identification of the main defect species in ZnO and characterisation of their properties; but the definitive picture, particularly for intrinsic defects in the bulk, has not yet emerged (see ref. 1–3 and references therein).

We have undertaken a series of embedded molecular cluster, hybrid quantum mechanical/molecular mechanical (QM/MM) studies of localised states in the ZnO system.[4] Previously our work focused on surface properties of ZnO and its interface with Cu.[5] Here we turn our attention to point defects in the bulk.

ZnO is a wide-gap, n-type semiconductor. The excess of electron charge carriers is commonly attributed to excess zinc present in typical nonstoichiometric samples of ZnO; it is usually assumed that the excess Zn is accommodated as interstitials.[6] Complex luminescence spectra (including ultraviolet, violet–blue, green, yellow–orange and red bands in its visible part) characterise ZnO samples of different origin subject to various treatments, and are usually associated with ubiquitous Cu, Li, Na, Fe and other impurities as well as elementary point defects and their complexes.

[a] The Royal Institution of Great Britain, 21 Albemarle Street, London, UK W10 5LP. E-mail: alexey@ri.ac.uk; Fax: 44 (0)20 7670 2958; Tel: 44 (0)20 7409 2992
[b] Johnson Matthey Technology Centre, Blount's Court, Sonning Common, Reading, UK RG4 9NH
[c] Departament de Química Física, Universitat de Barcelona, Martí i Franquès 1, E-08028, Barcelona, Spain. E-mail: s.bromley@qf.ub.es; Fax: + 34 93 402 1231; Tel: +34.93.403.9266
[d] Department of Chemistry, UCL, 20 Gordon Street, London, UK WC1H OAJ. E-mail: c.r.a.catlow@ucl.ac.uk; Fax: 44 (0)20 7679 7463; Tel: 44 (0)20 7679 7482
[e] CCLRC Daresbury Laboratory, Keckwick Lane, Daresbury, Warrington, UK WA4 4AD. E-mail: .
E-mail: p.sherwood@dl.ac.uk; Fax: 44 (0)1925 603634; Tel: 44 (0)1925 603553

Fig. 1 ZnO structure with hexagonal (ab) sheets on the left and the contents of the primitive unit cell (side view). Here and in the next figures, light grey colour is reserved for Zn and dark grey for O.

Furthermore, ZnO is an ionic dielectric with a wurtzite structure in ambient conditions, which can be represented as an AB stacking of hexagonally arranged ZnO sheets in the c direction of the crystallographic axes (see Fig. 1). Both Zn^{2+} and O^{2-} ions tetrahedrally coordinate to each other. Three neighbours of any given ion are situated in the same sheet. Coordination to the fourth ion in the adjacent sheets alternates in direction up and down the c axis between Zn^{2+} and O^{2-}, respectively. This spatial arrangement determines that a bulk termination normal to the c axis necessarily results in two complementary, Zn and O rich surfaces. Both surfaces are routinely observed and well characterised by various surface specific experimental techniques.[7] However, the dipole moment between the two *polar* surfaces is physically unsustainable in any system of macroscopic dimensions and is therefore compensated. The mechanism of dipole quenching is still a matter of debate; herein as previously we postulate an ionic reconstruction mechanism, in which about a quarter of the surface ions are removed from each surface. Thus, both the macroscopic charge and dipole in our model are set to zero from the onset of the calculations. From the alternating arrangement of parallel Zn and O rich layers stem the piezoelectric properties of ZnO.

Defect properties are either directly exploited or strongly influence numerous applications of ZnO in electronics and catalysis. In particular we are concerned with intrinsic and extrinsic point defects in bulk and at the surfaces of ZnO. In this work, we have further investigated Li, Cu and Al impurities, which represent typical donor and acceptor centres that control electric and optical properties of this material. In one of perhaps the most important applications, Cu supported on ZnO is used as a catalyst for methanol synthesis from syngas, a mixture of CO, CO_2 and H_2. Cu both naturally present and intentionally introduced in ZnO is also a very efficient electron scavenger, greatly improving resistivity of this material. Furthermore, Cu doped thin films of ZnO have recently been shown to exhibit ferromagnetism at and above room temperature, which makes it a good candidate for application in the emerging field of spintronics.

2 Hybrid QM/MM approach

Most advanced methods of modern quantum chemistry are available within hybrid QM/MM approaches, of which ours is one of many. However, these approaches have not been used extensively for the study of ZnO. Our embedded molecular cluster, hybrid QM/MM approach to model localised states in bulk and at the surface of ionic solids has been described in much detail elsewhere.[8] So we present here only a short outline of the idea and pertinent details.

We confine the subject of our study to the localised states in ionic solids. When a point defect is introduced, it breaks the periodic symmetry of a solid and gives rise either to localised states within the forbidden gap or to resonance states in the valence and conduction bands. By their nature, resonance states are not reliably treated within the current approach. Hence, when we calculate defect levels outside of the gap, we consider our description of the corresponding defect only as a guide, but not as a proper predictive result. With this word of caution we also note that although point defects have long-range effects on the crystal matrix, in which they are embedded, these effects are well accounted for as polarisation of the environment and can be dealt with using appropriate MM methods. The electronic states of the defect in contrast require a QM treatment. Hence, a large number of techniques have been developed in recent years, which treat a small region around a defect site in a solid with QM techniques and the rest with suitable MM approaches (see ref. 8 and references therein).

In our approach, a cluster of atoms including a defect site is defined as the QM region, which is described at the density functional theory (DFT) level. The remainder of the crystal is represented by a large finite cluster, all atoms of which are treated with pair wise interatomic potentials using formal ionic charges and the shell model.[9] The effect of the remainder of the crystal on the QM cluster and *vice versa* includes both long-range electrostatic and short-range interactions. The electrostatic terms are included in the QM Hamiltonian as Coulomb contributions from point charges centred on cores and shells in the MM region. The short-range contributions enter the QM Hamiltonian as semi-local pseudopotentials centred on the cations within an interface between the QM and MM regions, where atoms take part both in QM and MM interactions. The artificial effects of lattice termination that arise from excising the large cluster are counteracted by placing a group of compensating point charges around the cluster. The values of these point charges are fitted to reproduce the Madelung potential and field at lattice sites (calculated for an MM model of the system of interest). The outer layer of MM centres is frozen in our calculations; its main purpose is to provide an appropriate short-range embedding potential at the defect site. The thickness of this frozen region has been determined by the cut-off of the short-range interatomic potentials employed in our MM model of ZnO. This approach implements an approximate form of the Mott–Littleton approach. In order to account for the missing polarisation effects outside the active region, we include an *a posteriori* correction to the total energy of the cluster with a defect using the Jost formula (see detailed derivation in ref. 8 and further references therein): $E_{pol} = -Q^2/2r_a (1-1/\varepsilon)$, where Q is the charge of the defect, r_a is the radius of the active region and ε is the dielectric constant. For single-point calculations, we used high-frequency dielectric constants, whilst for geometry optimised configurations static dielectric constants calculated with our 3D MM model: $\varepsilon = (\varepsilon_{11} + \varepsilon_{22} + \varepsilon_{33})/3$, were employed.

To analyse the energetics of defect states, we employed the total energy of an embedded cluster rather than one-electron Kohn–Sham states of a QM region, in contrast to our earlier reports. The reason for this approach is not only that the many-electron energies we obtain are more reliable than one-electron energies of a given QM system, but also that the balance of energies in an embedded approach is shifted, with polarisation corrections not included self-consistently in the QM Hamiltonian.

This approach has been implemented in the computational chemistry environment software ChemShell.[10] The QM part of the calculation is performed using the GAMESS-UK package,[11] while the MM part is implemented by the GULP code.[12]

Structural models for the idealised bulk of ZnO have been obtained using MM with GULP. Parameters of the interatomic potentials employed have been previously described,[13] and we note only that they allowed us to achieve not only an excellent reproduction of the atomic structure but also of the physical properties of ZnO, including dielectric, elastic and piezoelectric constants, phonon spectra and the low–high pressure (wurtzite-to-rock salt) phase transition.

All QM calculations have been performed with a B97-1 exchange and correlation functional.[14] This second generation hybrid functional is based on the generalised gradient approximation (GGA) terms augmented by a (20%) nonlocal exchange contribution in the Hartree–Fock form, which allows for a very accurate reproduction of structure and binding energies in a wide range of systems. This level of theory, however, has been shown to be insufficiently accurate in the way that it deals with hole polarons trapped at negatively charged impurities in oxides: Al in silica and Li in MgO.[15] When experiment and accurate post-Hartree–Fock approaches predict hole localisation just on one oxygen site, the local density approximation (LDA) and GGA based calculations typically spread the hole over all nearest neighbour O sites, whilst hybrid methods such as the B3LYP and current B97-1 functionals split it equally between two O sites. Curiously, the B97-1 method performs significantly better, as we will see, when treating similar defects in ZnO, but the problem resurfaces in the calculation of neutral and singly charged Zn vacancies (see below). The over-delocalisation of effective one-electron systems can be traced to an incomplete cancelling of electron self-interaction by GGA based density functionals. A more accurate treatment of the hole defect states using alternative techniques is under way and will be presented elsewhere.

Throughout this work we made use of TZV basis sets[16] with large-core effective core potentials on cations.[17] An all-electron basis set, however, has been used for the central Zn and Cu atoms. Single optimised polarisation functions were placed on all centres in the QM region except the central Zn and substitutional metal atoms (Li, Al and Cu), where two functions were used. A full oxygen basis set has been retained at a vacant oxygen site, which significantly reduced the basis set superposition error (*ca.* 0.2 eV per electron per vacant site).

3 Point defects in bulk

3.1 Computational background

The first computational studies of point defects in ZnO exploited the method of interatomic potentials using the Mott–Littleton or supercell approaches.[18] Necessarily, only defects based on ions with formal ionic charges could be investigated directly. Electronic charge transfer processes, crucial for the understanding of many physical and chemical properties of ZnO, could only be studied at a semi-classical level.

More recently, electronic structure techniques have been applied to investigate the defect properties of ZnO.[19] This work mainly employed the LDA and GGA DFT levels of theory. Even though this approximation has worked reasonably well in numerous instances, as mentioned above it also has severe limitations when dealing with the problem of charge localisation, which is the key to the correct description of point defects. Furthermore, the typical errors in reproduction of binding energies by this method are also still too high, which puts in question many recent predictions as to the defect concentrations and related electrical properties of ZnO. (The same, of course, is true for other materials). To amend some of deficiencies of the GGA Janotti and Van de Walle[20] have used a semi-empirical LDA + U procedure, which allowed them to get a better reproduction of Zn 3d states and, remarkably, of the band gap in this material, with the value increased from 0.80 to 1.51 eV (*cf.* a similar study of Erhart *et al.*[21]). However, this procedure also led to deterioration in the reproduction of the structure and cohesive energy. The problematic part of this and similar studies is an unequal treatment of on-site electron correlation for anionic and cationic states, with O 2p electron states being particularly poorly reproduced. A more consistent approach is offered by hybrid exchange–correlation functionals such as B3LYP, which is implemented for example in the periodic boundary conditions code CRYSTAL (note, B3LYP and B97-1 functionals use equal fractions of Hartree–Fock exchange). Indeed, this technique has been applied successfully to study the structure and properties of perfect ZnO and its surface properties.[22] Important physical properties of ZnO including its structure and band gap (in Kohn–Sham one-electron states) are well reproduced by this method. Although, the method has not so far been applied to intrinsic defects in the bulk of ZnO, it has been successfully employed in a study of vibrational properties of H compensated Li impurity centre.[23]

Alternative embedded cluster approaches for the study of bulk defects have not yet been used to their full potential. A simple electrostatic version of such an approach has

This journal is © The Royal Society of Chemistry 2006

recently been realised by Fink, who studied oxygen vacancies in bulk and at the surfaces of ZnO (see discussion below).[24] We are not aware of any other work to date.

3.2 Structural model of embedded cluster

All of our bulk calculations have been based on one large excised cluster 27 Å in radius, centred on a Zn ion. Five QM regions, comprising of 1, 5, 17, 42 and 86 atoms, have been defined, based on the number of shells of neighbours, so that these regions are always terminated either by O or Zn ions. Here we report results obtained using the largest, Zn terminated cluster, in which boundary atoms are separated from the central Zn by four Zn–O bonds (at *ca.* 6.0–6.5 Å). (Electron poor cations at the outer boundary provide a faster converging series of results.) The smaller clusters have been used in this work to provide an initial guess of the defect configurations, which were then transferred to the largest cluster for production runs. All cations outside the QM region but within a 4 Å cut-off from any QM atom are included in the interface (96 centres). All centres in the embedded cluster, which are situated within a 15 Å cut-off distance from the centre, form an active region (978 atoms), where QM atoms, MM cores and shells are relaxed during geometry optimisation. In single-point calculations, MM shells in the active region are relaxed self-consistently with the electron density of the QM region. The remaining ions in the outer spherical layer, 12 Å across, remain fixed throughout (6440 atoms). A further 73 point charges have been fitted to reproduce the Madelung potential on all centres in the active region (QM atoms, MM cores and shells) to 10^{-4} V of their reference values.

3.3 Ideal material and electronic defects

We start these studies with the analysis of the defect free cluster, which represents an ideal ZnO material. Upon cluster relaxation we observed only minor movement of ions from their MM relaxed positions; with the displacements in the range 0.01–0.06 Å throughout the QM region and interface; and dying away when moving from the centre into the MM active region. One larger displacement of 0.08 Å is shown by an O ion bonded to the central Zn ion, which is described by an all-electron basis set. (In a small subset of calculations, which employed all-electron basis sets for all QM atoms, we observed smaller displacements more uniformly spread through the cluster; unfortunately such calculations are still too computationally expensive.) These displacements are obviously due to a slight mismatch of the potential energy surfaces originating from MM and QM levels of theory. Whereas parameters of the interatomic potentials were fitted to reproduce the experimentally known structure of ZnO, at the given level of QM description there are small errors in comparison with the experimental data. Importantly these displacements result in much less pronounced changes in the local structure of ZnO, with the Zn–O bond distances typically within a 0.05 Å margin of their experimental value.

Next we determined the vertical ionisation potential (IP) of ZnO with a single-point calculation of a charged cluster on the relaxed structure of the neutral cluster, which results in a hole state at the top of the valence band (VB) of this material. The calculated value of 7.71 eV for the IP is in excellent agreement with an experimentally determined value of 7.82 eV reported by Swank.[25] This agreement, however, should be treated with caution as good single crystal ZnO samples have appeared only very recently, which warrants further experimental investigation.

Furthermore, we find the hole state to be delocalised throughout the cluster with the dominant contributions from the four central oxygen ions (split into two pairs with the Mulliken spin populations, n_s, of 2×0.19 and 2×0.09 e). The localisation of the band state in the central part of the QM region could be argued to be an artefact of the embedding scheme used and is in contrast with typical periodic boundary condition calculations, which describe these states using periodic, Bloch functions. Within the periodic picture a band hole state could be expected to be evenly spread through a small periodic cell. Eigenstates obtained using embedded cluster approaches are, however, localised by their nature, and we should not expect one-to-one correspondence with the Bloch picture. (In the limit of large enough clusters the two sets of states should span the same space and be interconvertible.) Moreover, when using large enough supercells, a localisation of the solution should occur by breaking the initial full periodic symmetry with the hole trapping by the (high-frequency) dielectric response in the electron subsystem. In any case, the radius of localisation of the hole state is thus determined by our calculation to be in of the order of one Zn–O bond distance. As discussed above, such delocalised one-electron states are not necessarily described very well by the current technique; however, the associated error in energy can be as small as 0.1 eV, which is still well within the error margin of these calculations.

A test is provided by the simulation of a self trapping process (characterised by its energy E_{STh}) for the hole polaron state. Indeed, upon lattice relaxation, the hole becomes trapped on just one of the oxygens ($n_s = 0.83$ e and $E_{STh} = 0.67$ eV); tails of the spin density can still be observed in the spin populations of the next nearest neighbour oxygen ions ($n_s = 0.01–0.03$ e). Upon self trapping the central Zn ion and the oxygen, on which the hole is localised, move apart, with their interatomic distance dramatically increasing from 2.02 to 2.48 Å. (Note that both atoms belong to the same hexagonal ZnO sheet). In this configuration, the Zn ion displaces further, by 0.30 Å, than the O ion, which moves by 0.16 Å. In the final configuration, the pair of Zn^{2+} and O^- ions retains three nearest neighbours each with a decreased distance from the Zn to its neighbours of 0.08 Å while the O ion moves further away from its neighbours by 0.12 Å. The problem considered might seem to be somewhat artificial as no self-trapped polaronic states were reported for ZnO to our knowledge. However, closely related self-trapped and bound excitons as well as bound hole polarons are prominent features of this material.

Electron states at the bottom of the conduction band (CB) and related polarons have been assessed in a similar fashion, by adding an electron to the neutral cluster and performing a single-point self-consistent calculation. We calculate the vertical electron affinity of ZnO as 1.24 eV with respect to vacuum. This result positions the bottom of the conduction band 3.03 eV too high compared to experiment. Indeed, our calculation of the optical band gap ($E_g = I- A$) yields 6.47 eV compared to the recently revised experimental value of 3.44 eV.[26] The root of this large overestimation is in the highly delocalised nature of the conduction states. Indeed, the spin population (as expected predominantly on zinc ions) dies away very slowly, from the centre to the periphery of the cluster $n_s = 0.12$ e for the central Zn ion, 0.07 e for the first coordination zinc shell and 0.03 e for the second. With the bottom of the conduction band situated far below the vacuum level a significantly larger QM region is needed to achieve a reasonable description. Upon self trapping ($E_{STe} = 0.31$ eV), we observe only a very small shift of the spin density from the cluster periphery to its centre ($n_s = 0.14$ e). This picture of the delocalised electron state is in agreement with the current opinion in the literature, which, however, also estimates the binding energy of the electron charge carriers to be as small as 0.03 eV, which shows that our calculations overestimate this quantity. We note that during the geometry optimisation atoms in our cluster undergo only minor displacements, none of which are larger than 0.02 Å. This type of behaviour is characteristic of a large polaron, which is in contrast with the small hole polaron we observed above.

Finally, we also attempted to gain insight into the near band edge excited properties of ZnO by studying a triplet exciton, which is the ground state of the system with spin 1. The calculated energy of a singlet to triplet transition is 5.35 eV, way above the experimental band gap value, which we again consider to be due to the problem with the description of the delocalised electron component of the exciton.

We conclude that while the valence states are reasonably well reproduced by our approach, the description of delocalised states in the conduction band is not reliable. The calculated energies of charge trapping can be considered as an upper bound, and are probably much more reasonable for the hole states. Now we turn our attention to the intrinsic atomic defects in ZnO.

3.4 Intrinsic defects

Four main point defects in ZnO have been considered in this study in their principal oxidations states (both neutral and charged species): (i) zinc vacancy, V_{Zn}, (ii) zinc interstitial, Zn_i, (iii) oxygen vacancy, V_O and (iv) oxygen interstitial, O_i (using the Kröger–Vink notation, with the dot over the chemical symbol reserved for the positive charge, prime—for the negative charge; the cross—for neutral species is omitted for clarity). Antisites (Zn at the O site and O at the Zn site) have not been considered in this work (*cf.* refs. 19 and 20).

Defect structures. All neutral defect configurations have been generated from the idealised symmetric lattice positions and geometry optimised. When appropriate all relevant spin states have been considered. The charge has then been introduced at the relaxed configuration and the system optimised again.

The interstitial space in ZnO includes octahedral sites, which can be seen in the middle of the 6-ring channels running in the *c* direction, and tetrahedral sites, which are situated directly on the lines of the Zn–O bonds connecting the hexagonal sheets (see Fig. 1). Another way of looking at these sites is provided by analogy with covalent, tetrahedrally coordinated semiconductors, where antibonding sites are defined. Their location is determined by following the direction of the bond through the terminal atoms into the interstitial space. In this case, one can easily visualise the three symmetry-equivalent octahedral sites around any lattice site, obtained by following the three ZnO bonds within the hexagonal sheet, and the fourth, tetra-hedral site on the line of continuation of the fourth bond as explained above.

Fig. 2 shows defect structures obtained by the geometry optimisation procedure. The oxygen and zinc vacant sites are highlighted in the figure by placing the missing atom back at its original lattice site in the unrelaxed configuration. The figure shows all ions within a 3.8 Å separation distance from the main site of interest.

Zinc vacancy. In our calculations, the Zn vacancy (Fig. 2, 1st row) proved to be a powerful charge trap, which is stable in five charge states with respect to the lattice: $+2$, $+1$, 0, -1, and -2. A detailed account of all the Zn vacancy states and their properties will be given elsewhere, whilst here we will concentrate on the neutral and negatively charged states, which are of more interest for the electron rich materials (see, however, our discussion of the defect levels below).

Formation of the neutral vacancy creates two holes on the nearest neighbour O ions. These two holes can couple with total spin 0, in a diamagnetic (closed-shell singlet) or antiferromagnetic (open-shell singlet) configuration; alternatively a ferromagnetic coupling (triplet state) is realised with spin of 1. The diamagnetic case proves to be irrelevant (as the resultant defect species is a higher energy defect complex including an O vacant site and a peroxy split interstitial), whilst spin polarised calculations give preference to the triplet state with a triplet–singlet energy splitting of 1.09 eV in the relaxed atomic configuration of the triplet. On relaxation the singlet state lowers its energy by nearly 1 eV, but still remains 0.11 eV higher in energy than the ground state triplet. In the triplet state the spin density is predominantly localised on two oxygen ions from the same the hexagonal sheet ($n_s = 0.89$ e) with the remaining small contribution ($n_s = 0.10$ e) on the third. Our description of the singlet state here causes some concern as in the relaxed config-uration, one hole remains strongly localised on one of the oxygen ions ($n_s = 0.93$ e), whilst the other spreads evenly between the other two oxygen ions ($n_s = 0.48$ e). In the relaxed triplet configuration the O^{2-} ion moves away from the vacancy, along the *c* axis, by 0.48 Å, the two main hole bearing O^- ions also relax outward, by 0.31 Å, and the third O ion with a tail of the spin density on it, relaxes weakly, by

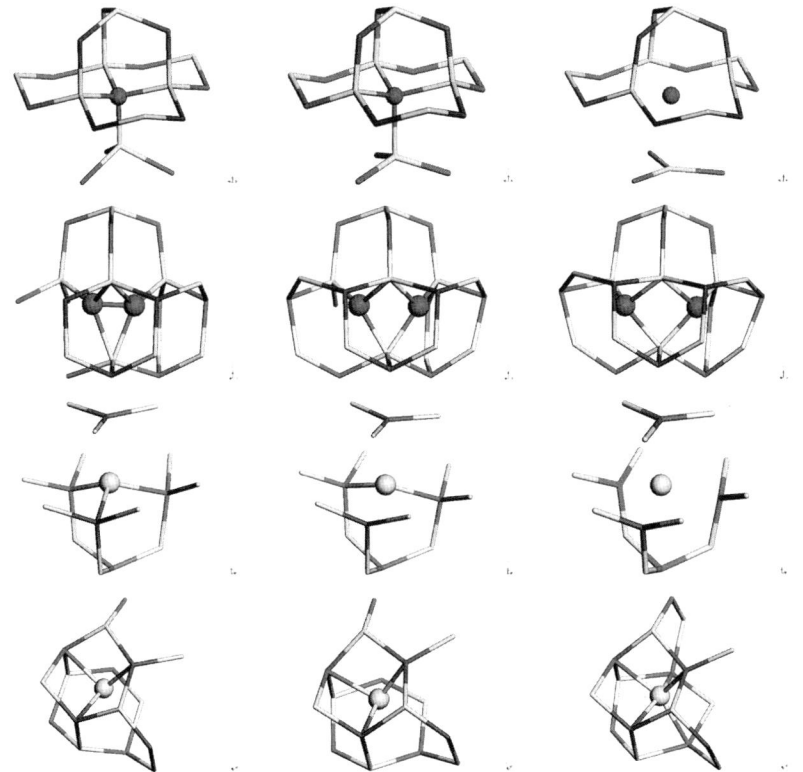

Fig. 2 Intrinsic defect structures in ZnO. 1st row: O vacancies; 2nd row: O interstitials, 3rd row: Zn vacancies; 4th row: Zn interstitials. The position of the vacant site is indicated by a ball. 1st column shows neutral, 2nd: singly charged and 3rd: doubly charged defects.

0.11 Å. In an alternative configuration, one of the holes occupies the axial O ion, this state lying 0.047 eV higher in energy.

As electron charge carriers can recombine with one of the holes, a V'_{Zn} centre is formed (sometimes called in the literature V^- with many analogous centres in oxide materials, see ref. 27). The remaining hole remains localised on one of the oxygens of the hexagonal sheet ($n_s = 0.90$ e), with the tail shared equally by the other two oxygen ions ($n_s = 0.03$ e). The oxygen ion, with a trapped electron, moves toward the vacancy by 0.10 Å, and the third O ion relaxes now away from the vacancy into a symmetry equivalent configuration, with an approximate local symmetry C_{2v}. Again, the corresponding axial hole configuration is 0.099 eV higher in energy.

Finally, yet another free electron carrier can be trapped by the vacancy and recombine with the remaining hole, resulting in the doubly charged defect. The electronic configuration of the defect is now of a closed-shell character with all three nearest neighbour O^{2-} ions, which belong to the same hexagonal ZnO sheets as the vacancy, having relaxed outward from their perfect lattice positions (by 0.31 Å) into a symmetric configuration.

Zinc interstitial. Our calculations showed that tetrahedral Zn interstitials are not local minima, but that they rather provide barriers for diffusion between stable octahedral sites in the (*ab*) crystallographic plane *via* a split-interstitial configuration (most probably Zn ions are too big compared to the available space). Therefore, we concentrate here on the octahedral sites (Fig. 2, 2nd row), whilst the diffusion processes with relevant configurations will be considered elsewhere. The neutral interstitial has a

This journal is © The Royal Society of Chemistry 2006

lower coordination by electron rich O^{2-} ions and forms a trigonal pyramid with the closest Zn_i–O separation distance of 2.15 and the two other distances of 2.20 Å. On ionisation this nearly symmetric configuration is broken, with the Zn^+ ion moving towards one of the lattice oxygens (1.96, 2.10 and 2.17 Å). The next nearest O ions move now towards the interstitial Zn (by about 0.3 Å), but do not approach close enough to coordinate to this ion directly (by a dative bond). However, with a loss of the second electron we observe a strong local rearrangement with Zn occupying a 5-coordinated site. This configuration can be considered to result from an off-centre displacement from an ideal octahedral site, as the Zn^{2+} ion is too small.

Oxygen vacancy. In its neutral state, the O vacancy (Fig. 3, 3rd row) is a conventional (colour) F-centre, identified in many halide and oxide materials. Its behaviour is well understood and theoretically described (see, for example, a recent discussion in ref. 28). Vacancy formation results in the two electrons from the removed O anion becoming trapped at the vacancy site and occupying a hydrogenic state in the band gap. We discuss this energy level later, but will concentrate here on the effect of the vacancy formation on the atomic structure. We observe a strong cation relaxation inwards, with the three Zn ions of the hexagonal ZnO sheet displaced by 0.08–0.09 Å, whilst the fourth Zn ion shifts along the c axis of ZnO by 0.25 Å. On ionisation this effect is reversed and now all four nearest neighbour Zn ions move away from the vacancy with equal displacements of 0.12 Å (compared to ideal bulk positions). This configuration is in line with the electron occupying a strongly localised s-like orbital in the vacancy. When the second electron is removed from the vacancy, we again observe a strong outward relaxation: the three Zn ions in the hexagonal sheet displace further by 0.12 Å, whilst the fourth cation moves drastically away by 0.48 Å in a trigonal configuration.

Oxygen interstitial. In agreement with periodic boundary conditions DFT calculations,[20] we find that the ground state configuration of the neutral O interstitial (Fig. 2, 4th row) is not at an octahedral or tetrahedral interstitial site, but a split interstitial. When placing the neutral O atom, in the triplet state (the ground state of atomic oxygen in the gas phase), at an octahedral site, it stays there assuming a trigonal pyramidal configuration. However, when excited in a singlet state or on electron trapping, this atom relaxes without a barrier into a split interstitial position. The ground state electronic configuration of the neutral split interstitial oxygen, in

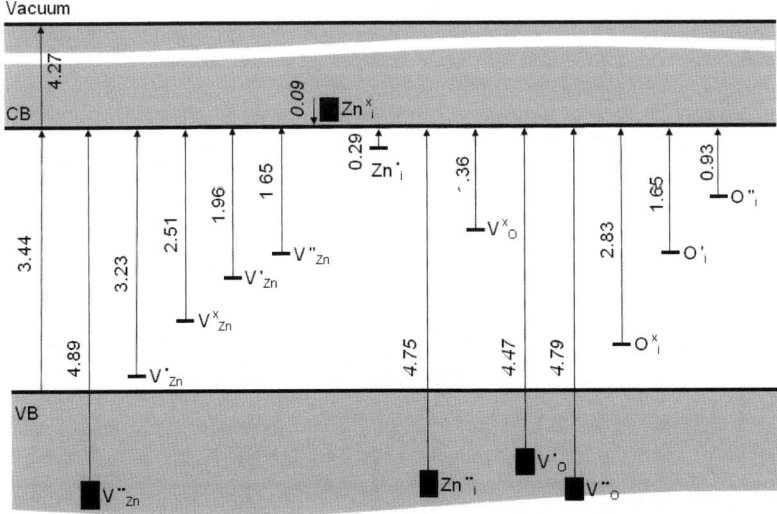

Fig. 3 Calculated donor defect levels in ZnO. Energies in eV. Resonance states are shown as filled rectangles.

contrast to the octahedral configuration, is a singlet. The defect relaxes to become a typical dioxygen peroxy species, O_2^{2-} with a bond length of 1.51 Å. Furthermore, the same pattern of relaxation has been shown by an O atom placed in a tetrahedral initial configuration. Two of the nearest neighbour Zn ions thus become 5-coordinated; correspondingly the distance to the two bonded O ions is larger by 0.10 and 0.13 Å compared to the defect free material. The singlet ground state of the defect is of a closed-shell nature. Electron trapping at the split interstitial site causes the distance between the two split ions to increase to 2.13 Å while the distance to the higher coordinated Zn ions returns to normal. The other Zn–O bond distances, however, become reduced by about 0.10 Å compared to their normal values, which should be simply determined by the Coulomb attraction to the charged centre. The hole is practically completely localised on the peroxy and shared equally by the two oxygen ions. On addition of the second electron, the two O^{2-} (closed-shell) ions move further apart, with the distance between them now being 2.59 Å.

Energies of formation. Below we provide a brief account of the defect energetics, summarised in Table 1.

As cohesion (or binding) energies of solids cannot be obtained consistently in our approach, standard thermodynamic data were used from ref. 29. Defect energies have been calculated here with respect to solid metal Zn (-124.203 kJ mol^{-1}), wurtzite ZnO (-357.393 kJ mol^{-1}) and molecular O_2 (also compare calculated and experimental values of -486.429 kJ mol^{-1} and -485.966 kJ mol^{-1} for the enthalpies of formation from atomic O). For the hole formation energy we used the value calculated above of the IP for the ideal cluster of 7.71 eV, which corresponds to a hole state at the top of the VB (see discussion of the electronic defects above). The electron formation energy has been estimated as -4.27 eV, which corresponds to the positioning of the bottom of the CB, 3.44 eV over the VB.

Assuming the self compensating mechanism for the defect formation (*i.e.* starting with the perfect stoichiometric material) we predict that under reducing conditions ZnO would be O deficient with O vacancy and Zn^{2+} interstitial formation having similar energies (as can be seen from the differences between the formation energies of Frenkel and Schottky pairs). We note that under different chemical conditions, which are realised during syntheses, sample preparation and exploitation in catalysis or electronic devices, zinc is in fact usually present in excess while oxygen can be both excessive and deficient. Owing to the close balance in the defect formation energies, the stoichiometry in this system is likely to be controlled by the history of the samples and working conditions, but clearly both Zn^{2+} interstitial and O vacancy formation are feasible in ZnO. Indeed, depending on these conditions, most defects considered can be present in significant concentrations and are therefore of interest.

A striking feature of calculated reaction energies is the stabilisation of electrons in the conduction band when compensated by zinc interstitials and oxygen vacancies. This is in agreement with all common samples of ZnO being of n-type, unless they have intentionally been doped with acceptor impurities. To obtain a more direct measure of the ease (or difficulty) of stabilisation of charge carriers by point defects in ZnO, we now consider defect levels in the band gap.

Defect levels. Vertical ionisation potentials of all defect species in this work have been used to align defect states in their main oxidation states with the band states of ZnO as shown in Fig. 3.

Zinc vacancy. In good quality undoped ZnO the photoluminescence spectra show a high energy sharp line at 3.22 eV.[1] Thonke et al. identified it with a shallow donor–acceptor transition (DAP), in which the acceptor binding energy is 0.195 ± 0.010 eV.[23] The authors proposed nitrogen impurities at O sites to be responsible. Here we will suggest a different assignment, but first let us consider how we can connect the calculated defect levels with the observed luminescence spectra or at least some of their features.

Table 1 Calculated defect formation energies

Defect reactions[a]	QM/MM energy/eV	MM energy[b]/eV
Oxygen vacancy		
$O_O^\times \rightarrow V_O^\times + 1/2\ O_2(g)$	5.296	
$O_O^\times \rightarrow V_O^\bullet + e' + 1/2\ O_2(g)$	5.275	
$O_O^\times \rightarrow V_O^{\bullet\bullet} + 2\ e' + 1/2\ O_2(g)$	5.403	
Oxygen interstitial		
$V_i^\times + 1/2\ O_2(g) \rightarrow O_i^\times$	1.652	
$V_i^\times + 1/2\ O_2(g) \rightarrow O_i' + h^\bullet$	6.129	
$V_i^\times + 1/2\ O_2(g) \rightarrow O_i'' + 2\ h^\bullet$	10.345	
Zinc vacancy		
$Zn_{Zn}^\times \rightarrow V_{Zn}^{\bullet\bullet} + 2\ e' + Zn(s)$	10.895	
$Zn_{Zn}^\times \rightarrow V_{Zn}^\bullet + e' + Zn(s)$	8.776	
$Zn_{Zn}^\times \rightarrow V_{Zn}^\times + Zn(s)$	7.280	
$Zn_{Zn}^\times \rightarrow V_{Zn}' + h^\bullet + Zn(s)$	9.514	
$Zn_{Zn}^\times \rightarrow V_{Zn}'' + 2\ h^\bullet + Zn(s)$	12.077	
Zinc interstitial		
$V_i^\times + Zn(s) \rightarrow Zn_i^\times$	4.722	
$V_i^\times + Zn(s) \rightarrow Zn_i^\bullet + e'$	3.563	
$V_i^\times + Zn(s) \rightarrow Zn_i^{\bullet\bullet} + 2\ e'$	2.183	
Frenkel pair in O sublattice		
$O_O^\times + V_i^\times \rightarrow V_O^\times + O_i^\times$	6.948	
$O_O^\times + V_i^\times \rightarrow V_O^\bullet + O_i'$	7.964	
$O_O^\times + V_i^\times \rightarrow V_O^{\bullet\bullet} + O_i''$	8.868	8.797
Frenkel pair in Zn sublattice		
$Zn_{Zn}^\times + V_i^\times \rightarrow V_{Zn}^\times + Zn_i^\times$	12.002	
$Zn_{Zn}^\times + V_i^\times \rightarrow V_{Zn}' + Zn_i^\bullet$	9.637	
$Zn_{Zn}^\times + V_i^\times \rightarrow V_{Zn}'' + Zn_i^{\bullet\bullet}$	7.380	7.501
$Zn_{Zn}^\times + V_i^\times \rightarrow V_{Zn}'' + Zn_i^{\bullet\bullet} + 2\ e'$	9.463	
Schottky pair		
$Zn_{Zn}^\times + O_O^\times \rightarrow V_{Zn}^\times + V_O^\times + ZnO(s)$	8.872	
$Zn_{Zn}^\times + O_O^\times \rightarrow V_{Zn}' + V_O^\bullet + ZnO(s)$	7.645	
$Zn_{Zn}^\times + O_O^\times \rightarrow V_{Zn}'' + V_O^{\bullet\bullet} + ZnO(s)$	6.896	6.749

[a] Letters g and s in parentheses indicate the gas phase and solid standard reference states. [b] MM energies calculated using the Mott–Littleton approach with current interatomic potentials.

As we discussed before the exciton structure in ZnO should involve a relatively localised hole state and a rather delocalised electron. Upon interaction with a defect, a process of recombination of the localised electron of the defect with a hole component of the exciton could occur. For such a process, we expect that the luminescence will be characterised by energies close to the defect levels we calculated. On the other hand, on ionisation of the defect (loss of an electron to the CB), the electron component of the exciton could recombine with a hole, which is localised now on the defect. In this case, we should observe some reduction in the emission energy compared to the ionisation energy (Stokes shift). In our approach, the problem is with the extent of this reduction. For most defects our calculations show large relaxation energies, in some cases comparable with the band gap, which suggests that nonradiative mechanisms of recombination take place. Quantitative predictions of these effects are difficult in particular due to the problems of positioning the conduction states. With this in mind, we start our analysis with higher energy bands.

We have calculated the donor level for a hole trapped on a neutral Zn vacancy at 0.215 eV above the VB, which corresponds to the ionisation energy to the CB of 3.225 eV. These numbers are very close to the values reported by Thonke, which suggests that positive $V_{Zn}^\bullet$ could be the acceptors responsible for the DAP. In contrast, our preliminary calculations suggest that the neutral nitrogen substitutional level lies *ca.* 0.9 eV above the VB, whilst the positive species is a resonance state, which could not serve as an acceptor in the DAP considered.

Recent applications of positron annihilation lifetime spectroscopy coupled with photoluminescence measurements were able to identify both near band edge (3.36 eV) and deep level (broad band with a maximum shifting between 2.2 and 2.4 eV) emission with negatively charged defects in ZnO.[30] (These measurements have also identified higher energy line at 3.346 eV with complexes of Zn vacancies.[31]) Magnetic resonance studies by two different groups uncovered a green (a broad band at 2.5 eV) photoluminescence centre with the initial triplet state.[32,33] Leiter *et al.* proposed that this signal involves intra-defect transitions in O vacancies.[33] In our view, an exciton coupling and recombination with a V_{Zn}^{x} triplet centre, which can also involve transitions to and from its singlet state is a plausible, alternative explanation of the green band.

Zinc interstitial. These calculations identify the interstitial zinc as the main intrinsic source of n-type conduction in ZnO, at least based on the shallow donor mechanism. (When present in sufficient concentrations, extrinsic defects, of course, still determine this behaviour—see also our results for the Al impurity below.)

As noted before our predictions for the resonance states should be treated with caution. In particular, we find that the level of the neutral Zn interstitial lies 0.09 eV above the bottom of the conduction band, which should result in a loss of its outer electron to the conduction band. In a more realistic calculation the valence states of Zn atoms would hybridise with the continuum states at the bottom of the CB, which would lead to some stabilisation of this species (compared with our calculation) probably making it a shallow donor. The shallow nature of the Zn interstitial is in agreement with a tentative assignment to this defect of a 0.031 eV donor state by Look *et al.*[34]

The donor level of singly ionised Zn interstitial, in contrast, lies 0.29 eV below the bottom of the conduction band, which is in remarkable agreement with electrical measurements by Auret *et al.*[35] These researchers identified four donor bands at E1 = 0.12 eV, E2 = 0.10 eV, E3 = 0.29 eV and E4 = 0.57 eV. The E3 band has been shown to be very similar to another band at *ca.* 0.3 eV reported by other authors, who also used capacitance measurement based techniques and samples of ZnO prepared differently. Based on the universal character of the band, the excess of Zn commonly observed in ZnO and that the experimental and calculated values are practically identical, we suggest that this band is due to the $Zn_i^{.}$ defect. Moreover, Auret *et al.* reported an unusual behaviour of E3 centres: external electric field slowed down their emission, which is in contrast to behaviour of typical shallow donors (such as E1), whose emission is enhanced. This effect, however, fits with our assignment of the band to a positive centre, an acceptor defect in agreement with an earlier proposal.[35] Auret *et al.* also report that the E3 band is possibly due to two different defects with closely overlapping levels. With present calculations we found so far only one possible source of this band.

Oxygen vacancy. Recent identification of green luminescence bands with O vacancies by Leiter *et al.* has been supported by other optical experiments.[36] However, our calculations do not support this assignment: the neutral F-centre, which could exhibit singlet to triplet transitions, lies too high in the band gap (1.36 eV in our calculation), so no simple absorption of emission process could occur with such a high energy that is required (2.5–3.0 eV). A similar conclusion has been reported by Fink.[24] An alternative explanation could be found in the luminescence of an F^{+} centre ($V_O^{.}$) due to intra-defect transitions, as predicted long ago by Wei.[37] The presence of positively charged O vacancies in ZnO has also been suggested based on positron annihilation spectroscopic data.[38] However, F^{+} is a doublet centre which is incompatible with the triplet and singlet nature of the initial and final states. Thus the oxygen vacancies might be responsible for some form of the green luminescence, but not that which has been observed and characterised by Leiter *et al.*

This journal is © The Royal Society of Chemistry 2006

As we place the neutral vacancy level at 2.08 eV above the VB, we can, however, offer a plausible explanation for a red–orange peak from a He-implanted ZnO reported by Hamby et al.[39] This peak is rather broad and centred at 1.86 eV at low temperatures. As the temperature rises, the peak shifts towards 2 eV, which is explained by the DAP nature of the transition. On heating the sample the shallow donor states lose their electrons into the conduction band, and the transition changes its origin from DAP to a free electrons/acceptor recombination. This latter mechanism can be modified to include an exciton recombination on an F centre. Another possibility is offered by a V'_{Zn} defect.

Oxygen interstitial. Currently, there is no convincing spectroscopic or other assignment to oxygen interstitials in the experimental literature. With the low energy of formation that we calculated and under oxidising conditions, this defect could be expected to be present in appreciable concentrations. Furthermore, this defect should be prominent in high-energy radiation damaged material. The calculated defect levels could be used to link this defect in its neutral state (at 2.83 eV below the bottom of the CB) with the blue[40] (2.78 eV) and blue-green (2.5–2.3 eV) luminescence bands; the latter would also require an intermediate-deep donor such as $Zn_i^\cdot$ discussed above.

Finally, we conclude this section by making reference to the work of Bylander[41] on cathodoluminescence, who observed a (blue-)green band at 2.55 eV, green at 2.33 eV and red at 2.00 eV. On the basis of the work of Kröger,[42] Bylander assigned the 2.55 eV band to electron recombination with V'_{Zn}, the 2.33 eV band to a transition from $Zn_i^\times$ to V'_{Zn} and the 2.00 eV band to an electron transition to a $V_O^\cdot$ defect level. As we consider correspondence between excitation and recombination levels, we find a very good agreement with our results, but one should be very careful here when considering that the relaxation of the state does not seem to be included by Bylander, and the charge state assignment is not identical to ours. The level assignment still might be valid, but independent spectroscopic evidence would be of great help here. Therefore we are currently performing calculations of spectroscopic signatures of point defects, aimed at establishing a closer link with experiment.

3.5. Extrinsic defects

Here we present only a short outline of our results on impurities in ZnO; defect levels of which are summarised in Fig. 4. Below, we consider Li, Cu and Al substitutionals in the Zn sublattice, for which we outline particularly interesting features. As noted in our discussion of the 3.22 eV DAP transition, N commonly proposed as an acceptor for this process, cannot play this rôle as clearly seen from the diagram. Further discussion of N in the O sublattice will be presented elsewhere.

Lithium. On substitution in ZnO at a Zn site, in the charge neutral state, there is a very strong lattice relaxation around the resulting $Li_{Zn}^\times$ species. In agreement with the detailed analysis of Schirmer,[43] we find two closely related forms of this defect. In the axial defect, the hole introduced by Li is localised on the axial O ion, while in the nonaxial centre the hole is on one of the three nearest neighbour O ions in the hexagonal sheet. Both O, bearing the hole, and Li ions move from their idealised lattice positions, and the distance between them increases by 0.74 and 0.72 Å, respectively. We also confirm that the axial form of the defect is more stable; curiously, the opposite is true for an analogous V'_{Zn} (or V^-) defect considered above. This centre proves to be a deep acceptor with the calculated value of the Li_{Zn}' defect level 0.807 eV above the VB in very good agreement with experiment.[44] A yellow–orange luminescence band (with a peak at 2.2 eV) is associated with this centre; it has been shown to be of the DAP character with the nature of the donor still unresolved.

Copper. On substitution, we observe that the neutral Cu substitutional fits nearly perfectly in the Zn site, with the distances to the nearest neighbour O ions only slightly decreased. In contrast, the Cu_{Zn}' defect shows three bonds elongated by

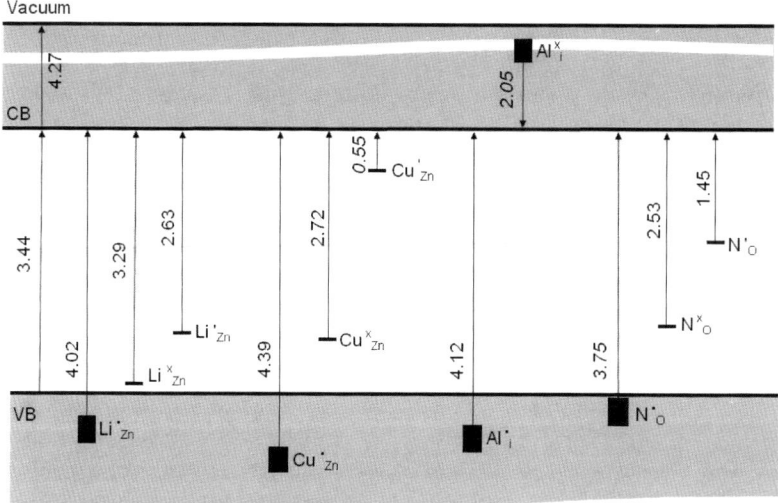

Fig. 4 Calculated extrinsic donor defect levels in ZnO. Energies in eV. Resonance states are shown as filled rectangles.

0.12 Å, while the fourth distance to an O ion, which is in the hexagonal ZnO sheet, increases by 0.20 Å. Copper impurities have been firmly identified as a source of a different green luminescence band characterised by a zero-phonon absorption at 2.86 eV (see ref. 1 and further references therein), which can be compared with the 2.89 eV we calculated as the corresponding donor level (in the acceptor state) above the VB. Significantly, this level is 0.55 eV below the bottom of the CB, which is in excellent agreement with the binding energy of 0.54 eV reported by Auret et al.[35] for Ep2 charge carriers in ZnO after bombardment with high energy protons. This band was identified by the authors with the E4 band (0.57 eV) in regular samples. It was also suggested that this is acceptor band. Thus the calculated energies strongly suggest that the E4 band is in fact due to the Cu impurities, which commonly contaminate the samples of ZnO. (Both metals have similar atomic masses and are difficult to separate.) In the neutral state, Cu assumes a d^9 configuration with a donor defect level 0.72 eV above the top of the VB and 2.72 eV below the CB bottom. These numbers are in good agreement with the sharp absorption by Cu at 0.72 eV.

Aluminium. In a recent work,[45] Sabioni proposed that in undoped and Al doped ZnO, oxygen diffusion occurs by an interstitial mechanism, with interstitial oxygen being neutral or singly charged. This statement is in contrast to other opinions in the literature,[46,47] which gives preference to the doubly charged interstitial O. One of the key points in Sabioni's analysis is based on electric measurements which unambiguously show Al to be a donor. In our calculations, not surprisingly, we find that neutral Al introduces a resonance state in the CB, from which we conclude that it must be a shallow donor. On ionisation the defect levels drops below the top of the VB.

Further work on extrinsic defect energetics and properties is now in progress and will be reported.

Conclusions

We have established the structure and energetics of the main point defects in ZnO. In particular, O interstitials found to have the lowest energies of formation among all intrinsic defects, which suggests their dominance under oxidising conditions. Furthermore, Zn interstitials and O vacancies have similar and relatively low energies of formation. Hence, the dominant defect species under reducing conditions should be decided upon by the sample history and working conditions.

A number of observations from electric and optical measurements are rationalised by computational models reported here:

- We propose that neutral and singly positively charged Zn interstitial defect is responsible for E1 and E3 donor bands from electric measurements.

- Zn vacancies are confirmed as majority acceptor in agreement with experimental assignment based on positron annihilation spectroscopic and other studies. This defect is found to be stable in five charged states. Exciton recombination on this defect species is proposed as a source for main photoluminescence bands: ultraviolet (an acceptor level at 3.2 eV in a DAP transition); green (a triplet level at 2.5 eV), red (at 1.9–2.0 eV).

- Neutral O interstitial in a split-interstitial peroxy configuration (at 2.8 eV) should also contribute to blue and green luminescence via an exciton recombination and DAP transition from donor Zn interstitials.

- O vacancies could not be a source of green luminescence, but could contribute to near-gap (UV) and red-orange luminescence bands (at 2.1 eV and below) via the exciton recombination mechanism.

- Our calculations confirm that Cu, which is stable in ZnO in two charge states, is an efficient electron scavenger. The singly negatively charged Cu impurity is proposed as an E4 donor. Calculated defect levels (at 2.7 and 0.55 eV) are in good agreement with experiment, which established Cu as a distinct source of green luminescence from ZnO.

Our calculations on Li, Al and N impurities show their donor and acceptor properties in agreement with experiment where available. Curiously, our calculations suggest that Li and N can also contribute to the green luminescence band.

We conclude that while our work, modelling spectroscopic signatures of the defect species, is in progress, more direct experimental measurements aiming at the atomic structure of the defects are highly desirable.

Acknowledgements

We are indebted to J. L. Gavartin and S. M. Woodley for many stimulating discussions. Computational resources for this work have been provided by the HPCx service via our membership of the Materials Chemistry Consortium and funded by the grant GR/S 13422. We gratefully acknowledge the use of visualisation software by Accelrys.

References

1   Ü. Özgür, Ya I. Alivov, C. Liu, A. Teke, M. A. Reshchikov, S. Doğan, V. Avrutin, S.-J. Cho and H. Markoç, *J. Appl. Phys.*, 2005, **98**, 041301-1-103.
2   S. J. Pearton, D. P. Norton, K. Ip, Y. W. Heo and T. Steiner, *Prog. Mater. Sci.*, 2005, **50**, 293–340.
3   D. C. Look and B. Claflin, *Phys. Status Solidi B*, 2004, **241**, 624–630.
4   (a) S. A. French, A. A. Sokol, S. T. Bromley, C. R. A. Catlow, S. Rogers, F. King and P. Sherwood, *Angew. Chem.*, 2001, **113**, 4569–4572; (b) S. A. French, A. A. Sokol, S. T. Bromley, C. R. A. Catlow and P. Sherwood, *Top. Catal.*, 2003, **24**, 161–172.
5   (a) S. T. Bromley, S. A. French, A. A. Sokol, C. R. A. Catlow and P. Sherwood, *J. Phys. Chem. B*, 2003, **107**, 7045–7057; (b) C. R. A. Catlow, S. A. French, A. A. Sokol, M. Alfredsson and S. T. Bromley, *Faraday Discuss.*, 2003, **124**, 185–203; (c) C. R. A. Catlow, S. A. French, A. A. Sokol and J. M. Thomas, *Philos. Trans. R. Soc. London, Ser. A*, 2005, **363**, 913–936; (d) N. S. Phala, G. Klatt, E. van Steen, S. A. French, A. A. Sokol and C. R. A. Catlow, *Phys. Chem. Chem. Phys.*, 2005, **7**, 2440–2445.
6   K. Lott, S. Shinkarenko, T. Kirsanova, L. Türn, A. Grebennik and A. Vishnjakov, *Phys. Status Solidi C*, 2005, **2**, 1200–1205.
7   V. E. Heinrich and P. A. Cox, *The Surface Science of Metal Oxides*, Cambridge University Press, Cambridge, UK, 1996.
8   A. A. Sokol, S. T. Bromley, S. A. French, C. R. A. Catlow and P. Sherwood, *Int. J. Quantum Chem.*, 2004, **99**, 695–712.
9   C. R. A. Catlow and W. C. Mackrodt, in *Computer Simulation of Solids*, ed. C. R. A. Catlow and W. C. Mackrodt, Springer-Verlag, Berlin, 1982, p. 3.
10 P. Sherwood, A. H. de Vries, M. F. Guest, G. Schreckenbach, C. R. A. Catlow, S. A. French, A. A. Sokol, S. T. Bromley, W. Thiel, A. J. Turner, S. Billeter, F. Terstegen, S.

Thiel, J. Kendrick, S. C. Rogers, J. Casci, M. Watson, F. King, E. Karlsen, M. Sjøvoll, A. Fahmi, A. Schäfer and Ch. Lennartz, *J. Mol. Struct. (THEOCHEM)*, 2003, **632**, 1.

11 M. F. Guest, J. M. H. Thomas, P. Sherwood, I. J. Bush and H. J. J. van Dam, *Mol. Phys.*, 2005, **103**, 719.

12 J. D. Gale, *J. Chem. Soc., Faraday Trans.*, 1997, **93**, 629.

13 L. Whitmore, A. A. Sokol and C. R. A. Catlow, *Surf. Sci.*, 2002, **498**, 135–146.

14 (a) A. D. Becke, *J. Chem. Phys.*, 1997, **107**, 8554–8560; (b) F. A. Hamprecht, A. J. Cohen, D. J. Tozer and N. C. Handy, *J. Chem. Phys.*, 1998, **109**, 6264–6271.

15 (a) G. Pacchioni, F. Frigoli, D. Ricci and J. A. Weil, *Phys. Rev. B*, 2000, **63**, 054102; (b) R. Catlow, R. Bell, F. Cora, S. A. French, B. Slater and A. Sokol, *Annu. Rep. Prog. Chem., Sect. A*, 2005, **101**, 513–547; (c) J. To, A. A. Sokol, S. A. French, N. Kaltsoyannis and C. R. A. Catlow, *J. Chem. Phys.*, 2005, **122**, 144704.

16 A. Schäfer, C. Huber and R. Ahlrichs, *J. Chem Phys.*, 1994, **100**, 5829–5835.

17 M. Dolg, U. Wedig, H. Stoll and H. Preuss, *J. Chem. Phys.*, 1987, **86**, 866.

18 (a) W. C. Mackrodt, R. F. Stewart, J. C. Campbell and I. H. Hillier, *J. Phys. (Paris), Colloq.*, 1980, **41**, 64; (b) D. J. Binks and R. W. Grimes, *J. Am. Ceram. Soc.*, 1993, **76**, 2730.

19 (a) A. F. Kohan, G. Ceder, D. Morgan and C. G. Van de Walle, *Phys. Rev. B*, 2000, **61**, 15019–15027; (b) J.-L. Zhao, W. Zhang, X.-M. Li, J.-W. Feng and X. Shi, *J. Phys.: Condens. Matter*, 2006, **18**, 1495–1508.

20 A. Janotti and C. G. Van de Walle, *J. Cryst. Growth*, 2006, **287**, 58–65.

21 P. Erhart, K. Albe and A. Klein, *Phys. Rev. B*, 2006, **73**, 205203-1-9.

22 A. Wander and N. M. Harrison, *J. Chem. Phys.*, 2001, **115**, 2312.

23 G. A. Shi, M. Stavola and W. B. Fowler, *Phys. Rev. B*, 2006, **73**, 081201-1-.

24 (a) K. Fink, *Phys. Chem. Chem. Phys.*, 2005, **7**, 2999; (b) K. Fink, *Phys. Chem. Chem. Phys.*, 2006, **8**, 1482–1489.

25 R. K. Swank, *Phys. Rev.*, 1967, **153**, 844.

26 K. Thonke, Th. Gruber, N. Teofilov, R. Schönfelder, A. Waag and R. Sauer, *Physica B*, 2001, **308–310**, 945.

27 (a) O. F. Schirmer, *Z. Phys. B*, 1976, **24**, 235; (b) N. M. Mott and A. M. Stoneham, *J. Phys. C: Condens. Matter*, 1977, **10**, 3391.

28 A. L. Shluger, A. S. Foster, J. L. Gavartin and P. V. Sushko, in *Nano and Giga Challenges in Microelectronics*, ed. J. Greer, A. Korkin and J. Labanowski, Elsevier, Amsterdam, 2003, pp. 151–222.

29 *CRC Handbook of Chemistry and Physics*, ed. D. R. Lide, CRC Press, Boca Raton, FL, 86th edn, 2005.

30 M. A. Hernández-Fenollosa, L. C. Damonte and B. Marí, *Superlattices Microstruct.*, 2005, **38**, 336–343.

31 A. Zubiaga, J. A. Garcia, F. Plazaola, F. Tuomisto, K. Saarinen, J. Z. Pérez and V. Muñoz-Sanjosé, *J. Appl. Phys.*, 2006, **99**, 053516-1-6.

32 W. E. Carlos, E. R. Glaser and D. C. Look, *Physica B*, 2001, **308–310**, 976.

33 F. H. Leiter, H. R. Alves, A. Hofstaetter, D. M. Hofmann and B. K. Meyer, *Phys. Status Solidi B*, 2001, **226**, R4–R5.

34 (a) D. C. Look, D. C. Reynolds, J. R. Sizelove, R. L. Jones, C. W. Litton, G. Cantwell and W. C. Harsch, *Solid State Commun.*, 1998, **105**, 399; (b) D. C. Look, J. W. Hemsky and J. R. Sizelove, *Phys. Rev. Lett.*, 1999, **82**, 2552.

35 (a) F. D. Auret, S. C. Goodman, M. Hayes, M. J. Legodi and E. Wendler, *Superlattices Microstruct.*, 2006, **39**, 17–23; (b) F. D. Auret, J. M. Nel, M. Hayes, L. Wu, W. Wesch and E. Wendler, *Superlattices Microstruct.*, 2006, **39**, 17–23.

36 K. Sakai, T. Kakeno, T. Ikari, S. Shirakata, T. Sakemi, K. Awai and T. Yamamoto, *J. Appl. Phys.*, 2006, **99**, 043508-1-7.

37 W. F. Wei, *Phys. Rev. B*, 1977, **15**, 2250–2253.

38 F. Tuomisto, K. Saarinen, K. Grasza and A. Mycielski, *Phys. Status Solidi B*, 2006, **243**, 794.

39 D. W. Hamby, D. A. Lucca, J.-K. Lee and M. Nastasi, *Nucl. Instrum. Methods Phys. Res., Sect. B*, 2006, **242**, 663–666.

40 D. H. Zhang, Z. Y. Xue and Q. P. Wang, *J. Phys. D: Appl. Phys.*, 2002, **35**, 2837–2840.

41 E. G. Bylander, *J. Appl. Phys.*, 1978, **49**, 1188–1195.

42 F. A. Kroger, *The Chemistry of Imperfect Crystals*, North Holland, Amsterdam, 1964, p. 691.

43 O. F. Schirmer, *J. Phys. Chem. Solids*, 1968, **29**, 1407.

44 O. F. Schirmer and D. Zwingel, *Solid State Commun.*, 1970, **8**, 1559.

45 A. C. S. Sabioni, *Solid State Ionics*, 2004, **170**, 145148.

46 H. Haneda, I. Sakaguchi, A. Watanabe, T. Ishigaki and J. Tanaka, *J. Electrochem.*, 1999, **4**, 41.

47 A. C. S. Sabioni, M. J. F. Ramos and W. B. Ferraz, *Mater. Res.*, 2003, **6**, 173–178.

Formation of, and ion-transport in, low-dimensional crystallites in carbon nanotubes

Mark Wilson

Received 20th February 2006, Accepted 28th March 2006
First published as an Advance Article on the web 21st July 2006
DOI: 10.1039/b602488b

The formation of low-dimensional crystal structures, obtained by filling carbon nanotubes from the molten salts, is considered for three stoichiometries (the MX, MX_2 and MX_3). For the MX stoichiometry, general classes of inorganic nanotube (INT) are predicted to exist whose morphology depends both on the low-energy (bulk) crystal structure and the encasing carbon nanotube diameter. These INTs are generally found to have no direct bulk analogues. For both the MX_2 and MX_3 stoichiometries crystal structures are predicted which either have a direct bulk analogue, or whose structure can be considered as a distortion of a bulk fragment. In both of these stoichiometries unusual (high anion coordination) crystallites are predicted. For the MX stoichiometry the ion transport mechanism is investigated and discussed, whilst for the MX_3 the vibrational densities of states are analysed with respect to both the pure liquid and idealized crystallites.

1. Introduction

Over the past decade, experimental investigations using high resolution transmission electron microscopy (HRTEM) have demonstrated that many molten salts fill simple carbon nanotubes (C-NTs). Materials as diverse as NiO, PbO, Bi_2O_3, V_2O_5 and MoO_3,[1–5] UCl_4,[6] AgCl/AgBr,[7,8] KI,[9,10] BaI_2,[11] CoI_2[12] and Sb_2O_3[13] have been successfully inserted. Significantly, the observed HRTEM patterns show that the structures formed within the C-NTs tend to be *crystalline* in nature despite the surrounding materials remaining in the molten state (*i.e.* above the bulk salt melting point). Molten salts represent an attractive class of filling materials since their surface tensions are typically low enough so as to not crush the immersed C-NT (as would most molten metals). Atomistic control of the internal crystallite structure is crucial if their potentially useful mechanical and electronic properties are to be fully exploited. Typical experimentally-based analysis of the formed internal structures relies on identifying a relatively simple correspondence between the internal crystallite structure and known bulk structures. However, computer simulation models, in which the atom–atom interactions are accounted for through relatively simple potential models, not only account for these "simple" bulk-related structures, but also *predict* the existence of new classes of inorganic nanotubes (INTs).[14–19] These INTs, although structurally related to the known bulk structures, have no direct bulk analogue. Furthermore, these simulation models not only predict complex INT structures but also allow for an understanding of their atomistic formation

Department of Chemistry, University College London, 20 Gordon Street, London, UK WC1H 0AJ

mechanisms (and, as a result, how material is transported along the C-NT). Such atomistic mechanistic information is not directly available from experimental studies.

In this paper we shall construct a range of simulation models in order to help understand the structures formed by salts across three different stoichiometries: MX, MX_2 and MX_3 (M = metal cation, X = halide anion). In all cases typical (static) crystal structures will be highlighted and their relationship to bulk crystal structures discussed. For the MX stoichiometry we focus on the dynamics of ion motion (atomistic filling mechanism) for a specific INT. For the MX_3 stoichiometry potentially interesting features in the vibrational densities of states are highlighted and discussed.

2. Potential models

In order to construct an overview picture of the factors controlling the INT formation, simplified atomistic models are employed in which the C-NTs are modelled as rigid superstructures of fixed atoms. The carbon nanotubes are generated in the usual way by folding a single graphene sheet along the chiral vector $C_h = n\mathbf{a}_1 + m\mathbf{a}_2$,[20] where $\mathbf{a}_1$ and $\mathbf{a}_2$ are the unit cell vectors.

The ion–ion interactions are accounted for via a polarizable-ion model (PIM) which incorporates a description of the (dipolar) anion polarization (see ref. 21 for details). The ion–ion interactions are modelled using a Born–Mayer potential,

$$U_{ij}(r_{ij}) = B_{ij}\exp(-a_{ij}r_{ij}) + \frac{Q_iQ_j}{r_{ij}} - \frac{C_6}{r_{ij}^6},$$ (2.1)

where $Q_{i(j)}$ is the (valence) charge on ion $i(j)$, B_{ij} and a_{ij} control the short-range repulsion, and C_6 is the dipole–dipole dispersion term. The PIM description of the anion polarization requires two additional parameters, the anion polarizability, α, and the short-range damping parameter, b, which control the anion response to the electric field and short-range (nearest-neighbour) effects, respectively.[22]

The ion–carbon interactions are modelled by simple Lennard-Jones potentials

$$U_{iC}(r_{iC}) = 4\varepsilon_{iC}\left\{\left(\frac{\sigma_{iC}}{r_{iC}}\right)^{12} - \left(\frac{\sigma_{iC}}{r_{iC}}\right)^{6}\right\},$$ (2.2)

where r_{iC} is the ion–carbon separation. The parameters are derived from isoelectronic noble gas potentials coupled with a model for graphitic carbon[23] using standard mixing rules. All of the potential parameters used in the present work are listed in Table 1.

Two sets of calculation procedures are employed throughout. In the first, direct filling of the C-NTs is observed. A cylindrical section is extracted from an equilibrated liquid configuration and a C-NT of the required morphology inserted. The two ends of the nanotube are initially covered with graphene sheets in order to allow the surrounding liquid to re-equilibrate without entering the C-NT. Once re-equilibrated the graphene end-groups are removed, allowing molten salt ingress. In the second method, specific INT structures are systematically constructed and energy minimisations performed using a steepest-descent algorithm. The latter methodology allows the underlying phase diagram (as a function of C-NT diameter and morphology) to be constructed. The inclusion of the direct filling methodology, however, suggests classes of INT structures for further energy minimisation calculations.

Table 1 Potential parameters used in the present work. The ion–ion interactions are modelled using a Born–Mayer potential with the charge–charge interactions calculated using full formal valence charges. The short-range interaction is given by $U(r_{ij}) = B_{ij} \exp(-a_{ij}r_{ij}) - C_{ij}^6/r_{ij}^6$. The carbon–ion interactions are handled *via* a Lennard-Jones potential and are unchanged throughout. The parameters "MX" and "XX' refer to the idealised system having a four-coordinate crystal structure ground state

ij	B_{ij}/a.u.	a_{ij}/a.u.	C_{ij}^6/a.u.	ε_{ij}/a.u.	σ_{ij}/a.u.
MX	8.68	1.55	2.09	—	—
XX	61.66	1.55	115.99	—	—
KI	102.88	1.491	107.64	—	—
II	269.52	1.491	624.76	—	—
LaCl	100.0	1.53	137.6	—	—
ClCl	450.0	1.80	319.9	—	—
SrCl	328.8	2.078	58.8	—	—
ClCl	44.93	1.380	165.8	—	—
MC	—	—	—	57.9	3.40
XC	—	—	—	79.4	3.74

3. Calculations and results

3.1. MX stoichiometry

Experimental (HRTEM) information is available for both KI[9,10] and AgCl/AgBr.[7,8] In the former case two basic structures are observed to form within the C-NT depending upon the diameter in the enclosing nanotube. Both structures formed can be understood as being extracted directly from a section of bulk (rocksalt, B1) crystal. In smaller C-NTs a "2 × 2" structure forms in which a cross-section perpendicular to the C-NT major axis contains two MX molecular units (arranged approximately as an M–X–M–X square).[9] In the larger C-NTs a "3 × 3" structure forms in which there are nine ions in the cross-section.[10] For the AgCl/AgBr mixture a structure is formed that can be considered as extracted directly from the bulk (wurtzite, B4) crystal structure.[7,8]

Two potential models are considered for this stoichiometry which differ in the relative stabilities of bulk crystal polymorphs. In the first model (I) the rocksalt (B1, 6:6 coordination) is favoured over wurtzite (B4, 4:4 coordination), whilst in the second model (II) the reverse energetic ordering is stabilised. Model I reproduces a number of crystal and liquid properties for KI[14] whilst model II represents an archetypal system in which a four-coordinate ground state crystal structure is predicted.

3.1.1. Static structure. Fig. 1a shows two typical INT structures (one generated for each model I and II) formed within two C-NTs. For model I the structure formed is based on percolating 2 × 2 square net structures but is not related simply to known bulk structures. Furthermore, the INTs formed within the C-NTs fall into a general class of INT which can most usefully be described by reference to folding a sheet of 2 × 2 square nets (Fig. 1b). As a result, we can develop a nomenclature to describe the inherent morphology of the INTs which is analogous to that utilised to describe the C-NT morphologies themselves. Fig. 1b, for example, highlights the chiral vector labelled $(2,2)_{sq}$. The "sq" subscript indicates that the INT is formed from a plane of square nets. Note that the experimentally observed "2 × 2" structure, which was originally understood by reference to a section of bulk crystal, corresponds to a specific INT (of morphology $(2,2)_{sq}$ as highlighted in the figure) which happens to correspond to a section of bulk structure. Fig. 1a shows a square net INT of morphology $(3,2)_{sq}$. The structure is chiral and, although containing square net

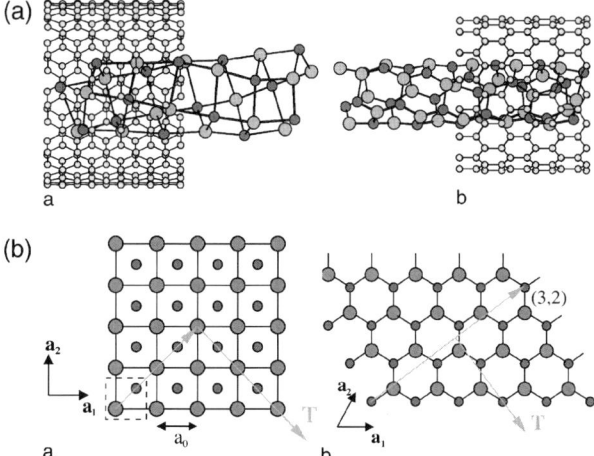

Fig. 1 (a) Inorganic nanotube (INT) structures formed from models I and II respectively. The left-hand panel shows a $(3,2)_{sq}$ INT formed within a $(11,11)$ C-NT, whilst the right-hand panel shows a $(3,2)_{hex}$ INT formed within a $(16,0)$ C-NT. In both cases a section of the C-NT has been cut away in order to reveal the internal INT structure. (b) The square net and hexagonal planes which can be folded to produce the INTs formed by models I and II. The left-hand panel shows the square net with unit cell (highlighted by the dashed box) length a_0, with the chiral vector shown to generate the $(2,2)_{sq}$ INT. The right-hand panel shows the analogous hexagonal plane with chiral vector appropriate to form the $(3,2)_{hex}$ INT. In both cases the direction of the resulting INT major axis, **T**, is shown.

structures which are common to the bulk (rocksalt) structure, clearly cannot be constructed purely by extracting a section of that bulk structure. The presence of the C-NT acts, therefore, to order the internal ionic structure. This ordering effect can be relatively easily understood by considering, as an example, the $(2,2)_{sq}$ INT structure. In this structure the cross-section area of the C-NT is able to accommodate four ions. Once a single ion is placed on the internal surface, however, then the locations of the remaining three ions are effectively defined as the energetic penalty to having nearest-neighbour like–like interactions which is significantly higher than the available thermal energy.

For model II, the structures formed are based on percolating hexagonal units. Again, we can classify these structures by reference to folding sheets of such hexagons (Fig. 1b), the unit cell of which is the direct analogue of that for a graphene sheet except that the two atoms in the unit cell have different chemical identities. The chiral vector shown in Fig. 1b is for the $(3,2)_{hex}$ INT which, when folded, corresponds to the structure shown in Fig. 1a. Again, the observed experimental structure (for the AgCl/AgBr system), originally rationalised in terms of extracting a fragment of the bulk wurtzite structure, can be understood equally well in terms of folding the hexagonal plane. In this case, a fold along the $(3,0)_{hex}$ chiral vector corresponds to the observed structure.

For both classes of structure, therefore, the observed structures are specific cases which coincide with bulk fragments but actually lie within more general classes of INT.

Fig. 2 shows the "phase diagrams" for models I and II as a function of C-NT diameter. The range over which specific INTs are (thermodynamically) favourable are generated by calculating the energies of specific INTs as a function of C-NT diameter (see, for example, ref. 15). Not surprisingly, the diameter of the enclosing C-NT appears highly correlated with the structure of the energetically-favoured INT.

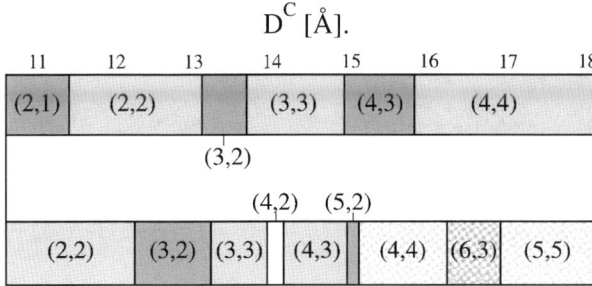

D^C [Å].

Fig. 2 Phase diagrams for models I (upper, square nets) and II (lower, hexagons) as a function of carbon nanotube diameter, D^C.

In order to understand any kinetic and/or entropic aspects of INT formation, a collection of filling simulations are performed in which both the starting liquid configurations and the C-NT morphology are varied. Equilibrated liquid configurations (obtained from constant pressure and temperature molecular dynamics simulations running at zero pressure and 1000 K) are extracted at time intervals of ~500 ps. The selected time interval is significantly greater than the characteristic relaxation time associated with the liquid dynamics at this temperature and so the extracted liquid configurations can be considered as distinct.

The carbon nanotubes considered are formed into a group according to nanotube diameter. Five tubes are used [the (11,11), (19,0), (12,10), (16,5) and (18,2)] with diameters ranging from 14.90–14.96 Å. For each carbon nanotube ten filling events are simulated. In addition, for each chiral nanotube (those whose indices are not (n, n) or $(n, 0)$) simulations are performed on each of the enantiomers. Overall, therefore, each chiral nanotube has twenty associated filling events whilst the achiral tubes have ten.

Fig. 3 shows the morphologies of the INTs formed in this group. Three distinct morphologies are observed to form given by indices $(4,3)_{hex}$, $(5,2)_{hex}$ and $(6,0)_{hex}$ (diameters 7.11, 7.30 and 7.01 Å, respectively). The most abundant INT formed is

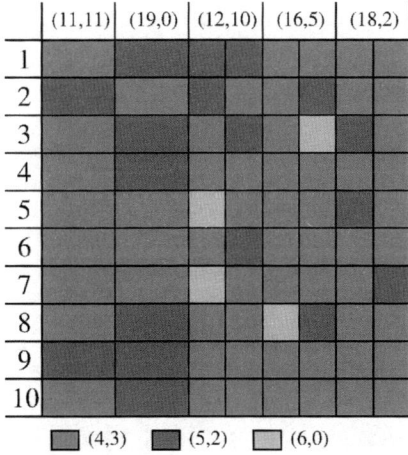

Fig. 3 Structure map for the INTs formed in the C-NTs of diameter range 14.90–14.96 Å. The coloured squares indicate the formation of the $(4,3)_{hex}$, $(5,2)_{hex}$ and $(6,0)_{hex}$ INT structures (as indicated by the key) from liquid configurations labelled '1' to '10'. Both enantiomers are filled for the three chiral C-NTs [the (12,10), (16,5) and (18,2)] giving twenty filling events for each of these tubes.

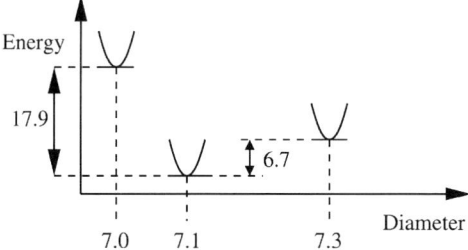

Fig. 4 Schematic energy landscape for the hexagonal INTs formed with C-NTs of diameter 14.90–14.96 Å. The three energetically-accessible INTs are the $(6,0)_{hex}$-diameter 7.0 Å, the $(4,3)_{hex}$-diameter 7.1 Å and the $(5,2)_{hex}$-diameter 7.3 Å. The energy differences (in kJ mol^{-1}) are calculated assuming a Boltzmann population distribution for the formation of these three structures with the populations taken from Fig. 3.

the $(4,3)_{hex}$ accounting for 72.5% of the structures observed, with the $(5,2)_{hex}$ and $(6,0)_{hex}$ accounting for 22.5 and 5.0%, respectively. Previous work has shown that a factor determining the class of INT formed is the difference in diameter, ΔD, between the INT (diameter D^{INT}) and the enclosing carbon nanotube (diameter D^C).[19] The INT energy can be usefully expressed as the sum of two terms, namely the energy required to create the INT from the hexagonal plane, and the C-NT–INT interaction energy. These two energies balance to give an energy minimum at ΔD of ~ 7.9 Å. The diameters for the observed INTs are in the range 7.01–7.30 Å compared with a C-NT diameter range of only 14.90–14.96 Å. The confining environment offered by the C-NT can be considered as generating an effective energy landscape in which only specific INT structures have the appropriate properties (*i.e.* diameter) to form. Considering these systems further in terms of an energy landscape, we can usefully imagine the three observed structures as energy basins on this landscape. The natural "reaction coordinate" for the landscape is the INT diameter (given that the diameter of the C-NT is approximately constant). The energy scale can be obtained (approximately) by linking the observed frequencies of formation of the INTs with an effective population in a Boltzmann distribution. At 800 K, therefore, the energy differences $E_{(4,3)} - E_{(5,2)}$ and $E_{(4,3)} - E_{(6,0)}$ are 6.7 and 17.9 kJ mol^{-1}, respectively. Fig. 4 shows the energy landscape based on these considerations. The next smallest INT [the $(5,1)_{hex}$] and the next largest [the $(6,1)_{hex}$] have respective diameters 6.50 and 7.65 Å. As a result, the difference in diameter with respect to the enclosing C-NTs considered in this group imposes too great an energetic penalty for these structures to form.

3.1.2. Filling dynamics. In addition to determining the internal (static) structures formed inside the C-NTs, the molecular dynamics simulations of the direct filling allows for basic formation mechanisms to be understood. For the $(2,2)_{sq}$ INT, for example, molecular graphics analysis of the ion filling shows that the ions are transported along the C-NT *via* a "folding"/"unfolding" mechanism (Fig. 5). In order to highlight the role of the confining carbon nanotube, the energy pathway for an idealised movement of a KI molecule along the $\langle 110 \rangle$ direction (consistent with the observed diffusion mechanism) is calculated as a function of the confining nanotube radius. The anion–cation pair labelled "1", as shown in Fig. 6, is moved in the $\langle 110 \rangle$ direction, with the nearest-neighbour anion–cation pairs (labelled "2" and "3") allowed to relax (using a conjugate-gradient algorithm) with all other ions held fixed. The energy minimisations are performed for a range of C-NTs of morphology (n, n). In order to further probe the effect of the tube, additional calculations are performed, which exploit the fixed nature of the C-NT, in which the (n, n) C-NT coordinates are scaled in order to generate a confining tube of a more controllable radius.

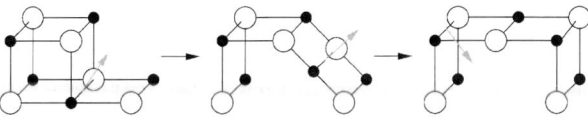

Fig. 5 "Folding/unfolding" mechanism responsible for ion transport in the $(2,2)_{sq}$ INT. The two ions indicated move along the $\langle 110 \rangle$ direction as shown to the final position given by the right-hand panel. The central panel is the effective "intermediate" at a local energy maximum.

Fig. 6a shows the energy pathway for the motion along the $\langle 110 \rangle$ vector shown in Fig. 6b. At the energy maxima the ion pair "1" is directly in between pairs "2" and "3" (and hence the 1–2 and 1–3 anion–cation bond lengths are at their most compressed). Fig. 7 (upper panel) shows the energy of interaction of the ions with the carbon nanotube as a function of the tube diameter. In the small diameter C-NTs the INT is highly compressed such that the ions are pushed against the repulsive Lennard-Jones wall (positive Lennard-Jones interaction energy). The most energetically favourable INTs are in the C-NTs of diameter ~ 12.5 Å in which the nearest-neighbour ion–carbon separations are around the Lennard-Jones energy minimum. Fig. 7 (lower panel) shows the activation barriers obtained from the energy minimisations (*i.e.* from Fig. 6) as a function of C-NT diameter. For $D \geq 13.6$ Å (corresponding to the $(10,10)$ C-NT) the activation barrier is approximately constant (corresponding to ~ 100 K) and equivalent to the barrier predicted from an unconfined INT. For $D < 13.6$ Å the activation barrier appears to rise near linearly. On the simulation time scales (of the order of nanoseconds) the smallest tube observed to fill is the $(9,9)$ ($D^C \sim 12.2$ Å) approximately corresponding to the location of the energy minimum in the figure. The absence of structures in smaller diameter C-NTs can be assigned as essentially a kinetic effect in which the low energy pathway mechanism described above is effectively removed by the presence of the small diameter C-NTs. Considering the "intermediate" structure (Fig. 6b) the critical relaxation is of the molecules "2" and "3" outwards (*i.e.* away from the INT major axis). As the confining C-NT diameter is reduced this relaxation is

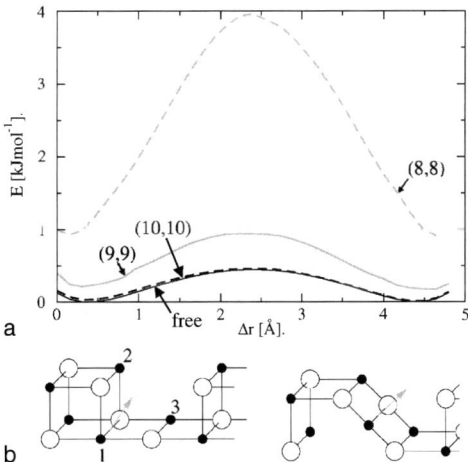

Fig. 6 (a) Direction of movement for energy minimisation calculations to highlight the energetics of the observed ion transport mechanism. Molecule '1' is moved along the $\langle 110 \rangle$ direction as shown in increments (Δr). A conjugate-gradient energy minimisation is applied allowing the locations of the molecules labelled '2' and '3' to relax, but holding all other ions fixed. (b) Energy pathways for the movement of the molecule '1' along the $\langle 110 \rangle$ direction as a function of the C-NT diameter. The C-NT morphologies are indicated.

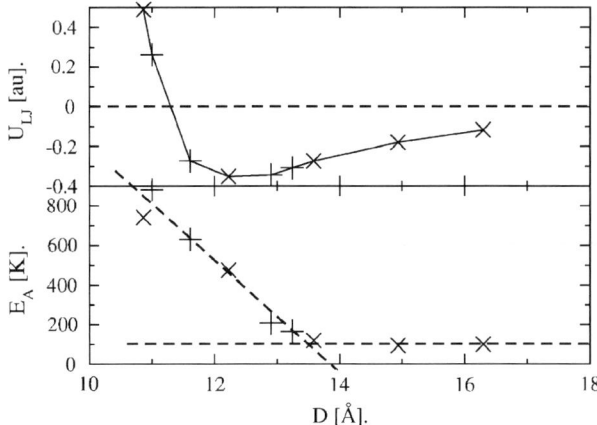

Fig. 7 (a) C-NT/INT Lennard-Jones energies, U^{LJ}, for a $(2,2)_{sq}$ INT as a function of C-NT diameter, D^C. (b) Activation barrier height, E_A (expressed as a temperature), as a function of C-NT diameter.

effectively removed as these ions are compressed against the C-NT repulsive wall. As a result, the M–X bond lengths in the intermediate structure are greatly reduced with respect to those in the intermediate structures in the free INT or in large diameter C-NTs corresponding to a significant rise in the height of the activation barrier. In the (12,12) C-NT, for example, the separation between an anion in pair '1' and a cation in pair '2' is compressed by ~ 0.22 Å from the starting configuration to the intermediate structure. In the (8,8) C-NT, this compression increases to ~ 0.56 Å as the ability for outward relaxation has been effectively lost.

3.2. MX₃ stoichiometry

$LaCl_3$ is chosen as an archetypal MX_3 salt. The central rationale behind this choice of system is the existence of a high quality PIM[24] which reproduces both crystalline and liquid state properties[24] (Table 1). Experimental HRTEM observations are focused on the metal triiodides[25–27] owing to their high electron beam visibility. However, ready contact can be made with the simulated patterns by noting that the chlorides and iodides share a similar crystal structure map.[28]

3.2.1. Static structure. Fig. 8 shows the crystallite structures formed within the (16,0) and (19,0) C-NTs (diameters 12.68 and 15.06 Å), respectively. Both systems are clearly highly crystalline in nature. In the smaller nanotube a structure forms which can usefully be considered as a percolating chain of face-sharing $LaCl_6$ octahedra which are distorted towards trigonal bipyramidal geometry. This fragment can be considered as extracted from a number of bulk MX_3 structures.[18] The fragment appears along the [001] direction of the ZrI_3 crystal,[29] the [100] direction in the $PuBr_3$ structure[30] and the [001] direction in the UCl_3.[31] In the larger C-NT the structure can be considered as having three cations in a cross-section perpendicular to the C-NT major axis. This structure has no direct bulk analogue, although it can be considered as extracted from the UCl_3 structure in which a series of metal–anion nearest-neighbour "bonds" point "inwards" (*i.e.* towards the centre of the C-NT) rather than outwards as in the bulk structure.[18] As a result, the anions, which occupy a chain that runs along the centre of the C-NT, have an unusually high coordination number of four compared with values of two and three observed in the typical bulk structures.[30]

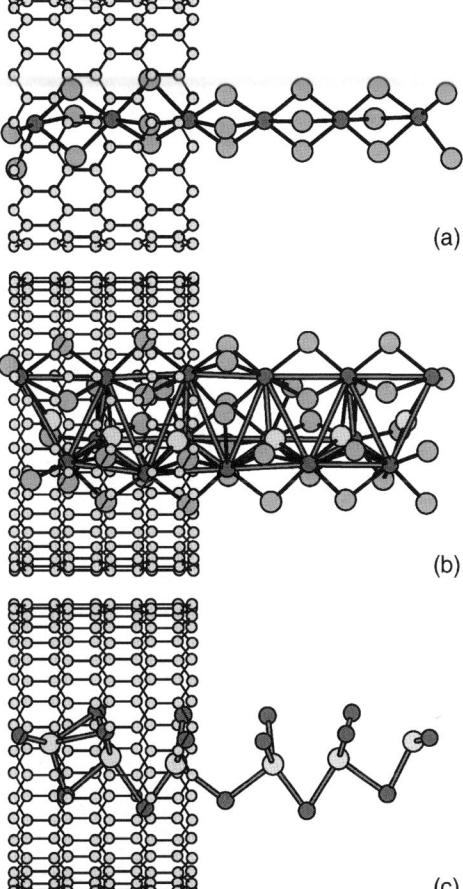

Fig. 8 Molecular graphics "snapshots" of the LaCl$_3$ crystal structures formed in the (a) (16,0) and (b) and (c) (19,0) C-NTs, respectively. In all cases the C-NT is cut away to highlight the internal structure. In (a), which can be considered as a percolating chain of face-sharing octahedra, "bonds" are drawn between nearest-neighbour anion (large, light circles)–cation (small, dark circles) pairs. Thick "bonds" are drawn between nearest-neighbour cation–cation pairs to highlight the cation super-structure. In (c) only the central anion chain and associated nearest-neighbour cations are drawn in order to highlight the unusually high anion coordination number (four) achieved for this stoichiometry.

3.2.2. Dynamics. In order to understand how the presence of the carbon nanotube affects the ion dynamics, the vibrational modes are probed. In one approach, the condensed phase vibrational density of states (VDOS) may be decomposed into contributions from the vibrations of modes associated with coordination complexes. The calculated VDOS may be used to make contact with experimental Raman and/or infrared spectroscopy.

The single-chain crystallite is constructed from linked (face-sharing) octahedra. For an isolated octahedral complex the Raman-active vibrational normal modes are of symmetry A_{1g}, T_{2g}, and E_g in addition to an infrared-active, Raman-forbidden mode of T_{1u} symmetry. Combinations of the relative velocities of the ions which make up the transient octahedral complexes present in the simulation may be used to construct the symmetry coordinates. The correlation functions associated with these velocities yield the associated oscillation frequencies which can then be compared directly with mode frequencies experimentally observed in the Raman (and infrared) spectra. At a given time step in the simulation, the six X^- neighbours, $i1 \rightarrow i6$, about a central M^{3+} cation, i are identified. The velocities $\boldsymbol{v}_{i1}$ and positions $\boldsymbol{r}_{i1}$ of these neighbours *relative to the cation* are then calculated and velocities parallel to the anion–cation "bond" projected out:

$$\boldsymbol{v}_{i1}^{\parallel} = \boldsymbol{v}_{i1} \cdot \hat{\boldsymbol{r}}_{i1}, \tag{3.1}$$

where $\hat{\boldsymbol{r}}_{i1}$ is the unit vector along $\boldsymbol{r}_{i1}$. As a result, for example, the velocity of the A_{1g} symmetric stretching vibration is

$$V_{A_{1g}}^i = \sum_{i\alpha=1 \rightarrow 6} \boldsymbol{v}_{i\alpha}^{\parallel} \tag{3.2}$$

The symmetry coordinates for the degenerate modes are given in Pavlatou *et al.*[32]
The spectra of each of these velocities is given by, *e.g.*

$$C_{V_{A_{1g}}}(\omega) = \boldsymbol{Re} \int_0^\infty \mathrm{d}t \, e^{i\omega t} \langle V_{A_{1g}}^1(t) \cdot V_{A_{1g}}^1(0) \rangle, \tag{3.3}$$

(where the average is taken over all cations, which are six-coordinated, and over time).

In addition to calculating the VDOS from these (isolated) octahedral vibrational modes, the instantaneous normal modes (INM) are calculated for the isolated octahedral chain from the Hessian matrix, $\boldsymbol{H}_{i\mu,j\nu} = \frac{1}{m}\frac{\partial^2 U}{\partial r_{i,\alpha}\partial r_{j,\beta}}$, where U is the total system energy for a given configuration $\{\boldsymbol{R}^N\} = \{\boldsymbol{r}_1, \boldsymbol{r}_2, \ldots, \boldsymbol{r}_N\}$. This derivative is calculated numerically and diagonalised to produce the vibrational density of states.[33] In addition, this procedure also generates the eigenvectors associated with each eigenmode and, as a result, the ion dynamics associated with specific features present in the VDOS may be understood.

Fig. 9 shows the vibrational modes calculated for liquid $LaCl_3$ at 1300 K and zero pressure and for the fully-filled $LaCl_3/(16,0)$ C-NT system under the same conditions. The latter system corresponds to the outer (liquid) salt in equilibrium with both the carbon nanotube and the fully grown single octahedral-chain internal

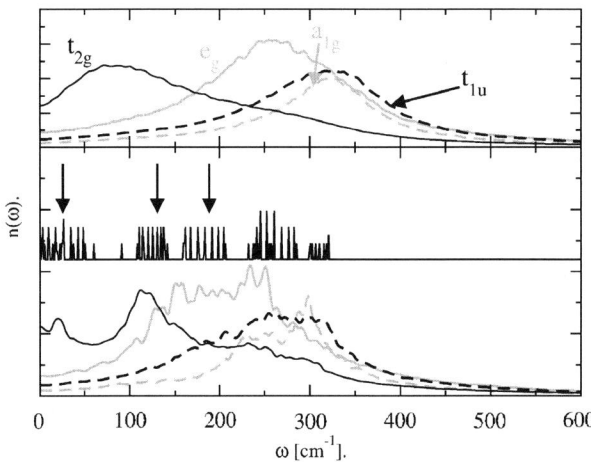

Fig. 9 Vibrational densities of states (VDOS) obtained *via* specific time correlation functions as discussed in the text. The upper panel shows the vibrational modes predicted in the pure $LaCl_3$ liquid at 1300 K. The lower panel shows the VDOS obtained for the fully-filled (16,0) C-NT (which contains a chain of face-sharing octahedra) encased in outer $LaCl_3$ liquid under the same conditions. The central panel shows the INMs obtained for an isolated face-sharing octahedral chain with the arrows indicating the groups of vibrational modes which correspond to features observed in the lower, but not the upper, panel.

crystallite. The spectra show potentially significant differences which may be attributed to both the presence of the C-NT (which partially orders the outer liquid over short length scales[15]) and the internal crystallite.

The T_{2g} mode, for example, shows features at ~ 25 cm^{-1} and ~ 120 cm^{-1} whilst the E_g mode shows a feature at ~ 200 cm^{-1}. The eigenvalues extracted from the normal mode analysis for the isolated chain are superimposed onto the VDOS shown in Fig. 9. This spectrum shows significant groups of eigenmodes at the frequencies associated with the "additional" features in the LaCl$_3$/C-NT VDOS. If we assume, therefore, that these additional features arise as a consequence of the cation-centred chains, then the associated eigenvectors in these frequency regimes will indicate the ion motions responsible for these features. The low frequency modes ($\nu \leq 50$ cm^{-1}) are associated with bending vibrations of the cation-centred chains with associated T_{2g}-like anion motion. The modes at $\nu \approx 120$ cm^{-1} are associated with rotations of the anions about the octahedra (with significant E_g character), and the high frequency modes ($\nu \approx 200$ cm^{-1}) are associated with stretching vibrations along the cation chain. Therefore, all three sets of vibrational models are present as a direct result of the ability of the C-NT to effectively disconnect the octahedral chain from the network liquid.

3.3. MX$_2$ stoichiometry

In the MX$_2$ stoichiometry, both filling by molten BaI$_2$ and CoI$_2$ have been studied by HRTEM.[11,12] For CoI$_2$, "twisted" crystal structures are observed which can be considered to be based on fragments of the underlying CdI$_2$ layered crystal structure in which a systematic twisting distortion is present. For BaI$_2$ a percolating chain of edge-linked polyhedra, in which the Ba^{2+} cations are (uniquely) five- and six-coordinate, are obtained.

In order to highlight the ubiquitous nature of internal crystallite formation we have performed a (preliminary) study on a test system of MX$_2$ stoichiometry. The system under study is SrCl$_2$, as a recently developed potential model is available (Table 1) which reproduces a range of crystal and liquid properties.[34]

Fig. 10 shows a molecular graphics "snapshot" of a structure formed within both the (18,0) and (19,0) C-NTs (diameters 14.27 and 15.06 Å, respectively). As for the MX and MX$_3$ stoichiometries considered above, a highly ordered structure has been formed directly from the filling liquid. The structure formed is based on a superstructure of a chain of edge-linked cation octahedra. In the (18,0) C-NT these octahedra are more compressed perpendicular to the C-NT major axis compared with the structure formed in the (19,0) C-NT. Fig. 10 shows the cations to be coordinated to seven nearest-neighbour anions compared with eight in the bulk fluorite structure. The anions on the outer surface of the crystallite are three-coordinate. The structure formed in both C-NTs again has no direct bulk analogue. In order to emphasize this Fig. 10 highlights the anion chain which runs along the centre of the C-NT. These anions have an unusually high (for this stoichiometry) cation coordination number of six compared with four in the bulk fluorite crystal and a mixture of four and five in the high pressure cotunnite form.[30] This high anion coordination has parallels with the MX$_3$ stoichiometry formed within the (19,0) C-NT. As with the BaI$_2$ structure (observed experimentally) the cations in the present structure have a lower coordination number than any observed for known bulk structures.

3.4. Summary and conclusions

In this work relatively simple metal–anion systems, covering three stoichiometries, have been inserted directly into carbon nanotubes which vary in diameter. In all cases crystalline structures are found to form within the C-NT. The presence of the confining nanotube can be considered as having effectively two major effects:

1. The ions are ordered by the presence of the C-NT. This can be rationalised in terms of the strong coulombic forces which preclude nearest-neighbour like–like

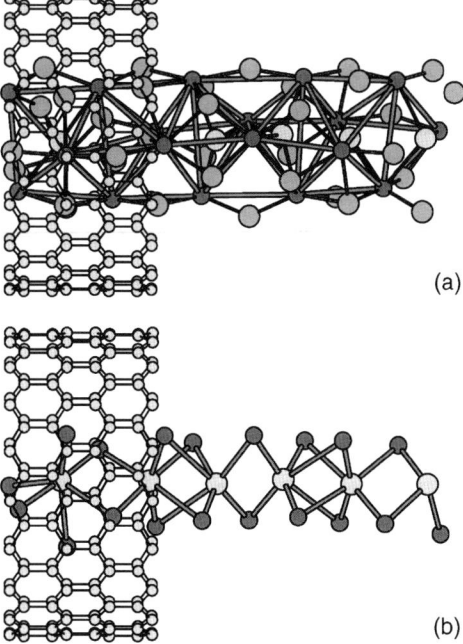

(a)

(b)

Fig. 10 Molecular graphics "snapshots" of the $SrCl_2$ crystal structure found in the (18,0) C-NT. In (a) thin "bonds" are drawn between nearest-neighbour anion (large, light circles)–cation (small, dark circles) pairs and thick "bonds" are drawn between nearest-neighbour cation–cation pairs to highlight the cation super-structure. In (b) only the central anion chain and associated nearest-neighbour cations are drawn in order to highlight the unusual high anion coordination number achieved for this stoichiometry.

interactions at these thermal energies, and extends well beyond the melting point of the bulk solid.

2. The energy landscape is significantly altered. The C-NT can be considered as effectively offering the system a highly simplified energy landscape (the C-NT acts as an energy landscape filter) in which only highly ordered (owing to the strong coulombic ordering) INTs within specific diameter ranges are energetically accessible.

However, despite the crystalline ordering imposed on the internal structure by the presence of the confining environment, relatively low energy (compared with the available thermal energy) diffusion pathways exist which allow the ions to be transported along the tubes and promote further crystal growth. The dynamics of the filled tubes can be probed by experimental techniques such as Raman spectroscopy which, as indicated in the present work, may highlight specific dynamic modes associated with the ion motion along the C-NTs.

Overall, therefore, the C-NTs offer formation pathways to crystal structures which may be novel both in their morphology and in their specific ion coordination environments. Potential (atomistic) control over these environments offers the possibility of tuning useful electronic and mechanical properties. The experimental investigation of these theoretical predictions will present challenges in the near future.

Acknowledgements

The work was supported by EPSRC grant GR/S06233/01. The author thanks Dr Steffi Friedrichs [TTP (The Technology Partnership plc), Melbourn Science Park, Cambridge Road, Melbourn, SG8 6EE, U.K.] for helpful discussions.

This journal is © The Royal Society of Chemistry 2006

References

1 S. C. Tsang, Y. K. Chen, P. J. F. Harris and M. L. H. Green, *Nature*, 1994, **372**, 159.
2 P. M. Ajayan, T. W. Ebbesen, T. Ichihashi, S. Iijima, K. Tanigaki and H. Hiura, *Nature*, 1993, **362**, 522.
3 P. M. Ajayan, T. Ichihashi and S. Iijima, *Chem. Phys. Lett.*, 1993, **202**, 384.
4 P. M. Ajayan, O. Stephan, Ph. Redlich and C. Colliex, *Nature*, 1995, **375**, 564.
5 Y. K. Chen, M. L. H. Green and S. C. Tsang, *Chem. Commun.*, 1996, 2489.
6 J. Sloan, J. Cook, A. Chu, M. Zwiefka-Sibley, M. L. H. Green and J. L. Hutchison, *J. Solid State Chem.*, 1998, **140**, 83.
7 J. Sloan, D. M. Wright, H.-G. Woo, S. R. Bailey, G. Brown, A. P. E. York, K. S. Coleman, J. L. Hutchison and M. L. H. Green, *Chem. Commun.*, 1999, 699.
8 J. Sloan, M. Terrones, S. Nufer, S. Friedrichs, S. R. Bailey, H.-G. Woo, M. Rühle, J. L. Hutchison and M. L. H. Green, *J. Am. Chem. Soc.*, 2002, **124**, 2116.
9 J. Sloan, M. C. Novotny, S. R. Bailey, G. Brown, C. Xu, V. C. Williams, S. Friedrichs, E. Flahaut, R. L. Callendar, A. P. E. York, K. S. Coleman, M. L. H. Green, R. E. Dunin-Borkowski and J. L. Hutchison, *Chem. Phys. Lett.*, 2000, **329**, 61.
10 R. R. Meyer, J. Sloan, R. E. Dunin-Borkowski, A. I. Kirkland, M. C. Novotny, S. R. Bailey, J. L. Hutchison and M. L. H. Green, *Science*, 2000, **289**, 1324.
11 J. Sloan, S. J. Grosvenor, S. Friedrichs, A. I. Kirkland, J. L. Hutchison and M. L. H. Green, *Angew. Chem., Int. Ed.*, 2002, **41**, 1156.
12 E. Philp, J. Sloan, A. I. Kirkland, R. R. Meyer, S. Friedrichs, J. L. Hutchison and M. L. H. Green, *Nat. Mater.*, 2003, **2**, 788.
13 S. Friedrichs, R. R. Meyer, J. Sloan, A. I. Kirkland, J. L. Hutchison and M. L. H. Green, *Chem. Commun.*, 2001, 929.
14 M. Wilson and P. A. Madden, *J. Am. Chem. Soc.*, 2001, **123**, 2101.
15 M. Wilson, *J. Chem. Phys.*, 2002, **116**, 3027.
16 M. Wilson, *Chem. Phys. Lett.*, 2002, **366**, 504.
17 M. Wilson, *Nano Lett.*, 2004, **4**, 299.
18 M. Wilson and S. Friedrichs, *Acta. Crystallogr., Sect. A*, 2006, **62**, 287.
19 M. Wilson, *J. Chem. Phys.*, 2006, **124**, 124706.
20 R. Saito, G. Dresselhaus and M. S. Dresselhaus, *Physical Properties of Carbon Nanotubes*, Imperial College Press, London, 1998.
21 (*a*) P. A. Madden and M. Wilson, *Chem. Soc. Rev.*, 1996, **25**, 339; (*b*) P. A. Madden and M. Wilson, *J. Phys.: Condens. Matter*, 2000, **12**, A95.
22 (*a*) P. Jemmer, P. W. Fowler, M. Wilson and P. A. Madden, *J. Chem. Phys.*, 1999, **111**, 2038; (*b*) C. Domene, P. W. Fowler, P. A. Madden, M. Wilson and R. J. Wheatley, *Chem. Phys. Lett.*, 2001, **333**, 403; (*c*) C. Domene, P. W. Fowler, P. A. Madden, Xu Jijun, R. J. Wheatley and M. Wilson, *J. Phys. Chem. A*, 2001, **105**, 4136.
23 W. A. Steele, *J. Phys. Chem.*, 1978, **82**, 817.
24 (*a*) Y. Okamoto and P. A. Madden, *J. Phys. Chem. Solids*, 2005, **66**, 448–451; (*b*) W. J. Glover and P. A. Madden, *J. Chem. Phys.*, 2004, **121**, 7293; (*c*) F. Hutchinson, M. Wilson and P. A. Madden, *Mol. Phys.*, 2001, **99**, 811; (*d*) F. Hutchinson, A. J. Rowley, M. K. Walters, M. Wilson, P. A. Madden, J. C. Wasse and P. S. Salmon, *J. Chem. Phys.*, 1999, **111**, 2028.
25 S. Friedrichs and M. L. H. Green, *Z. Metallkd.*, 2005, **96**, 419.
26 S. Friedrichs, U. Falke and M. L. H. Green, *ChemPhysChem*, 2005, **6**, 300.
27 S. Friedrichs, A. I. Kirkland, R. R. Meyer, J. Sloan and M. L. H. Green, *Electron Microsc. Anal.*, 2004, 455.
28 D. G. Pettifor, in *Intermetallic Compounds*, ed. J. H. Westbrook and R. L. Fleischer, J. Wiley and Sons., New York, 1994, ch. 18.
29 E. M. Larsen, J. S. Wrazel and L. G. Hoard, *Inorg. Chem.*, 1982, **21**, 2619.
30 A. F. Wells, *Structural Inorganic chemistry*, Clarendon, Oxford, 1984.
31 W. H. Zachariasen, *Acta Crystallogr.*, 1948, **1**, 265.
32 E. A. Pavlatou, M. Wilson and P. A. Madden, *J. Chem. Phys.*, 1997, **107**, 10446.
33 (*a*) M. C. C. Riberio, M. Wilson and P. A. Madden, *J. Chem. Phys.*, 1998, **108**, 9027; (*b*) M. C. C. Riberio, M. Wilson and P. A. Madden, *J. Chem. Phys.*, 1998, **109**, 9859; (*c*) M. C. C. Riberio, M. Wilson and P. A. Madden, *J. Chem. Phys.*, 1999, **110**, 4083.
34 M. Wilson, unpublished.

Screening and strain in superionic conductors

Angus Gray-Weale

Received 3rd March 2006, Accepted 10th May 2006
First published as an Advance Article on the web 27th July 2006
DOI: 10.1039/b603250h

I calculate the screening length and its relation to elastic constants from mean-field models of superionic conductors. The Debye formula for a continuous fluid's screening length is replaced by one that accounts for the discrete nature of the lattice. Interactions alter the screening length and in some cases give oscillations in the charge structure. The screening lengths derived here are exact not only at low temperature where there are few defects, but also at high temperature where the effective interaction strength is small. The mean-field treatment correctly gives the decrease in an elastic constant observed at the superionic transition for a variety of crystals. This decrease is best explained by a change in the defect creation entropy with density. The effect of doping on the elastic constant also agrees well with experiment. The extension to the mean-field models given here provides a consistent picture of the response of a superionic crystal to electrical, mechanical and thermal stresses, as well as to doping.

1. Introduction

The screening length in a superionic crystal is important for its relevance to the enhanced conductivity of nanostructures.[1–4] It describes the effective interactions between charge carriers, both the response to an applied field and the fluctuations in charge density. For example, changes in the charge density near an interface persist at most a few screening lengths from the interface.[5,6] In this case the 'applied field' is the interface itself. The screening length is often approximated as the Debye length for a continuous fluid,[7] but a more accurate treatment is possible.[8–11] This paper improves the calculation of the screening length for superionic crystals beyond the Debye formula for continuous fluids, taking account of the lattice's configurational entropy. The aim is to provide a better picture of the response of a superionic to a stress than is available within the assumptions of dilute point defects and the Poisson–Boltzmann equation.[7,12] The calculations given below also treat the coupling of density fluctuations to charge fluctuations, so that the responses in strain and charge density near an interface or under another stress may be compared.

The Debye length for a continuous fluid is often a good estimate of the screening length.[2,7] But even when coupling is so strong that the Debye length fails as such an estimate, it may be used to predict the true range and character of charge correlations.[1,3] For the Debye length to be used in this way in a superionic conductor, it should be modified to account for the nature of the lattice on which the ions move. This is also a goal of this paper: to establish the correct expression for the screening length in the limit of weak-coupling (a lattice-based Debye length) as a basis for further work on materials with stronger interactions and correlations.

The mean-field models used here consist of an approximation for the free energy. Minimising this free energy with respect to defect concentration gives the level of

School of Chemistry F11, University of Sydney, NSW 2006, Australia

disorder, heat capacity and other quantities.[13–15] Rice, Strässler and Toombs[9] and Huberman[10] gave the first such treatments. The most recent model is the 'cube-root model'[11,16,17] and it fits defect concentrations better than other models.[8,18] The more sophisticated models include correlations between ions and sometimes distortions of the crystal lattice.[19–21] All of these models could be used in the calculations reported here, but a few of the simpler ones have been chosen.

Fleming and Cohen described the large-scale fluctuations of strain and defect density fluctuations for a monatomic crystal in ref. 22. In their treatment the density of defects decays by the transport of defects, not by local rearrangements of ions, which is possible in the systems discussed here. Zeyher gives a treatment of the hydrodynamics of superionics in ref. 23. No explicit treatment of defects was given by Zeyher, who took as variables the densities of the two species of ion, fixed and mobile. These two works concerned the decay of slow modes rather than the screening, but illustrate different pictures of charge distribution in a crystal.

This paper is the first to examine screening as it changes from the low temperature, normal crystal with few defects to the high temperature superionic with many mobile ions, with explicit account of the lattice's effect on configurational entropy and a mean-field estimate of the effect of interactions on screening. The study of time-correlations is left to a later work.

2. Theory

The simplest model of a superionic is a lattice on which ions move freely. There are more sites than ions, so the ions move from site to empty neighbour site. Creation of a Frenkel defect pair involves moving an ion from a lattice of sites at lower energy, to a lattice of interstitial sites at higher energy. In the ground state, the lower-energy lattice is full and the interstitial empty, if the crystal is not doped. It is the presence of two lattices that distinguish this treatment of screening from studies of other Coulomb lattice fluids.[24]

At very low temperatures, the charge-carrying species are the vacancies and interstitials. They are isolated and migrate, but only when they meet does their number change. Their low density makes the interaction strength small. At high temperatures (*i.e.* above the superionic transition) the rate at which these species are created and destroyed is much higher and averaged over a small region of the crystal there may be substantial numbers of defects, though this region remains neutral.[8] Fluctuations in the number of Frenkel pairs decay rapidly compared to fluctuations in charge density, because local rearrangements can alter the number of defect pairs, but a long-wavelength charge fluctuation requires an actual flow of charge to decay. At such high temperatures, it is more natural to take the ions themselves as the charge carrying species. The concentration of Frenkel pairs is in local equilibrium while a large-scale charge fluctuation decays.

The screening length depends on the probability of a long-wavelength charge fluctuation, which in turn depends on the reversible work required to induce the fluctuation.[25] This reversible work has two contributions, one from the electrostatic energy and one from the change in free energy of the fluid of charged particles with density. The balance between the two couples Maxwell's equations to the microscopic features of the fluid. It is the second contribution which is calculated here with mean-field models. Section 2.1 described mean-field models as they have been previously used. Section 2.2 generalises mean-field models to allow variation in the local density of mobile ions.

2.1 Mean-field models

Vacancies and interstitials are randomly placed on the two lattices to estimate the configurational entropy. There is a free-energy cost associated with moving an ion from the lower energy lattice to the higher, creating a Frenkel pair of interstitial and vacancy. The final element is an interaction, typically varying as some power of the

This journal is © The Royal Society of Chemistry 2006

concentration of defect pairs. I assume there are N sites at a lower energy and αN sites at some higher energy. This free energy is, per ground state site,[9–11]

$$\beta g(c) = (\beta h_0 - s_0) c - s_r(\alpha,c) + \beta g_{ex}(c), \tag{1}$$

where s_r is the configurational entropy; $\beta = 1/k_B T$ is the reciprocal temperature in energy units; the defect concentration is $c = m/N$; the number of Frenkel pairs, in some volume Nl^3, is m; and l is the separation between the lower energy sites. The free energy in eqn (1) is denoted g because it is most often used at constant pressure, but it can be used at constant density[16,17] and is for a couple of the cases described below.

The excess free-energy describing the interactions is often of the form

$$g_{ex}(c) = \int_0^c \mu_{ex}(x)\mathrm{d}x, \tag{2}$$

$$\mu_{ex}(c) = -Jc^m. \tag{3}$$

The natural first choice is $m = 1$, so that eqn (1) is an expansion of the free energy in the defect concentration. It is also possible to take $m = 1/2$, a Debye–Hückel model for the chemical potential,[26] or to modify that expression with the extended Debye–Hückel theory that accounts for the finite size of ions.[14] A more recent approach uses $m = 1/3$, the 'cube-root model'.[11] It is argued that this interaction is due to an effective Madelung potential for the 'lattice' of defects; but it may also be obtained (approximately) by integrating the fraction of occupied neighbour pairs out of the 'QBW-2' model described by Vlieg et al.[19] The relation between the cube-root and Debye–Hückel models is discussed in the case of continuous fluids in ref. 27.

Other interactions include the exponential model described by Bouteiller,[15]

$$\beta g_{ex}(c) = \beta h_0 c[\exp(-\eta c) - 1] - B\frac{c^2}{2}. \tag{4}$$

The parameter B is an interaction entropy. Expanding eqn (4) to first-order in ηc gives a model similar to the work in ref. 9 and 10 but with energy and entropy interaction terms both proportional to the square of the defect concentration.

Fig. 1 compares the defect concentrations, $c = m/N$, predicted by various models and for various materials. Table 1 gives the model used for each material and the sources for parameters. Some of the parameters are taken from experimental work and some fitted to simulation. Most of the curves in Fig. 1 are obtained from the cube-root model, but the Debye–Hückel and Bouteiller models are included for comparison.

2.2 Configurational entropy with an excess, or lack, of ions

The number of mobile ions is normally the same as the number of lower energy sites and at low temperature the ions settle onto the lower energy lattice. Here I allow the number of mobile ions to differ, due either to a fluctuation or doping. The only similar treatment I am aware of is by Vlieg et al., who consider the concentration of defects and correlations between them and extend it to the doped compound $Sr_{1-x}Gd_xCl_{2+x}$. They study the variation of the superionic transition temperature with x, rather than the electrical screening under consideration here.

Let the number of ions in the volume $V = Nl^3$ of the crystal be n, where the number of sites on the lower-energy lattice in the volume is N and they are separated by a distance l. In the following it is assumed that the lower energy lattice is a simple

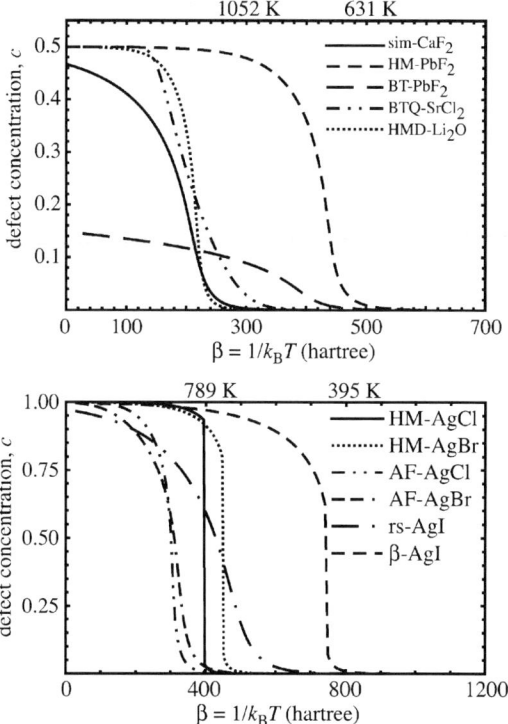

Fig. 1 Concentrations of Frenkel pairs of interstitials and vacancies, varying with temperature for a variety of mean-field models of superionic conductors. Details of the models and references are given in Table 1. The upper plot shows curves for superionics with the fluorite structure and the lower shows three crystals with the rock-salt structure and one with the wurtzite structure. AgCl and AgBr are each represented by two models, one cube-root (HM) and one Debye–Hückel (AF). The fluorites conduct fluoride ions, the others silver ions. Note that Bouteiller's exponential model (see eqn (4)) for PbF$_2$ gives a very low concentration of Frenkel pairs (upper plot BT-PbF$_2$). In the case of the wurtzite and two cube-root models of rock-salts the transition is first-order (lower plot). Mean-field models do not properly describe first-order superionic transitions, but we can still learn about the screening length from these curves (see sections 3.2 and 7).

cubic with spacing l. This is not really a loss of generality, as l only appears in the number density of lower energy sites, l^{-3}. Define an excess, or lack, of ions as,

$$\lambda = \frac{n}{N} - 1. \tag{5}$$

If $\lambda > 0$ the ground state contains ions on the higher energy lattice (interstitials) and if $\lambda < 0$, the ground state has vacancies on the lower energy lattice. These two cases are analogous to the n- and p-doping of semiconductors. The number of Frenkel pairs per lattice site is denoted c and the local number of ions per lattice site λ. If $\lambda \neq 0$ in some region of the crystal, then that region will carry a charge unless there are compensating charges on the immobile ions, perhaps due to doping. The number of ions on the higher energy interstitial lattice is proportional to $c + \lambda$, if $\lambda > 0$ and c otherwise; and the number of vacancies is proportional to $c - \lambda$ if $\lambda < 0$ and c otherwise. λ may vary slowly in space, vanishing only on the average. Such a variation in λ would constitute a fluctuation in the charge density. The screening length may be obtained by calculating the probability of such a fluctuation in charge density.[25]

This journal is © The Royal Society of Chemistry 2006

Table 1 The models and sources for the parameters taken from the literature as the basis of the screening length calculations

Abbre-viation	Model type	Structure	h_0/eV	s_0/k_B	Source (Ref.)	Note
Sim-CaF$_2$	Cube root	Fluorite	1.42	2.5	8	Fit to simulation data.
HM-PbF$_2$	Cube root	Fluorite	1.08	8.48	11	Parameters from ref. 31.
BT-PbF$_2$	Exponential	Fluorite	1.44	14.0	15	
BTQ-SrCl$_2$	Quadratic	Fluorite	2.67	23.98	15	Entropy interaction term included.
HMD-Li$_2$O	Cube root	Fluorite	2.1	8	36	Interaction strength estimated from other fluorites in ref. 11.
HM-AgCl	Cube root	Rock salt	1.48	8.37	11	
HM-AgBr	Cube root	Rock salt	1.15	7.67	11	
AF-AgCl	Debye–Hückel	Rock salt	1.45	9.37	31	
AF-AgBr	Debye–Hückel	Rock salt	1.1	6.54	31	
rs-AgI	Cube root	Rock salt	0.656	3.5	28	From neutron diffraction at high pressure.
β-AgI	Cube-root	Wurtzite	0.82	11.77	11	See also ref. 16

If the number of ways of arranging the interstitials and vacancies is W and the entropy given by Boltzmann's equation $S = k_B \log W$, where k_B is Boltzmann's constant, then the entropy per lower energy site in units of k_B is

$$s(\alpha, c, \lambda) = \frac{S}{k_B N} = \alpha \log \alpha - (c + \lambda_+) \log(c + \lambda_+) - (c - \lambda_-) \log(c - \lambda_-) \quad (6)$$

$$- (\alpha - (c + \lambda_+))\log(\alpha - (c + \lambda_+)) \quad (7)$$

$$- (\alpha - (c + \lambda_-))\log(1 - (c + \lambda_-)) \quad (8)$$

where Stirling's approximation has been used, and,

$$\lambda_+ = \lambda, \text{ if } \lambda > 0,$$
$$= 0, \text{ otherwise,}$$
$$\lambda_- = \lambda, \text{ if } \lambda < 0,$$
$$= 0, \text{ otherwise.}$$

Eqn (6) reduces to the more usual expression when λ vanishes.[9–11]

The mean-field free energy that uses the generalised configurational entropy from eqn (6) is

$$\beta g(c, \lambda) = (\beta h_0 - s_0)\left(c + \frac{|\lambda|}{2}\right) - s_r(\alpha, c, \lambda) - \beta g_{ex}\left(c + \frac{|\lambda|}{2}\right), \quad (9)$$

where the configurational entropy now depends on the local excess of ions λ. The change to the first term comes from regarding the two lattices as having free energies of $\pm(\beta h_0 - s_0)/2$ and the modification to the excess term is needed to make the free energy behave sensibly at $\lambda = 0$. If the screening length is calculated for $\lambda < 0$ and for $\lambda > 0$, then the limits as $\lambda \to 0$ taken, the same result for the screening length should be obtained. To make the change physically plausible, note that ions interact as do Frenkel pairs and while c represents the densities of two species, vacancies and interstitials, λ represents only one, the mobile ions themselves. This is the origin of the factor of $1/2$.

If interactions are neglected, the value of c that minimises βg is easily found:

$$c_+ = \frac{\sqrt{(\alpha + f\lambda + 1)^2 + 4f(\alpha - \lambda)} - 1 - \alpha - f\lambda}{2f}, \tag{10}$$

$$c_- = \frac{\sqrt{(\alpha - f\lambda + 1)^2 + 4f\alpha(\lambda + 1)} - 1 - \alpha + f\lambda}{2f}, \tag{11}$$

where $f = \exp(\beta h_0 - s) - 1$ and c_+ is the concentration of Frenkel pairs for $\lambda > 0$ and c_- for $\lambda < 0$.

2.3 Thermodynamic fluctuations and the Debye length

The simplest treatment of the screening length for a continuous fluid uses the isothermal compressibility of an ideal gas and balances the Coulomb energy of a sinusoidally varying charge density against the free energy cost of compressing and expanding the gas. This gives the entropy change needed to induce the fluctuation, the mean-squared magnitude of the fluctuation and in turn the screening length. The result is[7,25]

$$(\Lambda_D^{cts})^{-2} = 4\pi\beta \sum_i \rho_i Q_i^2 = \frac{8\pi Q^2 c}{l^3 k_B T}, \tag{12}$$

where the charge on and number density of species i are Q_i and ρ_i and 'cts' denotes a continuous, rather than a lattice-based, fluid. The second expression is the application of the continuous fluid formula to a superionic crystal and Q is the charge on the mobile species. This choice of Debye length is the one that is used in the Poisson–Boltzmann equation.[7,12]

Now suppose that on average $\lambda = 0$ throughout a cubic region of a crystal of a large volume $V = Nl^3$ and Fourier expand the local variations of λ, so that

$$\lambda(r) = \frac{1}{V} \sum_k \hat{\lambda}(k) \exp(-ik \cdot r). \tag{13}$$

The electrostatic energy is

$$U_{QQ} = \frac{1}{V} \sum_q \frac{4\pi}{q^2} \frac{Q^2}{l^6} |\hat{\lambda}(q)|^2, \tag{14}$$

where the sum is over all wavevectors commensurate with the cubic cell of volume $V = Nl^3$. If the variation in charge density is slow, then we can assume local equilibrium, so that the local concentration of defect pairs, c, minimises the free energy. The numbers of interstitials and vacancies are not separately conserved, they fluctuate in and out of existence rapidly.[8] A charge fluctuation (*i.e.* a fluctuation in λ), on the other hand, must decay by the actual transport of ions. If we choose the wavevector, q, to be small enough, the charge fluctuation will always be the slower mode and the assumption that c is in local equilibrium while λ decays will be justified.

Let the actual value of the free-energy at the minimum with respect to c at each value of λ be $\bar{g}(\lambda)$. Then the entropy change associated with a sinusoidal fluctuation in λ, of wavevector q, is

$$\frac{\Delta S}{k} = \frac{\beta}{V} \sum_q \left(\frac{1}{2} \frac{4\pi Q^2}{q^2 l^6} + \frac{1}{2l^3} \frac{\partial^2 \bar{g}}{\partial \lambda^2} \right) |\hat{\lambda}(q)|^2. \tag{15}$$

This last expression could be compared to the continuous fluid version in ref. 25. The probability of a fluctuation in λ is $\exp(\Delta S/k_B)$, and this probability lets us calculate the mean-squared fluctuation in charge density, and in turn the screening length.[2]

The screening length, Λ_S, is given by

$$\Lambda_S^2 = \frac{l^3}{4\pi Q^2} \frac{\partial^2 \bar{g}}{\partial \lambda^2}.$$

(16)

If the ideal gas compressibility is inserted into eqn (16) we obtain the usual continuous fluid Debye length, eqn (12).

Using the lattice-based free energy with no interaction (eqn (9) with $g_{ex} = 0$) in eqn (16) gives

$$(\Lambda_D^{\text{lat}})^2 = \frac{l^3 k_B T}{4\pi Q^2 c \left(2 - \frac{1+\alpha}{\alpha} c\right)}.$$

(17)

This is the Debye length for the fluid of mobile ions moving on the two lattices, ground state and interstitial.

Eqn (12) and (17) are closely analogous. Both are based on a non-interacting fluid, augmented by the Coulomb energy of a sinusoidal density fluctuation. The former is for a continuous fluid and so naturally uses the ideal gas compressibility, the latter for a lattice fluid and so naturally is based on non-interacting particles on two lattices. They agree, neglecting c^2 with respect to c. The difference between eqn (12) and (17) may be understood as follows. As the interstitial lattice fills, the number of available sites decreases, though defect interactions may cause the number of defects to continue increasing. As the number of configurations decreases there is less freedom in the distribution of defects and so less freedom in the distribution of charge. The system responds less efficiently to an applied field and so the screening length becomes longer. The expression for the lattice Debye length has been derived for superionics using the same physical principles as may be used to obtain the more usual Debye length for continuous fluids.[25] It is to a superionic as the usual Debye length is to a liquid electrolyte or molten salt and so has the same attractive features.

First, it will be a good estimate for the screening length whenever coupling is weak, i.e. at low density of charge carriers or at high temperature. The failure of the continuous version is illustrated below in sections 3 and 5 and about a factor of two at high temperature and possibly worse (the continuous fluid expression under-estimates the screening length).

Second, for continuous fluids the Debye length is an important quantity when interactions are so strong that it is no longer the same as the screening length. Even when the charge correlations have begun to oscillate, the range and wavelength of the correlations remain a simple function of Debye length. This is true of simple models of electrolytes,[1] but also more realistic ones that include the polarisability of the ions.[3] The continuous fluid Debye length is not necessarily the screening length for weak interactions in a superionic conductor, so it is very unlikely that it will be useful in this way at strong coupling. By analogy with the continuous fluid, it is the lattice Debye length that will play this role.

Fig. 2 shows the coefficients in the high temperature expansion of the two Debye lengths, using the non-interacting expressions for the defect concentrations 10 and 11. More specifically, the two are expanded in $f = \exp(\beta h_0 - s_0) - 1$ around $\exp(s_0) - 1$, for various values of s_0, the entropy of creation of a Frenkel pair in units of k_B. Published entropies are usually positive and often much larger than the ones chosen here (see Table 1), though there is some evidence that they may be negative.[18]

2.4 The effect of interactions

If interactions are included in the calculations, the screening length changes in two ways. First, the number of defects changes and that in turn changes the screening length even if eqn (12) and (17) are used. But there is also the direct influence of the interaction on the reversible work required to induce a charge fluctuation. This alters the expression for the screening length. In the case of power-law interaction (see eqn

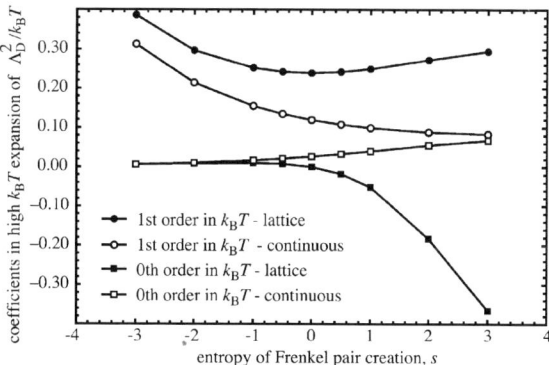

Fig. 2 Expanding the two expressions for the Debye length (that for a continuous fluid (eqn (12) and the one derived here for a two-lattice fluid, eqn (17)) at high temperature shows that they depend on the entropy of Frenkel pair creation in very different ways. Entropies of twice the magnitude illustrated here have been quoted in the literature.[36] The temperature dependence of the two expressions is qualitatively different for $s > 0$. For simplicity here $l = 1$.

(3)), the screening length is, following the same calculation as in section 2.3,

$$\Lambda_S^2 = \frac{l^3 k_B T \left(1 - \frac{1}{4k_B T}\frac{mJ}{c^m}\left(2 - \frac{1+\alpha}{\alpha}c\right)\right)}{4\pi Q^2 c \left(\left(2 - \frac{1+\alpha}{\alpha}c\right) - \frac{mJ}{k_B T}c^m(1-c)\left(1 - \frac{c}{\alpha}\right)\right)}, \quad (18)$$

where m is the power in the expression for the excess chemical potential and J is the strength of the interaction. Both the numerator and denominator are augmented with a term that depends on J and if J is zero this expression reduces to eqn (17), as it should. Eqn (18) is more elaborate than eqn (17), but all the parameters are available in the literature for a variety of superionics, in the work of Maier and co-workers[11,16,17] and in other works.[8,18,28] The cube-root model has $m = 1/3$ in eqn (18) and is capable of closely fitting the defect concentration and thermodynamic quantities from low temperature to melting. This suggests that eqn (18) will provide a good picture of screening up to melting. A similar calculation for the case of Bouteiller's exponential model is possible and some results are given below, but the formula for the screening length is too long to give here.

2.5 Strain fields

Suppose also that the crystal is strained. The strain field could be described by a tensor $\boldsymbol{u}$, but for the moment I take account only of $\psi = \mathrm{Tr}\, \boldsymbol{u}$, the fractional change in density. This approximation is supported by the Brillouin scattering data in ref. 29, where it is observed that most of the elastic constants do not change particularly at the superionic transition. For simplicity in this section the parameter α is taken as $1/2$ as for a fluorite-structured material.

In section 3 only the mobile ion density was allowed to vary in space. If the density of the crystal is allowed to vary also, the entropy change associated with a general fluctuation is (compare to eqn (15) above),

$$\frac{\Delta S}{k} = \frac{\beta}{V}\sum_q \frac{1}{2}\left(\frac{4\pi}{q^2}\frac{Q^2}{l^6} + \frac{g_{,\lambda\lambda}}{l^3}\right)|\lambda(\hat{q})|^2$$

$$+ \frac{1}{2}\left(\frac{1}{\kappa_0} + \frac{g_{,\psi\psi}}{l^3}\right)|\psi(q)|^2$$

$$+ \frac{g_{,\psi\lambda}}{l^3}\psi(\boldsymbol{q})\lambda(-\hat{q}),$$

where κ_0 is the isothermal compressibility of the crystal without any defects.

In order to obtain useful results, some coupling of density fluctuations to the defect concentration is necessary. Perhaps the simplest is to assume that the enthalpy required to create a Frenkel defect pair depends on the density change ψ:

$$h(\psi) = h_0(1 + \varepsilon\psi), \tag{19}$$

but we might also take the entropy of Frenkel pair creation to depend on the strain,

$$s(\psi) = s_0(1 + \varepsilon\psi). \tag{20}$$

The results of these two models for the effect of a density change on defect creation are similar, but the latter fits experimental data better (see section 4 and Fig. 8), so only results for the latter are given here. Where εs_0 appears here, replace it with $\varepsilon h_0/k_B T$ to get the result that would be obtained from eqn (19).

If eqn (20) is used, with $\alpha = 1/2$ as for a fluorite, the screening length is

$$\Lambda_S^2 = \frac{l^3 k_B T (1 - (2 - 3c)cs_0^2\varepsilon^2\kappa_0 k_B T)}{4\pi Q^2 c((2 - 3c) + (1 - c)(1 - c/2)2cs_0^2\varepsilon^2\kappa_0 k_B T)}. \tag{21}$$

This equation resembles eqn (18), but with interaction strength varying with temperature. That is, $m = 1$ and J a function of temperature would make the two identical.

The compressibility is modified by the disorder and charge fluctuations to give

$$\kappa^{-1} = \kappa_0^{-1} + \frac{(2c^2 - 3c + 1)}{(3c - 2)l^3}s_0^2\varepsilon^2 k_B T. \tag{22}$$

It is also possible to repeat the calculation with the average value of λ non-zero. The modification to the compressibility becomes

$$\kappa^{-1} = \kappa_0^{-1} + \frac{(2c^2 + (4\lambda - 1)c + \lambda(2\lambda - 1))}{(3c^2 + (4\lambda - 2)c + \lambda(2\lambda - 1))l^3}(c - 1)cs_0^2\varepsilon^2 k_B T. \tag{23}$$

3. Results

3.1 Fluorite-structured superionics

The upper part of Fig. 1 shows the defect concentration for various fluorites, with various models and parameters given in Table 1. Fig. 3 shows the corresponding screening lengths, calculated using the formulae from section 2.2. In fluorite-structured superionics the mobile species at low temperature occupies the tetrahedral interstices of the cubically close packed lattice of immobile ions. On heating they begin to occupy also the octahedral interstices, leaving vacancies on the tetrahedral lattice, so that $\alpha = 1/2$. The temperature axis in Fig. 3 is extended far beyond the physically sensible range to emphasise the limiting behaviour of the three screening lengths, but the effects described here are important between the superionic transition and melting. That range of temperature is experimentally accessible.[30] The pure materials described here tend to have fairly high superionic transitions, but doped or more complex materials have lower transitions.[19] The screening lengths are actually shown as $\Lambda^2/k_B T$, a quantity which becomes constant at high temperature in most cases. This makes the differences between the three screening length formulae clearer. (The intercept of the Debye length curves with the right hand axis are the high-temperature limits of $\Lambda_D^2/k_B T$ and correspond to the square symbols in Fig. 2.)

At low temperatures there are very few defects and the lattice- and continuous-fluid Debye lengths agree with the more sophisticated results and all are large, because fluctuations in charge are thermodynamically expensive and screening is not efficient. Here the charge carriers are vacancies and interstitials and with few defects the field of a test charge can only be screened over great distances.

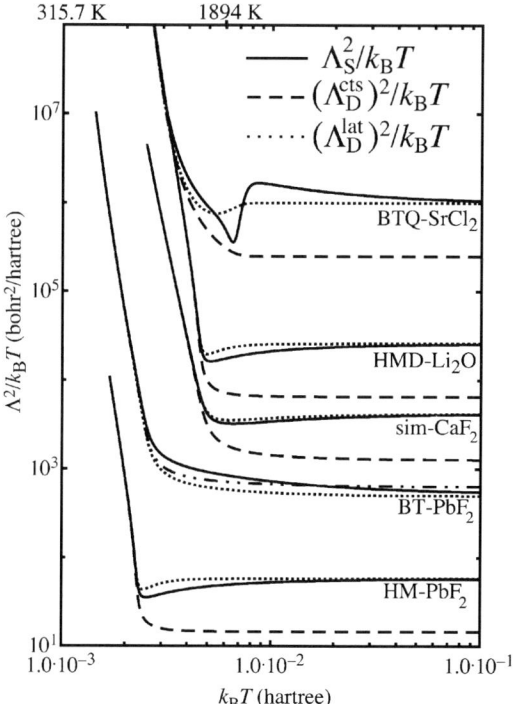

315.7 K   1894 K

Legend:
— $\Lambda_S^2/k_B T$
--- $(\Lambda_D^{cts})^2/k_B T$
⋯ $(\Lambda_D^{lat})^2/k_B T$

BTQ-SrCl$_2$
HMD-Li$_2$O
sim-CaF$_2$
BT-PbF$_2$
HM-PbF$_2$

y-axis: $\Lambda^2/k_B T$ (bohr^2/hartree)

x-axis: $k_B T$ (hartree), from $1.0 \cdot 10^{-3}$ to $1.0 \cdot 10^{-2}$ to $1.0 \cdot 10^{-1}$

Fig. 3 The screening length for five models with the fluorite structure. Two different models are examined for PbF$_2$. All three expressions for the screening length are shown (eqn (12), (17) and (16) with an interaction included in the free energy). The sets curves from bottom to top are each shifted by a factor of ten for clarity: the HM-PbF$_2$ curves are not shifted, the BTQ-SrCl$_2$ curves by a factor of 10^4. It is the screening length divided by $k_B T$ that is shown, because that makes the behaviour at high temperature simpler. The temperature axis is extended up to unphysically high temperatures to make the limiting behaviour clear, but the effects discussed here are important below the melting points. The lattice Debye length (dots) is much closer to the expression that includes interactions (solid line). The main strength of the lattice formula is that it is necessarily correct whenever interactions become weak, at low or high temperature.

Near the sharp increase in disorder at the superionic state the screening length is small and begins to increase with temperature (*i.e.* $\Lambda^2/k_B T$ is constant). In this region the mean-field treatment given here is not really valid, because the screening length becomes similar to the size of a unit cell. Still, the continuous fluid Debye length is much less than a unit cell length, so the lattice-based formulae are still superior. Experience with continuous fluids teaches us that a mean-field treatment probably works beyond conditions under which it is strictly valid and in any case is enough to characterise the crossover between the low temperature, normal crystal and a high temperature disordered material.[1,2] In both these cases the screening length is larger than the size of a unit cell and interactions become weak.

Fig. 4 illustrates that the lattice Debye expression for the screening length approaches the interacting one in the limits of both high and low temperature. More specifically it shows $\Lambda_D/\Lambda_S - 1$, which vanishes as $\Lambda_D \to \Lambda_S$ with weakening interactions. Here I am assuming that Λ_S is a much more accurate estimate of the screening length than either Λ_D and this seems reasonable for the reasons discussed in section 3. The lattice Debye length is a much more useful quantity than the continuous fluid equivalent, as it follows Λ_S more closely. The error in using the continuous fluid Debye length is typically a factor of two.

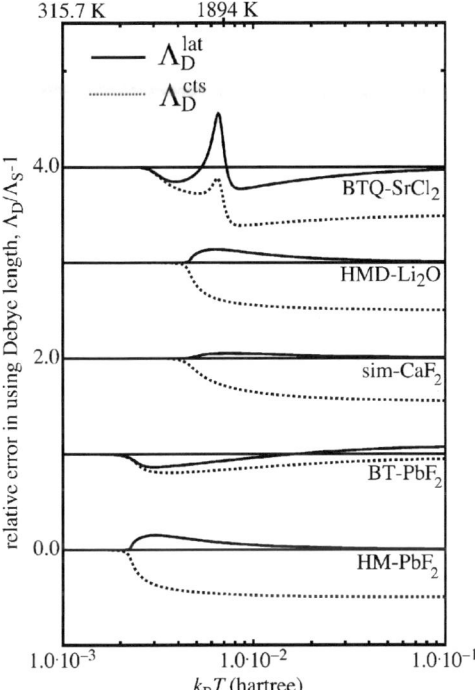

Fig. 4 The two expressions for the Debye length are compared to the mean-field interacting screening length, for each of the models of fluorite superionics used already in Fig. 1 and 3. The sets of curves are shifted by 1.0 for clarity: the HM-PbF$_2$ curves are not shifted, the BTQ-SrCl$_2$ curves by 4. The continuous fluid Debye length (dashes, eqn (12)) fails by a factor of ∼2 above the superionic transition. The two models proposed by Bouteiller show unusual behaviour, for PbF$_2$ and SrCl$_2$. For the exponential model the concentration of defects is low so the discrepancy between the three expressions is smaller.

3.2 Rock-salt structured superionics

The rock-salt structure has the mobile species occupying the octahedral holes in the cubically close packed lattice of fixed ions and the tetrahedral interstices are the higher energy sites ($\alpha = 2$). Defect thermodynamic parameters for several superionics of the rock-salt structure are available,[11,31] but these materials often have a first-order transition to the superionic state, an effect clearly not included in this mean-field treatment. We can, however, take these parameters as reasonable ones for the study of the effects that interactions have on screening. Note also that in at least one case it was possible to inhibit the first-order structural phase transition with mild doping.[32] That study revealed a first-order superionic transition without a change of crystal structure, like the type of first-order transition that occurs in the mean-field models.

Fig. 5 shows the screening length for three superionics with the rock-salt structure, with models and parameters and their sources listed in Table 1. The cusps in the screening length correspond to the first-order transitions visible in Fig. 1. The defect concentrations in these models are small below the transitions, so only above does the difference matter, for those two materials. The rock-salt structured phase of AgI is more like the fluorites in Fig. 3, having no first-order transition and shows a range of temperatures where disorder is high enough for the effects described here. See ref. 28 for more information on this material and the conditions under which it was studied.

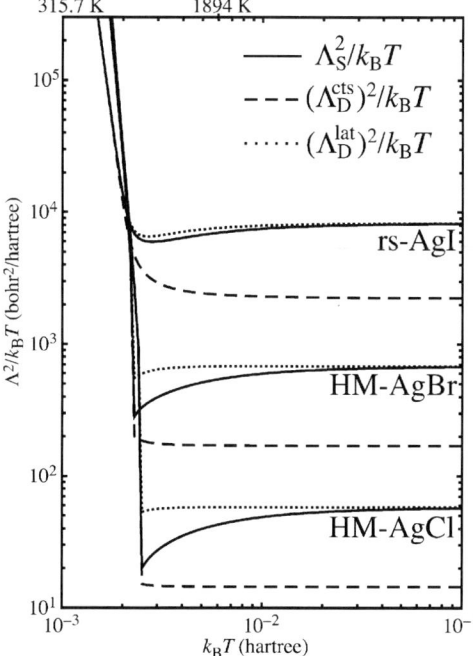

Fig. 5 As for Fig. 3, but for three superionics with the rock-salt structure, using the cube-root model. The corresponding concentration profiles are shown in Fig. 1 and details of the models in Table 1. The cusps in the curves correspond to the first-order superionic transitions predicted by these models. The successive sets of curves are shifted by factors of ten: the HM-AgCl curves are not shifted, the rs-AgI curves by a factor of 10^2 The curve for rs-AgI has no first-order transition and is at high pressure[28] and in this case the continuous fluid Debye length does fail.

Fig. 6 shows the screening lengths for AgCl and AgBr now calculated from the Debye–Hückel model. This model is based on the continuous fluid Debye length[31] and perhaps a better one would use the new lattice Debye length. The remarkable new feature is that Λ_S^2 becomes negative, implying an imaginary screening length and oscillatory correlations in the distribution of charge.[2] Such correlations do still decay over a characteristic distance, but a more sophisticated treatment is necessary to make an estimate of this distance. Guided by work on continuous fluids,[1,3] I expect the prediction of oscillations to prove accurate, though the strength of interaction necessary to induce oscillations is likely to be incorrect. The results indicate a better treatment like that in ref. 1 is worthwhile. Both computer simulation and neutron scattering experiments indicate the presence of clusters of defects in fluorite-structured superionics,[18,33] so the prediction of oscillations is plausible. This mean-field treatment gives no oscillations in the fluorite-structured superionics, so it seems likely that the mean-field treatment suppresses oscillations.

3.3 Wurtzite structure

The β phase of silver iodide undergoes a first-order transition to a superionic state with a different crystal structure. Parameters to describe this material are given in ref. 11 and are used in generating the concentration profile in Fig. 1 and the screening length in Fig. 7. A more sophisticated treatment of this material is given in ref. 16, but the simpler version serves as a good illustration of the behaviour of the screening length. The feature of this model that makes it worth examining is that

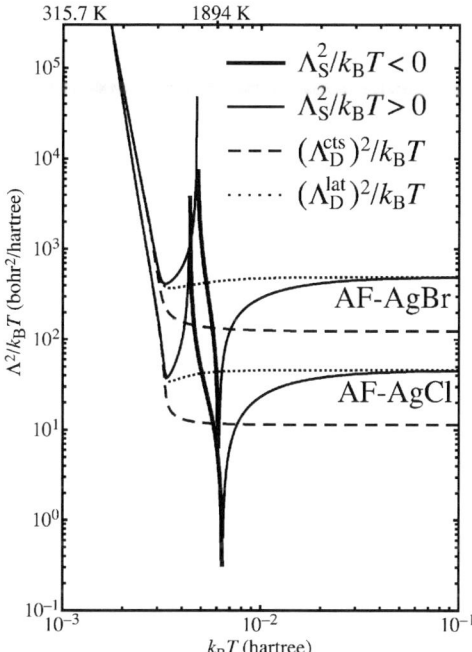

Fig. 6 The screening lengths for AgCl and AgBr now calculated using the Debye–Hückel model and parameters from ref. 31. This model shows oscillations in the charge structure, a feature that can appear for any of the mean-field models, for certain choices of the parameters. A better theoretical treatment is necessary to establish the exact extent of the oscillations. It seems likely that these oscillations are present in more materials than the mean-field treatment indicates.

the same number of sites are available on the lower and higher energy lattices (*i.e.* $\alpha = 1$).

Fig. 7 shows the screening length calculated using the three formulae from section 2.3, just as for Fig. 3 and 5. The continuous fluid Debye length behaves much as it did for the other materials, becoming proportional to temperature (note that $\Lambda^2/k_B T$ is plotted). The lattice and interacting results agree very well with each other, but behave very differently. They both become large as the temperature is increased, much more quickly than in proportion to temperature. This is because there are the same number of sites on the lower and higher energy lattices. The high temperature state, like the low temperature state, has one lattice nearly full and the other nearly empty. This explains both the unimportance of interactions and the very high screening lengths at high temperature: the number of charge carriers is small. This example is a little unnatural as the real material changes its structure at its first-order transition, but it does very clearly show how naturally the lattice Debye length accounts for the influence of the number of available sites.

4. Elastic constants of fluorite crystals

Fig. 8 compares the results on the compressibility of a superionic with the fluorite structure to the Brillouin scattering results on calcium fluoride reported in ref. 29. It is mainly the elastic constant C_{11} that shows a change at the superionic transition and it is this change that I treat with the formulae obtained in section 2.5.

The elastic constants are reported as changes in the squares of the Brillouin frequencies[29] and they show a linear decrease with temperature followed by a faster

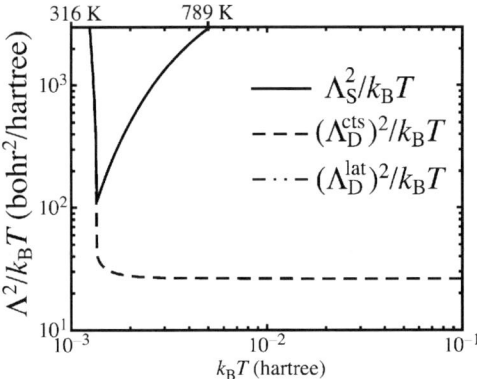

Fig. 7 Defect concentration and screening length in β-AgI, a superionic with the wurtzite structure. This model illustrates the greatest discrepancy between the lattice and continuous Debye lengths. The lower and higher energy lattices have the same number of sites in this simple treatment and so at high temperature the system looks much as it did at low temperature—one lattice nearly full, the other with only a few ions.

decrease near the superionic transition where the crystal becomes disordered. The upper half of Fig. 8 shows the change in the square of the Brillouin frequency near the superionic transition. The linear variation visible at low temperatures has been subtracted. The predictions for the disorder-induced change based on eqn (19) and

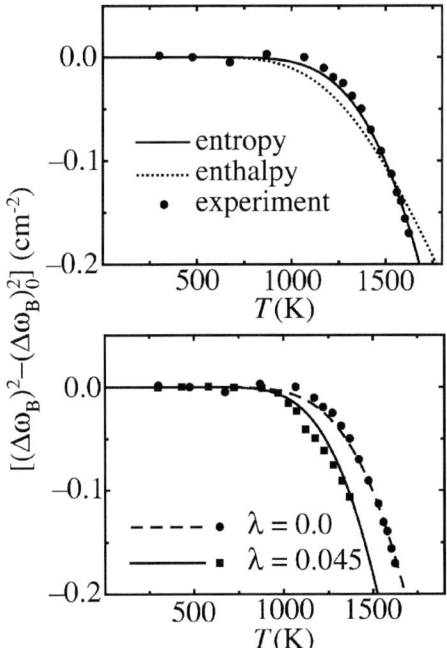

Fig. 8 Comparison of the predictions of section 2.5 to experimental data on CaF_2, taken from Fig. 9 of ref. 29. The upper plot shows the deviation of the square of the Brillouin frequency as disorder increases with temperature. The model in which density fluctuations couple to the entropy of Frenkel pair creation fits the shift better than one with coupling to the enthalpy. Only the parameter ε is varied in the fit (see eqn (22)). In the bottom half of the figure, the Brillouin data for doped calcium fluoride are also shown. Eqn (23) correctly gives the shift in Brillouin frequencies of doped CaF_2 without further fitting.

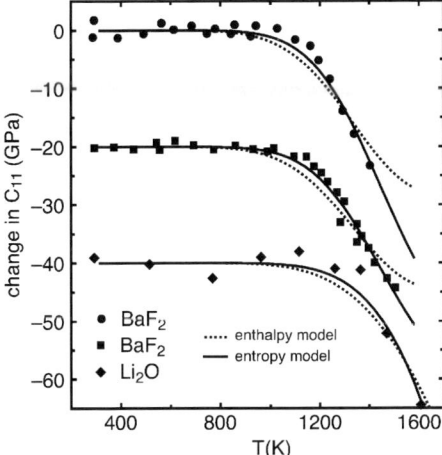

Fig. 9 The changes in compressibility predicted by eqn (22) for barium fluoride and lithium oxide. The three sets have been shifted vertically for clarity. The first set of barium fluoride data (circles, top) is from ref. 29, the second (squares, middle) is from ref. 37. The lithium oxide results are from ref. 38. In each case the model with strain varying the entropy of pair formation is the better fit, though only marginally in some cases.

(20), which couple the density fluctuation to the Frenkel pair creation enthalpy and entropy, respectively, are also shown. The necessary parameters are taken from ref. 8 and are based on a remarkably accurate simulation model for calcium fluoride.[34,35] The parameter ε is the only one allowed to vary in matching formula (22) to the experimental data. Fig. 9 compares the fit of the two possible expressions for the change in elastic constant for barium fluoride and lithium oxide. The fit of the expression with entropy varying with strain is consistently better, though only marginally so in some cases. The effect of defect interactions is usually to make the transition sharper and this is likely to account for the remaining discrepancy between the theoretical and experimental elastic constants.

Further evidence that section 2.5 gives the correct variation of elastic constants with disorder is given by the shift in Brillouin frequencies for calcium fluoride doped with 9% yttrium fluoride. The pure and doped data are shown in the bottom half of Fig. 8. Now the value of ε is taken from the fit to the pure CaF_2 data and used with $\lambda = 0.045$ to obtain the match to the experimental data on the doped material. The lower superionic transition and the different shape of the curve are both captured by eqn (23) without further fitting. More work is needed, as the effect in calcium fluoride is not very great, but it is very promising that with a single parameter the shapes of the superionic effects on the elastic constants are captured. This result suggests an origin for the change in elastic constants in the vibrational entropy of a Frenkel pair.

5. Conclusions

I have presented an analysis of screening in superionic conductors based on the various mean-field models available in the literature, in particular the cube-root model. Evidence for the accuracy of the cube-root model for a wide range of superionics is available and suggests that estimates for the screening length based on it will be useful up to melting. On the other hand, neglecting interactions gives a lattice version of the ordinary Debye length, which can be guaranteed to become exactly the screening length in any limit of low interaction strength, either low temperature and low defect density, or high temperature. This expression is derived for the lattice fluid using the same physical principles that lead to the more familiar Debye length for a continuous fluid.

The lattice version of the Debye length is just as easy to use and better founded in a non-interacting model. Using studies of continuous Coulomb fluids as a guide, it is likely that the lattice version of the Debye length will be an important parameter in more accurate descriptions of the charge structure. The prediction of oscillations in the charge correlations when mean-field interactions are included suggests that a thermodynamically consistent treatment is worthwhile to establish more reliably the conditions under which such oscillations appear.

Finally, it is shown that the method of calculating the probability of a long-wavelength fluctuation may be extended to couple the density fluctuations to the disorder. A first comparison to experiment shows that the result for the elastic constant fits well and without further fitting reproduces the effect of 9% doping on the superionic change in elastic constants, including the lowering of the superionic transition temperature. Bringing together the treatment of screening and strain in the same tractable mean-field theory makes it possible to estimate the contributions of electrical and mechanical response to stresses, such as those near interfaces in nanostructures.

References

1   R. J. F. L. Decarvalho and R. Evans, *Mol. Phys.*, 1994, **83**, 619.
2   J. Hansen and I. McDonald, *Theory of Simple Liquids*, Academic Press, London, 1986.
3   A. Gray-Weale and P. A. Madden, *Mol. Phys.*, 2003, **101**, 1761.
4   N. Sata, K. Eberman, K. Eberl and J. Maier, *Nature*, 2000, **408**, 946.
5   N. Sata, N. Jin-Phillipp, K. Eberl and J. Maier, *Solid State Ionics*, 2002, **154–155**, 497–502.
6   N. Jin-Phillipp, N. Sata, J. Maier, C. Scheu, K. Hahn, M. Kelsch and M. Rühle, *J. Chem. Phys.*, 2004, **120**, 2375–2381.
7   S. Kim, J. Fleig and J. Maier, *Phys. Chem. Chem. Phys.*, 2003, **5**, 2268.
8   A. Gray-Weale and P. Madden, *J. Phys. Chem. B*, 2004, **108**, 6624–6633.
9   M. J. Rice, S. Strässler and G. A. Toombs, *Phys. Rev. Lett.*, 1974, **32**, 596.
10 B. A. Huberman, *Phys. Rev. Lett.*, 1974, **32**, 1000.
11 N. Hainovsky and J. Maier, *Phys. Rev. B*, 1995, **51**, 15789–15797.
12 S. Kim and J. Maier, *J. Eur. Ceram. Soc.*, 2004, **24**, pp, 1919–1923.
13 D. O. Welch and G. J. Dienes, *J. Phys. Chem. Solids*, 1977, **38**, 311.
14 J. Oberschmidt, *Phys. Rev. B*, 1981, **23**, 5038.
15 Y. Bouteiller, *Phys. Rev. B*, 1992, **45**, 8734.
16 F. Zimmer, P. Ballone, J. Maier and M. Parinello, *J. Chem. Phys.*, 2000, **112**, 6416.
17 F. Zimmer, P. Ballone, M. Parinello and J. Maier, *Solid State Ionics*, 2000, **127**, 277.
18 A. Gray-Weale and P. Madden, *J. Phys. Chem. B*, 2004, **108**, 6634–6642.
19 E. Vlieg, H. W. den Hartog and M. Winnink, *J. Phys. Chem. Solids*, 1986, **47**, 521.
20 T. Ishii and O. Kamishima, *J. Phys. Soc. Jpn.*, 1999, **68**, 3580.
21 T. Ishii and O. Kamishima, *J. Phys. Soc. Jpn.*, 2001, **70**, 159.
22 P. D. Fleming and C. Cohen, *Phys. Rev. B*, 1976, **13**, 500.
23 R. Zeyher, *Z. Physik B*, 1978, **31**, 127.
24 A. C. G. Stell, *Phys. Rev. E*, 2004, **70**, 016144.
25 P. Vieillefosse and J. P. Hansen, *Phys. Rev. A*, 1975, **12**, 1106.
26 T. Kurosawa, *J. Phys. Soc. Jpn.*, 1957, **12**, 338.
27 E. Pines, *Chem. Phys. Lett.*, 1998, **288**, 270.
28 D. A. Keen, S. Hull, W. Hayes and N. J. G. Gardner, *Phys. Rev. Lett.*, 1996, **77**, p4914.
29 J. D. Comins, P. E. Ngoepe and C. R. A. Catlow, *J. Chem. Soc., Faraday Trans.*, 1990, **86**, pp1183–1192.
30 B. M. Voronin, *J. Phys. Chem. Solids*, 1995, **56**, 839.
31 J. K. Aboagye and R. J. Friauf, *Phys. Rev. B*, 1975, **11**, 1654.
32 D. A. Keen, S. Hull, A. C. Barnes, P. Berastegui, W. A. Crichton, P. A. Madden, M. G. Tucker and M. Wilson, *Phys. Rev. B*, 2003, **68**, 014117.

33 J. Goff, W. Hayes, S. Hull and M. Hutchings, *J. Phys.: Condens. Matter*, 1991, **3**, 3677.

34 M. Gillan, *J. Phys. C: Solid State Phys.*, 1986, **19**, 3391–3411.

35 M. Gillan, *J. Phys. C: Solid State Phys.*, 1986, **19**, 3517–3533.

36 Marc Hayoun, Madeleine Meyer and Aurélie Denieport, *Acta Mater.*, 2005, **53**, 2867.

37 K. E. Rammutla, J. Comins, R. M. Erasmus, T. T. Netshisaulu, P. E. Ngoepe and A. V. Chadwick, *Radiat. Eff. Defects Solids*, 2002, **157**, 783.

38 S. Hull, T. W. D. Farley, W. Hayes and M. T. Hutchings, *J. Nucl. Mater.*, 1988, **160**, 125.

General Discussion

Professor Haile opened the discussion of Dr De Souza's paper: Your analysis doesn't give an explicit dependence of the grain boundary capacitance on ground boundary thickness. Yet many of us use the brick layer model, from which one can extract the grain boundary thickness (assuming $\varepsilon_{bulk} = \varepsilon_{gb}$). Can this thickness be related to any quantities in your model? Is your model compatible with the brick layer model?

Dr De Souza replied: One can identify the grain boundary thickness obtained from the brick layer model with the extent of the space-charge layers at the grain boundary ($= 2 \lambda^*$, where λ^* is the extent of one of the space-charge layers). Thus, setting $C_{gb} = \varepsilon_0 \varepsilon_r / 2\lambda^*$ equal to the C_{gb} derived in the paper {eqn (8)}, we find

$$\lambda^* = \sqrt{\frac{2\varepsilon_0\varepsilon_r(\Phi_0 - \frac{1}{2\beta})}{e[\text{Fe}'_{\text{Ti}}]_b}}.$$

If Φ_0 is greater than 0.5 β^{-1}, then the above equation can be approximated by

$$\lambda^* = \sqrt{\frac{2\varepsilon_0\varepsilon_r\Phi_0}{e[\text{Fe}'_{\text{Ti}}]_b}}.$$

This is exactly the form of λ^* that is obtained by solving completely the Poisson–Boltzmann equation, eqn (4), having neglected the depleted oxygen vacancies.[1]

1 E. H. Rhoderick and R. H. Williams, *Metal–Semiconductor Contacts*, Clarendon Press, Oxford, UK, 1988.

Professor Kilner asked:
(1) Is there any trapping of the vacancies and, if so, could there be a dipolar reorientation component to the increase in relative permittivity with PO_2 (*i.e.* vacancy concentration)?
(2) Are the grain boundary capacitances you see comparable to those observed in polycrystalline material?

Dr De Souza answered:
(1) The oxygen vacancies could be trapped by Fe'_{Ti} and $\text{Fe}^\times_{\text{Ti}}$. From the study by Merkle and Maier[1] we could rule out $\{\text{Fe}'_{\text{Ti}}\text{–}V^{\cdot\cdot}_{\text{O}}\}$ associates being present at the temperature of interest. We believe that the binding energy of $\{\text{Fe}^\times_{\text{Ti}}\text{–}V^{\cdot\cdot}_{\text{O}}\}$ is lower than that of $\{\text{Fe}'_{\text{Ti}}\text{–}V^{\cdot\cdot}_{\text{O}}\}$, as there is no Coulomb attraction between the defects and the elastic strain resulting from size mismatch is low: in six-fold coordination $r_{\text{Ti}^{4+}} = 0.605$ Å and $r_{\text{Fe}^{4+}} = 0.585$ Å.[2]
(2) The boundary capacitance per boundary obtained for a polycrystal is comparable to those we measured. See also ref. 3 and my response to Professor Haile's question.

1 R. Merkle and J. Maier, *Phys. Chem. Chem. Phys.*, 2003, **5**, 2297.
2 R. D. Shannon, *Acta Crystallogr., Sect. A*, 1976, **32**, 751.
3 S. Rodewald, J. Fleig and J. Maier, *J. Am. Ceram. Soc.*, 2001, **84**, 521.

Professor Maier asked: What misorientation angle would correspond to the transition from a homogeneous boundary effect to a constriction behaviour?

Dr De Souza replied: The misorientation angle at which the current flow across the interface is constricted between the one and two dimensional cases is unclear. It will depend on the ratio d_{dis}/λ^*_{dis} and thus on dopant concentration and, *via* ε_r, on temperature. In Fig. 5 $\theta = 3.3°$ was employed, which was obtained from an arbitrary limit of $d_{dis}/\lambda^*_{dis} = 0.5$ (for $T = 623$ K and $[Fe_{Ti}]_b = 4.2 \times 10^{24}$ m^{-3}).

One may expect that current constriction becomes relatively more important, as the space-charge tubes surrounding each dislocation become effectively isolated. In a simple approximation this occurs at $d_{dis}/\lambda^*_{dis} = 2$; the corresponding misorientation angle is $\theta = 0.8°$. However, the resistance ratio for $\theta < 1°$ is $R_{gb}/R_b < 10^{-4}$, and therefore I believe that such a nanoscale constriction effect will not be observable.

1   R. A. De Souza, J. Fleig, J. Maier, Z. Zhang, W. Sigle and M. Rühle, *J. Appl. Phys.*, 2005, **97**, 053502.

Professor Haile asked: In your measurements you observe a remarkable variation in grain boundary resistance as a function of grain boundary structure. In a polycrystalline material how will these variations be averaged?

Dr De Souza answered: From studies on Fe-doped SrTiO$_3$ bicrystals[1,2] we showed that R_{gb}/R_b at $T = 623$ K can vary by six orders of magnitude. In a polycrystalline sample, the measured resistance ratio will be biased towards lower values. This is because the current will seek the path of least resistance, and thus avoid extremely blocking grain boundaries. For the same reason, even examination of individual boundaries in a polycrystalline sample with microcontacts (see, for example, ref. 3) will not probe highly resistive boundaries.

1   R. A. De Souza, J. Fleig, J. Maier, Z. Zhang, W. Sigle and M. Rühle, *J. Appl. Phys.*, 2005, **97**, 053502.
2   Z. Zhang, W. Sigle, R. A. De Souza, W. Kurtz, J. Maier and M. Rühle, *Acta Mater.*, 2005, **53**, 5007.
3   S. Rodewald, J. Fleig and J. Maier, *J. Am. Ceram. Soc.*, 2001, **84**, 521.

Professor Haile asked: Can you comment on the differences between YBCO and metal electrodes for these measurements?

Dr De Souza answered: YBa$_2$Cu$_3$O$_{6+x}$ is preferred for use as electrodes on Fe-doped SrTiO$_3$ for two reasons. First, it allows the equilibration of the SrTiO$_3$ samples with gaseous oxygen at the temperatures of interest within a reasonable timescale.[1,2] Second, YBa$_2$Cu$_3$O$_{6+x}$ electrodes give rise to a tiny, generally insignificant, contribution to the impedance spectra of SrTiO$_3$.[2,3] For bicrystals, the space-charge layers at metal electrodes would give rise to an impedance arc that would obscure the response of the grain boundary.

1   Denk, F. Noll and J. Maier, *J. Am. Ceram. Soc.*, 1997, **80**, 279.
2   R. A. De Souza, J. Fleig, J. Maier, O. Kienzle, Z. Zhang, W. Sigle and M. Rühle, *J. Am. Ceram. Soc.*, 2003, **86**, 922.
3   I. Denk, J. Claus and J. Maier, *J. Electrochem. Soc.*, 1997, **144**, 3526.

Dr Zhukovskii opened the discussion of Mr Björketun's paper:

(1) Mr Björketun complains in his paper about the marked underestimate of the band gap when using the VASP code. Why did he use it at all in such a case? To improve the calculated band gap, the semi-empirical option of the Hubbard potential has been implemented recently in this code (DFT + U method). In any case, first principles codes which implement localised orbital formalism (CRYSTAL, periodic Gaussian, SIESTA) should be also used for the comparative calculations on defective crystals.

(2) For a single defect approximation, a 3 × 3 × 3 supercell is still not enough, at least a 4 × 4 × 4 supercell should be used to avoid interdefect interactions (the criterion is a straight line of defect level or neglecting dispersion).

(3) It is known that for perovskites, VASP calculations overestimate relaxation of the nearest atoms around defect, probably it would be better to compare it with calculations using other codes.

Mr Björketun replied:

(1) In Section 3 we emphasize that the underestimation of the electronic band gaps of insulators and semiconductors is an artefact of traditional DFT (and thus not specifically associated with the particular code we have used). This is likely to affect the calculated formation energy of donor states derived from the conduction band. Since we in most cases (throughout Sections 4.1 and 4.3) consider a predominantly p-type material, where the problematic donor states are empty, the band gap error will generally not cause us any trouble. Difficulties arise in Section 4.2 when we study hydrogen interstitials in charge states $q = 0$ (H^0) and $q = -1$ (H^{-1}). The defect levels close to the conduction band are then no longer empty. In Section 3 we mention two possible remedies: the first one is a simple correction scheme where the conduction band is rigidly shifted upwards to match the experimental band gap and the defect levels are assumed to follow this shift. Making such a correction will increase the formation energies of H^0 and H^{-1}. Our main conclusion, that $+1$ is the most stable hydrogen state regardless of the actual position of the Fermi level in the band gap, will therefore remain valid. The second remedy is to include a better treatment of the exchange part of the total-energy functional. Using L(S)DA + U is one way to do this and is, as you mention, supported by VASP.4.6 and on. Unfortunately it was not implemented in VASP.4.5, which we used when the project was first initiated. Regarding the localized orbitals, we are aware that atom centred basis sets probably improve the description of point defects. Our main objective has however not been to obtain the best possible representation of single defects, but rather to discern general trends, in particular relative energy differences between differently doped $BaZrO_3$. For that purpose we think a plane wave formalism should be good enough. Still we admittedly have problems with an incorrect description of H^0, where localized orbitals might have improved our results.

(2) We agree that there is probably some spurious interactions of the defect levels left when using 3 × 3 × 3 supercells (see, for instance, ref. 1). This remaining dispersion should, however, hardly exceed 0.1 eV. One could also assume the defect levels to be shifted as the supercell size is increased, but that should also be a minor effect.[1] Consequently, the total-energy errors associated with the usage of this somewhat too small cell should be in the 0.1 eV range. All conclusions concerning trends in formation energies and differences in dopant solution energies should therefore remain valid.

(3) In principle, we should have used other codes to check the relaxation around defects. However, we have compared our calculated structures and binding energies of dopant–proton clusters in $BaZrO_3$ with the theoretical results obtained by other groups (e.g., M. S. Islam et al.[2]) and found good agreement.

1 R. A. Evarestov, E. A. Kotomin and Yu. F. Zhukovskii, *Int. J. Quantum Chem.*, 2006, **106**, 2173–2183
2 M. S. Islam, P. R. Slater, J. R. Tolchard and T. Dinges, *Dalton Trans.*, 2004, **19**, 3061–3066

Professor Islam asked:

(1) You have examined, as we have,[1] the proton–dopant binding (association) energetics in $BaZrO_3$. Do your energetic trends link up with the observed proton conductivities for the same series of dopants?

(2) You use the term "hydrogen interstitial". Is it really an interstitial defect or an $OH_O^{\cdot}$ (hydroxyl) defect (as in eqn (1)) since the H^+ species is very small?

1   M. S. Islam, P. R. Slater, J. R. Tolchard and T. Dinges, *Dalton Trans.*, 2004, 3061

Mr Björketun responded:

(1) Yes, they do. First of all, the experimental trend in hydration enthalpies (hydration *via* dissociative absorption of water from a surrounding atmosphere)— which determines the proton concentration in the dilute limit at low temperatures— is reproduced when we account for both $M–(OH_O)^{\cdot}$ and $M–(V_O)^{\cdot\cdot}$ interactions. Secondly, the trend in activation energy for proton conduction (or more correctly proton self-diffusion), is reproduced by taking the $M–(OH_O)^{\cdot}$ interactions into account.

(2) We chose the term "hydrogen interstitial" instead of "proton" because in Section 4.2 we investigated not only protons (H^+), but also hydrogen in the $q = 0$ and -1 charge states (H^0 and H^-). This was mainly done to show that hydrogen preferentially dissolves in the $q = +1$ charge state, *i.e.* as protons (H^+). Whereas H^+ and H^0 always reside deep within the oxygen charge cloud, forming a hydroxyl defect, H^- is instead found to occupy a true interstitial site due to repulsive interactions with the surrounding lattice oxygens.

Professor Beck addressed Mr Björketun and Professor Islam: I come back to the question put to Dr Slater earlier this morning. I asked him whether it is right to speak of interstitials in the case of apatites, because it is known that the chain of anions in the channel is very often buckled due to the space needs of the anion, so the position called "interstitial" would be occupied due to a "normal" shift of the anion out of its place in 0 0 Z by the influence of its neighbours.

I ask Professor Islam the same question when he suggests not to speak of interstitial hydrogen in the case of $BaZrO_3$ but rather an oxygen defect species. This discussion highlights the need to (re)define the expression "interstitial".

Professor Islam replied: In terms of definition, the "interstitial" is viewed as a defect at a crystal lattice site that is *not* normally occupied. As is well known, one of the best examples is the silver ion (Ag^+) interstitial defect in rock-salt structured silver halides (AgX); here the interstitial position is a site not normally occupied at the centre of the eight-fold cube. In this context, the oxygen-ion interstitial in apatite-silicates discussed by Dr Slater is also at a lattice site not normally occupied as indicated by atomistic simulations[1] and neutron diffraction studies.[2] However, for the protonic defect in perovskite oxides such as $BaZrO_3$ this is less clear: the proton is a very small species in comparison with the O^{2-} ion, and not in a "free" unbound state, but sits within the electron cloud of an oxygen ion as a hydroxide defect ($OH_O^{\cdot}$) at a normal oxygen site.[3] This is indicated from spectroscopic[4] and DFT-based computational studies of proton-conducting perovskite oxides.[4,5]

1   J. R. Tolchard, M. S. Islam and P. R. Slater, *J. Mater. Chem.*, 2003, **13**, 1956.
2   L. León-Reina, E. R. Losilla, M. Martínez-Lara, S. Bruque, A. Llobet, D. V. Sheptyakov and M. A. G. Aranda, *J. Mater. Chem.*, 2005, **15**, 2489.
3   T. Norby, M. Widerbe, R. Glockner and Y. Larring, *Dalton Trans.*, 2004, 1.
4   K. D. Kreuer, *Annu. Rev. Mater. Res.*, 2003, **33**, 333.
5   M. S. Islam, R. A. Davies and J. D. Gale, *Chem. Mater.*, 2001, **13**, 2049.

Professor Haile addressed Mr Björketun:

(1) Does your analysis treat $BaZrO_3$ as a material with a fixed Ba : Zr = 1 : 1 composition, or does it allow for variability in the Ba : Zr ratio?

(2) In $BaCeO_3$, we have done a lot of studies of dopant partitioning particularly as a result of BaO evaporation at high temperature, and have found that the proton

uptake and transport properties can be readily explained by the presence of dopants on both the A and B sites. The situation in very different, however, for barium zirconate. A slight Ba deficiency can lead to a reduction in conductivity by two orders of magnitude, far greater than can be explained by dopant partitioning.

Mr Björketun responded:

(1) We consider a solid phase of stoichiometric $BaZrO_3$ in equilibrium with an oxygen atmosphere and a solid phase of either stoichiometric BaO or stoichiometric ZrO_2 in order to assign values for the atomic chemical potentials of Ba, Zr, and O.

(2) The reason why we decided to investigate dopant partitioning in the first place was the fact that the proton concentration in $BaZrO_3$ after hydration sometimes is considerably smaller—down to 60% of—than the nominal value.[1] This lowering of the proton content would, however, only reduce the conductivity by a factor of two. I wonder if it is possible that the remaining reduction in conductivity is caused by proton–barium vacancy association, effectively increasing the activation energy for proton diffusion.

1 K. D. Kreuer, St. Adams, W. Münch, A. Fuchs, U. Klock and J. Maier, *Solid State Ionics*, 2001, **145**, 295–306.

Dr Kohanoff asked:

(1) Do you know for certain that the charged states of the hydrogen ($+1$, 0, -1) correspond to H^+, H^0 and H^-? Have you checked that the excess or defect charge is effectively localised on the H? Using a plane wave method, this charge could have gone anywhere in order to minimise the energy.

(2) Have you got an estimate for the barrier for proton jumping between equivalent positions, associated to the various oxygen atoms in the structure (4 equivalent locations)? Could the H be tunnelling between them?

Mr Björketun replied:

(1) A Bader analysis[1] reveals, first of all, that the $+1$ state (one electron removed from the supercell) corresponds to H^+. Moreover, it shows that when one electron is added to the supercell (the -1 state) almost all additional charge associates with the hydrogen; *i.e.* H^- is formed. However, it seems like the neutral state resembles that of H^+. In this case the charge that was assumed to be localized on H has been transferred to the Zr atoms (distributed quite evenly over the Zr atoms in the supercell) so that talking about hydrogen in the $q = 0$ state is therefore not really appropriate.

(2) In pure $BaZrO_3$ we obtain a classical barrier of approximately 0.2 eV for proton transfer between neighbouring oxygens. Since zero-point motion effects appear to reduce it even further, the possibility of proton tunnelling at low temperatures cannot be excluded (*cf.* the formation of quantum-mechanically delocalized hydrogen states in bcc metal hydrides). However, preliminary results from our calculations using the Flynn–Stoneham model of incoherent phonon-assisted hopping for the proton transfer step indicate an activation energy that is comparable with the classical barrier but with a much smaller prefactor.

1 G. Henkelman, A. Arnaldsson and H. Jónsson, *Comput. Mater. Sci.*, 2006, **36**, 354–360

Dr Lord commented: It should be possible to use muon spin relaxation to identify the various (possibly metastable) charge states of H in $BaZrO_3$, and the electron distribution in the H^0 state—either a shallow donor or an atomic muonium state.

Professor Ishihara asked: From experiment, this material is not stable under highly humidified conditions or a high CO_2 atmosphere. According to your calculations, the solubility of the dopant or trapping energy of H^+ is strongly affected by the size of the dopant. Can you expand your calculation to the chemical stability, which depends on the dopant? According to your calculations, which system is the most stable?

Mr Björketun answered: It has been suggested that $BaZrO_3$ should be rather stable in CO_2-containing atmospheres, in contrast to, for instance, $BaCeO_3$ (see ref. 1 and 2). In principle, we could predict the stability of $BaZrO_3$ doped either on the Ba- or Zr-site by calculating the heats of formation of the binary oxides BaO, ZrO_2 and M_2O_3 (M = any trivalent dopant). That, however, requires the total energy of M_2O_3, which we have not calculated. Instead we decided to study the difference in dopant solution energy on the Ba- and the Zr-site (eqn (3) and (4)). This eliminates the contribution from the total energy of M_2O_3.

1   K. D. Kreuer, *Solid State Ionics*, 1997, **97**, 1–15.
2   K. H. Ryu and S. M. Haile, *Solid State Ionics*, 1999, **125**, 355–367.

Professor Navrotsky commented: The question of stability with respect to CO_2 or H_2O is a case in point of the comment I made earlier—namely that one has to take the calculated defect energies, combine them with other steps in a thermochemical cycle, to get an energy for a reaction that can be measured directly.

Professor Ishihara replied: I think the stability against CO_2 or H_2O is strongly related to the lattice energy, which is also influenced by the dopant as mentioned. As pointed out, defect energy is also important for considering the lattice energy, but I think that the size of the dopant is more important for determining the stability of materials. This means that in spite of the same amount of defects, the stability of the materials may change depending on the dopant. These effects of the dopant on the lattice energies are also better analysed from calculations based on quantum chemistry as well as thermodynamics.

Professor Yashima asked:
(1) How did you estimate the force constant in the vibrational part of eqn (13)?
(2) What is the effect of the dynamical matrix and density of states (DOS) of vibration?
(3) What was the effect of the vibration term?

Mr Björketun replied:
(1) With the lattice frozen in the fully relaxed configurations, the vibrational frequencies of the single atoms under consideration were obtained by calculating and diagonalizing the corresponding one atom dynamical matrices; that means we applied an Einstein model. The general expression (13) does not in itself imply an Einstein formulation. This restriction is introduced in the more explicit expressions (16) and (17).
(2) It is difficult to estimate the error we make when not accounting for the whole vibrational DOS, but our results regarding hydration at finite temperatures and pressures should not be affected that much. In our analysis we never consider the whole vibrational contents of the lattice, only the difference between pure and defective lattices. This suggests that a significant amount of error cancellation can be expected. Moreover, Fig. 7 clearly indicates that the contribution to the hydration energy from changes in vibrational free energy of the lattice is relatively small compared to the part associated with absorption of water from the atmosphere.
(3) The effect of neglecting the vibrations of all metal atoms and the oxygen atoms far from defects in expression (17) should also be small. These atoms are never exchanged with the surrounding atmosphere and therefore do not change the

number of modes. Their only contribution to the vibrational free energy comes from small frequency differences between the pure and defective lattices. Fig. 7 shows that this type of contribution is small.

Professor Maier opened the discussion of Dr Sokol's paper: What is your opinion on the compensating point defect to the excess electrons in pure ZnO (which is oxygen efficient, relatively speaking)?

Dr Sokol answered: Our results are not accurate enough for a reliable thermo-dynamical analysis, for example, to predict concentrations of particular defect species. Therefore, based solely on our calculations, I cannot give you a definitive answer, and you found some ambiguity in our discussion in the paper. I should stress that the problem of accuracy occurs in current periodic density functional studies, moreover, it is complicated by the deficiencies of commonly used functionals. Further, in the discussion of the charge carriers and point defects in the bulk, we usually neglect surface effects. As shown in my short presentation to this meeting, the band offset for the dominant polar reconstructed surfaces in this material is about 1.35 eV, which makes it imperative to include boundary conditions. The oxygen and zinc terminated polar surfaces are the essential source and sink for electron and hole charge carriers as well as for atomic point defects as confirmed by the presence of surface charge accumulation layers (space charge) in ZnO.

Professor Islam asked:
(1) You investigated Cu substitution. Do you find Jahn–Teller distortion, as expected for d^9 Cu^{2+}?
(2) For Cu^+ substitution, what was your charge-compensation mechanism? What is the implication for optical and catalytic behaviour of ZnO?
(3) Is ZnO deliberately doped with Cu or are they impuritites (or both)?

Dr Sokol responded:
(1) We did not observe nor expect the "typical" Jahn–Teller distortion as found in weakly bound systems. See for comparison a good account in Figgis and Hitchman.[1] However, although we found that the dopant's environment was not symmetrical, upon closer analysis after your question, we note that there is a lengthening of one of the four Cu–O bonds, which may be due to a form of Jahn–Teller distortion.
(2) As observed in experiments, we considered Cu as an electron acceptor with Cu^+ formed on trapping an electron from the conduction band. In this sense, the Cu^+ species would be compensated in real material by any electron donating defect, including Zn interstitials and O vacancies or other impurity species such as Al. As with all centres studied, Cu has been considered here as an isolated centre with Cu^{2+} playing the rôle of a deep acceptor and Cu^+ as a deep donor. The corresponding optical transitions are in close agreement with available experimental data as outlined in the paper. The catalytic properties of this material are dominated rather by surface Cu, which we investigated previously.[2]
(3) Copper is present in ZnO as a ubiquitous impurity. However, ZnO can be deliberately doped with Cu for various applications, for example, because it can dramatically decrease the n conduction in ZnO. Moreover, in one of the main industrial applications of ZnO, a mixture of Cu (at 60%) with ZnO and Al_2O_3 is used for the production of methanol from syngas, which of course leads to very heavy doping. Finally, it has been recently discovered that Cu doped ZnO thin films exhibit ferromagnetism at and above room temperature, which makes it a good candidate for application in the emerging field of spintronics.

1   B. N. Figgis and M. A. Hitchman, *Ligand Field Theory and Its Applications*, Wiley-VCH, New York, 2000, p. 158

2 S. T. Bromley, S. A. French, A. A. Sokol, C. R. A. Catlow and P. Sherwood, *J. Phys. Chem.*, 2003, **107**, 7045–7057.

Professor Beck commented: The fact that especially polar surfaces, such as the (0001) in ZnO, are not stable and should be stabilised by a certain number of defects puts into question whether the terminating (0001) surfaces could be alternatively stabilised by substitution, *e.g.*, OH^- for O^{2-} or Ag^+ for Zn^{2+} instead of forming vacancies.

Dr Sokol agreed: You are absolutely right. Depending on the synthetic route, history of particular samples, various treatments and typical exploitation conditions, polar surfaces can be hydroxylated and hydrogenated, terminated by various impurities or metallic layers *etc.* The charge disproportionation can also be achieved by adatoms and *via* larger scale features such as triangular islands and depressions on the Zn terminated (0001) surface, which were recently identified and investigated by the group of Diebold and colleagues.[1]

1 O. Dulub, U. Diebold and G. Kresse, *Phys. Rev. Lett.*, 2003, **90**, 016102.

Dr Zhukovskii asked: Are you sure that there is no sense in using periodic first principles calculations for defects on ZnO surfaces? What are the main advantages of embedded cluster model calculations on the same system?

Dr Sokol replied: I do not want to give you an impression that the periodic calculations would not be of value, but we should clearly realise their limitations. Embedded cluster approaches have been developed exactly to address the problem of the description of an isolated defect in homogeneous or uniform extended media such as periodic solids. These approaches are therefore the most economical means to deal, for example, with defects in ZnO. Our calculations and simple estimates based on semiclassical calculations show that to get a reliable picture we have to model explicitly regions approximately 15 Å in radius. We can hardly afford to use supercells of this size in first principles calculations on a routine basis for many alternative configurations. Moreover, such calculations typically employ local density or generalised gradient approximations, which are in very large relative error when dealing with energies of defect formation and electron localisation processes.

Professor Navrotsky asked a general question: How does an experimentalist understand how accurate a particular calculation is for a given property such as structure, dynamics or thermodynamics?

Dr Sokol answered: It is very difficult to give a particular recipe as the answer will depend on the system of interest. For ionic systems, for example, semiclassical calculations are often very accurate in reproduction of both structure and energetics of charged defects. The so-called first principles, LDA and GGA periodic calculations are usually quite reliable for the structure and dynamics, or vibrational properties of closed-shell systems. The energetics, however, particularly when assessing the thermodynamics of extended defective systems, are too poor, and more accurate quantum chemical methods are essential. Actually, as we discussed previously, using modern hybrid exchange–correlation functionals, such as B97-1, as used here, approach the limit of chemical accuracy where we calculate, for example, the energies of desorption from well characterised surfaces within 1 kcal mol^{-1} of their experimental values (see, *e.g.*, ref. 1).

1 R. Catlow, R. Bell, F. Cora, S. A. French, B. Slater and A. Sokol, *Annu. Rep. Prog. Chem., Sect. A*, 2005, **101**, 513.

Professor Islam said: In general, all the computational techniques (*e.g.*, atomistic simulation, molecular dynamics, *ab initio*, DFT type) have strengths and weaknesses depending on the material and property being investigated. There is a strong body of work showing that atomistic simulation (interatomic potential) methods are valuable in probing defects and atomic transport in solid state ionics, and that DFT-based methods are useful in examining the electronic structure of materials. It is worth stressing that experimental techniques (*e.g.*, diffraction, NMR, conductivity, EXAFS) also have their strengths and weaknesses.

Professor Navrotsky said: I agree. We need to apply all techniques—computational, thermodynamics, structural, transport—simultaneously to develop models and understanding. The whole will be greater than the sum of its parts

Dr Kohanoff commented: Current approximations in DFT are indeed good at describing structure and dynamics, but they still lack the dispersive, or van der Waals, component. The consequence is the underestimation (LDA) or overestimation (GGA) of lattice parameters. In usual ionic crystals this amounts to a few percent in the equilibrium value, but this can be enough to destabilise one phase in favour of another (ferroelectricity in perovskites). In molecular ionic crystals, where dispersion is more important, the differences are even larger. Unfortunately, there are no clear indications that an improved functional including van der Waals attraction is going to be available soon.

Mr Björketun replied: There is a joint Rutgers/Chalmers initiative aimed at incorporating van der Waals interactions into DFT. Recently these groups proposed a functional that consistently accounts for the dispersive interactions in systems of general geometry.[1] Although not implemented in any public DFT software, this functional has been applied to, among others, polyethylene crystals and dimers of polycyclic aromatic hydrocarbons. Not only does it lead to stable structures, but it also predicts crystal parameter values and binding energies in promising agreement with experimental data and values obtained with advanced quantum chemistry methods.

1   M. Dion, H. Rydberg, E. Schröder, D. C. Langreth and B. I. Lundqvist, *Phys. Rev. Lett.*, 2004, **92**, 246401.

Professor Navrotsky commented: Comparing DFT results with measured calorimetric data may help delineate how important these long range forces are in terms of energetics.

Mr Martin opened the discussion of Dr Wilson's paper: At greater radii does the inorganic nanotube wall fold underneath the wall already formed, and so form spiralling layers?

Dr Wilson answered: This has not been observed in any of our simulations to date. All of our INTs have a circular, or near-circular, cross-section. One possibility is that the sorts of spiral structures suggested may form in larger diameter carbon nanotubes (C-NT). Alternatively, improvements in the potential models, such as the inclusion of a flexible model for the C-NT, may promote these types of structure. It is also possible that systems with MX_2 stoichiometry which favour layered structures ($CdCl_2$, CdI_2 . . .) may form such tubular structures. These systems are known to form complex structures in the bulk, of the sort outlined in the question, which result from the layered nature of the bulk crystals.

Professor Heitjans asked: What is the meaning of the residual activation barrier height at large carbon nanotube diameters (Fig. 7a), which is constant and said to be equivalent to the barrier predicted from an unconfined inorganic nanotube? Why is it non-zero?

Dr Wilson responded: The diffusion pathway under investigation involves moving an ion pair along the [110] direction (Fig. 5 and 6). If we consider in detail the ion motion shown in Fig. 6, then the ion pair which is moved undergoes a change in coordination environment. Initially, both ions are three-coordinate. In between the start and end points these ions have two nearest-neighbour ions and two next-nearest-neighbour ions (which are the ions to which the "bonds" are being simultaneously created and destroyed). As a result, even in the absence of the C-NT, there is an activation barrier resulting from this change in ion coordination environment. The C-NT acts to compress the internal structure and so hampers ions relaxation along the diffusion pathway and, as a result, kinetically controls the formation of the internal INT structure.

Professor Islam asked:

(1) You presented the electron microscopy (HRTEM) of Malcolm Green showing inorganic (metal–anion) structures within carbon nanotubes. From your simulations, do you predict any unusual or exotic structures of MX_2 or MX_3 within carbon nanotubes that have now been observed by HRTEM (or other structural techniques)?

(2) I agree with your view that you cannot assume the same structures as in the bulk; as found from experimental and simulation studies of solid surfaces, the real surfaces cannot be assumed to be a simple termination of the bulk solid.

Dr Wilson replied:

(1) This is very much an on-going question. My understanding of the experimental techniques is that many filling events are observed from a single filling experiment. Carbon nanotube bundles (which have a significant dispersion in nanotube diameter) are inserted into the molten salt and the filled INT structures observed by HRTEM. Early work[1] focused on the INT structures *directly* related to known bulk structures essentially because the HRTEM images were readily interpretable. The simulation work indicates that these structures are, in fact, special cases within more general INT frameworks. Furthermore, the simulation models offer the possibility of predicting the HRTEM images,[2] which can aid experimental interpretation and also highlight the relative complexity of these images even for relatively simple non-bulk structures (*i.e.*, they appear difficult to interpret in the absence of model structures). More experimental work is now emerging in which the INT structures do not directly relate to bulk. In the MX_2 stoichiometry, for example, CdI_2 forms a "twisted" structure.[3] In the MX_3 stoichiometry current experimental observations indicate that the structures can all be understood in terms of sections of known bulk structures.[4] However, recent simulations show that subtle, non-bulk, distortions may be present.[5]

(2) Absolutely. These structures can be viewed in very much the same way—single-walled INTs are essentially all surface (basically a folded surface). Clearly, one expects a strong thermodynamic connection between the structures formed by the INTs and the bulk crystals but factors associated with the low-dimensional confining environment have a significant effect on the underlying thermodynamics.[6]

1 R. R. Meyer, J. Sloan, R. E. Dunin-Borkowski, A. I. Kirkland, M. C. Novotny, S. R. Bailey, J. L. Hutchison and M. L. Green, *Science*, 2000, **289**, 1324.
2 M. Wilson, *Chem. Phys. Lett.*, 2002, **366**, 504.
3 E. Philp, J. Sloan, A. I. Kirkland, R. R. Meyer, S. Friedrichs, J. L. Hutchison and M. L. H. Green, *Nat. Mater.*, 2003, **2**, 788.
4 S. Friedrichs and M. L. H. Green, *Z. Metallkd.*, 2005, **96**, 419.
5 M. Wilson and S. Friedrichs, *Acta Crystallogr., Sect. A*, 2006, **62**, 287.
6 M. Wilson, *J. Chem. Phys.*, 2006, **124**, 124706.

Dr Zhukovskii asked:

(1) Do you consider carbon nanotubes to be rigid structures in your model of inorganic ionic transport? Could you expand your model to allow C-NTs to be more flexible?

(2) What chirality of C-NT is more appropriate for ionic transport: I (armchair) or II (zigzag-like)?

Dr Wilson answered:

(1) The C-NT is, indeed, considered as fixed and atomistic. The original remit was to determine whether the basic C-NT filling could be modelled with a relatively simple model. The effect of including C-NT flexibility (by using existing potential models for the nanotubes) is currently under investigation.[1] A question to answer, for example, concerns the role (if any) of the C-NT vibrational modes on the filling mechanism. We anticipate additional development of the potential models, for example, to incorporate the effect of the metallic or semiconducting nature of the C-NT.

(2) Within our simple model the typical rates of ion diffusion appear insensitive to the morphology of the confining nanotube. I suspect that a more flexible model, in which the C-NT vibrational modes are included, may highlight differences in morphology.

1 C. L. Bishop and M. Wilson, in preparation.

Professor Bruce asked: Have you considered using inorganic nanotubes as host structures? The polar surfaces should exhibit more structure directing properties for the growth of what are polar ionic crystals.

Dr Wilson replied: Not as yet! I agree that polar enclosing structures exhibit a greater influence on the filled structure. Some clues are present in the present simulation work. Recent work shows that, within a specific range of carbon nanotube diameters, double-walled INTs form.[1] In these simulations the inner INT forms after the outer and, as a result, any correlation between the inner and outer INT structures may uncover any role of the outer tube morphology in determining the structure of the inner INT. Work is on-going to uncover any such correlations.

1 M. Wilson, *J. Chem. Phys.*, 2006, **124**, 124706.

Professor Maier opened the discussion of Dr Gray-Weale's paper: How far is it consistent to confine the cube-root law that we promoted[1] for the description of bulk defect interaction with space charge concepts (in which one carrier is depleted, the other enriched). What about gradient effects? Given the validity of such an approach, what are the results for screening length or possible phase transformations at interfaces?

1 N. Hainovsky and J. Maier, *Phys. Rev. B*, 1995, **51**, 15789.

Dr Gray-Weale answered: One aspect of my work is a slight generalisation of the cube-root and other models to handle enrichment of one carrier and depletion of the other. I did this by requiring that the models' predictions were consistent in the limit of equal carrier concentrations. This, I think, makes my treatment consistent but does not prove the cube-root law works for interactions in space-charge layers.

One intriguing possibility is that the interaction described by the cube-root law is connected to relaxation of the density, and to the strain field. My calculations point to this, in giving similar modifications to the screening length by the cube-root interaction and by coupling defect creation to strain (see eqn (18) and (21)). More work is needed, but this suggests that simply by including strain, a superionic transition in a space-charge layer could be studied.

I note also that the screening length is a bulk property, and so interactions specific to the bulk may well be used to calculate it. To then use it to estimate the size of a space-charge layer is to treat the layer as a perturbed region of the bulk crystal.

Dr Gray-Weale then asked Professor Maier: Do you think it possible that in some cases at least, the cube-root interaction is describing a coupling of density to defect concentration?

Professor Maier answered: The cube-root law seems to come out as an approximation of an accurate description of electrolytic interactions.[1] Yet, an interpretation is difficult here. A reasonable explanation is through an effective mean distance between oppositely charged carriers, *i.e.*, in the simplest case through an effective Madelung superlattice of defects as a mean-field model,[2] which then becomes denser the higher the concentration.

1 A. R. Allnatt and M. H. Cohen, *J. Chem. Phys.*, 1964, **40**, 1860; A. R. Allnatt and M. H. Cohen, *J. Chem. Phys.*, 1964, **40**, 1871.
2 N. Hainovsky and J. Maier, *Phys. Rev. B*, 1995, **51**, 15789.

Professor Haile addressed Dr Gray-Weale: Your model is based on two lattices in which occupancies vary with temperature, but there is no change in the lattices themselves. How well would this model do in terms of describing a system in which there is a first order transition from, say, a monoclinic to a cubic phase?

Dr Gray-Weale answered: I have made no effort to describe first-order structural transitions, like the change from a cubic to a monoclinic phase. Such a change would be much harder to handle, and the model is really only intended to capture the various influences on the screening length within a given structure, including doping, strain and interaction. My approach also extracts more predictions from a mean-field model than are usual, so as to put it to harsher test against experiment.

In order to make comparisons for structurally different phases, a different set of parameters would be needed for each phase. You could apply my treatment as well to the monoclinic phase to the cubic, but it would not tell you anything about the transition between the two, only about the different character of charge correlations and screening in the two phases.

Professor Nazar opened a general discussion by addressing Dr Wilson: Are there examples of highly complex lattice structures that you have investigated in which crystallization within the nanotube fails to occur?

Dr Wilson replied: Not as yet. Our investigations to date have been limited to relatively small diameter C-NTs and relatively simple filling materials (two chemical species only). Both of these "constraints" would be expected to promote the formation of crystalline internal structures. In small diameter C-NTs, for example, the strong ion–ion interactions, coupled with the strongly confining environment, essentially force the formation of highly ordered structures.[1] As the C-NT diameter increases the confining effect is reduced and, since the whole system is above the bulk salt melting point, there must be a critical C-NT diameter above which the internal structure is liquid-like (with some structure owing to perturbation by the C-NT walls). In addition, it is possible that systems with more complex stoichiometries, in which bulk crystallisation is relatively sluggish, may favour the formation of internal disordered structures. Both areas are currently under investigation.

1 M. Wilson, *Chem. Phys. Lett.*, 2004, **397**, 340.

Professor Bruce asked:

(1) Why do the inorganic structures not grow as scrolls? I can understand that wetting of the C surface nucleates the process but given that the atomic spacing is unlikely to be an integral number of carbon tube diameter there must be strain where the ionic tube closes unless it scrolls.

(2) Are the inorganic structures defect free?

Dr Wilson replied:

(1) This links directly with the answer I gave to Mr Martin. The lack of scrolling structures in the INTs observed in the simulations can be understood in terms of the number of possible structures which the INTs may form. The critical factor (in these simple models) is the difference between the C-NT diameter and that if forming INT (*i.e.* the INT forms essentially by internally wetting the C-NT). If we consider folding either a hexagonal or square-net sheet then a wide range of INT diameters are readily accessible simply by folding along different chiral vectors (Fig. 1). Furthermore, the high temperatures at which these systems form (above the bulk salt melting point) means that these INT structures are kinetically accessible.

(2) To date the INTs formed are defect free. We can account for this in two ways. Firstly, the high temperatures at which the INTs form means that the ionic network is flexible enough to remove potential defects. Secondly, the heavily-confined environment may effectively increase the defect formation energies by stabilising the ideal crystal structures.

Professor Navrotsky asked: Is there any evidence of discontinuity or stepwise formation of the concentric nanotubes, in terms of rates of formation?

Dr Wilson answered: Simulation studies show that, over a specific range of C-NT diameters, double-walled INT structures are formed. Fig. 1 shows the number of ions inside a nanotube (a (14,10) C-NT, diameter 16.4 Å) as a function of time during a typical filling simulation. The first plateau corresponds to the almost-complete formation of the outer INT and the second to the formation of the overall double-walled structure. Thus, for relatively short sections of C-NT filled in these simulations, the outer walls form prior to the inner (each INT effectively forms by wetting the outer tube). One might imagine that, as the C-NT being filled is made longer, the inner wall may form behind the outer with the latter not needing to be complete.

Professor Maier asked: How reliable is modelling in such highly interfacial systems with varying chemical bond situations using empirical potentials of constant functionality?

Dr Wilson responded: The models chosen for our studies to date have been deliberately kept simple. The original simulation work focused on KI[1] as this was the system that had been most investigated experimentally and represented a system for which highly reliable potential models were available. Since then, the systems studied have fallen into two groups. Firstly, specific systems (*i.e.* LaCl$_3$[2]) have been pursued in order to promote on-going collaborations with experimental investigations. Secondly, models have been investigated in which the underlying potential parameters simply produce potentially interesting structures. For example, the INT structures which form based on a percolating hexagonal motif are predicted using a general ionic model in which the ion radius ratio is chosen so as to promote the formation of a four-coordinate crystal structure.

Overall, therefore, the models act to highlight possible INT structures and, to date, have been highly successful when considering specific systems. However, these systems have exclusively been those considered as highly ionic and, as a result, the

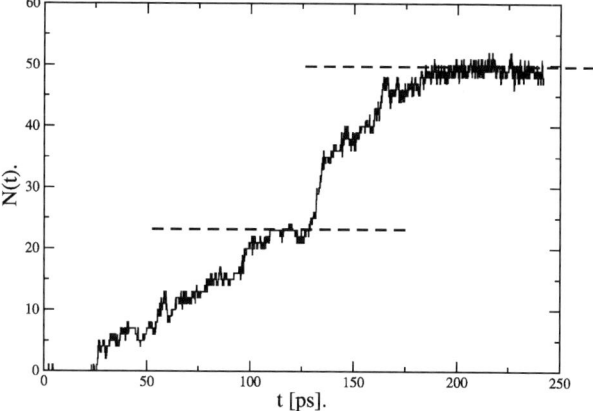

Fig. 1 The number of ions inside a nanotube (a (14,10) C-NT, diameter 16.4 Å) as a function of time during a typical filling simulation.

systems for which the bulk models are most appropriate. It remains to be seen if the models are as successful when considering systems in which additional, non-ionic, bonding components may be present.

1   M. Wilson and P. A. Madden, *J. Am. Chem. Soc.*, 2001, **123**, 2101.
2   M. Wilson and S. Friedrichs, *Acta. Crystallogr., Sect. A*, 2006, **62**, 287.

Professor Kilner commented: If the computational methods are good at highlighting trends in materials properties then a very useful application would be a sort of computational combinatorial chemistry to predict compositions with interesting properties.

Professor Nazar added: It seems though that experiment tends to drive theory at present, rather than the other way around.

Dr Wilson replied: It is important to stress that simulations should not be seen as an alternative to experimental investigation. Experiments remain very important for determining material properties. A significant question concerns the relationship between experiment and simulation. Traditionally, simulations operate in a responsive mode (in that they are driven by experimental observation). However, a more cost-effective approach is for simulations to behave in a predictive mode in which properties of interest are identified prior to experimental investigation. The limiting factor to this approach appears to be the construction of truly transferable potential models, in which parameter sets retain a physical meaning and may transmute between systems in a physically transparent manner. The construction of such potentials remains the 'holy grail'.

Professor Chadwick asked: You are probably aware of the work of Nick Quirke at Imperial College, London, who found that van der Waals liquids, like argon, fill nanotubes at supersonic speeds. Is this also true for ionic melts?

Dr Wilson answered: In the simulations performed to date, once the ions begin to enter the C-NT, then the filling proceeds at high filling velocities. What is not known, however, is if these velocities remain high when filling significantly longer nanotubes. As ever, we are limited by both the time- and length-scales accessible to atomistic computer simulation.

Professor Islam asked: Do you find any unusual ion transport mechanism down the carbon nanotubes that is different from the bulk? I presume the long-range electrostatic terms will be different and hence modify the potential energy surface.

Dr Wilson replied: Our investigations of the transport mechanisms are at the early stages. If we consider the ion movement discussed in the paper (of an ion pair in a $(2,2)_{sq}$ INT) then differences with respect to the bulk are expected. In this case the ions that move are charge neutral pairs promoted by the relatively low activation barrier associated with this process. The relatively facile nature of this diffusion pathway is critically related to the low dimensionality of the INT. In a bulk environment such a motion would require the breaking of more bonds and, as a result, have a significantly higher activation barrier. A critical factor has to be the role of the C-NT in effectively disconnecting the internal structure from the outer system (the molten salt). This can be rationalised in several ways, one of which is in terms of the underlying potential energy surface. Recent work,[1] for example, shows how the C-NTs can be considered as acting as an "energy landscape filter".

1 M. Wilson, *J. Chem. Phys.*, 2006, **124**, 124706.

Professor Haile asked: Is this a system in which one might observe significant nonstoichiometry with electrostatic charge balance being provided by the carbon nanotube? This would be the kind of phenomenon described earlier by Professor Maier in his discussion of excess mass storage in nano systems.

Dr Wilson answered: Quite possibly. As stressed in the paper and in answers to previous questions, the present models are deliberately simple. As a result, at present, there is no possibility of charge transfer between the filling material and the carbon nanotube. One possibility is to perform high level electronic structure calculations in order to assess the magnitude of any such transfer. These calculations could form a symbiotic relationship with the simpler models, which could provide test configurations for analysis. Furthermore, extended models, in which ion–carbon charge transfer is incorporated, could be developed by reference to the electronic structure calculation.

Professor Nazar asked: My recollection is that oxides such as MoO_3 and V_2O_5 do not completely fill carbon nanotubes (although I may recall this incorrectly). If so, however, it might be expected due to the weaker interaction of the oxide with the carbon surface that would result in less "wetting". Can your calculations take this into account?

Dr Wilson replied: My recollection is that some oxides have been made to fill the nanotubes. To date, NiO, PbO, Bi_2O_3, V_2O_5 and MoO_3 have been inserted. I agree that the subtle differences in ion–carbon interactions may have profound effect upon the formation thermodynamics and hence lead to different filling properties. We do not, as yet, have any insight from the simulation models.

A self-consistent mean field theory for diffusion in alloys

Maylise Nastar* and Vincent Barbe

Received 24th April 2006, Accepted 9th May 2006
First published as an Advance Article on the web 24th July 2006
DOI: 10.1039/b605834e

Starting from a microscopic model of the atomic transport *via* vacancies
and interstitials in alloys, a self-consistent mean field (SCMF) kinetic theory
yields the phenomenological coefficients L_{ij}. In this theory, kinetic
correlations are accounted for through a set of effective interactions within
a non-equilibrium distribution function of the system. The introduction of a
master equation describing the evolution with time of the distribution
function and its moments leads to general self-consistent kinetic equations.
The L_{ij} of a face centered cubic alloy are calculated using the kinetic
equations of Nastar (M. Nastar, *Philos. Mag.*, 2005, **85**, 3767, ref. 1)
derived from a microscopic broken bond model of the vacancy jump
frequency. A first approximation leads to an analytical expression of the L_{ij}
and a second approximation to a better agreement with the Monte Carlo
simulations. A change of sign of the L_{ij} is studied as a function of the
microscopic parameters of the jump frequency. The L_{ij} of a cubic centered
alloy obtained for the complex diffusion mechanism of the dumbbell
configuration of the interstitial (V. Barbe and M. Nastar, *Philos. Mag.*,
2006, in press, ref. 2) are used to study the effect of an on-site rotation of
the dumbbell on the transport.

1 Introduction

When an alloy is irradiated with particles, the atomic transport is not only supported
by the vacancy mechanism but also by the interstitials, both concentrations being
fixed by the irradiation conditions. The estimation of the transport coefficients
associated with both mechanisms was therefore intensively studied in the past
decades, mainly to predict the variation of atomic fluxes and point defect fluxes
with the local concentration and temperature.[3]

Although there is an experimental procedure to determine the L_{ij}, experiments are
usually performed on non-irradiated alloys without interstitial defects, at high
temperature, for a few compositions, and only some of the total set of the transport
coefficients characterizing the diffusion properties of an alloy are measured. The link
between these partial data and the complete set of the L_{ij} is not obvious. The
previous attempts relating the diffusion coefficients of isotopes with the L_{ij} turn out
to be less and less valid as the Monte Carlo simulations become a more reliable test.[4]
Moreover relations like the famous Manning's ones[5] never predict a negative sign of

*CEA Saclay, Service de Recherches de Métallurgie Physique, bât. 520, 91 191, Gif-sur-Yvette,
France. E-mail: maylise.nastar@cea.fr; Fax: 331 6908 6867; Tel: 331 6908 2767*

an off-diagonal L_{ij} although it has been observed between some solutes and the solvent atoms in aluminium alloys.[6] It is a case where a rough estimation of the L_{ij} may affect not only the rate of reaction but also the path of reaction. For the interstitial diffusion, with the exception of a few values of effective migration energies deduced from resistivity measurements at low temprerature there are almost no experimental data.

On the other hand, *ab initio* methods are now able to provide accurate values of jump frequencies in pure and dilute binary alloys not only for the vacancy diffusion mechanism[7] but also for the intertitials in the dumbbell configuration.[8,9] Therefore an appropriate solution to estimate the L_{ij} is to start from an atomic diffusion model for which the parameters are fitted on *ab initio* calculations and/or the available partial diffusion data completed by thermodynamic data.[10,11] The time evolution of the system is then described by a master equation. Either the L_{ij} are extracted from the equilibrium fluctuations[12] or, like within the self-consistent mean field (SCMF) theories[1,2,13–15] and the path probability method (PPM)[16,17] they are estimated for a system subject to an external force. Apart from the SCMF theory, the methods that take into account short-range order explicitly in a concentrated alloy with vacancy transport use the jump frequency model of the PPM and derive approximate expressions for the correlation coefficients, but the different versions of the PPM were limited to body centered cubic (bcc) alloys.[16,17] Although the model of Stolwijk[18] was verified to be in excellent agreement with the Monte Carlo simulations in bcc binary alloys, it was restricted to the calculation of tracer coefficients. The later model of Qin, Allnatt and Allnatt[19] was, on the contrary, limited to the Onsager matrix of a binary cubic alloy.

The SCMF theory is based on a more recent atomic diffusion model introduced by Martin;[20] extended by Nastar *et al.*,[21] Athènes and Bellon[22] and Lebouar and Soisson,[23] including a variation of the saddle point energy as well as equilibrium energy with the surrounding nearest neighbours (nn) of the exchanging vacancy-atom pair; and adapted to the complex interstitial diffusion mechanism[2] both in bcc and face centered cubic (fcc) alloys. The SCMF theory based on a different formalism leads to kinetic equations very similar to the PPM ones but opposite to the PPM it distinguishes two independent hierarchies of approximation. When the PPM considers a statistical pair approximation both for the equilibrium averages and for the calculation of the most probable path between two successive times, the SCMF theory uses a statistical point approximation for the equilibrium averages and a pair approximation for the kinetic correlations[2,13] or a higher approximation with N-body effective interactions (N > 1) to study percolation effects in alloys with high jump frequency ratios.[15] It was also extended[1] to a pair approximation for both the equilibrium averages and the kinetic correlations to obtain the same L_{ij} as the multi-frequency ones available in dilute binary alloys[24] and to suggest a new formulation of the effect of solutes on the self-diffusion coefficient of the solvent tracer. This flexibility allows us to tackle complex diffusion mechanisms and allows this to be the first theory to propose a calculation of the transport coefficients for the dumbbell mechanism in bcc concentrated alloys.[2]

We present here a new extension of the SCMF theory devoted to the calculation of the L_{ij} of a concentrated fcc alloy based on a pair approximation for the equilibrium averages and the correlation part. Attention will be paid to the possible change of sign of the off-diagonal coefficients with respect to the composition and atomic jump frequencies. We will also use the kinetic equations of ref. 2 to study the effect of the on-site rotation mechanism of the dumbbell on the L_{ij}. Both studies will be supported by exact results of Monte Carlo simulations run with the same atomic diffusion models.

After a short introduction of the diffusion atomic models and the principles of the SCMF theory, we provide an analytical formulation of the L_{ij} for the vacancy and the dumbbell and finish with a discussion.

2   The self-consistent mean field theory

2A   Diffusion models of vacancies and interstitials

2A.1   Vacancy jump frequency. The exchange frequency between a vacancy and an atom has the classical thermally activated form

$$W_\alpha = \nu_\alpha \exp(-E_\alpha^{act}/k_B T), \tag{1}$$

where k_B is the Boltzmann's constant and T the temperature. The term ν_α is the attempt frequency which is assumed to depend only on the jumping atom, and the term in the exponential is the 'migration enthalpy', which is the difference between the total energy of the system in the initial configuration and when the jumping atom is at the saddle point. If we assume that both energies follow a broken bond model limited to nn pair interactions, the difference depends on the chemical species of the jumping and surrounding atoms

$$E_\alpha^{act} = \frac{1}{k_B T} \left(\sum_{k,\beta} \gamma_{i'k}^s V_{\alpha\beta}^s n_k^\beta - \sum_{k,\beta} \gamma_{ik} V_{\alpha\beta} n_k^\beta \right), \tag{2}$$

where the first term in the exponential corresponds to the new binding energies created by the atom at the saddle point and the second term to the binding energies to be cut by the exchanging species at the initial state. The term $\gamma_{i'k}^s$ is equal to 1 if site k is a nn site of atom α at the saddle point and zero if else; and γ_{ik} is equal to 1 if lattice sites i and k are nn. Occupation of a site i is specified by a set of occupation numbers $(n_i^\alpha)_\alpha$: n_i^α is equal to one if site i is occupied by α and zero if else. Note that the mean occupancy of site i by species α is the ensemble average $\langle n_i^\alpha \rangle$. The equilibrium interactions $V_{\alpha\beta}$ are adjusted so as to reproduce thermodynamic properties of the system whereas the other parameters are fitted to reproduce diffusion coefficients. This type of model was applied to real systems like the austenitic steels[21] and ferritic steels.[23] On the other hand, *ab initio* calculations are now able to compute atomic jump frequencies, although mainly in dilute alloys.[7-9] A combination of both approaches can also be used, in a similar way to the multiscale modelling of precipitation in aluminium alloys.[10,11]

2A.2   Interstitial jump frequency. In most bcc alloys, the most stable configuration of the dumbbell is along the $\langle 110 \rangle$ direction, although the $\langle 111 \rangle$ direction is also possible for mixed dumbbells in alloys.[8,9] A dumbbell can rotate on its site or jump. In a jump, one atom, say B of the dumbbell AB migrates to a nn 'target site' and forms a new dumbbell with the species C which was already on this site ('target atom'). The remaining atom A of the initial dumbbell is then left in a substitutional position

$$AB + C \rightarrow A + BC \tag{3}$$

The associated jump frequency has the form $\omega_{AB/C}$. Unlike the vacancy model, the variation of jump frequencies with local composition is not taken into account. It has been looked at in the case of dilute alloys.[25] The present study is rather exploring the competition between the rotation and jumps of a dumbbell and its effect on the L_{ij}. It is restricted to the dumbbell with a $\langle 110 \rangle$ direction, which has also been the most extensively studied up to now. In this system, three jump mechanisms have been considered:[2] an on-site rotation R of 60° to another $\langle 110 \rangle$ direction with four possible final orientations; a translation T towards a nn target site; a translation and rotation of 60° towards the same nn target site (RT) with two equivalent final orientations (see Fig. 1).

In terms of jump frequencies, we differentiate the T and RT mechanisms by introducing a factor τ as:

$$\omega_{AB/C}^T = \tau \omega_{AB/C}^{RT} \tag{4}$$

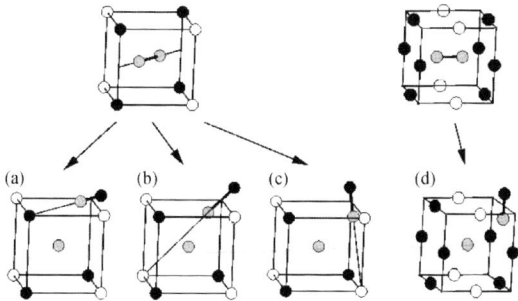

Fig. 1 Top: the geometries of the dumbbell in a bcc structure (left) and in a fcc structure (right). The nn sites of the dumbbell (represented in grey) are separated in target sites of one of the atoms forming the dumbbell (in black) and non-target sites (in white). Bottom: the final configurations after a jump of the dumbbell toward a target site, (a) by translation (bcc); (b–c) by rotation–translation (bcc); (d) by rotation–translation (fcc).

The limit $\tau = 0$ is equivalent to neglecting the simple translation, and $\tau = 1$ sets a degeneracy between the three possible nn jumps. The specificity of the bcc structure compared to the fcc structure is the possibility for a dumbbell to jump toward a same target site in two or three different jumps with two or three final orientations (Fig. 1). It is because of this complexity that Bocquet[26] was not able to derive the L_{ij} of a bcc alloy as he did for an fcc alloy.[27]

2B Expression of the L_{ij}

Under a gradient of chemical potential constant in time there is no variation of mean occupations but permanent fluxes of matter. Within the theory of linear non-equilibrium thermodynamics, the L_{ij} are defined as the constants coupling the fluxes to the external forces which here will be the gradients of chemical potentials. Following Vaks,[28] the SCMF theory introduces an effective Hamiltonian to calculate the probability of non-equilibrium configurations. One introduces then a master equation traducing the fact that transitions between a given configuration and the others are controlled by the vacancy or the interstitial jump frequencies. One obtains eventually a series of kinetic equations which describe the time evolution of mean occupancies of one chemical species on one site, two chemical species on a pair of sites, *etc.* The pair effective interactions (the N-body interactions) are calculated in such a way as to guarantee a stationary state of the mean occupations of pairs of sites (multiplets of N sites). This effective Hamiltonian represents the correlation contribution to the L_{ij}. The level of approximation of the SCMF theory is fixed by two hierarchies. The first concerns the statistical approximation used to describe the equilibrium correlations between the atoms, and the second is fixed by the number of effective interactions used to describe the kinetic correlations. The range of these latter interactions is controlling the level of description of the diffusion paths offered to a point defect after a first jump.

The last upgrade of the SCMF theory for alloys with vacancy transport is based on a statistical pair approximation and introduces pair effective interactions.[1] In the case of a dilute binary alloy it converges to the multi-frequency models of dilute alloys.[1] An extension of the model to concentrated alloys is presented below. The same theory was adapted to the interstitial diffusion mechanism.[2] Under the judicious assumption that the effective interactions are limited to pair interactions between the two atoms forming the dumbbell,[2] this theory provides simple analytical expressions of the L_{ij} (first shell approximation) that we recall below. A more complex calculation (second shell approximation) also considering pair effective

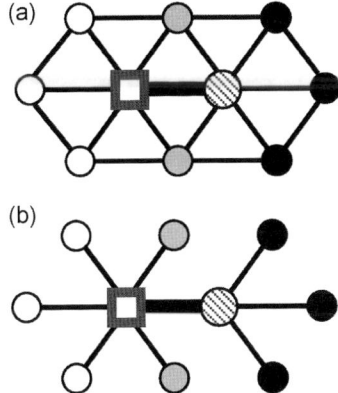

Fig. 2 A (111) projection of the 'cocoon' made up of the exchanging pair (the vacancy is represented by a square and the exchanging atom by a hatched disc) and its nn sites (the nn sites of the vacancy are represented by white discs and the nn of the exchanging atom by grey discs). Lines in black represent in (a) all the nn pairs and in (b) the ones that are considered in the calculation of the 'cocoon' probability in terms of pair probabilities.

interactions between nn sites is used to compare with new Monte Carlo simulations including an on-site rotation mechanism.

2B.1 Transport *via* vacancy. In ref. 1 general kinetic equations were written with an effective Hamiltonian restricted to pair interactions. The complete derivation of the L_{ij} was achieved in the case of a dilute fcc alloy. The same kinetic equations are used here to extend the calculation of the L_{ij} to fcc concentrated alloys. When only nn pair effective interactions are considered, they lead to analytical expressions of the L_{ij}. This first approximation is called the first shell approximation. The second shell approximation considers nn sites and nn of nn sites. Both approximations are plotted in order to show that the successive approximations of the SCMF converge rather well to the exact results of the Monte Carlo simulations.

In a fcc lattice with nn interactions, around an exchanging vacancy–atom pair, there are five categories of nn pairs (see Fig. 2):

1. the exchanging pair $\alpha \rightarrow v$,
2. the seven pairs connecting α to species on its nn sites,
3. the seven pairs connecting the vacancy to species on its nn sites,
4. the four pairs connecting the vacancy to a species on a nn site of both the vacancy and α,
5. the pairs connecting two nn sites of the exchanging pairs.

A determination of the L_{ij} requires the calculation of equilibrium averages of a jump frequency multiplied by the occupation numbers of the exchanging pair and in some cases by one or two occupation numbers of their nn sites.[1] A jump frequency depends upon the configuration formed by the exchanging pair on sites i and j and the pairs of categories 2 to 4. An average of a jump frequency then requires the calculation of the probability of every configuration and to allocate to it the corresponding value of the jump frequency. An estimation of the probability of a configuration of such a large cluster, also called a 'cocoon', is beyond the usual applications of the cluster variation method (CVM) (see ref. 16 and references therein). In the case of a Bethe type lattice, *e.g.* one not containing closed circuits of nn sites, it is possible to deduce a total density probability from the partial pair densities. It is simply equal to the product of pair densities comprised in the system.[29] The present 'cocoon' is not a Bethe lattice since it contains triangles of nn sites. It is approximated by a Bethe lattice by neglecting some nn pairs (*cf.* Fig. 2). For example, a probability of exchange is then decomposed into a product of pair terms of the categories 1–4:

$$L_{\alpha\alpha}^{(0)} = \left\langle \omega_{ij}^{\alpha\,v} n_i^\alpha n_j^v \right\rangle^{(0)} = \nu_\alpha s_\alpha \, {}_v c_\alpha \left(\sum_\beta s_{\alpha\,\beta} \right)^7 \left(\sum_\beta t_{v\beta} \right)^4 \left(\sum_\beta s_{v\beta} \right)^7, \qquad (5)$$

with

$$s_{\alpha\beta} = p_{\alpha\beta} c_\beta \exp\{V_{\alpha\beta}/k_B T\} \text{ and } t_{v\beta} = p_{v\beta} c_\beta \exp\{(V_{\alpha\beta} + V_{v\beta} - V_{\alpha\beta}^{(s)})/k_B T\}, \qquad (6)$$

where $s_{\alpha\beta}$ (resp. $s_{v\beta}$) is an 'averaged pair' of category 2 (resp. 3) made up of the exchanging atom α on site i and a species β on one of the seven nn sites of site i (resp. j) for which the corresponding interaction to be cut is $V_{\alpha\beta}$ (resp. $V_{v\beta}$) and the probability of a nn site of i (resp. j) to be occupied by species β is $p_{\alpha\beta} c_\beta$ (resp. $p_{v\beta} c_\beta$). The latter is calculated using a statistical pair approximation of the CVM.[16]

Neglecting the pairs of category 5 at the periphery of the 'cocoon' in eqn (5) is not a drastic approximation. Indeed, jump frequencies do not depend on these pairs, nor on the relative positions of the other pairs, but on the global occupation of pairs of category 2, 3 and 4, so that pairs at the periphery are summed over several configurations and should not play the same role as that of the other pairs. Note that the bcc diffusion model of Kikuchi skipped such interactions too.[16,17] A more serious approximation specific to the fcc structure concerns the pairs of category 4 forming a triangle of nn sites with the sites of the exchanging pair. This type of triangle is treated as if it was an open three-point circuit with the pair of category 4 transformed into a pair of category 3 (vβ). Such an 'approximate pair' (vβ) is represented by $t_{v\beta}$ in eqn (5). Note that although the nn distance between α and β is neglected in the probability term it is considered in the broken bond model of the jump frequency both at the saddle and initial position ($V_{\alpha\beta}$ and $V_{\alpha\beta}^{(s)}$ of eqn (6)). The resulting probability of the 'cocoon' is written in eqn (5) as a product of the selected pair probabilities illustrated in Fig. 2. Like Kikuchi et al.[16,17] and Stolwijk[18] the pairs are grouped by category and the number of equivalent configurations are directly calculated by the sum over pairs raised at the power of the number of pairs in the category.

The kinetic equations of ref. 1 involve other averages of jump frequency. For example in fluxes some jump frequencies are multiplied by an extra occupation number on a nn site s of the exchanging atom or on a nn site t of the exchanging pair. The corresponding average is easily deduced from the average jump frequency defined in eqn (5)

$$\left\langle \omega_{ij}^{Av} n_i^A n_j^v n_s^B \right\rangle^{(0)} = s_{AB} \Omega_A^s \text{ with } \Omega_A^s = \left\langle \omega_{ij}^{Av} n_i^A n_j^v \right\rangle^{(0)} \Big/ \left(\sum_\beta s_{A\beta} \right),$$

$$\left\langle \omega_{ij}^{Av} n_i^A n_j^v n_t^B \right\rangle^{(0)} = t_{vB} \Omega_A^t \text{ with } \Omega_A^t = \left\langle \omega_{ij}^{Av} n_i^A n_j^v \right\rangle^{(0)} \Big/ \left(\sum_\beta t_{v\beta} \right), \qquad (7)$$

where the average jump frequency is divided by an averaged pair and multiplied by the true occupation of the pair built from the occupation number of site s or t. The next kinetic equations associated with pair probabilities involve averages of jump frequencies multiplied by two additional occupation numbers:

$$\left\langle \omega_{ij}^{Av} n_i^A n_j^v n_{s1}^B n_{s2}^C \right\rangle^{(0)} = s_{AB} s_{AC} \Omega_A^{ss} \text{ with } \Omega_A^{ss} = \left\langle \omega_{ij}^{Av} n_i^A n_j^v \right\rangle^{(0)} \Big/ \left(\sum_\beta s_{A\beta} \sum_\beta s_{A\beta} \right),$$

$$\left\langle \omega_{ij}^{Av} n_i^A n_j^v n_t^B n_s^C \right\rangle^{(0)} = t_{vB} s_{AC} \Omega_A^{ts} \text{ with } \Omega_A^{ts} = \left\langle \omega_{ij}^{Av} n_i^A n_j^v \right\rangle^{(0)} \Big/ \left(\sum_\beta t_{v\beta} \sum_\beta s_{A\beta} \right).$$

$$(8)$$

This journal is © The Royal Society of Chemistry 2006

Because the detailed balance is satisfied, an average including a frequency multiplied by an occupation number of a nn site of the vacancy is found to be equivalent to averages of the type of eqn (7) and (8). Thus, it is easy to replace the averages of the kinetic equations (24)–(26) of ref. 1 by expressions of eqn (5), (7) and (8) and by means of eqn (21) of ref. 1 to derive analytical expressions of the L_{ij} as a function of composition in a binary alloy AB:

$$
\begin{aligned}
L_{\mathrm{AA}} &= L_{\mathrm{AA}}^{(0)} - 2(-3s_{\mathrm{AB}}\Omega_{\mathrm{A}}^{s} + 2t_{\mathrm{AB}}\Omega_{\mathrm{A}}^{t})^{2}/D, \\
L_{\mathrm{AB}} &= 2(-3s_{\mathrm{AB}}\Omega_{\mathrm{A}}^{s} + 2t_{\mathrm{AB}}\Omega_{\mathrm{A}}^{t})(-3s_{\mathrm{BA}}\Omega_{\mathrm{B}}^{s} + 2t_{\mathrm{BA}}\Omega_{\mathrm{B}}^{t})/D,
\end{aligned}
\tag{9}
$$

where

$$
\begin{aligned}
D = & [11s_{\mathrm{AB}}{}^{2}\Omega_{\mathrm{A}}^{ss} - 12s_{\mathrm{AB}}\,t_{\mathrm{AB}}\Omega_{\mathrm{A}}^{st} + 6t_{\mathrm{AB}}{}^{2}\Omega_{\mathrm{A}}^{tt}] + [7s_{\mathrm{AB}}\Omega_{\mathrm{A}}^{s} + 2t_{\mathrm{AB}}\Omega_{\mathrm{A}}^{t}] \\
& + [11s_{\mathrm{BA}}{}^{2}\Omega_{\mathrm{B}}^{ss} - 12s_{\mathrm{BA}}\,t_{\mathrm{BA}}\Omega_{\mathrm{B}}^{st} + 6t_{\mathrm{BA}}{}^{2}\Omega_{\mathrm{B}}^{tt}] + [7s_{\mathrm{BA}}\Omega_{\mathrm{B}}^{s} + 2t_{\mathrm{BA}}\Omega_{\mathrm{B}}^{t}]
\end{aligned}
\tag{10}
$$

The coefficient L_{BB} is obtained by inverting A and B in the above expressions.

To study the sign of L_{AB} as a function of concentration it is convenient to use the notations of the CVM and express the probability of a pair $(\alpha\beta)$ as a product of three contributions:

$$
p_{\alpha\beta}c_{\alpha}c_{\beta} = q_{\alpha}q_{\beta}K_{\alpha\beta} \quad \text{with} \quad K_{\alpha\beta} = \exp\{-V_{\alpha\beta}/k_{\mathrm{B}}T\},
\tag{11}
$$

If the vacancy concentration is considered to be negligible with respect to the concentrations of A and B, q_{α} and q_{β} are solution of the following equation:[16]

$$
\begin{aligned}
& K_{\mathrm{BB}}^{2}\,K_{\mathrm{AA}}\,q_{\mathrm{B}}^{4} - (c_{\mathrm{A}}K_{\mathrm{AB}}^{2} - 2c_{\mathrm{B}}K_{\mathrm{AA}}K_{\mathrm{BB}} - K_{\mathrm{BB}})q_{\mathrm{B}}^{2} + c_{\mathrm{B}}(1 + K_{\mathrm{AA}}c_{\mathrm{B}}) = 0, \\
& q_{\mathrm{A}} = (c_{\mathrm{B}} - q_{\mathrm{B}}^{2}\,K_{\mathrm{BB}})/(q_{\mathrm{B}}\,K_{\mathrm{AB}}),
\end{aligned}
\tag{12}
$$

To conclude, we propose for the first time a diffusion model for the fcc alloys that is analytical within this first shell approximation. Note that these expressions converge to the first shell approximation of the five frequency model of Allnatt and Lidiard[24] in the limiting case of the dilute alloy: a small amount of B in A and *vice versa*. The same model has been extended to the second shell approximation but details of the calculation will be given in a future publication.

2B.2 Transport *via* interstitial. The SCMF has been adapted to calculate the transport coefficients in a concentrated alloy for diffusion by the dumbbell mechanism. The calculation was performed in both fcc and bcc multicomponent alloys for simple sets of jump frequencies. Under the assumption of an effective Hamiltonian limited to pair interactions between the two atoms forming the dumbbell, the phenomenological coefficients in a binary concentrated bcc alloys simply writes:[2]

$$
\begin{aligned}
L_{\mathrm{AA}} &= L_{\mathrm{AA}}^{(0)}[1 - ((2 + \tau)(W_{\mathrm{AB}/X})^{2}/W_{\mathrm{A}})/D], \\
L_{\mathrm{AB}} &= L_{\mathrm{AA}}^{(0)}(2 + \tau)(W_{\mathrm{AB}/X}\,W_{\mathrm{BA}/X}/W_{\mathrm{A}})/D,
\end{aligned}
\tag{13}
$$

where

$$
L_{\mathrm{AA}}^{(0)} = 3(2 + \tau)W_{\mathrm{A}},
\tag{14}
$$

and

$$
D = (2 + \tau)(W_{\mathrm{BA}/X} + W_{\mathrm{AB}/X}) + W_{\mathrm{AB}}^{R} + (1 + \tau)(W_{\mathrm{BA}/B} + W_{\mathrm{AB}/A}),
\tag{15}
$$

and the general frequencies are defined as:

$$
W_{\mathrm{AB}/X} = c_{\mathrm{AB}} \sum_{\sigma} c_{\sigma}\omega_{\mathrm{AB}/\sigma}, \ W_{\mathrm{A}} = \sum_{\sigma,\sigma'} c_{\sigma'\mathrm{A}}c_{\sigma}\omega_{\sigma'\mathrm{A}/\sigma}, \ W_{\mathrm{AB}}^{R} = c_{\mathrm{AB}}\omega_{\mathrm{AB}}^{R},
\tag{16}
$$

where ω_{AB}^{R} corresponds to an on-site rotation frequency of a dumbbell AB. The way to calculate the concentration of a dumbbell AB c_{AB} and c_{σ} is detailed in Appendix B of ref. 2.

3 Results

3A A possible inversion of the atomic flux in relation to the vacancy flux

The SCMF theory accounts for the short-range order and the variation of jump frequencies with the surrounding trough pair interactions at the saddle and equilibrium positions. As the first shell approximation of the five frequency diffusion model of a fcc binary alloy AB dilute in B,[24] the SCMF theory predicts a change of sign of L_{AB} for $W_1 > 13/2W_3$. Here, W_1 and W_3 belong to the set of jump frequencies of an exchanging pair A–vacancy depending on the position of atom B: W_1 describes a vacancy jump from a nn site of atom B towards another nn site of atom B, and the jump frequency W_3 describes a vacancy jump which leaves the nn shell of atom B. As shown by Anthony,[6] the resulting amount of a solute interface segregation induced by a gradient of vacancy chemical potential is directly proportional to the sum, $L_{AB} + L_{BB}$. It allowed him to measure positive and negative L_{ij} values associated with different aluminum alloys quenched from high temperature. The SCMF theory shows that even in concentrated alloys L_{AB} can be negative and the sign of the sum $f_{BB} + f_{BA}^{(B)} = (L_{AB} + L_{BB})/L_{BB}^{(0)}$ tells you if the flux of B is in the same direction as the vacancy flux or in the opposite direction. To obtain a criteria for the change of sign of L_{AB} as a function of the alloy composition it is interesting to express the numerator of L_{AB} (eqn (9)) by means of the scalar notation of the pair probability p_{AB} introduced in eqn (11)

$$L_{AB} = 2L_{AA}^{(0)} q_B q_A [q_A(-3Q_{AA} + 2Q_{AB}) - q_B Q_{AB}][q_B(-3Q_{BB} + 2Q_{AB}) - q_A Q_{AB}]/D',$$

where $Q_{AB} = \exp\{(V_{AB} - V_{AB}^{(s)}/k_B T)\}$

and $D' = D(\sum_\beta q_\beta)^2 \sum_\beta Q_{A\beta} \sum_\beta Q_{B\beta}$ (17)

so that L_{AB} is equal to zero if

$$c_B/q_B^2 = K_{AA}K_{BB} + K_{AB}Q_{AB}/(2Q_{AB} - 3Q_{AA}) \text{ or}$$
$$c_A/q_A^2 = K_{AB}K_{AA} + K_{AB}Q_{AB}/(2Q_{AB} - 3Q_{BB}), \quad (18)$$

There is no analytical solution of the threshold composition but there is a numerical solution of eqn (13) combined with eqn (12). Fig. 3 presents an example of correlation coefficients where the variation of jump frequencies at the origin of the off-diagonal coefficient is mainly controlled by the energy interactions at the saddle point between a jumping atom A and its neighbours B.

The agreement between the predictions of the SCMF theory and the Monte Carlo simulations performed on the same atomic diffusion model is good at least for the off-diagonal collective correlation factor $f_{BA}^{(B)}$. The composition at which the change of sign of $f_{BA}^{(B)}$ occurs is well predicted by the SCMF theory and in this particular case can be easily related to the parameters of the atomic diffusion model

$$f_{BA}^{(B)} < 0 \Leftrightarrow c_B < \frac{3 - 2e^{-V_{AB}^s/kT}}{3(e^{-V_{AB}^s/kT} - 1)}. \quad (19)$$

In addition to the effect of interactions at the saddle point, Fig. 4 shows the effect of thermodynamic interactions on the correlation coefficients of a binary fcc alloy with a clustering tendency. The correlation coefficients which are the result of a complex combination of thermodynamic and kinetic correlations are well reproduced by the second shell approximation of the SCMF theory.

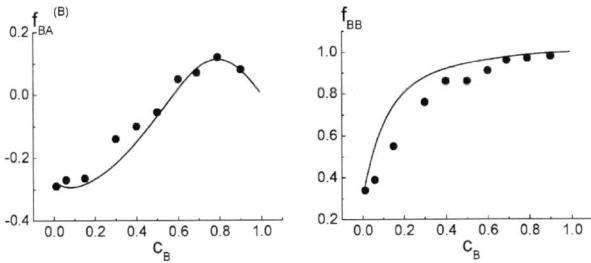

Fig. 3 Collective correlation factors $f_{BA}^{(B)}$ and f_{BB} as a function of composition in a fcc solid solution AB with vacancies. Interaction energies are all equal to zero except for the interaction of a migrating A atom at the saddle point with a nearest neighbour B atom: $V_{AB}^{(s)}/k_B T = -1.55$. The attempt frequency ratio ν_B/ν_A is equal to 16. Dots are the Monte Carlo simulations of Allnatt and Allnatt,[4] solid lines refer to the first shell approximation (second shell being similar) of the SCMF theory.

3B An acceleration of diffusion *via* the interstitial by the rotation mechanism

An on-site rotation frequency does not intervene in the non-correlated displacement of the dumbbells estimated by the diagonal coefficients $L_{\alpha\alpha}^{(0)}$ (see eqn (14)). We understand it clearly if we consider a diffusion model where all the jump frequencies are equal to zero except the rotation frequency: there is no displacement of the dumbbells. On the other hand the rotation mechanism modifies the correlation coefficients. To study this effect, we vary the on-site rotation frequency and keep the same values of the jump frequencies. For simplicity Fig. 5 represents the correlation coefficients f_{AA} and f_{BB} only. Note that the other coefficients can be deduced from them by means of the Sharma relations between the four collective correlation coefficients, proving that only one coefficient out of four is independent in a non-interacting alloy.[2,30] One can see that the second shell approximation is required to achieve a good agreement of the correlation coefficients with the Monte Carlo simulations. It should be noted that the first shell approximation was sufficient as long as the on-site rotation frequency was set to zero.[2] With the rotation, the first shell approximation is not quantitative anymore but predicts the reduction of the correlation effects with the increase in the rotation frequency well.

From the analytic expressions of eqn (13) it is straightforward to deduce that such a result seems to be general, whatever the model of jump frequencies, since the rotation frequencies show up in the denominator of the collective correlation coefficients. This effect was already predicted by Barbu and Lidiard[31] in a dilute bcc alloy using a multi-frequency approach equivalent to the first shell approximation of the SCMF.[2]

To confirm this result, additional calculations using a second shell approximation of the SCMF were performed starting from a different set of jump frequencies and a larger range of rotation frequencies (Fig. 6). We can understand how the rotation affects the migration path of a defect. After an on-site rotation the probability that the defect jumps back to its previous site is smaller because a previous site might become a non-target site. Therefore a high value of the rotation frequency makes the defect lose its memory of the previous jumps, decreases the correlation effects and enhances the transport *via* the dumbbells.

4 Conclusion

The SCMF theory based on an atomic model of the atom–vacancy exchange and the jump–rotation of an interstitial in the dumbbell configuration yields general expressions of the phenomenological coefficients L_{ij} in a multi-component alloy. The limitations and future improvements of the SCMF approach are easily related to the

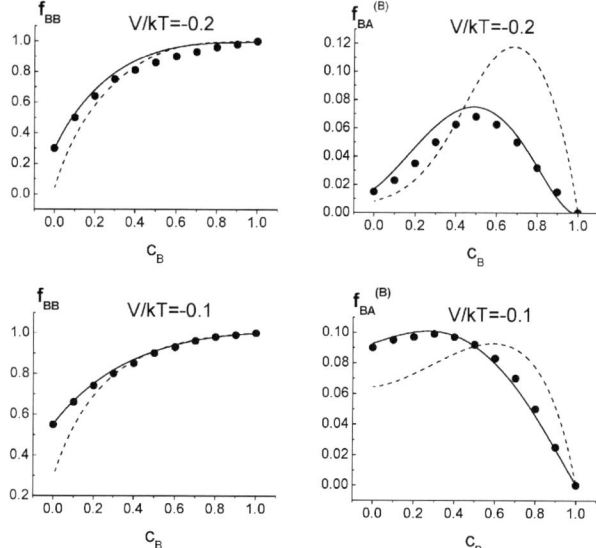

Fig. 4 Collective correlation factors $f_{BA}^{(B)}$ and f_{BB} as a function of composition in a fcc solid solution AB with vacancies. Interaction energies are all equal to zero except for the interaction between A atoms and nearest neighbour B atoms so that the order energy V normalized by k_BT is equal to $V/k_BT = -V_{AB}/k_BT$. Dots are Monte Carlo simulations of Allnatt and Allnatt,[4] solid lines refer to the second shell approximation and dashed lines to the first shell approximation of the SCMF theory.

statistical approximation of the thermodynamic correlations and to the time-dependent effective interactions used to describe the kinetic correlations.

In contrast to the previous models devoted to fcc concentrated alloys with vacancy transport, the present formulation makes the whole set of exchange frequencies associated with a given atom depending on the chemical species of the atoms nearby appear in the final result. As a first example of application, it tackles the variation of sign of an off-diagonal L_{ij} and predicts the composition of the alloy at which the change of sign will occur. By using the first existing diffusion model of a bcc alloy

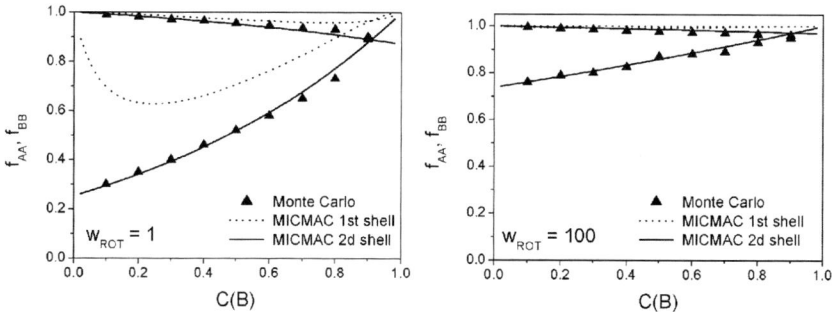

Fig. 5 Correlation coefficients in a AB concentrated bcc alloy as a function of the concentration $c(B)$ for different values of the dumbbell on-site rotation frequency. Jump frequencies come from a two-frequency model, where a single jump frequency ω_α is attributed to a function of the atomic species α which actually moves from one site to another: $\omega_A = 1$ and $\omega_B = 10$. Simple translations are accounted for with the ratio $\tau = 1$ and the on-site rotation W_{ROT} is equal to 1 (left) and 100 (right). The triangles stand for the Monte Carlo simulations. The SCMF results are plotted in dotted lines (first shell approximation) and solid lines (second shell approximation).

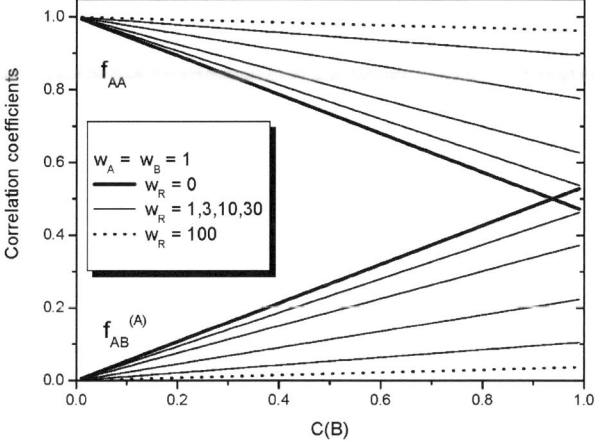

Fig. 6 Correlation coefficients in an AB concentrated bcc alloy as a function of the concentration $c(B)$ for different values of the dumbbell on-site rotation frequency. Jump frequencies come from the same two-frequency model of the previous figure except that $W_A = W_B = 1$. The on-site rotation is denoted by W_R. The lines correspond to the second shell approximation of the SCMF.

with dumbbell transport,[2] we could observe, in contrast to what we could imagine at first sight, that an on-site rotation mechanism systematically enhances the transport.

Acknowledgements

The authors are grateful to J. L. Bocquet for stimulating discussions. Support by F. Soisson and E. Clouet for the MC simulations was also appreciated. Part of this work has been funded by the joint research program 'SMIRN' between EDF, CNRS and CEA and by the European PERFECT project supported by the European Commission (F160-CT-2003-508840).

References

1   M. Nastar, *Philos. Mag.*, 2005, **85**, 3767.
2   V. Barbe and M. Nastar, *Philos. Mag.*, 2006, **86**, 3503.
3   M. Nastar, *Philos. Mag.*, 2005, **85**, 641.
4   A. R. Allnatt and E. L. Allnatt, *Philos. Mag. A*, 1991, **64**, 341.
5   (*a*) J. R. Manning, *Diffusion Kinetics for Atoms in Crystals*, Van Nostrand , Princeton, NJ, 1968; (*b*) J. R. Manning, *Phys. Rev. B*, 1971, **4**, 1111.
6   (*a*) T. R. Anthony, *J. Appl. Phys.*, 1970, **41**, 3969; (*b*) T. R. Anthony, *Phys. Rev. B*, 1970, **2**, 264.
7   C. Fu and F. Willaime, *Phys. Rev. B*, 2005, **72**, 064117.
8   C. Fu, F. Willaime and P. Ordejón, *Phys. Rev. Lett.*, 2004, **92**, 175503.
9   C. Domain and C. S. Becquart, *Phys. Rev. B*, 2005, **71**, 214109.
10 E. Clouet, M. Nastar and C. Sigli, *Phys. Rev. B*, 2004, **69**, 064109.
11 E. Clouet, L. Laé, T. Epicier, W. Lefebvre, M. Nastar and A. Deschamps, *Nat. Mater.*, 2006, **5**, 482.
12 L. K. Moleko, A. R. Allnatt and E. L. Allnatt, *Philos. Mag. A*, 1989, **59**, 141.
13 M. Nastar, V. Yu Dobretsov and G. Martin, *Philos. Mag. A*, 2000, **80**, 155.
14 M. Nastar and E. Clouet, *Phys. Chem. Chem. Phys.*, 2004, **6**, 3611.
15 V. Barbe and M. Nastar, *Philos. Mag.*, 2006, **86**, 1513.
16 R. Kikuchi and H. Sato, *J. Chem. Phys.*, 1969, **51**, 161.
17 H. Sato and R. Kikuchi, *Phys. Rev. B*, 1983, **28**, 648.
18 N. A. Stolwijk, *Phys. Status Solidi*, 1981, **105**, 223.
19 Z. Qin, A. R. Allnatt and E. L. Allnatt, *J. Phys.: Condens. Matter*, 1998, **10**, 5295.
20 G. Martin, *Phys. Rev. B*, 1990, **41**, 2279.

21 M. Nastar, P. Bellon, G. Martin and J. Ruste, *Mater. Res. Soc. Symp. Proc.*, 1998, **383**, 1749.
22 M. Athènes and P. Bellon, *Philos. Mag. A*, 1999, **79**, 2243.
23 Y. Le Bouar and F. Soisson, *Phys. Rev. B*, 2002, **65**, 094103.
24 A. R. Allnatt and A. B. Lidiard, *Atomic Transport in Solids*, Cambridge University Press, Cambridge, UK, 1993.
25 V. Barbe and M. Nastar, *Philos. Mag.*, 2006, in press.
26 J. L. Bocquet, *Technical Report*, 1990, CEA-R-5531.
27 J. L. Bocquet, *Res. Mech.*, 1987, **22**, 1.
28 (*a*) V. G. Vaks, *Pis'ma Zh. Eksp. Teor. Fiz.*, 1996, **63**, 447; English translation: ; (*b*) V. G. Vaks, *JETP Lett.*, 1996, **63**, 471.
29 F. Ducastelle, *Order and Phase Stability in Alloys*, North-Holland, Amsterdam, 1991.
30 S. Sharma, D. K. Chaturvedi, I. V. Belova and G. E. Murch, *Philos. Mag. A*, 2000, **80**, 65.
31 A. Barbu and A. B. Lidiard, *Philos. Mag. A*, 1996, **74**, 709.

Li$^+$ ionic diffusion and vacancy ordering in β-LiGa

Koichi Nakamura,*[a] Keisuke Motoki,[a] Yoshitaka Michihiro,[a] Tatsuo Kanashiro,[a] Masahito Yahagi,[b] Hiromi Hamanaka[c] and Kazuo Kuriyama[c]

Received 17th February 2006, Accepted 15th May 2006
First published as an Advance Article on the web 8th August 2006
DOI: 10.1039/b602445a

^{7}Li and ^{71}Ga NMR measurements have been performed to study the Li$^+$ ionic motion and vacancy ordering in the lithium semimetal β-LiGa. The temperature dependence of the spin–lattice relaxation rate, T_1^{-1} of the ^{7}Li nuclei in the 50 atom% Li sample shows an asymmetric broad peak around 175 K and is interpreted in terms of fast Li$^+$ ionic diffusion. The activation energy of hopping is estimated as 0.11 eV using a non-Debye type relaxation model. In the temperature dependence of T_1^{-1} of the ^{7}Li nuclei in 44 and 47 atom% Li samples, steep peaks are observed at 225 and 195 K, respectively. The origin of these anomalous peaks is attributed to the order–disorder transformation of Li$^+$ vacancies. The temperature dependence of T_1^{-1} of the ^{71}Ga nuclei measured above 200 K is interpreted in terms of the relaxation originating from the fluctuation of the electric field gradient at the ^{71}Ga nuclei due to mobile Li$^+$ ions. The activation energy for the Li$^+$ ionic diffusion estimated from T_1^{-1} of the ^{71}Ga nuclei is comparable with that obtained from T_1^{-1} of the ^{7}Li nuclei.

1. Introduction

Most lithium ion rechargeable batteries consist of layered LiCoO$_2$ (or other lithium transition metal oxides) as the positive electrode and graphite as the negative electrode. To improve the efficiency of the negative electrode there have been many studies which utilize a lithium alloy instead of graphite. In particular, some binary semimetals LiM (M = Al, Ga etc.) have attracted much attention because of the high diffusion coefficient for mobile Li$^+$ ions.[1–3]

LiGa is a mixed conductor with carriers of Li$^+$ ions and holes, which are caused by an electron transfer from Li to Ga through sp^3-like bonding.[4] The β-phase of LiGa exists over a wide range of Li content between 44 and 54 atom% and is a stable single phase. β-LiGa has an NaTl structure (space group Fd3m) and is composed of two interpenetrating sublattices consisting of Li and Ga atoms, each forming a diamond lattice as shown in Fig. 1.[5] It is known that the defect structure in β-LiGa and isostructural β-LiAl consists of two types of defects, the Li vacancy in the Li sublattice (V$_{Li}$) and the Li antisite defect in the Ga(Al) sublattice induced by the invasion of excess Li atoms (Li$_{Ga(Al)}$).[6,7] In the Li deficient region (less than

[a] Department of Physics, Faculty of Engineering, The University of Tokushima, Tokushima 770-8506, Japan. E-mail: koichi@pm.tokushima-u.ac.jp; Fax: +81 88 656 7577
[b] Faculty of Engineering, Aomori University, Aomori 030-0943, Japan
[c] College of Engineering and Research Center of Ion Beam Technology, Hosei University, Tokyo 184-8584, Japan

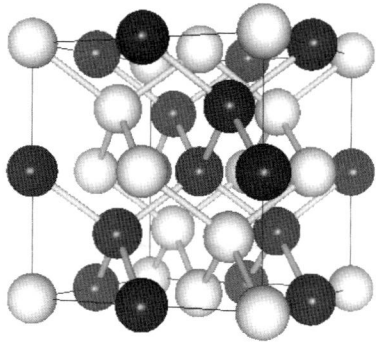

Fig. 1 The crystal structure of β-LiGa. Black and white spheres represent Li and Ga atoms, respectively. Solid lines represent the unit cell.

∼50 atom% Li) of β-LiGa, V_{Li} is dominant, while Li_{Ga} is dominant in the Li excess region (more than ∼50 atom% Li). It was reported that the concentration of V_{Li} and Li_{Ga} in the β-LiGa varies with decreasing Li content from 2.8 to 11.4% and 5.1 to 0.0%, respectively.

Some characteristic properties such as electrical resistivity and heat capacity in β-LiGa and β-LiAl are affected by such a complicated defect structure.[6–12] The electrical resistivity of β-LiGa depends largely on the Li content as shown in Fig. 2.[8] The large residual resistivity observed in β-LiGa is not interpreted by the carrier scattering for a single defect V_{Li} according to Linde's rule,[13] that is, the residual resistivity of a metal containing charged impurities is proportional to the square of the valence difference between the impurities and the matrix atoms. Therefore, the carrier scattering for the Li_{Ga} or V_{Li}–Li_{Ga} complex defect, which are created with increasing Li content, was proposed to interpret the large residual resistivity, because the valencies of these defects are larger than that of V_{Li} and their scattering is expected to have a greater effect than that of V_{Li}.[8] Besides, an anomalous jump around 230 K observed in the resistivity of a sample of 43.9 atom% Li becomes small and shifts to lower temperature with increasing Li content. In the heat capacity of β-LiGa of 43.8 atom% Li there is a jump at 233 K, which corresponds to the anomaly in the resistivity around 230 K for β-LiGa of 43.9 atom% Li.[10] The anomalous heat capacity was also observed at 95 K for β-LiAl.[11,12] However, such anomalies in heat capacity and resistivity are not observed in the LiGa and LiAl with excess Li.

It was pointed out that the anomalous resistivity and heat capacity in the β-phase result from the order–disorder transformation of the Li+ vacancies and the V_{Li}–Li_{Ga} defects play an important role in this vacancy ordering.[10] The increase in the number of the V_{Li}–Li_{Ga} defects obstructs the transformation and leads to the disappearance of the anomaly as the Li content increases.

The Li+ ions can easily diffuse into β-LiGa, including a large number of defects. The ionic conduction is rather small and usually masked by the electron/hole conduction. In such a case, NMR is a powerful tool for the detailed investigation of Li+ ionic diffusion. So far, some NMR studies have been performed on binary Li semimetals. Schone and Knight obtained an activation energy of 0.15 eV for the diffusion of Li+ ions in nearly stoichiometric LiGa from an 7Li NMR study of the line width.[14] Willhite et al. reported Li+ ionic diffusion with an activation energy of 0.12 eV in LiAl.[15] However, no anomaly in the NMR data corresponding to the anomalies in resistivity and heat capacity has been reported in these studies. A recent NMR study has revealed that LiGa is a fast ionic conductor with an activation energy of 0.11 eV and that there are anomalous peaks in the spin–lattice relaxation rate of the 7Li nuclei in the Li deficient samples. Such anomalies have been attributed to the order–disorder transformation of Li+ vacancies.[16]

   This journal is © The Royal Society of Chemistry 2006

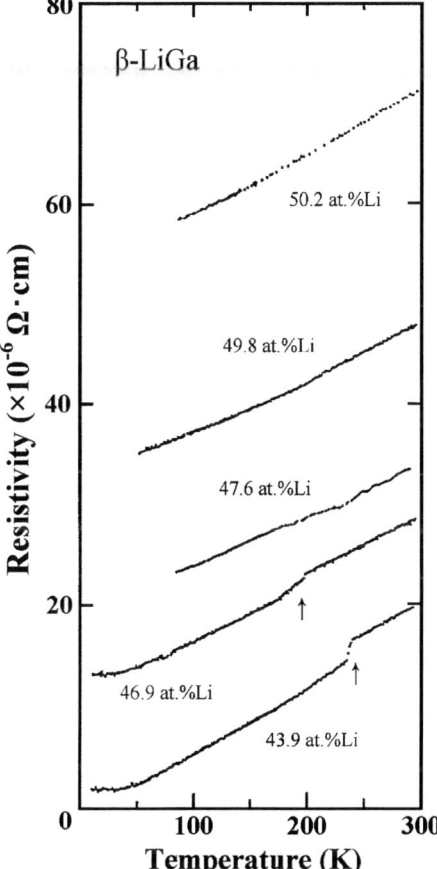

Fig. 2 The temperature dependence of electrical resistivities of β-LiGa samples with various Li content reported in ref. 8. The arrows represent the V_{Li} ordering temperatures.

Thus, anomalous spin–lattice relaxation, resistivity and heat capacity are closely related to the defect structure. Details of these anomalous properties and the ordering mechanism are unknown at present. We discuss here the Li^+ ionic diffusion and order–disorder transformation of the Li^+ vacancies from NMR relaxation measurements of the ^7Li nucleus as a carrier and the ^{71}Ga nucleus as a frame atom in β-LiGa.

2. Experimental

The polycrystalline β-LiGa samples used in this study were prepared by direct reaction of desired amounts of Li (99.9% pure) and Ga (99.999% pure).[8] The successful crystal growth of β-LiGa was achieved by adding a few percent excess Li to compensate for evaporation loss of Li and gravity segregation of Ga. A tantalum crucible was used as a container for β-LiGa. The charged crucible was set in a vertical resistance furnace in a chamber under a vacuum of 10^{-3} Torr. The chamber was filled with Ar gas of about 500 Torr prior to crystal growth. The furnace was heated to 790 °C at a rate of about 25 °C min^{-1}. After keeping the temperature constant for about 45 min to promote crystallization, the furnace was cooled to 350 °C at a rate of 70 °C h^{-1} for the growth of crystals with less than 50 atom% Li (or at a rate of 90 °C h^{-1} for those with more than 50 atom% Li). The furnace was then cooled to room temperature at a rate of 35 °C h^{-1}.

The polycrystalline samples obtained were confirmed to consist of a single phase exhibiting the NaTl structure using XRD measurements. The lattice parameters have been estimated to be 6.167, 6.192 and 6.220 Å for the 44, 47 and 50 atom% Li samples, respectively.

Each sample was crushed into a powder and packed into an evacuated Pyrex tube (9 mm in outer diameter) to prevent oxidation and water absorption. NMR measurements were performed using a conventional phase-coherent pulse NMR spectrometer constructed in house. The spin–lattice relaxation rate, T_1^{-1} of the ^{7}Li nuclei ($I = 3/2$) was measured at 10.03 MHz in the temperature range of 80–300 K for the 50 atom% Li sample and 100–300 K for the 44 and 47 atom% Li samples. The T_1^{-1} of the ^{71}Ga nuclei ($I = 3/2$) was measured at 10.03 MHz in the temperature range of 200–400 K for the 44 and 47 atom% Li samples and 200–300 K for the 50 atom% Li sample. The T_1 measurements were carried out by observing the recovery of the integrated spin echo intensity after a saturating pulse. Below the ordering temperature, the integrated intensity of the free induction decay (FID) was also used to measure T_1. The broad line NMR spectra of ^{7}Li and ^{71}Ga were measured at 10.03 MHz using the Fourier transform technique.

3.   Results and discussion

3.1   Broad line NMR spectrum of the ^{7}Li nuclei

Fig. 3 shows the broad line NMR spectrum of the ^{7}Li nuclei in the 44 atom% Li sample at room temperature. A central line is observed in β-LiGa. The satellite lines were broadened beyond observation because of the large distribution of the electric field gradient (EFG) caused by the large number of defects included in β-LiGa. Similar spectra were observed in the 47 and 50 atom% Li samples. A small resonance shift in the spectrum is observed and estimated to be less than 0.002%.

The second moment, $\langle \Delta \omega^2 \rangle$ of the ^{7}Li line arising from the dipole interaction among ^{7}Li and the nearest neighbor like (^{7}Li)/unlike (^{71}Ga and ^{69}Ga) nuclei is estimated to be about 3.7×10^8 rad s^{-2} for our samples and yields a line width of about 7.2 kHz. Here, we neglect the contribution from ^{6}Li nuclei because of their low abundance. This width is much broader than the observed width, which indicates that the motional narrowing has already occurred at room temperature.

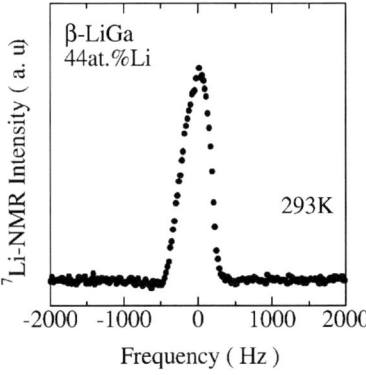

Fig. 3   Broad line NMR spectrum of ^{7}Li in the 44 atom% Li sample at 293 K.

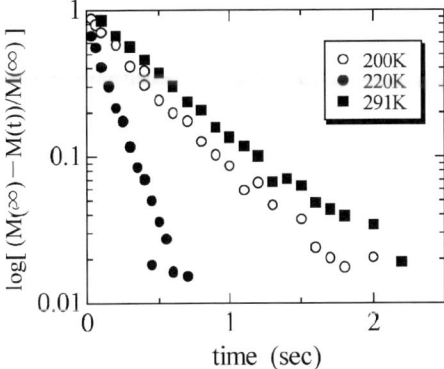

Fig. 4 Recovery curves of the nuclear magnetization of the 44 atom% Li sample at various temperatures.

3.2 Spin–lattice relaxation rate T_1^{-1} of the ^{7}Li nuclei

Fig. 4 shows the recovery curves of nuclear magnetization of the ^{7}Li nuclei in the 44 atom% Li sample in the temperature range from 200 K to room temperature. A single exponential decay is observed in each sample. This means that the spin–lattice relaxation is characterized by a single spin–lattice relaxation time, T_1, which is connected with the nuclear magnetization, $M(t)$ as $M(t) = M(\infty)\{1- \exp(-t/T_1)\}$ where $M(\infty)$ is the nuclear magnetization under thermal equilibrium.

The temperature dependence of T_1^{-1} of the 44, 47 and 50 atom% Li samples is shown in Fig. 5. Sharp peaks are observed at 225 and 195 K for the 44 and 47 atom% Li samples, respectively. An asymmetric broad peak is observed at about 175 K for the 50 atom% Li sample. If the spin–lattice relaxation is dominated by the Fermi contact interaction between the conduction electron and Li nucleus through a Li(2s) orbital, the relaxation time T_1 is connected with the Knight shift by Korringa's relationship, $T_1 T K_s^2 = (\hbar/4\pi k_B)(\gamma_e/\gamma_n)^2$ where K_s, γ_e, and γ_n are the

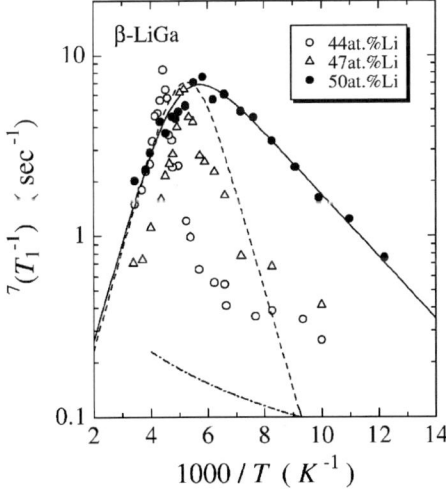

Fig. 5 Arrhenius plot of T_1^{-1} of the ^{7}Li nuclei in β-LiGa. A solid and dashed line for the 50 atom% Li sample represent the best fitting results calculated by eqn (2) and (3) and eqn (1) and (2), respectively. A dashed and single dotted line represents Korringa's relationship, $T_1^{-1} = 2.3 \times 10^6 K_s^2 T$ for β-LiGa.

Knight shift due to s-electrons, and the gyromagnetic ratio of the electron and nucleus, respectively.[17] The numerical calculation of T_1^{-1} yields $T_1^{-1} = 2.3 \times 10^6 k_s^2 T$ s^{-1}. Since the shift of the ^{7}Li resonance line of the 44 atom% Li sample is estimated to be less than 0.002% at room temperature, the contribution to T_1^{-1} from the Fermi contact interaction is estimated to be less than $9 \times 10^{-4} T$ s^{-1}. Clearly, these calculated T_1^{-1} values denoted by a dashed and single dotted line are apart from the measured values in Fig. 5. Thus, we neglect such a contribution here.

As mentioned above, the temperature dependence of T_1^{-1} observed in the 50 atom% Li sample is distinct from the steep peaks observed in the 44 and 47 atom% Li samples. As seen in Fig. 5, the broad peak around 175 K observed in the 50 atom% Li sample reminds us of the temperature dependence of T_1^{-1} arising from ionic motion. Willhite et al. and Clausen et al. also reported a similar temperature dependence of T_1^{-1} in LiAl.[15,18] Usually, a simple Debye-type relaxation model is used to analyze the spin–lattice relaxation originating from the fluctuating internal field due to ionic motion (Bloembergen–Purcell–Pound (BPP) model).[19] Here, we try to discuss the activation energy and the temperature dependence of T_1^{-1} in LiGa assuming a quadrupole relaxation. In the BPP relaxation model for $I = 3/2$, the relaxation rate, T_1^{-1} is expressed in the following notation,

$$ T_1^{-1} = C_q \left\{ \frac{\tau}{1 + (\omega_0 \tau)^2} + \frac{4\tau}{1 + (2\omega_0 \tau)^2} \right\}, \tag{1} $$

where ω_0 is the angular Larmor frequency and τ is the correlation time of the hopping of the ion. The coupling constant C_q associated with the quadrupole interaction is given by applying a point charge model[20] as follows

$$ C_q = \frac{8\pi}{75} \left(\frac{Nc}{V} \right) \frac{1}{l^3} \left\{ \frac{(1+\gamma)e^2 Q}{\hbar} \right\}^2, \tag{2} $$

where γ is the Sternheimer coefficient, Q is the quadrupole moment, c is the concentration of diffusing vacancies, N/V is the number of Li$^+$ ions included in unit volume and l is the distance of the closest approach between two spins. The correlation time τ, being the jumping time, has the form $\tau = \tau_0 \exp(E_m/k_B T)$ with the activation energy E_m required for the hopping of Li$^+$ ion.

The temperature dependence of T_1^{-1} observed in the 50 atom% Li sample at low temperatures deviates from the behavior deduced from the simple BPP-type relaxation model as denoted by the dashed line in Fig. 5. Such a deviation in the relaxation behavior has often been observed in superionic conductors. We use a phenomenological relaxation model, which was applied to the diffusion of Ag$^+$ ions in Ag$_3$SBr, to discuss the behavior at low temperatures.[21] In this case, T_1^{-1} is phenomenologically described by the following non-BPP type expression

$$ T_1^{-1} = C_q \left\{ \frac{\tau}{1 + (\omega_0 \tau)^n} + \frac{4\tau}{1 + (2\omega_0 \tau)^n} \right\}, \tag{3} $$

where n is the parameter which indicates the degree of correlation among Li$^+$ ions. The value of n is less than 2 and taking $n = 2$ leads to the simple BPP expression.

This model reproduces well the temperature dependence of T_1^{-1} over the whole measured temperature range as denoted by the solid line in Fig. 5. Here we have obtained the best fitted parameters, $\tau_0 = 2.12 \times 10^{-11}$ s, $c = 0.011$, $C_q = 2.53 \times 10^8$ s^{-2}, $E_m = 0.11$ eV and $n = 1.31$, applying eqn (2) and (3) to the T_1 data of the 50 atom% Li sample with a parameter set of $Q = -0.04 \times 10^{-24}$ cm^2, $N/V = 3.32 \times 10^{28}$ m^{-3}, $l = 2.694$ Å and $\gamma = -0.256$. Therefore, the jump time of the Li$^+$ ions is in the form of $\tau = 2.12 \times 10^{-11} \exp(+0.11/k_B T)$ s. The activation energy, 0.11 ± 0.02 eV, is comparable to that in LiAl. The vacancy concentration c estimated is comparable with the concentration of 4% reported for 50 atom% Li[8] and 2% estimated in LiAl of 50 atom% Li.[15] Thus, it seems that the spin–lattice relaxation of

^{7}Li nuclei in β-LiGa is reasonably explained by the quadrupole relaxation. We note that the values of the parameters of E_m and n obtained in this analysis are reasonable for interpreting T_1 in LiGa regardless of the relaxation mechanism originating from the quadrupole or dipole interaction.[15,18]

Fast ionic diffusion would also be expected in the Li deficient LiGa samples. In the 44 and 47 atom% Li samples, however, the temperature dependence of T_1^{-1} does not show the BPP-like relaxation behavior, that is, the T_1^{-1} peaks are too sharp. We consider that this behavior is related to the ordering of the Li^{+} vacancies, because the peak temperatures of these samples are in good agreement with the vacancy ordering temperatures for the anomalies in the electrical resistivity and heat capacity. On the other hand, a much narrower line width is observed around room temperature. These results suggest that the anomalous relaxations associated with the vacancy ordering at low temperatures changes into BPP-like relaxation with increasing temperature. The narrowing of the line width and decrease in T_1^{-1} at high temperatures is thought to correspond to the BPP-type relaxation behavior in the high temperature range ($\omega_0\tau \ll 1$). The calculation with the BPP-type relaxation model yields an activation energy of around 0.14 eV in both samples. It is noted that there is an ambiguity in this estimation because of the observation of incomplete BPP-like relaxation behavior. These activation energies estimated in samples of 44, 47 and 50 atom% Li are lower than the values reported in the Li^{+} ionic conductors Li$_3$N (0.62 eV),[22] LiI (0.35 eV),[23] LiNaSO$_4$ (0.70 eV),[24] and the Li transition metal oxides LiCoO$_2$ (0.30 eV) and LiNiO$_2$ (0.59 eV).[25,26]

The diffusion coefficient for the motion of Li^{+} ions is discussed below. The diffusion coefficient D obeys the Arrhenius-type equation $D = D_0 \exp(-E_m/k_BT)$ with the hopping energy E_m. If the Li^{+} ionic diffusion takes place by a vacancy mechanism in which V_{Li} hops every τ_v, the hopping time of V_{Li}, τ_v and the Li^{+} ion, τ_{Li} are related by $\tau_v = c\tau_{Li}$ with the vacancy concentration, c. In this case, the diffusion coefficient D_{Li} for mobile Li^{+} ions is expressed as

$$D_{Li} = \frac{Zl^2}{12\tau_{Li}}, \tag{4}$$

where l is the closest distance between the Li sites in the Li diamond sublattice and Z is the mean number of jumps. Using $Z = 1.7929$, which was used for LiAl,[15] we obtain $D_{Li} = 5.1 \times 10^{-6} \exp(-0.11/k_BT)$ cm^2 s^{-1} in the 50 atom% Li sample. At 300 K, D_{Li} is estimated to be 7.6×10^{-8} cm^2 s^{-1} and D_v for the vacancy is larger than D_{Li} by a factor of $(1/c)$ and is 7.6×10^{-6} cm^2 s^{-1} for $c = 0.01$. This value is comparable to that in LiAl and the value estimated at 415 °C is also comparable to the Li self-diffusion coefficient of about 3.0×10^{-7} cm^2 s^{-1} for the Li-excess LiAl.[2]

3.3   Order–disorder transition

Fig. 6 shows the temperature dependence of T_1^{-1} in the 44 and 47 atom% Li samples. As mentioned above, anomalous peaks in T_1^{-1} are closely related to the ordering of V_{Li}. In β-LiGa, V_{Li} are irregularly situated at high temperatures and Li^{+} ions can easily diffuse through V_{Li}. At low temperatures, the electrostatic correlation among these vacancies becomes stronger because of the slowing of the motion of vacancies with decreasing temperature and the vacancies are orderly arrayed by the Coulomb interaction. Thus, Li^{+} ionic motion through V_{Li} would be strongly suppressed at low temperatures. The fact that the peak temperature T_p decreases from 225 (for 44 atom% Li) to 195 K (for 47 atom% Li) means that such a correlation among V_{Li} declines because of the decrease in V_{Li} and the increase in V_{Li}–Li$_{Ga}$ with increasing Li content. The hopping frequency of Li^{+} ions through V_{Li} would become lower with decreasing temperature. Therefore, we suppose that the slow critical motion of V_{Li} around T_p causes the fluctuation of the EFG at the Li sites and enhances T_1^{-1} as the frequency of the fluctuation gets close to the Larmor frequency. Here we apply

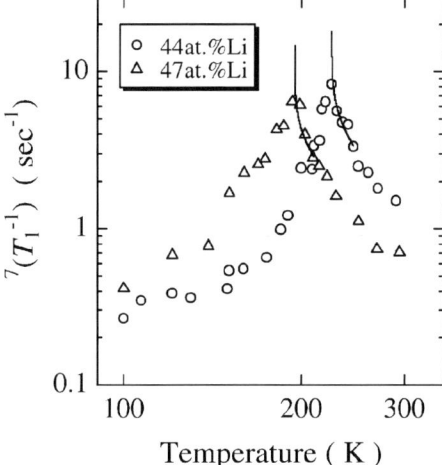

Fig. 6 Temperature dependence of T_1^{-1} of ^{7}Li nuclei in the 44 and 47 atom% Li samples. Solid lines denote the best fitting lines calculated by eqn (5).

the following phenomenological logarithmic divergence form to explain the T_1^{-1} enhancement immediately above T_p,

$$T_1^{-1} = -\zeta \ln\left(\frac{T - T_P}{T_P}\right) - \eta, \tag{5}$$

analogous to the T_1^{-1} enhancement due to the order–disorder phase transition in ferroelectric materials.[27] Applying eqn (5) to the data above T_p, we obtain adjustable parameters of $\zeta = 1.78$, $\eta = 0.92$ and $\zeta = 1.63$, $\eta = 0.96$ for the 44 and 47 atom% Li samples, respectively. The results of these calculations are denoted by solid lines in Fig. 6 and are in good agreement with the data immediately above T_p. Thus, the T_1^{-1} enhancement observed in the Li deficient samples could be qualitatively expressed as an order–disorder type relaxation. It is found that in the Li deficient phase (44 and 47 atom% Li), anomalous T_1 is strongly related to the change in the diffusion of Li$^+$ ions, which is caused by the order–disorder transformation.

3.4   Spin–lattice relaxation rate T_1^{-1} of the ^{71}Ga nuclei

A NMR spectrum of ^{71}Ga is observed above 200 K in each sample. ^{71}Ga, with its large quadrupole moment, is more sensitive to the motion of the defects than ^{7}Li. The line width in each sample is about 6 kHz at room temperature. The dipole width of about 3 kHz calculated from the second moment of the ^{71}Ga line is one half as large as the measured width, which includes a large quadrupole width.

Fig. 7 shows the temperature dependence of T_1^{-1} of ^{71}Ga. In the measured temperature range, T_1^{-1} of each sample decreases with increasing temperature and anomalous T_1^{-1} peaks similar to those observed for ^{7}Li in the 44 and 47 atom% Li samples are not observed. In LiGa the Li$^+$ ions are diffusive, while the Ga sublattice is rigid. The fluctuation of the EFG caused by mobile ions (vacancies) would affect the relaxation of the nucleus in the rigid lattice.[28] This means that the decrease observed in T_1^{-1} of ^{71}Ga would be related to the relaxation caused by the motion of Li$^+$ ions. We apply the BPP-type relaxation model to the T_1^{-1} data of ^{71}Ga, assuming that the decrease in T_1^{-1} corresponds to the incomplete behavior of the BPP-type relaxation in the range $\omega_0\tau \ll 1$. From the linearity in this Arrhenius plot, the activation energies are estimated as 0.14, 0.09 and 0.09 ± 0.02 eV for the 44, 47

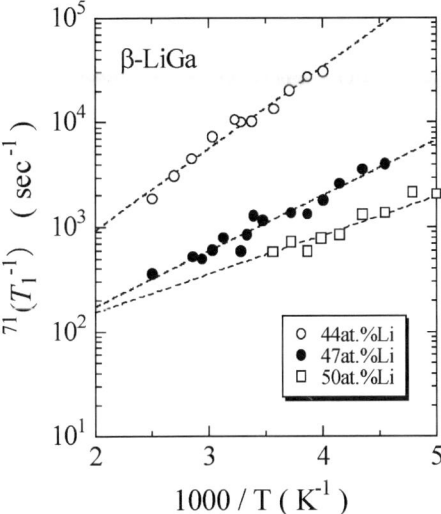

Fig. 7 Temperature dependence of T_1^{-1} of ^{71}Ga in the 44, 47 and 50 atom% Li samples. Dashed lines show the best fitting lines calculated using the BPP model for $\omega_0\tau \ll 1$.

and 50 atom% Li samples, respectively. These values are in reasonable agreement with the activation energy for mobile Li$^+$ ions. It is obvious that the relaxation of ^{71}Ga above 200 K indirectly reflects the motion of Li$^+$ ions.

Conclusions

The temperature dependence of T_1^{-1} in the 50 atom% Li sample, which shows no anomalous heat capacity and resistivity, shows an asymmetric broad peak around 175 K and is explained by the non BPP-like relaxation model. This analysis yields an activation energy of 0.11 eV. The anomalous peaks in the T_1^{-1} values of ^{7}Li nuclei are observed at 225 and 195 K in the 44 and 47 atom% Li samples in the Li deficient region, respectively, and are related to the order–disorder transformation of the Li vacancy, which would be associated with the correlation among the vacancies. Thus, the T_1 anomaly depends on the complicated defect structure in β-LiGa. The T_1^{-1} of ^{71}Ga decreases with increasing temperature above 200 K and would be related to the change in the fluctuation of the EFG caused by the Li$^+$ ions moving near Ga sites. The activation energy obtained from T_1^{-1} of the ^{71}Ga nuclei is in reasonable agreement with that from the T_1^{-1} of the ^{7}Li nuclei.

References

1　C. J. Wen, B. A. Boukamp, R. A. Huggins and W. Weppner, *J. Electrochem. Soc.*, 1979, **126**, 2258.
2　C. J. Wen and R. A. Huggins, *J. Electrochem. Soc.*, 1981, **128**, 1636.
3　J. Saint, M. Morcrette, D. Larcher and J. M. Tarascon, *Solid State Ionics*, 2005, **176**, 189.
4　E. Zintl and G. Brauer, *Z. Phys. Chem. B*, 1933, **20**, 245.
5　J. Hafner and W. Jank, *Phys. Rev. B*, 1991, **44**, 11662.
6　K. Kuriyama, K. J. Volin, J. O. Burn and S. Susman, *Phys. Rev. B*, 1986, **33**, 7291.
7　K. Kuriyama, T. Kamijoh and T. Nozaki, *Phys. Rev. B*, 1980, **22**, 470.
8　K. Kuriyama, H. Hamanaka, S. Kaidou and M. Yahagi, *Phys. Rev. B*, 1996, **54**, 6015.
9　T. O. Burn, S. Susman, R. Dejus, B. Guranelli and K. Skold, *Solid State Commun.*, 1983, **45**, 721.
10　H. Hamanaka, S. Kaidou, K. Kuriyama and M. Yahagi, *Solid State Ionics*, 1998, **113–115**, 69.
11　K. Kuriyama, S. Yanada, T. Nozaki and T. Kamijoh, *Phys. Rev. B*, 1981, **24**, 6158.

12 H. Sugai, M. Tanase, M. Yahagi, T. Ashida, H. Hamanaka, K. Kuriyama and K. Iwamura, *Phys. Rev. B*, 1995, **52**, 4050.
13 For example, N. F. Mott and H. Jones, *The Theory of the Properties of Metals and Alloys*, Dover, New York, 1958.
14 H. E. Schone and W. D. Knight, *Acta Metall.*, 1963, **11**, 179.
15 J. R. Willhite, N. Karnezos, P. Cristea and J. O. Brittain, *J. Phys. Chem. Solids*, 1976, **37**, 1073.
16 K. Nakamura, K. Motoki, Y. Michihiro, T. Kanashiro, K. Kuriyama, H. Hamanaka and M. Yahagi, *J. Phys. Soc. Jpn.*, 2002, **71**, 1409.
17 A. Abragam, *The Principles of Nuclear Magnetism*, Oxford University Press, Oxford, 1961.
18 D. Clausen, I. Burmester, P. Heitjans and A. Schirmer, *Solid State Ionics*, 1994, **70/71**, 482.
19 N. Bloembergen, E. M. Purcell and R. V. Pound, *Phys. Rev.*, 1948, **73**, 679.
20 M. H. Cohen and F. Reif, in *Quadrupole Effects in Nuclear Magnetic Resonance Studies of Solids Solid State Physics 5*, ed. F. Seitz and D. Turnbull, Academic Press, New York, 1957.
21 H. Huber, M. Mali, J. Roos and D. Brinkmann, *Phys. Rev. B*, 1988, **37**, 1441.
22 D. Brinkmann, M. Mali, J. Roos, R. Messer and H. Birli, *Phys. Rev. B*, 1982, **26**, 4810.
23 M. Mali, J. Roos, D. Brinkmann, J. B. Phipps and P. M. Skarstad, *Solid State Ionics*, 1988, **28–30**, 1089.
24 T. Kanashiro, T. Yamanishi, Y. Kishimoto, T. Ohno, Y. Michihiro and K. Nobugai, *J. Phys. Soc. Jpn.*, 1994, **63**, 3488.
25 K. Nakamura, H. Ohono, K. Okamura, Y. Michihiro, T. Moriga, I. Nakabayashi and T. Kanashiro, *Solid State Ionics*, 2001, **135**, 143.
26 K. Nakamura, M. Yamamoto, K. Okamura, Y. Michihiro, I. Nakabayashi and T. Kanashiro, *Solid State Ionics*, 2006, **177**, 821.
27 A. Rigamonti, *Adv. Phys.*, 1984, **33**, 115.
28 P. M. Richards, in *Physics of Superionic Conductors*, ed. M. B. Salamon, Springer-Verlag, Berlin, 1979.

Core structures and kink migrations of partial dislocations in 4H–SiC

Gianluca Savini,*[a] Malcolm I Heggie[a] and Sven Öberg[b]

Received 16th March 2006, Accepted 2nd May 2006
First published as an Advance Article on the web 7th August 2006
DOI: 10.1039/b603920k

First-principles calculations are used to investigate the Shockley partial dislocations in 4H–SiC. We show that both dislocations can sustain the asymmetric and symmetric reconstructions along the dislocation line. The latter reconstructions are always electrically active. In particular, the Si(g) 30° partials can explain the optical activation energy for the dislocation glide at ∼2.4 eV above the VB, the narrow peak at 2.87 eV and the broadband at ∼1.8 eV found in photoluminescence spectra. Further, we propose a new model to explain the stability of the symmetric reconstructions and the enhancement of the dislocation velocity in SiC.

Introduction

The first step in the discovery of semiconductors took place in 1883 when Michael Faraday observed that in silver sulfide the conductivity increased with an increase in temperature, in contrast to metals which show a decrease when heated.[1] Over the years many other semiconductors have been found and nowadays these materials are the basic elements of every electronic device.

Semiconductor crystals always contain numerous defects which perturb the perfect lattice periodicity and influence its electrical and mechanical properties. Some defects have useful effects and are crucial for the device design, (e.g. doping with impurity atoms to achieve n- or p-type conductivity), while others have a destructive effect on devices and their performance.

Dislocations are often associated with the degradation of the electrical and optical properties of the devices. It is commonly found that dislocations scatter charge carriers, decreasing the carrier lifetime and increasing the resistivity of the materials. Furthermore, dangling bonds can be present along the dislocation line giving rise to energy levels in the forbidden band gap. These levels act as electron–hole recombination centers degrading the optical properties of the device and giving a large leakage current when biased.

Since experiments alone often cannot yield information about the exact origin of defect-related effects or about how to suppress unwanted effects or to promote those which are desirable, we need theoretical models and calculations to interpret the experimental data.

Silicon carbide (SiC) has received special attention in recent years because of its suitability for electronic devices operating under high temperature, high power, high frequency and/or strong radiation conditions where conventional semiconductor materials, like Si and GaAs are considered to have reached their limits. In

[a] Department of Chemistry, University of Sussex, Falmer, Brighton, UK BN1 9QJ. E-mail: g.savini@sussex.ac.uk
[b] Department of Mathematics, University of Luleå, SE-97187, Luleå, Sweden

comparison with the other semiconductor materials silicon carbide distinguishes itself by a combination of superior properties, such as high thermal conductivity, high thermal stability, high critical breakdown field, a hardness second to diamond and high resistance to radiation. This semiconductor can exist in more than 200 different polytype structures with a wide band gap ranging from 2.3 eV for the 3C–SiC polytype to 3.3 eV for the 4H–SiC polytype.[2] The latter properties make SiC a suitable material for light emitting devices in the visible range, where silicon is useless. The different SiC polytypes are usually describe by the Ramsdell notation, where the first character indicates the total number of formula units contained in the unit cell (reflecting the different stacking sequences) and then the letter C, H or R denotes the lattice type as being cubic, hexagonal or rhombohedral, respectively.

Despite its outstanding properties, several studies report a drawback: an increase in the voltage drop of p–i–n diodes under forward bias. In a common diode the forward voltage drop is almost independent of the amount of current passing through the device, so they have a very steep characteristic in the current/voltage (I/V) graph. In SiC the I/V diode characteristics change under forward bias, reflecting an increase in the resistance in the device. Typically the initial forward drop of 3.5 V at 100 A cm^{-1} increased to 3.8 V after a few hours and over 15 V in a few days of constant operation.[3]

Such behaviour renders the SiC diode much less attractive than its real potential characteristics. In particular high-power systems are frequently designed with several diodes connecting in parallel in order to increase the total current rating. The increase in the resistance of one diode would cause an increase in the current flow in the remaining stable components. At some point, the current flowing through the stable diodes could exceed the threshold maximum current of the devices and the system could fail catastrophically.

Recent experiments have shown that the forward voltage drop is due to expansion of stacking faults (SFs) in the active region of the diodes.[3–5] The SF regions are always bounded by Shockley partial dislocations; therefore the SF expansions are strongly connected with the mobility of the partials.[6–9] Depending on the angle between the Burgers vector and the dislocation line the Shockley partial dislocations are 30° or 90° partial dislocations either with Si or C termination along the dislocation line in the glide plane, hence the respective labels Si(g) and C(g) (see Fig. 1).

Under forward bias a broad band gap transition at approximately 1.8 eV is observed simultaneously with the glide of the partials.[4] Latest photoluminescence (PL) spectra[10] obtained from the mobile dislocations confirm the broad radiative band at ∼1.8 eV, and reveal a further narrow peak at 2.87 eV. Furthermore, electron beam-induced current (EBIC) experiments indicate that both types of 30° partial act as nonradiative centers,[11] while optical studies show that the radiative transition rate on the Si(g) partials is much higher than that on the corresponding C(g) core dislocations.[12]

Several experimental studies show that not all the partials glide under electron–hole plasma injection, but rather that only the Si(g) 30° partials move rapidly, while the others are almost immobile.[12–14] Under electrical stress (forward bias) or optical excitation (laser beam) the activation energy for the kink migration of the partials is 0.27 ± 0.02[4] or 0.25 ± 0.05 eV,[10] respectively. These values are ∼10 times lower than the estimated value of 2.5 eV obtained from the temperature dependence of the yield stress.[15] It is commonly believed that recombination-enhanced dislocation glide (REDG)[16–18] is responsible for the rapid propagation of these planar defects. According to the 'phonon-kick' mechanism, a nonradiative electron–hole recombination center should be present along the dislocation line and the released transition energy is transferred into the reaction coordinate for glide migration of the Shockley partials.

The latest excitation spectrometry experiments reveal a nonradiative center at ∼2.4 eV above the valence band. Furthermore, these experiments clearly show that

This journal is © The Royal Society of Chemistry 2006

the latter deep level is responsible for the REDG mechanism on the Si(g) 30° partials.[10] Theoretical studies in 3C– and 2H–SiC have pointed out that the C(g) partials are electrically inactive, while the Si(g) gives rise to a band gap level at $E_v + 0.4$ eV. In addition, the activation energy for the glide of the C(g) 30° partial is calculated to be 4.95 eV while that for the Si(g) is 4.60 eV for short dislocation segments.[19] So far, the electrical activity of the 30° partials and the deep level required by the REDG process still remains unclear.

In this work, we investigate the dislocation core effect on electrical activity and kink migration. We show that the symmetric reconstructions along the dislocation line are always electrically active and have glide activation energies lower than the respective asymmetric reconstructions. Further, we propose a new model that can explain why the symmetric reconstructions become dynamically more stable under electron–hole plasma and suggest why only the Si(g) 30° partials are mainly involved in the enhancement of the dislocation mobility. This model can be applied to any semiconductor material in order to predict the behaviour under electron–hole plasma and could inspire new experimental techniques to reduce the degradation mechanism or in a more fascinating way to change the destructive effect in to a new useful property of semiconductor materials.

Computational methods

The calculations are based on density functional theory (DFT) in the local density approximation (LDA) using the exchange–correlation functional as parametrized by Perdew and Wang.[20] The basis sets employed consist of s, p, and d Gaussian orbital functions with four exponents, centered at the atomic sites.[21] Norm-conserving pseudopotentials based on the Hartiwigsen–Goedecker–Hutter[22] scheme were used. The charge density is represented by a plane-wave basis in reciprocal space expanded up to 300 Ryd. To perform the Brillouin zone integrations we use a Monkhorst–Pack (MP) scheme[23] with 8 k-points along the dislocation line. In order to take into account the possible dispersion of the levels inside the band gap, a metallic filling is used, where the number of electrons at each k-point can differ. The initial atomic positions have been produced within isotropic elastic theory and then relaxed using DFT and the conjugate gradient algorithm.

Dislocation core structures

In hexagonal SiC the partial dislocations lie in the basal plane with Burgers vector $b_p = \frac{1}{3}[1\bar{1}00]$ and dislocation line $l = [11\bar{2}0]$. In order to investigate separately the properties of the single partials, the dislocations have been modelled using the cluster–supercell hybrid. The periodicity of the lattice is kept only along the dislocation line (supercell component) while within the $\{11\bar{2}0\}$ plane, the unit cell is repeated to create a cluster keeping an empty space between the cluster and its images in the neighbouring unit cells of $\Delta = 8$ a.u. The silicon and carbon bonds along the surface are saturated by hydrogen-like atoms. The stoichiometry of the 90° partial unit cells are $Si_{84}C_{84}H_{52}$ with a total of 220 host atoms, while the stoichiometry of the 30° partials are $Si_{90}C_{92}H_{54}$ for the C(g) core and $Si_{92}C_{90}H_{54}$ for the Si(g) core, with a total of 236 host atoms. The lattice parameters of the bulk structure are $a = 3.06$ Å, $c/a = 3.27$ in excellent agreement with experimental values ($a = 3.07$ Å, $c/a = 3.27$),[24] while the Si–C bond lengths are 1.88 Å.

The possible reconstructions along the dislocation line are shown in Fig. 1. The symmetrical reconstructions (SRs) are characterized by dangling bonds or quasi-fivefold bonded atoms along the dislocation line. In the asymmetrical reconstructions (ARs) the core atoms in pairs form a bond along the dislocation line. The band structure analysis shows that the dangling/weak bonds of the SR give rise to a half-occupied band. In the ARs the strong covalent bond splits the half-filled band,

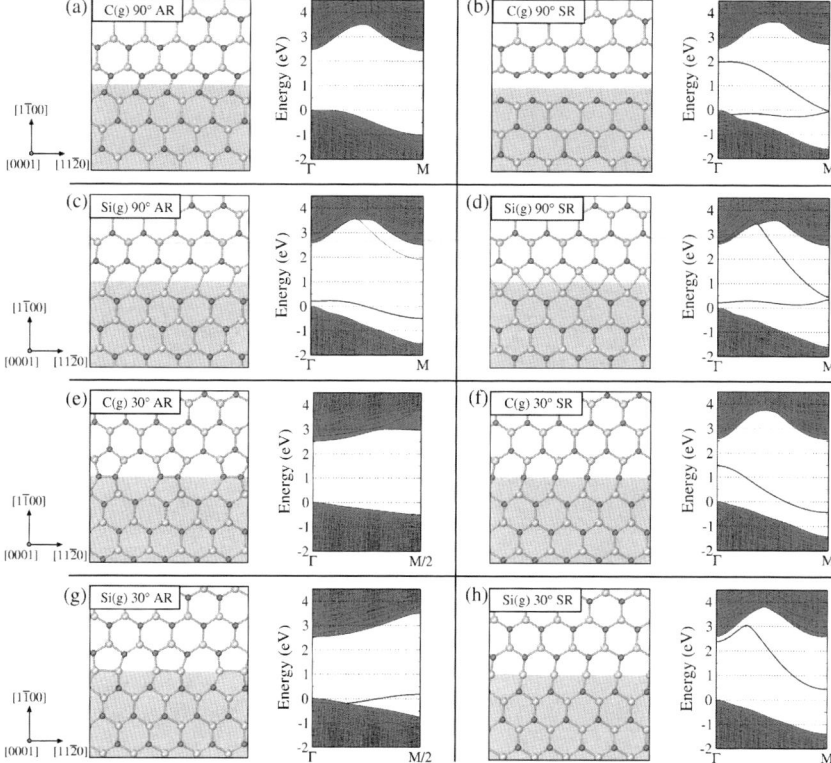

Fig. 1 Projections on the glide plane {0001} of the partial dislocations, and the respective Kohn–Sham band structures in the neutral charge state. Left: the asymmetric reconstruction (AR); right: the symmetric reconstruction (SR). The intrinsic stacking fault regions accompanying the partials are shaded; (a, b) C(g) core 90° partial; (c, d) Si(g) core 90° partial; (e, f) C(g) core 30° partial; (g, h) Si(g) core 30° partial.

generating a fully occupied band near the top of or inside the bulk valence band (VB), and an empty band near the bottom of or inside the conduction band (CB).

Depending on the position of the Fermi level in the bulk band gap, both partials can sustain either the AR or SR reconstructions along the dislocation line. The electron chemical potential or Fermi level E_F is the energy at which the Fermi–Dirac probability of occupation by an electron is exactly one half. The Fermi level represents an important quantity in the analysis of semiconductor behaviour. For intrinsic material the Fermi level lies at the middle of the band gap. In n-type the Fermi level lies closer to the bottom of the conduction band E_c, and the energy difference $(E_c - E_F)$ gives a measure of how strongly n-type the material is. In p-type materials the Fermi level lies closer to the top of the valence band E_v, and the energy difference $(E_F - E_v)$ gives a measure of how strongly p-type the material is. The formation energies of the respective reconstructions in different charge states are defined by[25]

$$E_{\text{form}}^q = E_{\text{tot}}^q - E_{\text{bulk}} + q\,[E_F + E_v + \Delta V] \qquad (1)$$

where E_{form}^q is the formation energy of the dislocation in the q charge state, E_{tot}^q is the total energy of the unit cell containing a dislocation with charge q, E_{bulk} is the energy of the bulk with the same stoichiometry and E_F is the Fermi energy with respect to the top of the valence band E_v. The potential ΔV is a correction term in order to line up the band structures of the different charge states with the bulk.[26] Fig. 2 shows the

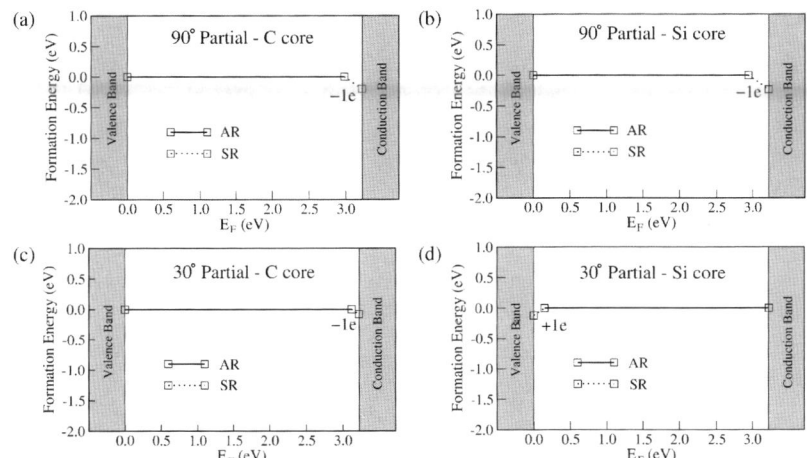

Fig. 2 Formation energy as a function of the Fermi level. The solid line indicates the asymmetric reconstruction (AR) while the dashed line indicates the symmetric reconstruction (SR). (a) C(g) core 90° partial; (b) Si(g) core 90° partial; (c) C(g) core 30° partial; (d) Si(g) core 30° partial. The figure shows that the AR is always favourable in intrinsic bulk and moderately doped semiconductors, while the SR may become stable in strongly doped material when the Fermi level E_F is approaching the conduction band (a), (b), (c) or valence band (d).

formation energy of the most stable charge states for each configuration as a function of Fermi energy. When E_F is approaching the valence band ($E_F = 0$) we are in a strongly p-type regime while when E_F is approaching the conduction band ($E_F = \Delta E_{gap} = 3.26$ eV) we are in a strongly n-type regime. In general we found that the AR (solid line in Fig. 2) is always favourable in intrinsic bulk and moderately doped semiconductors, while the SR may become stable in strongly doped material (dashed line in Fig. 2).

C(g) core 90° partial

In the AR the C–C bond length is 1.65 Å, longer by 7.8% than in bulk diamond (1.53 Å), while the C–Si backbond of the core atoms range from −3.2 to 5.5% compared with the bulk SiC. This reconstruction is not electrically active. In strongly n-type doping the SR may become more stable. In the neutral charge state, the dangling bonds along the dislocation line are separated by 2.56 Å, expanded by 67% from the bulk diamond bond length. As a consequence, the backbonds of threefold coordinate core atoms are shortened (by 2.6%). The dangling bonds localized on the C atoms give rise to a half-filled band ranging from inside the VB to $E_v + 1.97$ eV. The AR is higher in energy than the SR by 1.03 eV/a_0 ($a_0 = 3.06$ Å is the bulk parameter lattice). In the negative charge state the distance between dangling bond atoms is 2.69 Å, 5.1% longer than the respective neutral charge state, while the backbonds of the C atoms are shortened by 5.1% with respect to the bulk SiC. The band structure analyses show a deep band ranging from $E_v + 0.06$ eV to inside the CB.

Si(g) core 90° partial

In the AR the Si–Si bond length is 2.38 Å, expanded by 1.7% with respect to bulk silicon (2.34 Å), while the backbonds are distorted by ±1.6%. The band structure analysis gives a donor level at $E_v + 0.21$ eV at the Γ-point, and opens a little gap at the boundary of the BZ. In heavily n-type doped material the SR may become favourable. In the neutral charge state the Si atoms along the dislocation line are quasi-fivefold coordinate with a bond length of 2.69 Å, expanded by 15.0% with respect to the bulk silicon. The backbond of the Si core atoms range between -1.0 and 0.4% compared to the bulk SiC. The band structure analysis shows a half-filled band ranging from $E_v + 0.21$ eV to inside the conduction band (CB). The AR is 0.34 eV/a_0 higher in energy than the respective SR. In the negative charge state the bond lengths between like atoms along the dislocation line are 2.58 Å, 4% smaller than in the neutral state. The backbonds are slightly shortened (by 1.9%) with respect to the bulk SiC. The band structure analyses show a deep band ranging from $E_v + 0.38$ eV to inside the conduction band (CB).

C(g) core 30° partial

The AR shows a C–C bond length of 1.67 Å, expanded by 9.2% with respect to bulk diamond, while the backbonds are distorted by between -2.66 and $+3.72\%$. Due to the strong reconstruction, the band structure analysis shows that this dislocation has no deep state. In heavily n-type doped materials the SR can become stable. In the neutral and negative charge states the distance between like atoms along the dislocation line is 3.06 Å, *i.e.* double the bulk diamond bond length. The backbonds of the threefold coordinate atoms are shortened by 3.7% for the neutral SR and by 5.8% for the negative charge state. The dangling bonds give rise to a half-filled band ranging from VB to $E_v + 1.46$ eV for the neutral charge state and from $E_v + 0.21$ eV to the CB for the negative charge state. The neutral AR is 0.51 eV/a_0 higher in energy than the respective SR.

Si(g) core 30° partial

The AR shows a Si–Si bond length of 2.35 Å expanded by 0.4% with respect to bulk silicon, while the backbonds are shortened by 2.0%. The band structure shows a deep band ranging between the VB to $E_v + 0.19$ eV. In heavily p-doped material the SR can become favourable. In the neutral and positive charge states the distance between dangling bonds along the dislocation line is 3.06 Å, expanded by 31% with respect to the bulk silicon. The respective backbonds are shortened by about 2.4 and 4.3% for the neutral and positive charge state, respectively. These structures are electrically active with a deep band ranging between $E_v + 0.44$ eV at the M-point to $E_v + 2.36$ eV at the Γ-point for the neutral charge state and between the VB and $E_v + 2.01$ eV at the Γ-point for the positive charge state. The neutral AR is 0.55 eV/a_0 higher in energy than the SR.

We observe that the AR of the 90° partials requires a core shear between the unfaulted and stacking fault regions of about 0.8–1.2 Å for the C(g) core and 1.0–1.3 Å for the Si(g) core dislocations. In the case of the 30° partials the AR does not require a long ranged shear, but rather only requires flipping of alternate atoms in the core. Table 1 summarises the deep bands found for each dislocation in the neutral charge state. Later, in the discussion section, we will propose that under electron–hole plasma injection the free energy of the SRs are dynamically lowered by continuous electron–hole transitions between their respective deep levels and valence/conduction bands.

In conclusion for the AR, only the Si(g) core partials give rise to a deep narrow band at $E_v + 0.2$ eV in substantial agreement with those found previously ($E_v + 0.4$ eV).[19] The small difference between the two values can be attributed to the different sizes of the unit cells: 120 host atoms with a $1 \times 1 \times 2$ MP set of k-points in ref. 19, against 220–236 atoms with $1 \times 1 \times 8$ MP grid in the present work. Furthermore, the bond length between like atoms at the core of the Si(g) partials is 2.35 Å, while at the core of the C(g) partial it is 1.67 Å. These results are in very good agreement with the values found by Bernardini and Colombo (2.37 Å for the Si(g) partial, and 1.68 Å for the C(g) partial).[27] In the SR all the partials are electrically active with a half-filled deep band. In particular, the 30° dislocations in the neutral

Table 1 Deep band localized along the dislocation line, in the neutral charge state. The top of the valence band is at the Γ-point of the BZ, while the bottom of the CB is at the M-point. Later, in the discussion section, we propose that the free energy of the SRs would be dynamically lowered by continuous electron–hole transitions between the respective deep levels and valence/conduction bands. The energy units are in eV

	Asymmetric Reconstruction (AR)		Symmetric Reconstruction (SR)	
	Γ-point	$M/2$-point	Γ-point	M-point
90° partials C(g) core	—	—	$E_v + 1.97$	—
90° partials Si(g) core	$E_v + 0.21$	—	$E_v + 0.21$	—
30° partials C(g) core	—	—	$E_v + 1.46$	—
30° partials Si(g) core	—	$E_v + 0.19$	$E_v + 2.36$	$E_v + 0.44$

charge state give rise to a deep band ranging between the top of VB to $E_v + 1.46$ eV for the C(g) partial and a deep band ranging between $E_v + 0.44$ eV and $E_v + 2.36$ eV for the Si(g) partial.

Kink migration

Following the Hirth–Lothe model, the mobility of the dislocations is determined by the formation and migration of kinks. The dislocation velocity is given by[28]

$$v_{dis} \propto e^{-\frac{Q-TS}{kT}} \qquad (2)$$

where Q is the activation energy and S is an entropy term. The latter factor will not be calculated in this work. The activation energy for short dislocation segments is the sum of the formation energy $2F_k$ of a kink pair, and the kink migration energy W_m. The formation energy controls the density of kinks in thermodynamic equilibrium, while W_m determines the expansion of the kinks along the dislocation line. When the dislocation length is bigger than the mean separation between kinks or between strong obstacles, the activation energy becomes $Q = F_k + W_m$. The latter expression controls the migration velocity for long dislocation segments. Kinks can be formed only in pairs and the formation energy of a kink pair when the separation of the single kinks in nb is defined as

$$2F_k = \Delta E_{kink\ pair} + E_{int}(n) \qquad (3)$$

where F_k is the formation energy of a single kink, $\Delta E_{kink\ pair}$ is the formation energy of a kink pair with the smallest possible separation and E_{int} is the kink–kink interaction term. The latter term is approximately given by elasticity theory. For the 90° partials it is defined by[78]

$$E_{int}(n) = -\frac{\mu b_p^2 h^2}{8\pi n a_0} \frac{1 - 2v}{1 + v} \simeq -\frac{0.24}{n} eV \qquad (4)$$

while for 30° partials the kink–kink interaction term is given as[28]

$$E_{int}(n) = -\frac{\mu b_p^2 h^2}{32\pi n a_0} \frac{4 + v}{1 - v} \simeq -\frac{0.49}{n} eV \qquad (5)$$

where μ is the shear modulus (1.23 eV/Å),[29] b_p the modulus of the Burgers vector $(a_0/\sqrt{3})$, v is Poisson's ratio (0.21),[30] h is the height of the kink $(a_0/\sqrt{3})$ and na_0 is the separation between single kinks ($a_0 = 3.06$ Å). The term $\Delta E_{kink\ pair}$ is found by introducing a kink pair along the dislocation line for both the C(g) and Si(g) core dislocations. The formation energies are then found by subtracting the energies of the corresponding straight dislocations.

To model the single kink and kink pair we have used a hybrid cluster–supercell approach, with several layers along the dislocation line (supercell component). The stoichiometry of the unit cells used is $Si_{35}C_{35}H_{24}$ per layer for the 90° partials, while the stoichiometry of the 30° partials is $Si_{27}C_{28}H_{21}$ per layer for the C(g) core and $Si_{28}C_{27}H_{21}$ per layer for the Si(g) core. In order to check the convergence of the calculated energies we have increased the number of layers along the dislocation line up to 9. Test calculations have shown that 6–7 layers along the dislocation line are large enough to describe the quantum-mechanically bonds of the kinked dislocations. In the following discussion, we describe the kink migration of the AR and SR reconstructions in their more stable charge state, *i.e.* the neutral charge state for the AR, the negative charge state for the SR 90° partials and SR C(g) core 30° partial, and the positive charge state for the SR Si(g) core 30° partial. In this way the reconstructions under investigation are always the global minimum energy reconstruction.

The elementary step of single kink migration was found by rotating the central Si and C atoms (arrow in Fig. 3,5,6) by about 90° along an axis normal to the glide plane. This causes a kink migration along the dislocation and a consequent expansion of the stacking fault region associated with the partials. To investigate the intermediate structures between the initial and migrated kink we defined two variables, or constraints, $c_{silicon}$ and c_{carbon} associated with the two central atoms, R_1 and R_2

$$c_{silicon} = |R_1 - R_4|^2 - |R_1 - R_3|^2, \quad c_{carbon} = |R_2 - R_3|^2 - |R_2 - R_4|^2 \qquad (6)$$

where R denotes the coordinate of the atom and the subscript indicates the atom as depicted in Fig. 3,5,6. In some of the SRs it is only one atom that is mainly involved

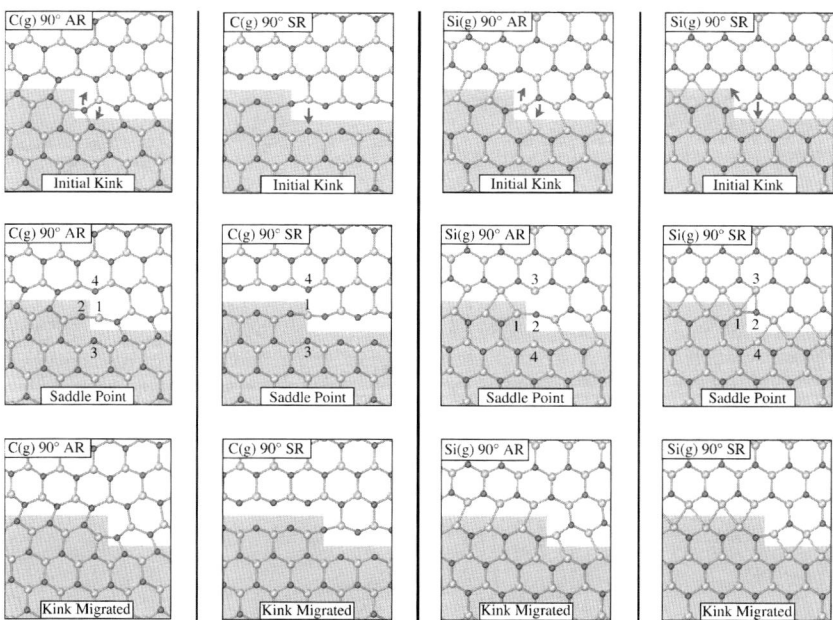

Fig. 3 Kink migration for the C(g) core 90° partial dislocation (first column: asymmetric reconstruction, second column: symmetric reconstruction) and Si(g) core 90° partial (third column: asymmetric reconstruction, fourth column: symmetric reconstruction). The shading region underline the stacking fault expansion associated with the kink migrations. (Top) Initial kink. The arrows indicate the atoms mainly involved during the kink migration. (Centre) Saddle point. The numbers show the atom positions used by the constraints. (Bottom) Kink migrated.

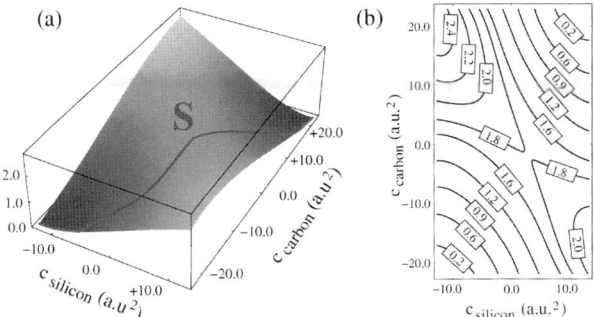

Fig. 4 (a) Surface energy and (b) contour plot of the kink migration for the C(g) core 90° partial dislocation. The two global minima represent the energy of the initial and migrated kink, while the maximum of the migration path (indicated with S) represents the saddle point.

in the migration process, therefore only one constraint is used. Then meshes of 10 × 10 intermediate points were used to model the single kink migrations. For each set of fixed values of the two constrains all the structures were relaxed using the conjugate gradient method. Fig. 3 shows the projection into the basal plane of the elementary kink migration steps for the respective 90° partial dislocations. For the 90° partial dislocations the initial and kinked structures are quantitatively equivalent in terms of bond lengths and strains. Fig. 4 shows the two-dimensional energy surface for the kink migration of the C(g) core 90° partial.

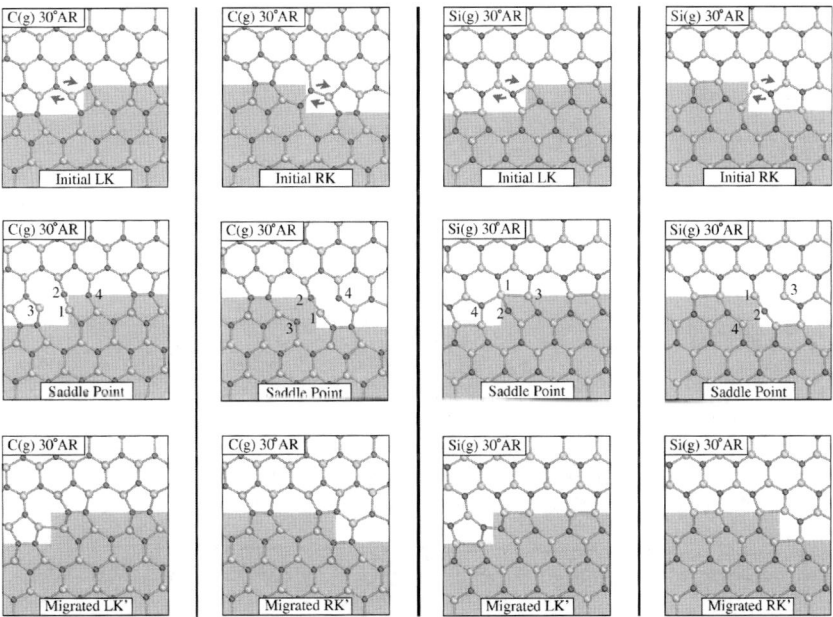

Fig. 5 Kink migration paths for the 30° partial in the asymmetric reconstructions. The migration path of the C(g) core dislocations (first column: left kink, second column: right kink) and the migration path of the Si(g) core dislocations (third column: left kink, fourth column: right kink). The shading regions underline the stacking fault expansion associated with the kink migrations. (Top) Initial kink. The arrows indicate the atoms mainly involved during the kink migration. (Centre) Saddle point. The numbers show the atom positions used by the constraints. (Bottom) Kink migrated.

For the AR 30° partials four topologically different types of kinks can be generated. Fig. 5 shows the structure of the left kink (LK) and the right kink (RK) in the asymmetric reconstructions. Each of the two kinks has its alternate configurations, depending on the kink position along the dislocation line. In general the alternative kinks, named LK′ and RK′, have higher formation energies than the normal kinks and act as intermediate steps during the kink migrations. All atoms in these different type of kinks are fourfold coordinate. We observe that the migrated left kinks form a reconstructed bond, of the opposite type to the reconstruction bonds of the respective partial, *i.e.* the C(g) partial LK′ presents an alien Si–Si bond and the Si(g) partial LK′ presents an alien C–C bond. (see Fig. 5).

For the 30° partials SR, the normal and alternative kinks are topologically equivalent, reflecting their single periodicity along the dislocation line (see Fig. 6). Considering that the dislocation motion is dominated by the kinks that migrate at the fastest rate,[31] we have chosen the lowest migration energy W_m between the LK and RK saddle point energies.

C(g) core 90° partial

In the AR the C–C bonds closest to the kink step are slightly compressed with lengths of 1.64 Å, while the others reproduce the bond length of the straight dislocation (1.65 Å). The backbonds range between 1.81–2.04 Å, representing strains of up 9%. We found a saddle point near the origin of the constraint $c_{silicon}$, c_{carbon} with migration energy W_m of 1.78 eV. The formation energy $2F_k$ for the corresponding kink pair is 0.36 eV. This yields an activation energy Q of 2.14 eV for short segment dislocations. In the SR the distance between dangling bond atoms closest to the kink step are 2.83 Å, while the others far from the kink step reproduce the value

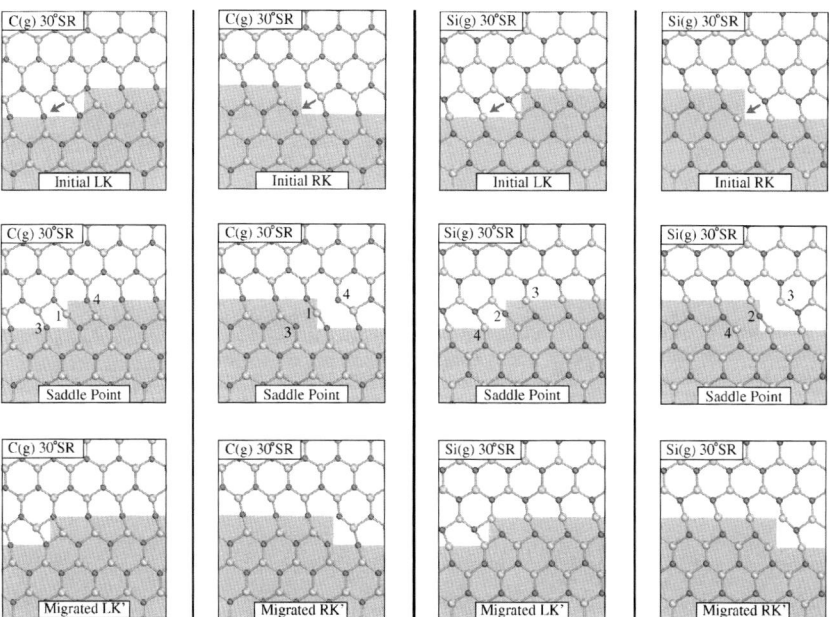

Fig. 6 Kink migration for the 30° partial in the symmetric reconstructions. The migration path of the C(g) core dislocations (first column: left kink, second column: right kink) and the migration path of the Si(g) core dislocations (third column: left kink, fourth column: right kink). The shading region underline the stacking fault expansion associated with the kink migrations. (Top) Initial kink. The arrows indicate the atoms mainly involved during the kink migrations. (Centre) Saddle point. The numbers show the atom positions used by the constraints. (Bottom) Kink migrated.

of the straight dislocation (2.69 Å). The migration barrier W_m is found exactly at the origin of the constraint with saddle point energy of 0.21 eV. The formation energy $2F_k$ for the corresponding double kink is 0.40 eV. Therefore the activation energy Q for the migration of short segment dislocations is 0.61 eV.

Si(g) core 90° partial

In the AR, the Si–Si bonds close to the kink step are 2.37 Å (only 0.4% compressed with respect to the straight dislocation) and 2.45 Å (3% stretched with respect to the straight dislocations). The backbonds have lengths ranging between 1.84–1.92 Å, representing strains of up to 2%. The difference in energy between the double kink and the straight dislocation is 0.10 eV, and the corresponding formation energy $2F_k$ is 0.34 eV. The saddle point barrier for the kink migration is 1.85 eV, which gives an activation energy Q of 2.19 eV for short dislocation segments. For the SR, the lengths between dangling bond like atoms close to the single kink are 2.57 Å (expanded by 8% with respect to the straight dislocation). The backbonds have lengths between 1.84–1.92 Å, representing strains of up to 2% (exactly the same of the AR). The saddle point energy W_m is 0.34 eV while the formation energy $2F_k$ is 1.21 eV. This yields the corresponding activation energy Q of 1.55 eV for short dislocation segment.

C(g) core 30° partial

In the AR structures, the C–C bond lengths close to the kink steps range between 1.62–1.70 Å, representing strains of up to 11%, whereas the backbonds have lengths ranging between 1.80–2.15 Å, representing strains of up to 14%. The alien Si–Si bonds present in the LK′ has roughly the same length of the bulk Si. The intermediate LK′ and RK′ structures are 2.51 and 0.33 eV higher in energy than the initial LK and RK structures. The barrier heights are calculated to be 2.98 and 1.47 eV for the LK and RK migrations, while the formation energy of the LK + RK kink pair, $2F_k$, is 1.78 eV. Then the corresponding activation energy Q for the fastest kink (RK) is 3.25 eV. In the case of the SR, the C–Si bonds range between 1.76–2.00 Å, representing strains of up to 6%. The saddle point energies are found to be 0.09 and 0.22 eV for the LK and RK migrations, while the formation energy of a kink pair is found to be 1.48 eV. Therefore both the LK and RK have fairly equivalent activation energies, with values of 1.57 eV for the LK and 1.70 eV for the RK.

Si(g) core 30° partial

In the ARs, the Si–Si bond lengths close to the kink steps range between 2.31–2.41 Å, representing strains of up to 3%, while the backbonds have lengths ranging between 1.79–2.02 Å, representing strains of up to 7%. The alien C–C bond length in the LK′ structure is 1.62 Å, expanded by 6% with respect to the bulk diamond. The intermediate structure LK′ is 2.07 eV higher in energy than the LK structures, while the structures RK and RK′ have roughly the same energy (the difference between the two formation energies is less than 0.03 eV). The saddle point energies are 2.47 and 2.17 eV for the LK and RK, respectively, while the formation energy of the LK + RK kink pair is 2.26 eV. The corresponding activation energy for the fastest kink (LK) is 4.43 eV. For the SR, the C–Si bonds range between 1.78–2.02 Å, The saddle point energies are calculated to be 0.08 and 0.06 eV for the LK and RK migrations, while the formation energy of the LK + RK kink pair is 2.12 eV. Therefore both the LK and RK have fairly equivalent activation energies, with values of 2.20 eV for the LK and 2.18 eV for the RK.

For all the partials, the dislocation dynamics of the ARs is controlled by the kink migration barrier W_m ($W_m > F_k$), while for the SRs the dislocation dynamics is governed by the kink formation energy F_k ($F_k > W_m$) (see Table 2). In general we

Table 2 Kink formation energies $2F_k$ and migration barrier W_m for the 90° partial disloca-tions in the AR (neutral charge state) and SR (negatively or positively charge state). The resulting glide activation energies $Q = 2F_k + W_m$ are relevant for short segment dislocations

Dislocation	Core structure	$\Delta E_{kink\ pair}$/eV	$2F_k$ /eV	W_m eV	$Q = 2F_k + W_m$/eV
C(g) core 90° partials	AR (0 e)	0.12	0.36	1.78	2.14
	SR (−1 e)	0.16	0.40	0.21	0.61
Si(g) core 90° partials	AR (0 e)	0.10	0.34	1.85	2.19
	SR (−1 e)	0.97	1.21	0.34	1.55
C(g) core 30° partials	AR (0 e)	1.29	1.78	1.47	3.25
	SR (−1 e)	0.99	1.48	0.09	1.57
Si(g) core 30° partials	AR (0 e)	1.77	2.26	2.17	4.43
	SR (+1 e)	1.63	2.12	0.06	2.18

have found that the SR dislocations are always more mobile than the AR ones, but in both reconstructions, the C(g) partial possesses the lower activation energy. Furthermore, in all the dislocations the 90° partials present higher mobility than the 30° partials. Therefore at high temperature and when large obstacles are not present, the C(g) core 90° partial is clearly the most mobile dislocation, which is in agreement with the results found by Blumenau et al.[19]

Discussion

Before describing the new model we briefly summarise what at present is suggested to explain the enhancement of the dislocation velocity in SiC. To the best of our knowledge two main models have been proposed:

(a) The model of Blumenau et al.:[19] REDG between Si(g) core partial and SF band level.

These authors show that only the Si(g) core partials are electrically active with a band gap level at $E_v + 0.4$ eV. Because the SFs formed under electron–hole plasma are predominantly of the single-layer type[7,8] with a narrow band at $E_c - 0.2$ eV,[32,33] this model predicts a nonradiative recombination center at about ∼2.7 eV (with a band gap of 3.3 eV). This model is not able to explain the electrical activity of both the 30° partials, and in particular the 1.8 eV radiative emission found under forward bias remains inexplicable.

(b) The model of Pirouz et al.:[10] Soliton model

A soliton is always associated with a dangling bond along the dislocation line and can act as a preferential site for nucleation of kink pairs (Heggie and Jones).[34] Since the Si(g) core dislocation gives rise to a band gap at $E_v + 0.4$ eV and assuming that the soliton dangling bond gives rise to a deep level $E_v + 2.4$ eV acting as both a radiative and a nonradiative site, this model can explain the radiative transition at ∼2.0 eV and provide a level deep enough for the REDG theory (∼2.4 eV). The possible weakness of this model is the soliton formation energy. For example, in silicon the formation energy of a soliton[35–37] is 1.2–1.4 eV, and the Boltzmann probability of having a density of soliton sites n_S along the dislocation line of length l becomes

$$n_S = e^{-\frac{F_S}{kT}} \frac{l}{a_0} \qquad (7)$$

where a_0 is the unitary repeat distance along the dislocation line (3.06 Å). With the temperature range of the measured[4,10] dislocation glide velocity in SiC of 300–500 K ($k_B T = 0.026 - 0.043$ eV), and a formation energy of $F_S = 1.2$ eV, we need a dislocation line of length ranging between ∼10^8–1 km in order to have ∼3–2 solitons in thermal equilibrium (Boltzmann constant $k_B = 8.617 \times 10^{-5}$ eV K^{-1}). In a later study we will show that the Si(g) soliton formation energy in 4H–SiC is less than a soliton in Si but still too large to explain the enhancement of the dislocation velocity alone.

Our theoretical study shows that both dislocations can support the symmetric and asymmetric reconstructions. In the AR only the Si(g) dislocation is electrically active with an energy level of ~ 0.2 eV above the VB. The SRs characterized by dangling bonds on like atoms along the dislocation line are always electrically active. In the neutral charge state the C(g) 90° partial gives rise to a deep band ranging from the top of the VB to $E_v + 1.97$ eV, while the Si(g) gives rise to a band ranging from $E_v + 0.21$ eV to the top of the CB. In a similar way, the C(g) 30° partial also gives rise to a deep band ranging from the bottom of the VB to $E_v + 1.46$ eV, while the Si(g) core gives rise to a deep band ranging from $E_v + 0.44$ eV to $E_v + 2.36$ eV.

The kink migration analysis shows that the SRs are always more likely to move. For C(g) 90° partials, the activation energy Q is lowered from 2.14 eV for the AR to 0.61 eV for the SR, while for the Si(g) core dislocations, the activation energy Q is lowered from 2.19 eV for the AR to 1.55 eV for the SR. For 30° partials, the activation energy Q is lowered from 3.25 eV for the AR to 1.57 eV for the SR, while for the Si(g) core dislocations, the activation energy Q is lowered from 4.43 eV for the AR to 2.18 eV for the SR.

Our results show that the AR does not possess band gap levels deep enough, as required by the REDG mechanism. Therefore the AR dislocations cannot explain the enhancement of the dislocation mobility. For SR dislocations the C(g) 90° partial and both the 30° partials present deep levels that can act as electron–hole recombination center, as required by REDG theory. At this stage we can only observe that Si(g) 30° partials possess a $E_v + 2.36$ deep level at the Γ-point, very close to the nonradiative center revealed by excitation spectrometry experiment (~ 2.4 eV above the valence band). Therefore, the following model is presented:

(c) New model: Savini model†

Under electron–hole plasma injections, the free energy of the SR 30° partials are dynamically lowered by continuous electron–hole transitions between the respective deep levels and valence/conduction bands.

To stabilize the SR 90° partials, a shear between the unfaulted and stacking fault regions at the core of the partials along the dislocation line is required, while for the 30° partials the AR does not require a long ranged shear, but rather only requires flipping of alternate atoms in the core. Therefore only for the 30° partials does the SR dislocation line become more stable than the AR with a strong dynamic charge screening provided by the continuous electron–hole plasma injections. The deep levels provided by the SR, are dynamically positive (hole recombination) and negatively (electron recombination) charged. However, the strong charge screening of the dislocation line surrounded by an electron–hole plasma freezes the deep levels inside the band gap, *i.e.* the 30° partial deep levels correspond to the respective neutral band structures.

The Si(g) 30° partials provide a deep band ranging from $D_1 = E_v + 2.36$ eV at the Γ-point of the Brillouin zone to $D_2 = E_v + 0.44$ eV at the M-point of the Brillouin zone. The level D_1 explains the latest optical activation energy for the dislocation glide at ~ 2.4 eV above the VB.[10] This deep level acts as the electron–hole recombination center as required by the REDG theory and explains why the Si(g) dislocations move under forward bias/optical excitation. A radiative transition of 2.82 eV between the bottom of the conduction band ($E_c - E_v = 3.26$ eV) and D_2 explains the narrow peak at ~ 2.87 eV found by the photoluminescence spectra,[10] while the radiative transition of 1.9 eV between D_1 and D_2 (here called T_{Si}) explains the ~ 1.8 eV electroluminescence peak found during the growing of the stacking faults.[4,10]

---

† Developed by Gianluca Savini.

The C(g) 30° partials provide a deep band ranging from $D_{C(g)} = E_v + 1.46$ eV to the top of the VB. This level can provide a radiative/nonradiative transition of 1.8 eV between the bottom of the CB to $D_{C(g)}$ (here called T_C) and a nonradiative transition of 1.46 eV between the top of the VB to $D_{C(g)}$. These results are in agreement with the latest EBIC experiment reporting that both types of 30° partials act as nonradiative centers.[11]

However, optical studies show that the radiative transition rate of the Si(g) partials is much higher than the corresponding C(g) core dislocations.[12] Here we observe that both the respective radiative transitions T_{Si} and T_C are of indirect type, *i.e.* with electron–phonon coupling. In particular, the two transitions involve the creation of phonons at different points of the Brillouin zone. We suggest that the reason why the radiative transition rate on the Si(g) partials is higher than the corresponding C(g) core dislocations is due to the different kind of phonons created that could hinder/increase the stability of the SR dislocation line. The same reason can explain why only the Si(g) core 30° partials are mobile under electron–hole plasma injection.

Conclusions

First-principle calculations show that both the dislocations can support the symmetric and asymmetric reconstructions. In the AR only the Si(g) core partials present a band gap level, while all the SR dislocations are electrically active. In particular, we have shown that the Si(g) 30° partials can explain the optical activation energy for the dislocation glide at ~ 2.4 eV above the VB, the narrow peak at 2.87 eV and the broad band at ~ 1.8 eV found in photoluminescence spectra. We have also proposed a new model that can explain why under electron–plasma injections the symmetric reconstruction become dynamically more stable, and suggest why the Si(g) 30° partials are the most mobile dislocations. This model can be applied to any semiconductor material in order to predict the behaviour under electron–hole plasma injection.

Acknowledgements

The authors thank the Swedish National Research Council for financial support. S. Gianluca thanks Serguei I. Maximenko for fruitful advice. S. Gianluca thanks his future wife Erminia Carillo for her patience during the process of developing the new model.

References

1 F. A. Stahl, *Am. J. Phys.*, 2003, **71**, 1170.
2 H. Morkoç, S. Strite, G. B. Gao, M. E. Lin, B. Sverdlov and M. Burns, *J. Appl. Phys.*, 1994, **76**, 1363.
3 J. P. Bergman, H. Lendenmann, P. Å. Nilsson, U. Lindefeldt and P. Skytt, *Mater. Sci. Forum*, 2001, **299**, 353–356.
4 A. Galeckas, J. Linnros and P. Pirouz, *Appl. Phys. Lett.*, 2002, **81**, 883.
5 S. Ha, M. Skowronski, J. J. Sumakeris, M. J. Paisley and M. K. Das, *Phys. Rev. Lett.*, 2004, **92**, 175504.
6 M. H. Hong, A. V. Samant and P. Pirouz, *Philos. Mag. A*, 2000, **80**, 919–935.
7 J. Q. Liu, M. Skowronski, C. Hallin, R. Söderholm and H. Lendenmann, *Appl. Phys. Lett.*, 2002, **80**, 749.
8 S. Ha, K. Hu, M. Skowronski, J. J. Sumakeris, M. J. Paisley and M. K. Das, *Appl. Phys. Lett.*, 2004, **84**, 5267.
9 S. Ha, M. Skowronski and H. Lendenmann, *J. Appl. Phys.*, 2004, **96**, 393.
10 A. Galeckas, J. Linnros and P. Pirouz, *Phys. Rev. Lett.*, 2006, **96**, 025502.
11 S. I. Maximenko, P. Pirouz and T. S. Sudarshan, *Appl. Phys. Lett.*, 2005, **87**, 033503.
12 S. Ha, M. Benamara, M. Skowronski and H. Lendenmann, *Appl. Phys. Lett.*, 2003, **83**, 4957.

13 M. Skowronski, J. Q. Liu, W. M. Vetter, M. Dudley, C. Hallin and H. Lendenmann, *J. Appl. Phys.*, 2002, **92**, 4699.
14 M. E. Twigg, R. E. Stahlbush, M. Fatemi, S. D. Arthur, J. B. Fedison, J. B. Tucker and S. Wang, *Appl. Phys. Lett.*, 2003, **82**, 2410.
15 P. Pirouz, M. Zhang, J. L. Demenet and H. M. Hobgood, *J. Appl. Phys.*, 2003, **93**, 3279.
16 J. D. Weeks, J. C. Tully and L. C. Kimerling, *Phys. Rev. B*, 1975, **12**, 3286.
17 H. Sumi, *Phys. Rev. B*, 1984, **29**, 4616.
18 K. Maeda, S. Takeuchi, in *Dislocation in Solids*, ed. F. R. N. Nabarro and M. S. Duesbery, North-Holland, Amsterdam, 1996, vol. 10, pp. 443–504.
19 A. T. Blumenau, C. J. Fall, R. Jones, S. Öberg, T. Frauenheim and P. R. Briddon, *Phys. Rev. B*, 2003, **68**, 174108.
20 J. P. Perdew and Y. Wang, *Phys. Rev. B*, 1992, **45**, 13244.
21 P. R. Briddon and R. Jones, *Phys. Status Solidi B*, 2000, **217**, 131.
22 C. Hartwigsen, S. Goedecker and J. Hutter, *Phys. Rev. B*, 1998, **58**, 3641.
23 H. J. Monkhorst and J. D. Pack, *Phys. Rev. B*, 1976, **13**, 5188.
24 G. L. Harris, in *Properties of Silicon Carbide*, ed. G. L. Harris, Institution of Electrical Engineers, London, UK, 1995, 4.
25 M. M. de Araújo, J. F. Justo and R. W. Nunes, *Appl. Phys. Lett.*, 2004, **85**, 5610.
26 D. B. Laks, C. G. Van de Walle, G. F. Neumark, P. E. Blöchl and S. T. Pantelides, *Phys. Rev. B*, 1992, **45**, 10965.
27 F. Bernardini and L. Colombo, *Phys. Rev. B*, 2005, **72**, 085215.
28 J. P. Hirth and J. Lothe, *Theory of Dislocations*, Wiley, New York, 2nd edn, 1982, p. 244.
29 G. M. Amulele, M. H. Manghnani, B. Li, D. J. H. Errandonea, M. Somayazulu and Y. Meng, *J. Appl. Phys.*, 2004, **95**, 1806.
30 S. Karmann, R. Helbig and R. A. Stein, *J. Appl. Phys.*, 1989, **66**, 3922.
31 N. Oyama and T. Ohno, *Phys. Rev. Lett.*, 2004, **93**, 195502.
32 U. Lindefelt, H. Iwata, S. Öberg and P. R. Briddon, *Phys. Rev. B*, 2003, **67**, 155204.
33 M. S. Miao, S. Limpijumnong and W. R. L. Lambrecht, *Appl. Phys. Lett.*, 2001, **79**, 4360.
34 M. I. Heggie and R. Jones, *Philos. Mag. B*, 1983, **48**, 365.
35 M. I. Heggie, R. Jones and A. Umerski, *Phys. Status Solidi A*, 1993, **138**, 383.
36 R. W. Nunes, J. Bennetto and D. Vanderbilt, *Phys. Rev. Lett.*, 1996, **77**, 1516.
37 C. P. Ewels, S. Leoni, M. I. Heggie, P. Jemmer, E. Hernández, R. Jones and P. R. Briddon, *Phys. Rev. Lett.*, 2000, **84**, 690.

Positional disorder and diffusion path of oxide ions in the yttria-doped ceria $Ce_{0.93}Y_{0.07}O_{1.96}$

Masatomo Yashima,*[a] Syuuhei Kobayashi[a] and Tadashi Yasui[b]

Received 8th February 2006, Accepted 19th May 2006
First published as an Advance Article on the web 3rd August 2006
DOI: 10.1039/b601806h

The scattering amplitude distribution of an yttria-doped ceria material ($Ce_{0.93}Y_{0.07}O_{1.96}$, space group: $Fm\bar{3}m$) has been investigated between 23 and 1434 °C by the maximum-entropy method (MEM) combined with a Rietveld analysis using neutron powder diffraction data. The refined unit cell and atomic displacement parameters increased with an increase in temperature. The results of the MEM analysis reveal that the oxide ions have a positional disorder spreading over a wide area. One possible diffusion path of the oxide ions lies on the tie line along the $\langle 100 \rangle$ directions. The other pathway of the oxide ions can be seen along the $\langle 110 \rangle$ directions. The curved feature in the diffusion path would be common in various ionic conductors.

1. Introduction

Atomic transport in solids is one of the central themes in contemporary solid state science and technology. Crystalline ionic conductors have been extensively studied by numerous researchers because of their many applications in solid oxide fuel cells (SOFCs), sensors, catalysts, and batteries. The development of better electrolyte materials requires a better understanding of the mechanism of ionic conduction, and crucial to this is a comprehensive knowledge of the crystal structure. Therefore, many researchers have studied the crystal structure and diffusion path of the ionic conductors.[1,2] It is well known that the ionic conductor has a positional disordering for the mobile ions. To describe the spatial distribution and disorder of the mobile ions, various techniques such as the split-atom model, anharmonic thermal motions, the probability density function, and Fourier synthesis have been applied. Recently the maximum-entropy method (MEM) has been developed for the determination of accurate structural features through the electron and nuclear density distribution in the crystalline materials.[3-8] We have investigated the crystal structure, structural disorder and diffusion path of mobile ions in the crystalline ionic conductors such as Bi_2O_3,[3] $Bi_{1.4}Yb_{0.6}O_3$,[4] $(La_{0.8}Sr_{0.2})(Ga_{0.8}Mg_{0.15}Co_{0.05})O_{3-\delta}$,[5] CeO_2,[6,7] and $La_{0.62}Li_{0.16}TiO_3$.[8] In the present work we focus on an oxide-ion conductor, yttria-doped ceria solid solution, $Ce_{0.93}Y_{0.07}O_{1.96}$.

Ceria (cerium dioxide, CeO_2) based compounds are attractive materials for potential use as electrolytes in SOFCs and gas sensors, because the materials have

[a] Department of Materials Science and Engineering, Interdisciplinary Graduate School of Science and Engineering, Tokyo Institute of Technology, Nagatsuta-cho 4259-J2-61, Midori-ku, Yokohama-shi, 226-8502, Japan. E-mail: yashima@materia.titech.ac.jp; Fax: +81-45-924-5630; Tel: +81-45-924-5630
[b] Daiichi Kigenso Kagaku Kogyo Company, Hirabayashi-Minami 1-6-38, Suminoe-ku, Osaka-shi 559-0025 Japan

a higher oxide-ion (O^{2-}) conductivity than that of stabilized zirconia[9] and a lower cost compared with lanthanum gallate-based phases.[10,11] Other promising applications of ceria-based materials include automotive catalysts, SOFCs anode materials, solid-electrolyte oxygen pumps, and mixed-conducting membranes for oxygen separation and partial oxidation of hydrocarbons.[12] Doped ceria with oxides such as yttria have high oxide-ion conductivity and have been extensively studied by a number of researchers.[13–16]

The development of ceria-based materials requires knowledge of the disorder and diffusion path of mobile ions in ceria-based materials at high temperatures where they work efficiently. The crystal structure of $CeO_{2-\delta}$ at high temperatures has been investigated by some researchers.[17–21] Berber et al.[20] studied the scattering amplitude distribution in $Ce_{1-x}Y_xO_{2-x/2}$ samples at high temperatures ($x = 0$–0.18). However, these previous works on the disorder of oxide ions were insufficient. Therefore we have studied the positional disorder of oxide ions in ceria at high temperatures using the MEM technique.[6,7] In the present work we investigate the crystal structure, disorder and diffusion path of the oxide-ion conductor, yttria-doped ceria solid solution, $Ce_{0.93}Y_{0.07}O_{1.96}$. We have chosen this composition, $Ce_{0.93}Y_{0.07}O_{1.96}$, because it exhibits the maximum electrical conductivity found for ceria–yttria solid solutions.[13]

2. Experiments and data processing

A ceria–yttria solid solution $Ce_{0.93}Y_{0.07}O_{1.96}$ sample was prepared from a precursor carbonate including Ce and Y atoms. The precursor carbonate was calcined at 1200 °C and then crushed and ground by a ball-milling technique. The $Ce_{0.93}Y_{0.07}O_{1.96}$ powders thus obtained were pressed into pellets, and then sintered at 1500 °C for 3 h in air. The sintered product was cylindrical with the size of 19 mm in diameter and 30 mm in height. These pellets were used for the high temperature neutron diffraction measurements. Inductively coupled plasma (ICP) emission spectroscopy indicated that the chemical formula of the sintered product was $Ce_{0.927}Y_{0.073}O_{1.962}$.† The measured density was 6.98 g cm^{-3}.

Neutron powder diffraction experiments were carried out in air in the temperature range 23–1434 °C (Table 1). Neutron powder diffraction measurements were conducted in air with a 150-detector system, HERMES,[22] installed at the JRR-3M reactor of the Japan Atomic Energy Association, Tokai, Japan. Neutrons

Table 1 Crystallographic parameters and reliability factors of Rietveld analysis of $Ce_{0.93}Y_{0.07}O_{1.96}$ measured at high temperatures[a]

Atom Site			Temperature/°C							
			23	178	394	594	798	998	1198	1434
$Ce_{0.927}$ CY (4a) $Y_{0.073}$	B(CY)/Å^2		0.36(3)	0.52(5)	0.60(5)	0.98(4)	1.15(6)	1.43(3)	1.62(4)	1.81(6)
O O (8c)	B(O)/Å^2		0.57(3)	0.77(5)	1.00(5)	1.50(4)	1.88(6)	2.36(3)	2.77(3)	3.26(6)
Unit-cell parameter/Å				5.4055(2)	5.4143(3)	5.4269(3)	5.4398(3)	5.4550(3)	5.4710(2)	5.4857(3)	5.5065(3)
Reliability	R_{wp}(%)		9.41	10.38	9.77	8.65	8.88	6.96	6.55	6.58
factors in the	R_I(%)		1.91	2.66	2.63	1.14	1.91	1.39	1.74	2.19
Rietveld analysis	R_F(%)		1.11	1.34	1.23	0.61	0.79	0.75	1.02	1.36

[a] Space group: $Fm\bar{3}m$; B(X): Atomic displacement parameter at the X site; g(O): Occupancy factor at the 8c O site, g(O) = 0.981 75.

---

† The ICP analysis also showed that the impurities in the sintered pellets were 0.006 wt% Fe_2O_3, 0.006 wt% SiO_2, 0.001 wt% Na_2O and 0.004 wt% CaO.

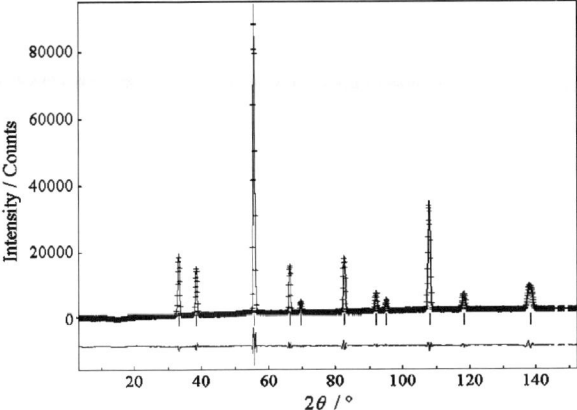

Fig. 1 Rietveld fitting pattern of $Ce_{0.93}Y_{0.07}O_{1.96}$ measured at 1434 °C. The crosses (+ symbols) and line denote observed and calculated intensities, respectively. Short verticals indicate the positions of possible Bragg reflections of the cubic fluorite-type structured $Ce_{0.93}Y_{0.07}O_{1.96}$. The line below the profile stands for the difference between observed and calculated intensities.

with wavelength 1.820 35 Å were obtained by the (331) reflection of a Ge mono-chromator. Diffraction data were collected in the 2θ range from 5–153° in the step interval of 0.1°, from 23–1434 °C (eight different temperature points). A furnace with $MoSi_2$ heaters was placed on the sample table, and used for neutron diffraction measurements at high temperatures.[23] Sample temperature was kept constant within ±1.5 °C during each data collection.

The neutron diffraction data were analyzed by a combination technique of the Rietveld analysis[24] and the MEM.[25,26] The Rietveld analysis was carried out using the computer program RIETAN-2000.[24] It is well known that the MEM can produce a scattering amplitude distribution map from the neutron diffraction data. In a MEM analysis, any kind of complicated scattering amplitude distribution is allowed as long as it satisfies the symmetry requirements. The MEM calculations were carried out using the computer program PRIMA,[26] with $128 \times 128 \times 128$ pixels.

3. Results and discussion

3.1 Temperature dependence of the neutron diffraction profile and Rietveld analysis of $Ce_{0.93}Y_{0.07}O_{1.96}$ at high temperatures

All the reflections in the neutron powder diffraction patterns of $Ce_{0.93}Y_{0.07}O_{1.96}$ in the whole temperature range of 23–1434 °C were indexed by a single phase with the cubic fluorite-type structure. Rietveld analysis was carried out assuming the fluorite-type structure with $Fm\bar{3}m$ symmetry where the cations (Ce^{4+} and Y^{3+}) and anions (O^{2-} and its vacancy) were put at $4a$ 0,0,0 and at $8c$ 1/4,1/4,1/4, respectively. The isotropic atomic displacement parameters were used for all of the atoms. The calculated profile agreed well with the observed data (Fig. 1). The refined crystallographic parameters and the reliability factors in the Rietveld analyses are shown in Table 1. The unit cell parameter increased with an increase in temperature as shown in Fig. 2. Both the isotropic atomic displacement parameters for the cations $B(CY)$ and anions $B(O)$ also increased with temperature as shown in Fig. 3. The $B(O)$ was larger than the $B(CY)$, suggesting a larger diffusion coefficient of oxide ions. These features are consistent with the previous results in Rietveld analyses for the non-doped and doped ceria materials.[6,7,17–21]

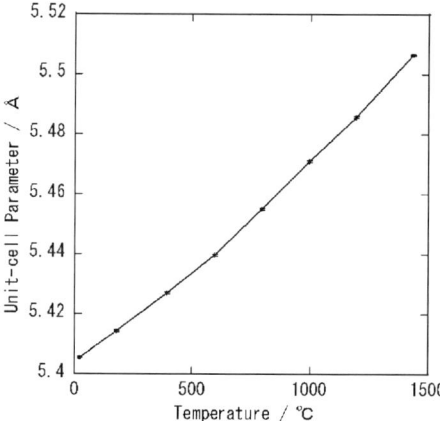

Fig. 2 Temperature dependence of the unit cell parameter of the $Ce_{0.93}Y_{0.07}O_{1.96}$ material. The unit cell parameter increases with an increase in temperature due to the thermal expansion.

3.2 Scattering amplitude distribution, disorder and diffusion path of oxide ions in the $Ce_{0.93}Y_{0.07}O_{1.96}$ at 1434 °C

MEM analyses were carried out using the structure factors obtained in the Rietveld analysis. The number of structure factors derived in the analysis was 11. In the MEM calculations we used the 111 peak intensity at the lowest 2θ position, which provides the most important information for the MEM analysis. The MEM map provided much information on the positional disorder of oxide ions (Fig. 4 and 5), compared with the structural model obtained in the Rietveld analysis. The conventional simple models consisted of atom spheres are no longer appropriate to describe the positional disorder of oxide ions at high temperatures. The MEM scattering amplitude distribution maps on the (110) plane are shown in Fig. 5 to visualize the positional disorder at 1434 °C. The present results reveal that the oxide ions in the $Ce_{0.93}Y_{0.07}O_{1.96}$ have a complicated disorder and spread over a wide area, compared with the cations. The spatial distribution of oxide ions at 1434 °C was larger than that at 23 °C (Fig. 4), corresponding to the larger atomic displacement

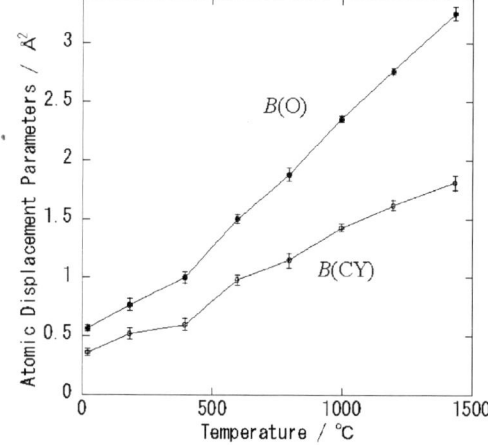

Fig. 3 Temperature dependence of the atomic displacement parameters of the $Ce_{0.93}Y_{0.07}O_{1.96}$ material. Open and filled circles denote the atomic displacement parameters of cations $B(CY)$ and anions $B(O)$, respectively.

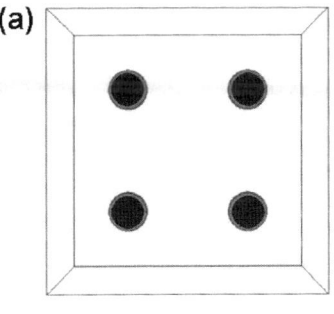

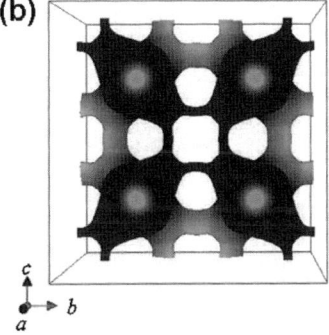

Fig. 4 Equicontour surface of the scattering amplitude distribution at 0.003 fm Å^{-3} with the scattering amplitude distribution on the b–c plane of $Ce_{0.93}Y_{0.07}O_{1.96}$ obtained by the combination technique of Rietveld refinement and the MEM of neutron powder diffraction data measured at (a) 23 °C and (b) 1434 °C ($0.15 < x < 0.3$).

parameters (Fig. 3 and Table 1). There were two types of bulges in the MEM distribution map (Fig. 5). One bulge leads to path (A) in Fig. 5(b). The other exists along the $\langle 110 \rangle$ directions. The oxide ions move away from the Ce and Y cations. The bulges are attributable to the repulsion between the cations and oxide ions. Such an anisotropic feature in the neutron scattering amplitude distribution has been observed in the ceria materials.[6,7,20] The shift of oxide ions in the $\langle 111 \rangle$ directions has been reported in the bismuth oxide-based compounds[3,4,27,28] and β′ phase in the Mg–Zr–O–N system.[29] These compounds have a defect fluorite-type structure, similar to the ceria-based materials. Thus, this feature might be common in fluorite-type structured materials.

The most striking feature in the scattering amplitude distribution map is the diffusion path shown by the arrows of (A) and (B) in Fig. 5(b). The connected distribution on path (A) between the two stable positions of oxide ions indicates the diffusion path (A) along the $\langle 100 \rangle$ directions. The bulges in the $\langle 110 \rangle$ directions suggests the diffusion path (B) along the $\langle 110 \rangle$ directions. It should be noted that the diffusion paths along the $\langle 100 \rangle$ directions were not seen in undoped CeO_2,[6,7] even for lower density levels. The difference in the scattering amplitude distribution between undoped and doped ceria compounds would reflect that in the conductivity between them. Namely, the observation of the diffusion pathway of oxide ions in $Ce_{0.93}Y_{0.07}O_{1.96}$ corresponds to the higher ionic conductivity compared with un-doped CeO_2. Similar diffusion paths in the $\langle 100 \rangle$ directions have been observed in the scattering amplitude distribution of the fluorite-type structured materials such as $Bi_{1.4}Yb_{0.6}O_3$.[4] Similar curved diffusion paths along the $\langle 100 \rangle$ and $\langle 110 \rangle$ directions were reported in some fluorite-type structured fluorides.[30,31]

It is interesting that the pathway along the $\langle 100 \rangle$ direction is not a straight line ((A′) in Fig. 5(b)), but a curved line ((A) in Fig. 5(b)). This is ascribed to the

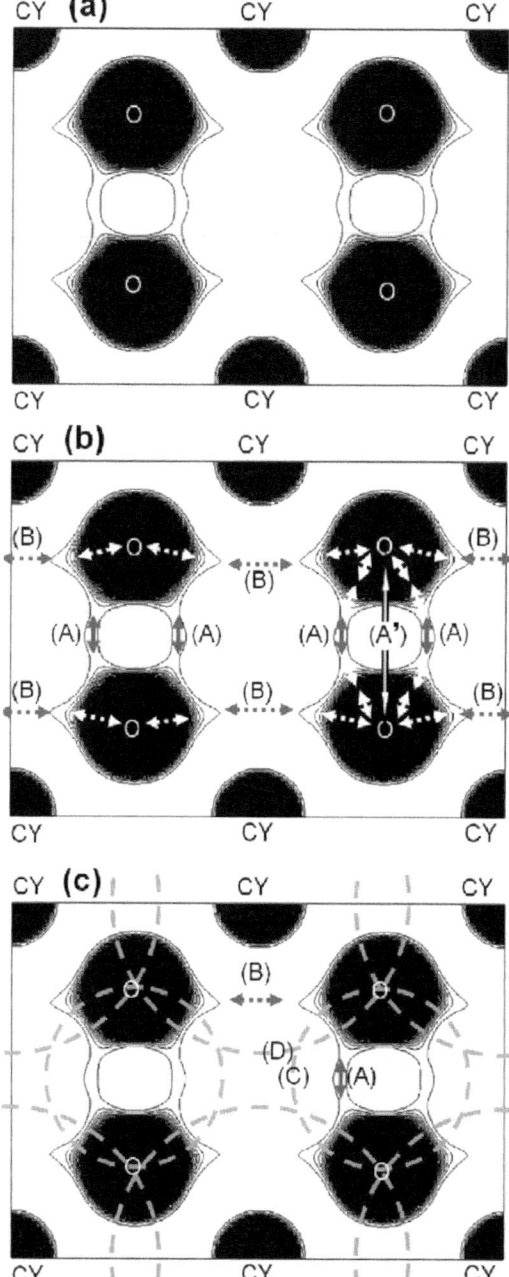

Fig. 5 Scattering amplitude distribution on the (110) plane of the $Ce_{0.93}Y_{0.07}O_{1.96}$ at 1434°C with contours in the range from 0.003 fm $Å^{-3}$ to the maximum (0.003 fm $Å^{-3}$ step). The (A) and (B) arrows in Fig. 5(b) stand for possible diffusion paths in the $\langle 100 \rangle$ and $\langle 110 \rangle$ directions, respectively. The (A') arrow denotes the most direct way between nearest neighbour sites of oxide ions. CY stands for the Ce and Y cations. The O denotes the oxide ions. In the Fig. 5(c), each dashed circle denotes the places where the distance from a cation is constant.

repulsion between the cations and anions. If an oxide ion diffuses through the straight line in the $\langle 100 \rangle$ directions ((A') in Fig. 5(b)), the interatomic distance between the cation and anion becomes too short. Therefore the oxide ion migrates through the curved line (A) in Fig. 5(b) keeping the interatomic distance between the cation and anion to some degree (Fig. 5(c)). Around the middle point of the diffusion pathway ((A) in Fig. 5(c)), the interatomic distance between the cation and anion is a little shorter than that between the stable positions of cations and anions ((C) in Fig. 5(c)). The positions of path (B) are also different from the circle ((D) in Fig. 5(c)). Such curved features of the diffusion pathway were observed in the scattering amplitude distribution in $Bi_{1.4}Yb_{0.6}O_3$,[4] $(La_{0.8}Sr_{0.2})(Ga_{0.8}Mg_{0.15}Co_{0.05})O_{3-\delta}$,[5] and $La_{0.62}Li_{0.16}TiO_3$.[8] This curved feature in the diffusion path would be common in various ionic conductors.

4. Conclusions

The scattering amplitude distribution of $Ce_{0.93}Y_{0.07}O_{1.96}$ (space group:$Fm\bar{3}m$) has been investigated between 23 and 1434 °C by MEM analysis combined with the Rietveld method using *in situ* neutron powder diffraction data. The refined unit cell and atomic displacement parameters increased with an increase in temperature (Table 1, Fig. 2 and 3). The results of the MEM analysis reveal that the oxide ions have a positional disorder spreading over a wide area. The disorder of the oxide ions is more significant at higher temperatures as shown in Fig. 4. One possible diffusion path of the oxide ions lies on the curved line along the $\langle 100 \rangle$ directions ((A) in Fig. 5(b)). The other pathway of the oxide ions can be seen along the $\langle 110 \rangle$ directions ((B) in Fig. 5(b)). The curved feature in the diffusion path would be common in various ionic conductors.

Acknowledgements

We thank Prof. K. Ohoyama and Mr K. Nemoto for the use of the HERMES diffractometer. This research was partially supported by the Ministry of Education, Science, Sports and Culture of Japan, Grant-in-Aid. Fig. 4 and 5 were drawn using the computer program *VENUS* developed by Dr R. Dilanian and Dr F. Izumi.

References

1 S. Hull, *Rep. Prog. Phys.*, 2004, **67**, 1233.
2 J. R. Tolchard, M. S. Islam and P. R. Slater, *J. Mater. Chem.*, 2003, **13**, 1956.
3 M. Yashima and D. Ishimura, *Chem. Phys. Lett.*, 2003, **378**, 395.
4 M. Yashima and D. Ishimura, *Appl. Phys. Lett.*, 2005, **87**, 221909.
5 M. Yashima, K. Nomura, H. Kageyama, Y. Miyazaki, N. Chitose and K. Adachi, *Chem. Phys. Lett.*, 2003, **380**, 391.
6 M. Yashima and S. Kobayashi, *Appl. Phys. Lett.*, 2004, **84**, 526.
7 M. Yashima, S. Kobayashi and T. Yasui, *Solid State Ionics*, 2006, **177**, 211.
8 M. Yashima, M. Itoh, Y. Inaguma and Y. Morii, *J. Am. Chem. Soc.*, 2005, **27**, 3491.
9 H. L. Tuller and A. S. Nowick, *J. Electrochem. Soc.*, 1975, **122**, 255.
10 H. Inaba and H. Tagawa, *Solid State Ionics*, 1996, **83**, 1.
11 M. Mogensen, N. M. Sammes and G. A. Tompsett, *Solid State Ionics*, 2000, **129**, 63.
12 A. Trovarelli, *Catalysis by Ceria and Related Materials*, Imperial College Press, London, 2002.
13 D. Y. Wang, D. S. Park, J. Griffth and A. S. Nowick, *Solid State Ionics*, 1981, **2**, 95.
14 P. Sarker and P. S. Nicholson, *Solid State Ionics*, 1986, **21**, 49.
15 G. B. Balazs and R. S. Glass, *Solid State Ionics*, 1995, **76**, 155.
16 C. Tian and S.-W. Chan, *Solid State Ionics*, 2000, **134**, 89–102.
17 J. Faber, Jr, M. A. Seitz and M. H. Mueller, *J. Phys. Chem. Solids*, 1976, **37**, 903.
18 J. Faber, Jr, M. A. Seitz and M. H. Mueller, *J. Phys. Chem. Solids*, 1976, **37**, 909.
19 J. Faber, Jr, *Physica B*, 1988, **150**, 241.
20 K. Berber, U. Martin, Z. Mursic, J. Schneider, H. Boysen and F. Frey, *Mater. Sci. Forum*, 1991, **79–82**, 685.

21   M. Yashima, D. Ishimura, Y. Yamaguchi, K. Ohoyama and K. Kawachi, *Chem. Phys. Lett.*, 2003, **372**, 784.

22   K. Ohoyama, T. Kanouchi, K. Nemoto, M. Ohashi, T. Kajitani and Y. Yamaguchi, *Jpn. J. Appl. Phys., Part 1*, 1998, **37**, 3319.

23   M. Yashima, *J. Am. Ceram. Soc.*, 2002, **85**, 2925.

24   F. Izumi and T. Ikeda, *Mater. Sci. Forum*, 2000, **321–324**, 198.

25   M. Sakata, T. Uno, M. Takata and C. H. Howard, *J. Appl. Crystallogr.*, 1993, **26**, 159.

26   F. Izumi and R. A. Dilanian, *Recent Research Developments in Physics*, Transworld Research Network, Trivandrum, 2002, vol. 3, part II, p. 699.

27   P. D. Battle, R. A. Catlow, J. Drennan and A. D. Murray, *J. Phys. C: Solid State Phys.*, 1983, **16**, L561.

28   P. D. Battle, C. R. A. Catlow, J. W. Heap and L. M. Moroney, *J. Solid State Chem.*, 1986, **63**, 8.

29   M. Lerch, H. Boysen and P. G. Radaelli, *J. Phys. Chem. Solids*, 1997, **58**, 1557.

30   H. Schulz, *Annu. Rev. Mater. Sci.*, 1982, **12**, 351.

31   K. Koto, H. Schulz and R. A. Huggins, *Solid State Ionics*, 1980, **1**, 355.

Oxygen transport in unreduced, reduced and Rh(III)-doped CeO$_2$ nanocrystals

Thi X. T. Sayle,[a] Stephen C. Parker[b] and Dean C. Sayle*[a]

Received 31st January 2006, Accepted 27th April 2006
First published as an Advance Article on the web 7th August 2006
DOI: 10.1039/b601521b

Ceria, CeO$_2$, based materials are a major (active) component of exhaust catalysts and promising candidates for solid oxide fuel cells. In this capacity, oxygen transport through the material is pivotal. Here, we explore whether oxygen transport is influenced (desirably increased) compared with transport within the bulk parent material by traversing to the nanoscale. In particular, atomistic models for ceria nanocrystals, including perfect: CeO$_2$; reduced: CeO$_{1.95}$ and doped: Rh$_{0.1}$Ce$_{0.9}$O$_{1.95}$, have been generated. The nanocrystals were about 8 nm in diameter and each comprised about 16 000 atoms. Oxygen transport can also be influenced, sometimes profoundly, by microstructural features such as dislocations and grain-boundaries. However, these are difficult to generate within an atomistic model using, for example, symmetry operations. Accordingly, we crystallised the nanocrystals from an amorphous precursor, which facilitated the evolution of a variety of microstructures including: twin-boundaries and more general grain-boundaries and grain-junctions, dislocations and epitaxy, isolated and associated point defects. The shapes of the nanocrystals are in accord with HRTEM data and comprise octahedral morphologies with {111} surfaces, truncated by (dipolar) {100} surfaces together with a complex array of steps, edges and corners. Oxygen transport data was then calculated using these models and compared with data calculated previously for CeO$_{1.97}$/YSZ thin films and the (bulk) parent material, CeO$_{1.97}$. Oxygen transport was calculated to increase in the order: CeO$_2$ nanocrystal < (reduced) CeO$_{1.95}$ nanocrystal $\approx$ Rh$_{0.1}$Ce$_{0.9}$O$_{1.95}$ nanocrystal < CeO$_{1.97}$/YSZ thin film < (reduced) CeO$_{1.97}$ (bulk) parent material; the mechanism was determined to be primarily vacancy driven. Our findings indicate that reducing one- (thin film) or especially three- (nanocrystal) dimensions to the nanoscale may prove deleterious to oxygen transport. Conversely, we observed dynamic evolution and annihilation of surface vacancies *via* surface oxygens migrating to the bulk of the nanocrystal; the vacancies left are then filled by other oxygens moving to the surface. Coupled with previous simulation studies, in which we calculated that oxygen extraction from the surface of a ceria nanocrystal was energetically easier compared with the bulk surface, our calculations predict that ceria nanocrystals would facilitate effective oxidative catalysis. This study describes framework simulation procedures, which can be used in partnership with experiment, to explore transport in nanocrystalline ionic systems, which include complex

[a] Dept. Environmental and Ordnance Systems, Cranfield University, Defence Academy of the United Kingdom, Shrivenham, Swindon, UK. E-mail: d.c.sayle@cranfield.ac.uk
[b] Dept. Chemistry, University of Bath, Claverton Down, Bath, Avon UK

microstructures. Such data can provide predictions for experiment or help reduce the number of experiments required.

1. Introduction

Yttrium stabilised zirconia (YSZ) can be used as an electrolyte in electrochemical devices such as solid oxide fuel cells (SOFC).[1] However, to ensure adequate conductivities, the system is necessarily operated at temperatures as high as 1000 °C, which has severe implications such as cost, accelerated ageing[2] and impracticality with respect to its use in electric vehicles. Accordingly, the search for a replacement, which can operate at much lower temperatures, is intense. Extrinsically doped Ceria, $M_{2x}Ce_{1-2x}O_{2-x}$, is a promising candidate because, at lower temperatures, its ionic conductivity is higher than YSZ.[3,4]

CeO$_2$ is also important with respect to exhaust catalysis.[5] In particular, the material can expose at its surface labile oxygen ions that can be extracted and used to oxidise, for example, CO into CO_2.[6–8] Moreover, ceria is able to release a high proportion of its oxygen whilst retaining its fluorite structure even at high temperatures. The surface oxygen vacancy can then move relatively easily to the bulk, to restore the surface oxygens to facilitate further CO oxidation. Oxygen can be replenished by extracting oxygen from another exhaust gas, such as NO_2 and reducing this material to N_2[9] The CeO$_2$ acts therefore as a dual oxidiser/reductor.

Oxygen vacancy formation and oxygen transport through the lattice is central to the exploitation of ceria in both SOFC and catalysis. Accordingly, considerable efforts focus on understanding such processes and how they may be improved or even controlled.

A milestone in this respect is the study of Sata and co-workers on anion transport in metal fluorides. In particular, the authors found that they could increase the fluoride ion conductivity in CaF$_2$ and BaF$_2$, compared with the parent materials, by fabricating heteroepixaxial BaF$_2$/CaF$_2$ layers with nanoscale thicknesses.[10] Moreover, rather than simply increase the conductivity the authors demonstrated tuneable conductivities by controlling the thickness of the heterolayers. A simulation study was performed to help understand this phenomenon[11,12] and revealed that the activation energy associated with the fluoride ion conductivity was much lower for ions in boundary regions compared with those ions in the bulk crystal. In particular, the grain-boundary regions offered energetically facile pathways for the fluoride ions to traverse.

This work prompted researchers to explore whether generating ceria thin films would similarly enhance oxygen transport.[13] Indeed, Azad and co-workers found an increase[14,15] for a system comprising gadolinia-doped CeO$_2$/ZrO$_2$ heterolayered thin films supported on Al$_2$O$_3$.

It is therefore apparent that a multitude of factors can influence transport properties. These include for example, grain-boundaries and dislocations, point and associated intrinsic and extrinsic dopants, epitaxy and the resulting lattice strain as the material accommodates misfit.

There is currently much interest in materials whose properties change significantly as one traverses to the nanoscale. Here, we use atomistic computer simulation to explore whether traversing down to the nanoscale will influence the transport properties of CeO$_2$. In particular, we generate atomistic models of unreduced (CeO$_2$), reduced (CeO$_{1.95}$) and doped (Rh$_{0.1}$Ce$_{0.9}$O$_{1.95}$) ceria nanocrystals and, using these models, calculate the oxygen transport to compare with the bulk parent material. Clearly, the ability to calculate accurate conductivities is dependent upon the atomistic models. However, generating the full (one that includes every atom explicitly) atomistic structure of a nanocrystal is not trivial—realistic atomistic structures cannot simply be cleaved from the bulk material. For example, the morphological structure will comprise many low-energy surfaces. And whilst there are many simulation codes that use the crystal symmetry to generate an individual surface, for a nanocrystal one also needs to consider the structure of edges where a pair of surfaces meet, and vertices where three or more surfaces meet together with the implications of dipolar surfaces and surface steps, edges, corners and niche sites. Moreover, for 'large' nanocrystals ($> 10\,000$ atoms) the system will likely comprise dislocations or grain-boundaries. And whilst there are recently developed simulation strategies to generate the structure of an individual dislocation[16,17] or grain-boundary[18] within the *bulk* material, these methods are not easily transferable to facilitate the introduction of such microstructural features into a finite polyhedral structure. In summary, each atomistic nanocrystal model requires:

- Morphology: Number of surfaces, Miller indices, edges and vertices, dipolar surfaces, steps, corners, niche sites.
- Microstructure: Grain-boundaries, dislocations, intrinsic/extrinsic defects: substitutionals, interstitials, vacancies.

It is certainly a challenge to generate full atomistic models that comprise all these features. However, if the simulator is to offer predictions pertaining to the atomistic structure of a nanocrystal that are both accurate and can be used reliably by the experimentalist, the models must include all of the above microstructural features because they are central to its chemical and physical properties. It is probably unfeasible to construct, in an automated way (perhaps using symmetry operators[16,17]), atomistic models that include all the structural features identified above and therefore we must use some kind of evolutionary method.

Fabrication of a material experimentally inevitably involves some kind of 'crystallisation' process. Indeed, crystallisation processes control the (micro)structure and hence the properties of the material. Moreover, by modifying the crystallisation process (whether crystallisation from flame or solution, vapour deposition, molecular beam epitaxy, ball milling *etc.*) one can exact some control over the microstructure and hence the properties of the material.

The best way of capturing, within a single atomistic model, all the microstructural features observed experimentally, is to simulate the crystallisation process itself. Here we use amorphisation and recrystallisation (A&R)[19] to generate the atomistic models for the ceria nanocrystals. This simulation approach has been shown to evolve accurate models of CeO_2 nanocrystals[20] and we use A&R to generate the atomistic models in this present study.

To generate realistic models for the reduced CeO_{2-x} nanocrystals we draw upon the findings of Zhang and co-workers who used X-ray photoelectron spectroscopy (XPS) and X-ray absorption near edge spectroscopy (XANES) to investigate the oxidation state of cerium ions in ceria nanocrystals.[21] For 10 nm particles they measured, using XPS, the concentration of Ce^{3+} ions, within 1.9 Å of the surface, to be 22.3% compared to 5.6% for the bulk. However, the authors cautioned that XPS yields a higher concentration of Ce^{3+} ions compared with XANES. Indeed, the XANES study on the same system suggests the concentration of Ce^{3+} in the bulk to be only 1%—about five times smaller compared with the XPS measurements. For nanocrystals 6 nm in size, the Ce^{3+} concentrations were found to be higher. Specifically, using XPS they found 25% of the Ce^{3+} at the surface (within a 2.2 Å shell) and 10% in the bulk region. Conversely, with XANES they found the Ce^{3+} concentration to be about 6%. Based upon this data we have generated models for ceria nanocrystals comprising 10% Ce^{3+} with charge neutrality facilitated by introducing oxygen vacancies. Specifically, $Ce_{0.1}^{3+}Ce_{0.9}^{4+}O_{1.95}$. In addition, the location of Ce^{3+} may be an important factor with respect to oxygen transport and therefore our simulations were designed to explore this possibility.

The ionic conductivity of ceria has also been extensively investigated with respect to different extrinsic dopants (for example see ref. 22). Accordingly, we have generated models where the Ce^{3+} (radius = 1.03 Å) have been replaced by Rh^{3+} (0.68 Å) to explore the influence of extrinsic dopants on the conductivity. Rh^{3+} was chosen because of the large size mismatch to Ce^{4+} (radius = 0.92 Å). In particular, the association between extrinsic dopants and oxygen vacancies has been shown to influence the conductivity,[23] which may be attributed to size effects.[24] Extrinsic dopants are likely to occupy Ce^{4+} lattice positions and therefore the structural relaxation around each dopant will be related to the disparity in size between the dopant ion and Ce^{4+}. We note that, experimentally, it has been shown that polycrystalline cerium oxide exhibits *decreasing* ionic conductivity upon reduction of the grain size (for example see ref. 25).

2. Theoretical methods

In this section, we describe briefly the potential models used to represent the interactions between the ions comprising the nanocrystals, the code used to perform the dynamical simulations and the A&R strategy used to evolve the atomistic structures for the nanocrystals. The final section describes the procedure for calculating various transport properties using the atomistic models.

2.1 Potential models and simulation code

The calculations presented in this study are based upon the Born model of the ionic solid in which the ions interact *via* attractive long-range Coulombic terms balanced by short-range repulsive interactions. The (rigid-ion) parameters are presented in Table 1 and have been used successfully to explore oxygen vacancy formation energies in CeO_2,[6,26] Ce^{4+}/Ce^{3+} reduction within CeO_2–ZrO_2 solid solutions,[33] ionic conductivity in nano-scale CeO_2/YSZ heterolayers[27] and segregation of metal ions in CeO_2.[28] The DL_POLY code[29] was used to perform all of the molecular dynamics (MD) simulations.

2.2 Nanocrystal generation

Seven 'cubes' of CeO_2 nanocrystals, comprising 15 972 atoms (5324 Ce and 10 648 O) and {100} surfaces were cut from the parent material. In the first three nanocrystals, 532 Ce^{4+} ions (about 10%) were substituted by Ce^{3+} and located at the surface (S), at the centre (C), and at mixed positions (M). Charge neutrality was facilitated by creating 266 oxygen vacancies (about 5%), which were located near to the Ce^{3+} species. The starting structures are shown in Fig. 1. The nanocrystals are thus $Ce^{3+}_{2x}Ce^{4+}_{1-2x}O_{2-x}$, $x \approx 0.05$. The second set of three nanocrystals were identical to the first three but with Rh^{3+} ions substituting for Ce^{4+} to give $Rh^{3+}_{2x}Ce^{4+}_{1-2x}O_{2-x}$, $x \approx 0.05$. Finally, for the seventh nanocrystal, no defects were introduced and we designate this unreduced or CeO_2 (U). The stoichiometry of the systems were therefore: unreduced nanocrystal: CeO_2 (U); reduced nanocrystals: $CeO_{1.95}$ (S), $CeO_{1.95}$ (C), $CeO_{1.95}$ (M); doped nanocrystals: $Rh_{0.1}Ce_{0.9}O_{1.95}$ (S), $Rh_{0.1}Ce_{0.9}O_{1.95}$ (C), $Rh_{0.1}Ce_{0.9}O_{1.95}$ (M). We shall use this formalism to identify each system in the manuscript.

Each of the seven nanocrystals were then amorphised and recrystallised. In particular, each nanocrystal was tensioned by 38.6% and rigid-ion, constant volume, MD simulation performed at 3750 K (timestep = 0.005 ps) for a duration sufficient

Table 1 Potential parameters of the form $V(r) = A\exp(-r/\rho) - Cr^{-6}$; all cation–cation interactions are set to zero

	A	ρ	C	Ref.
Ce^{4+}–O^{2-}	1986.83	0.351	20.40	6
Ce^{3+}–O^{2-}	1731.62	0.364	14.43	6
Rh^{3+}–O^{2-}	1404.43	0.365	13.12	28
O^{2-}–O^{2-}	22 764.30	0.149	27.89	30

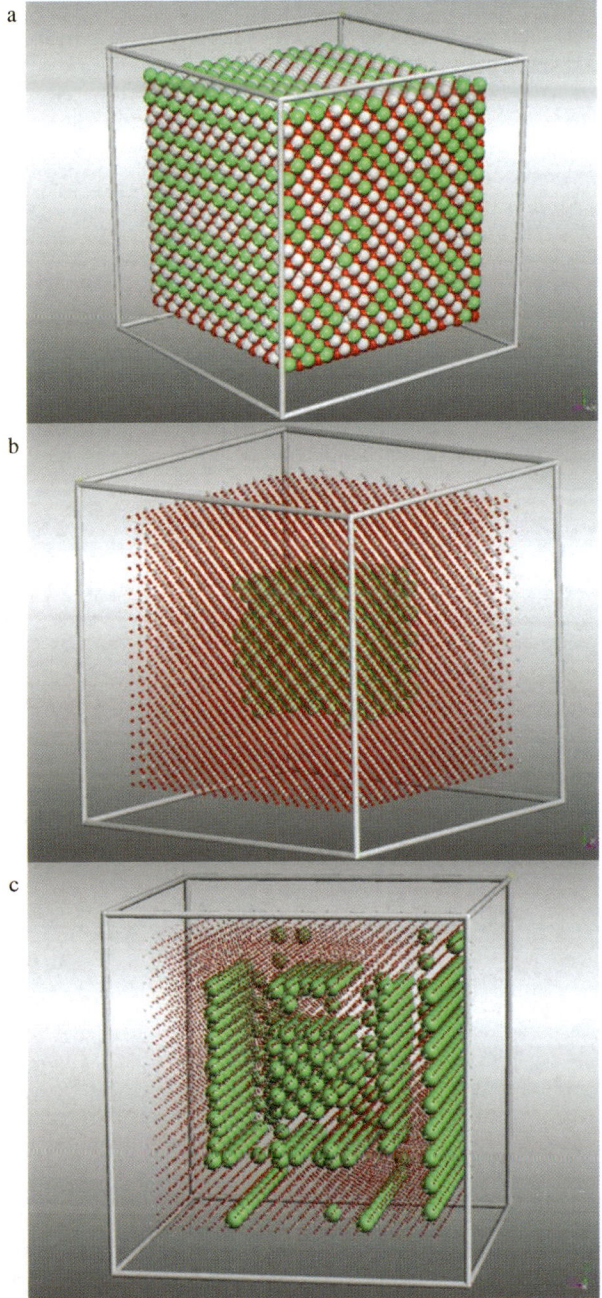

Fig. 1 Representations of the atom positions comprising the starting structures for the ceria nanocrystals. Each system comprises a cube of CeO_2 and differs only in the initial location of the 'defect' species (Ce^{3+}) specifically: (a) $Ce_{0.1}^{3+}Ce_{0.9}^{4+}O_{1.95}$ (Surface, S), (b) $Ce_{0.1}^{3+}Ce_{0.9}^{4+}O_{1.95}$ (Centre, C) and (c) $Ce_{0.1}^{3+}Ce_{0.9}^{4+}O_{1.95}$ (Mixed, M). Ce^{3+} species are coloured green, Ce^{4+} are white and oxygen is red. The sizes of the ions are modified to aid understanding of each structure and do not relate to the ionic radii.

to amorphise and then recrystallise completely the nanocrystals, typically this required about 4000 ps of MD simulation. Tensioning the lattice by 38.6% was achieved simply by multiplying all the coordinates comprising the nanocrystal by 1.386. The amorphisation and recrystallisation is illustrated in Fig. 2 for the Ce^{3+} doped system with Ce^{3+} located initially at the surface (S). Initially, the considerable tension in the system causes (under MD) the ions to accelerate back in an attempt to restore the lattice parameter. The collisions between the ions as they move results in the amorphisation of the system, Fig. 2(b), the nanocrystal then becomes spherical, Fig. 2(c), as the system attempts to minimise its surface energy by reducing the total area exposed. At some point in time during the course of the MD simulation, a (fluorite-structured) crystalline 'seed' evolves 'naturally' within the amorphous sea of ions. Ce and O ions surrounding this seed then start to condense onto its surface propagating the crystallization front, which traverses through the nanocrystal. Fig. 2(d) shows a region that has crystallised and is enlarged in Fig. 2(e) to show the structure more clearly. As the crystallization front impinges the surface, the crystallization was observed to evolve energetically stable {111} surfaces. We note that during the crystallisation, energy is liberated, which reflects the latent heat of

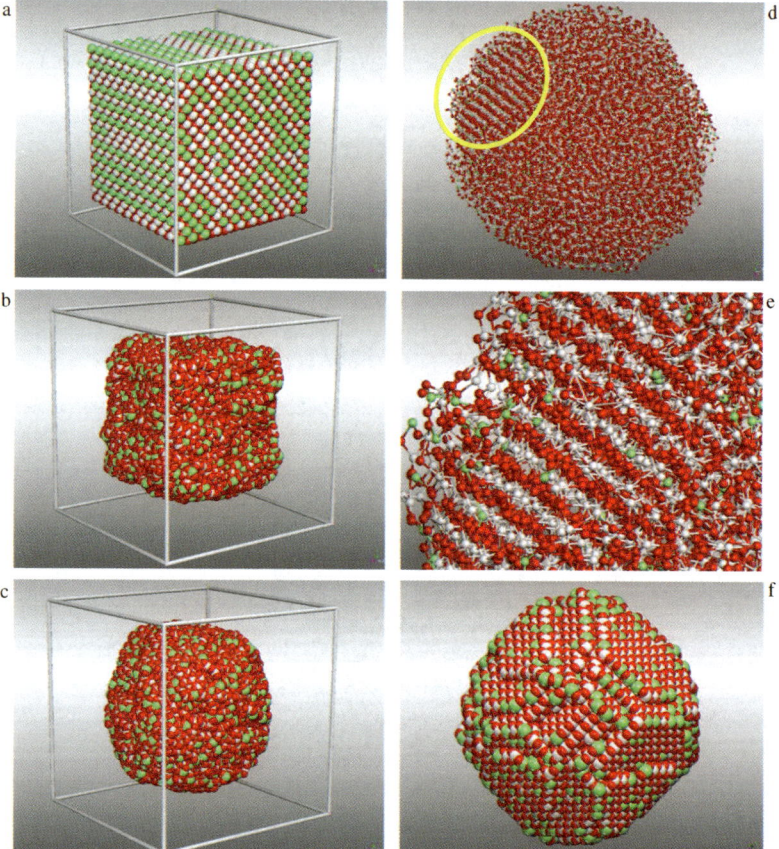

Fig. 2 Representation of the atom positions comprising the $Ce^{3+}_{0.1} Ce^{4+}_{0.9} O_{1.95}$ (S) system during MD simulation. Snapshots of the system were taken at intervals to show clearly the amorphisation and recrystallisation of the nanocrystal. In particular, (a) starting configuration; (b) after 25 ps; (c) after 250 ps (completely amorphous); (d) after 1000 ps showing a crystalline region within the amorphous sea of ions; (e) enlarged segment of '(d)' showing more clearly the crystalline region; (f) final, low (10 K) temperature, structure. Ce^{3+} species are coloured green, Ce^{4+} are white and oxygen is red.

crystallisation. After the system had recrystallised fully (as determined by a plateau in the configurational energy), the system was cooled gradually at intervals: 3500 K for 250 ps, 3000 K for 250 ps, 2000 K for 50 ps, 1000 K for 50 ps and 10 K for 10 ps. Finally, shell model MD was performed for 10 ps at 10 K, which acts effectively as a pseudo energy minimisation. The final structure is shown in Fig. 2(f). Further details describing the amorphisation and recrystallisation of nanocrystals can be found in ref. 19. We note that although the Ce^{3+} dopants originated at the surface, centre or at mixed positions, within the amorphisation state, the ions have high mobility and do not necessarily remain in the vicinity of their starting position.

2.3 Transport properties

Ionic self-diffusion, conductivity and the associated activation energies can be determined using the atomistic model and calculating the mean square displacements (MSD) of the ions at a particular temperature. Further details can be found elsewhere.[27] In particular, once a realistic atomistic model for each nanocrystal had been generated, MD simulation was performed on each system, for 500 ps at 1000, 1175, 1335, 1500, 1700, 2000 and 2500 K. Each simulation included a preliminary 250 ps equilibration to the simulation temperature. The simulations were performed within an NVT ensemble (constant Number of particles, constant Volume and constant Temperature) with thermostat and barostat relaxation times of 1 ps and a 0.005 ps timestep. The MSDs of the ions were then calculated as a function of time.

3. Results

In this section, we describe the atomistic structures comprising the nanocrystals, including their morphological shapes and surfaces exposed together with micro-structural features, such as grain-boundaries, dislocations and defects that the nanocrystals incorporate within their structure. Next, we report the calculated transport properties associated with these nanocrystals. Finally, we correlate the transport properties of the nanocrystals with their atomistic structure. For each section, structure, properties and correlation, we compare closely with available experimental data.

3.1 Structural analysis

The final, low temperature (10 K) structures of the three reduced nanocrystals are shown in Fig. 3 and comprise octahedral morphologies with {111} surfaces truncated by {100}. This is in accord with experimental data from Wang and Feng,[31] which are shown in Fig. 4 for comparison. It is perhaps surprising that the nanocrystals comprise {100} because these surfaces are dipolar.[32] Inspection of the structure of the {100} surfaces reveals that the dipole has been quenched. In particular, the surface layer comprises an equal number of Ce^{4+} species as oxygen. This is seen most easily by looking at Fig. 3(b)—the arrow points to the (100) surface.

Inspection and analysis of the nanocrystals using molecular graphics reveals that they are not all single crystals. In particular, the nanocrystal with Ce^{3+} species located initially at the centre, Fig. 3(b), comprises many grains. A slice cut through this nanocrystal is shown in Fig. 5, which depicts more clearly the grain-boundary structures.

A Ce^{3+} ion occupying a Ce^{4+} lattice site is associated with considerable local lattice strain because of the larger size of the Ce^{3+} ion (1.03 Å) compared with Ce^{4+} (0.92 Å). This strain can be alleviated *via* the formation of grain-boundaries because, for an equivalent number of ions, a grain-boundary is associated with an increase in volume compared with the bulk material and therefore provides more relaxational

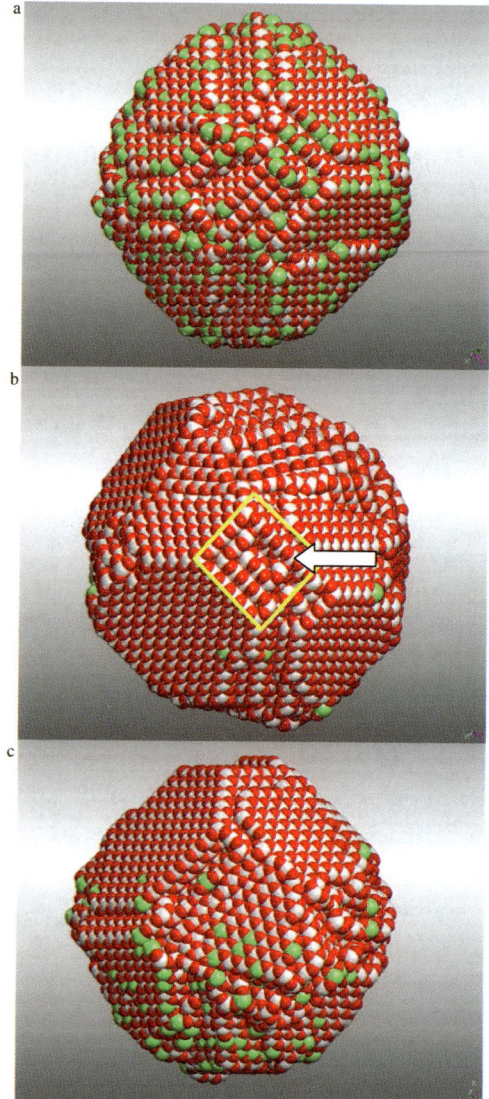

Fig. 3 Sphere model representations of the atom positions for the reduced nanocrystals comprising (a) $Ce_{0.1}^{3+}Ce_{0.9}^{4+}O_{1.95}$ (S); (b) $Ce_{0.1}^{3+}Ce_{0.9}^{4+}O_{1.95}$ (C); (c) $Ce_{0.1}^{3+}Ce_{0.9}^{4+}O_{1.95}$ (M). The graphical images are the final, low temperature structures; Ce^{3+} species are coloured green, Ce^{4+} are white and oxygen is red. The arrow in '(b)' points to a dipolar {100} surface showing that it comprises 50% Ce^{4+} and 50% O^{2-}, which facilitates quenching of the surface dipole.

freedom for the Ce^{3+} species thus reducing the strain energy. It was important to evolve grain-boundaries because they are likely to influence the oxygen transport. The mixed system, $CeO_{1.95}$ (M), also comprises a grain-boundary, which is shown in Fig. 6.

One might argue that the Ce^{3+} ions could have migrated in a way to facilitate a more even distribution of Ce^{3+} within the nanocrystal and therefore help alleviate the considerable strain at the central region. Indeed, one can observe from Fig. 3(b) that some of the Ce^{3+} species have migrated from the centre to the surface of the nanocrystal. However, we suggest that the high speed of crystallisation prevents the

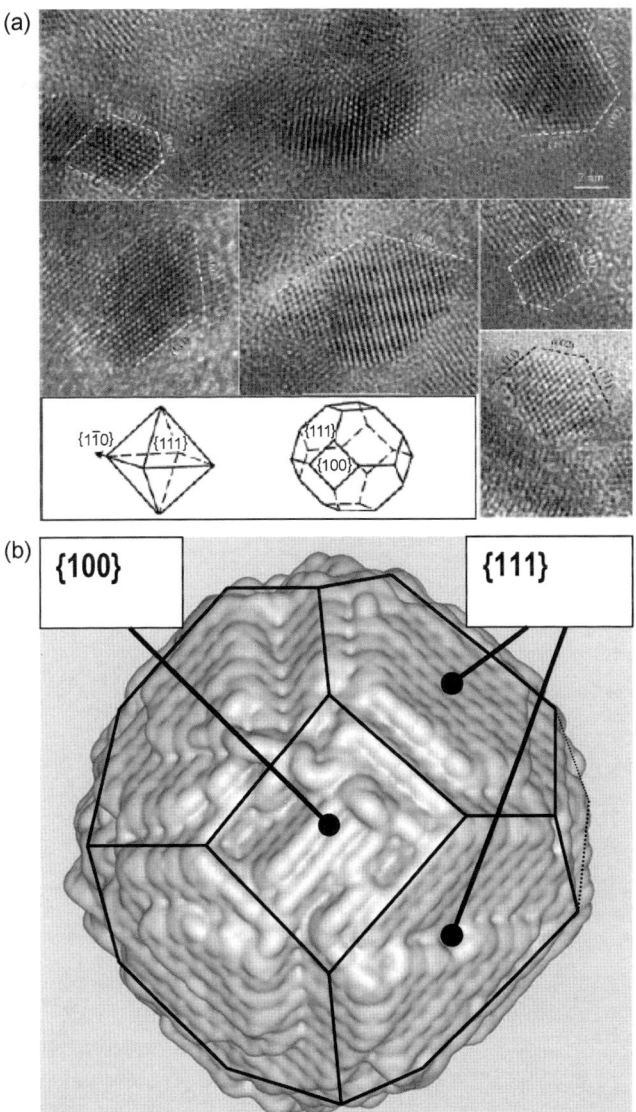

Fig. 4 (a) Experimental HRTEM images of CeO_2 nanocrystals about 10 nm in diameter and showing the octahedral morphologies with {111} surfaces truncated with {100} from the study of Wang and Feng.[31] Reprinted with permission from Z. L. Wang and X. D. Feng, *J. Phys. Chem. B*, 2003, **107**(49), pp. 13563–13566. Copyright 2003, American Chemical Society; (b) surface rendered atomistic model of the final, low temperature, $Ce_{0.1}^{3+}Ce_{0.9}^{4+}O_{1.95}$ (S) system indicating the {111} and {100} surfaces, which can be usefully compared with the HRTEM.

Ce^{3+} from distributing itself evenly throughout the nanoparticle. In particular, Ce^{3+} species have no possibility of migrating through *crystalline* regions within the timescale of the simulation and therefore once the crystalline front impinges upon Ce^{3+} species they are effectively trapped within the crystalline lattice. The only chance the ions have to rearrange significantly is within the amorphous state. For the simulations reported here, the amorphous precursors are *solid* and therefore the ions have less mobility compared to the molten equivalent. Moreover, if we had melted

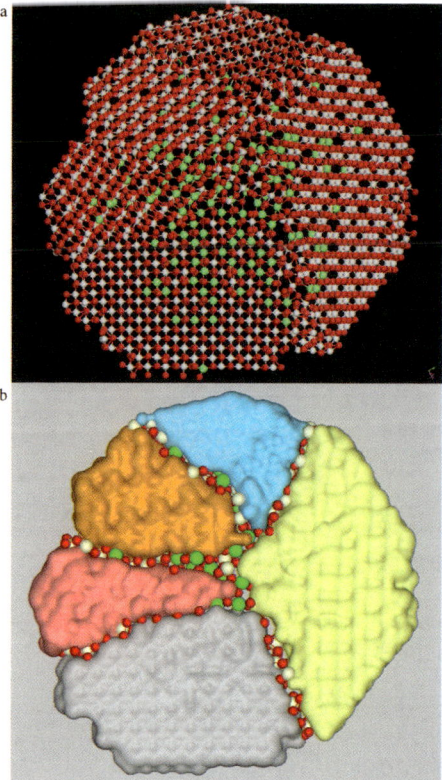

Fig. 5 Molecular graphics representations of the atom positions within a slice cut through the $Ce_{0.1}^{3+}Ce_{0.9}^{4+}O_{1.95}$ (C) system (final, low temperature structure) showing the multiple misoriented grains comprising the nanocrystal. (a) Ball and stick model showing the distribution of Ce^{3+} species in the centre of the nanocrystal and the grain-boundary and grain-junction structures; (b) surface rendered model of '(a)' in which each of the grains are coloured to show more clearly the individual grains. Ce^{3+} species are coloured green, Ce^{4+} are white and oxygen is red.

the nanoparticles, the Ce^{3+} may have formed a more even distribution prior to crystallisation, therefore preventing the evolution of grain-boundaries.

Grain-boundaries have been observed experimentally in CeO_2 nanocrystals and can result from the agglomeration of two nanocrystals, where the nanocrystals join to help reduce the total area of surfaces that are exposed and hence the surface energy, or grain-boundary formation within a single nanocrystal. In Fig. 7, several grain-boundaries are shown and were taken, with permission, from the study of Wang and Feng.[31] In particular, Fig. 7(a) shows two nanocrystals with a matched (111) plane and a twist parallel to the plane so that one nanocrystal is oriented along [$\bar{1}$10] while the other is slightly off the zone axis. The two nanocrystals in Fig. 7(b) have the same [$\bar{1}$10] orientation but there is a small relative twist and therefore the atoms at the interfacial region have to be strained to accommodate the local deformation. A twin boundary can be seen in Fig. 7(c) and edge dislocations, near the surface of the nanocrystal, in Fig. 7(d).

In Fig. 8, the nanocrystals, which have been doped with Rh^{3+} are shown. The morphologies are analogous to those of the (reduced) $CeO_{1.95}$ nanocrystals in that they comprise octahedral morphologies with {111} surfaces and truncated with {100}. The Rh^{3+} ions substitute for Ce^{4+} at lattice positions. In addition, the $Rh_{0.1}Ce_{0.9}O_{1.95}$ (M), system comprises a grain boundary with an edge dislocation.

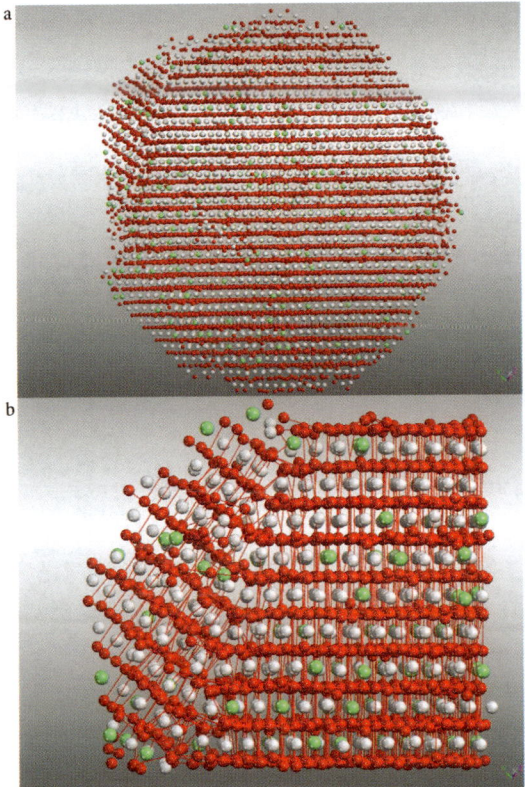

a

b

Fig. 6 Ball and stick representations of the atom positions comprising the $Ce_{0.1}^{3+} Ce_{0.9}^{4+} O_{1.95}$ (M) nanocrystal, final low temperature structure. (a) The ions have been reduced in size to enable one to view through the nanocrystal and see more clearly the grain-boundary, top left. (b) Enlarged segment showing more clearly the boundary plane. Ce^{3+} species are coloured green, Ce^{4+} are white and oxygen is red.

We note that the structure of $Rh_{0.1}Ce_{0.9}O_{1.95}$ (C) does not comprise grain-boundaries (in contrast to $CeO_{1.95}$ (C), Fig. 5). This is because the Rh^{3+} ion is smaller compared with Ce^{4+} ($Rh^{3+} = 0.68$; $Ce^{4+} = 0.92$ Å) and therefore does not require more space to relax the strain energy associated with its incorporation.

3.2 Ionic transport

The predominant mechanism for oxygen transport in reduced ceria is widely accepted to be *via* vacancy hopping.[33] Viewing animations of the MD simulations revealed that oxygen transport is also vacancy driven in the reduced and doped ceria nanocrystals: to illustrate visually the ionic transport, Fig. 9 shows the $Rh_{0.1}Ce_{0.9}O_{1.95}$ (M) nanocrystal at the start (Fig. 9(a)) and after 1800 ps of MD simulation performed at 2500 K, Fig. 9(b). The sphere sizes of the O, Ce^{4+} and Rh^{3+} in two perpendicular planes have been labelled (by enlarging the sphere size) to show how far they move. A schematic, illustrating the vacancy hopping mechanism, is shown in Fig. 9(c).

Inspection of the figure after 1800 ps Fig. 9(b) shows clearly that the oxygen anions move a considerable distance in this time whereas the cation mobility is negligible. In addition, we see no evidence of enhanced mobility at the surface of the nanocrystal.

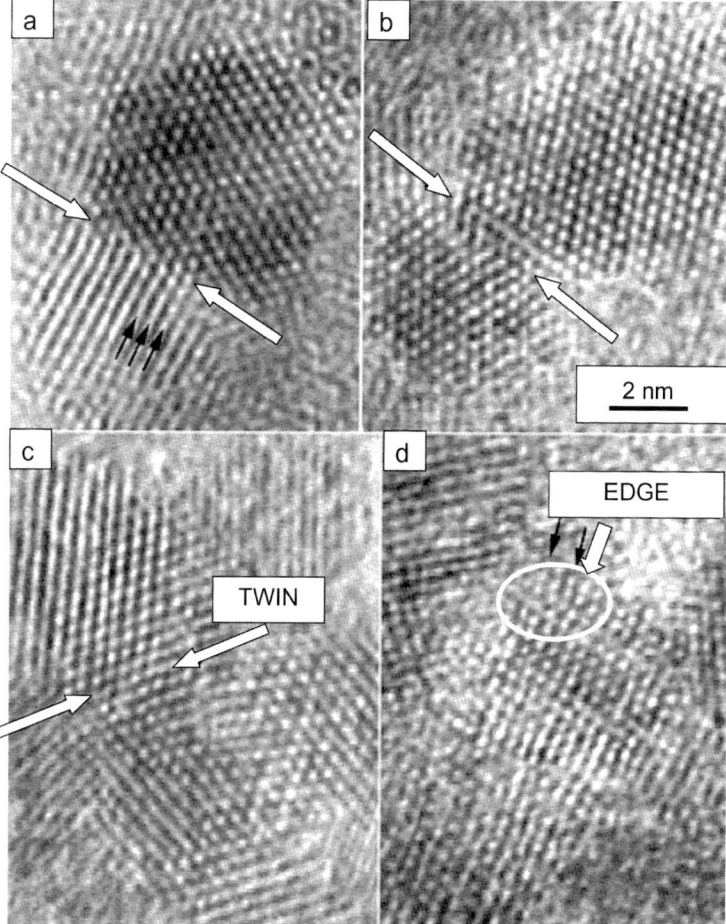

Fig. 7 Experimental HRTEM of CeO_2 nanocrystals synthesised by Wang and Feng.[31] Reprinted with permission from Z. L. Wang and X. D. Feng, *J. Phys. Chem. B*, 2003, **107**(49), pp. 13563–13566. Copyright 2003, American Chemical Society. Parts (a) and (b) show two aggregated nanocrystals that have a matched (111) plane but are twisted, (c) shows a twin-boundary and (d) an edge dislocation within a nanocrystal. These HRTEM can usefully be compared to our models to help validate the predicted atomistic structures.

To explore further the surface mobility, graphical techniques were used to observe the trajectories of the surface oxygen species in the $CeO_{1.95}$ (M) nanocrystal at the start, Fig. 10(a), and at the end, Fig. 10(b) of an MD simulation performed at 2500 K for 1800 ps. It is clear from the figures that throughout the simulation a high proportion of the surface oxygen ions move into the bulk of the nanocrystal rather than remain at the surface. However, the figure suggests that they do not move as far as oxygen ions in the centre of the nanocrystal indicating that oxygen transport at the surface of the nanocrystal is lower than in the bulk region. The arrow in Fig. 10(b) points to an area where there appears a particularly low mobility of surface oxygen. To understand why this is so, Fig. 10(c) shows the position of all the Ce^{3+} species. Clearly, they are well dispersed throughout the nanocrystal. However, it is evident that there are fewer Ce^{3+} species in the region indicated by the arrow. This suggests that oxygen transport is facilitated by the presence of Ce^{3+}.

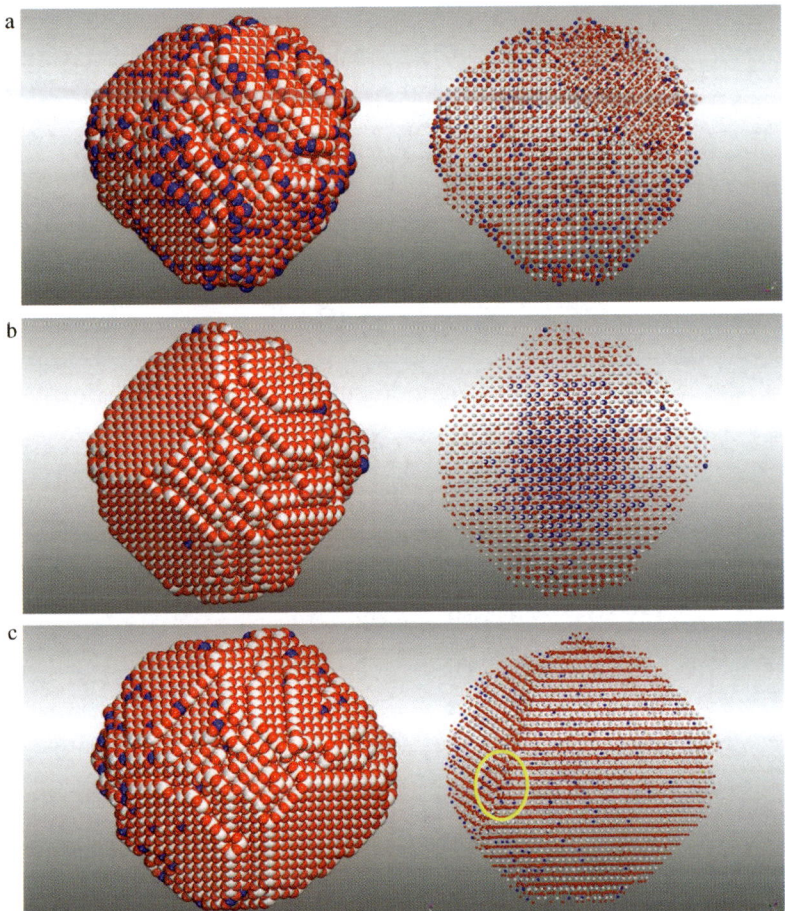

Fig. 8 Final, low temperature structure of the three Rh^{3+}-doped CeO_2 nanocrystals. (a) $Rh_{0.1}Ce_{0.9}O_{1.95}$ (S); (b) $Rh_{0.1}Ce_{0.9}O_{1.95}$ (C); (c) $Rh_{0.1}Ce_{0.9}O_{1.95}$ (M). To the left of each figure a sphere model representation of the atom positions is shown and to the right, the radii of the spheres has been reduced to enable one to look through the nanocrystal structure, which facilitates a closer comparison with HRTEM, enabling the grain-boundary structures (if present) to be seen. The yellow oval in Fig. '(c)' indicates the position of an edge dislocation, which can be usefully compared to a real dislocation in Fig. 7(d). Rh^{3+} species are coloured blue, Ce^{4+} are white and oxygen is red.

A plan view of a particular (111) surface, associated with the $CeO_{1.95}$ (M) nanocrystal, is shown in Fig. 11(a) at the start of the MD simulation, and Fig. 11(b) after 1800 ps at 2500 K. The figure shows that oxygen ions move from the outermost surface plane and into the bulk nanocrystal. However, the surface positions are not left vacant; rather oxygens from the bulk segregate to the surface to fill these positions. To aid visualisation, the oxygen ions that segregate to the surface to fill the vacancies are coloured purple in Fig. 11(b). The simulations show that surface anion vacancies are therefore continually being created and annihilated during the simulation.

Similar animations showing the MD simulations for the $CeO_{1.95}$ (C) system, which comprises multiple grain-boundaries, were made and then analysed (snapshot images not shown). Inspection of these movies revealed no obvious enhanced ionic mobility at grain-boundary regions.

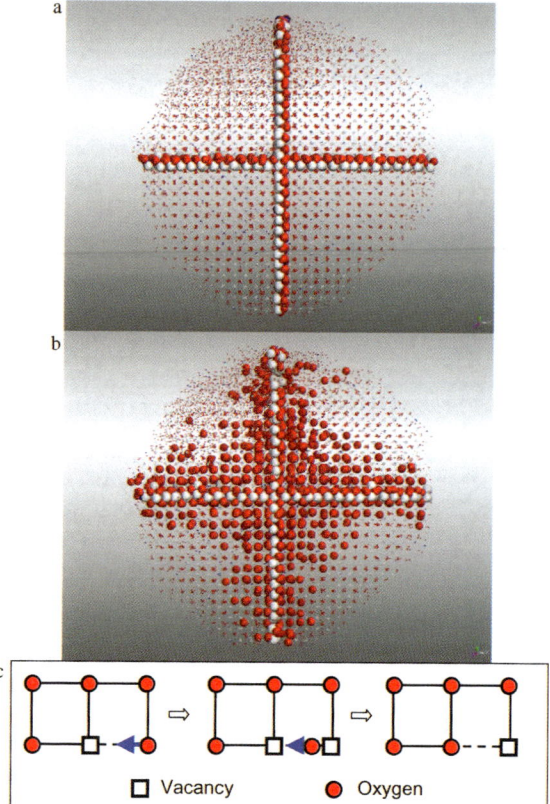

Fig. 9 Molecular graphics representation illustrating the mobility of oxygen ions in the $Rh_{0.1}Ce_{0.9}O_{1.95}$ (M) system. (a) Snapshot of the atomistic structure at the start of the (NVT) MD simulation; (b) Snapshot taken after 1800 ps of MD simulation, performed at 2500 K. In '(a)' the sizes of the oxygen and Ce^{4+} ions within two perpendicular planes through the nanocrystal are enlarged to show how far they move during the simulation. In '(b)' one can clearly see that the oxygen ions have moved a considerable distance, whereas in comparison, anion mobility is negligible. (c) Schematic illustrating oxygen transport, which proceeds *via* a vacancy mechanism. Rh^{3+} species (if visible) are coloured blue, Ce^{4+} are white and oxygen is red.

In Fig. 12, we show oxygen diffusion coefficients, calculated as a function of temperature, for nanocrystals CeO_2 (U), $CeO_{1.95}$ (M) and $Rh_{0.1}Ce_{0.9}O_{1.95}$ (M). On the same figure we have included, for comparison, oxygen diffusion coefficients calculated for $CeO_{1.97}/YSZ$ thin films, and in the bulk parent material, $CeO_{1.97}$; data taken from ref. 27. In Table 2, we report the mean square displacements of the oxygen ions calculated during MD simulation performed at 2000 K for 250 ps together with the activation energy barriers associated with ionic transport for the nanocrystals, thin films and parent materials. Fig. 11 shows graphical representations of the heterolayer $CeO_{1.97}/YSZ$ thin films and Fig. 12, the atomistic structure of the CeO_2 (U) nanocrystal.

It is clear from Fig. 12 that the calculated oxygen mobility increases in the order CeO_2(U) nanocrystal < $CeO_{1.95}$ (M) nanocrystal $\approx$ $Rh_{0.1}Ce_{0.9}O_{1.95}$ (M) nanocrystal < $YSZ/CeO_{1.97}$ thin film < $CeO_{1.97}$ parent material (bulk). The calculated activation energies, Table 2, *decrease* in the same order.

Molecular graphical images of the yttrium-stabilised zirconia/$CeO_{1.97}$ heterolayered system is shown in Fig. 13 and the structure of the CeO_2(U) nanocrystal is depicted in Fig. 14.

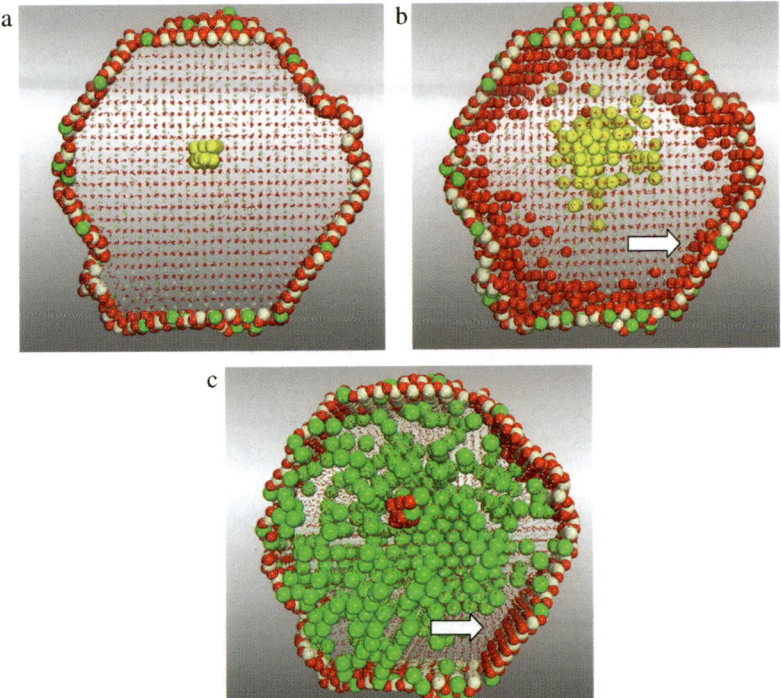

Fig. 10 Graphical images depicting the ion positions comprising the $Ce_{0.1}^{3+}Ce_{0.9}^{4+}O_{1.95}$ (M) nanocrystal system. (a) Atom positions at the start of the MD simulation and (b) after 1800 ps of MD simulation (NVT) performed at 2500 K. The two figures show the mobility of oxygen ions (coloured red), which move from the surface of the nanocrystal to deeper within the nanocrystal and compared with the oxygen ions at the centre of the nanocrystal (coloured yellow). In (c), the positions of the Ce^{3+} ions (coloured green) throughout the nanocrystal are highlighted and show that regions of low oxygen mobility correlate to regions relatively devoid of Ce^{3+} species.

4. Discussion

Our atomistic model for the CeO_2 (U) nanocrystal should, theoretically, comprise no oxygen vacancies. However, the amorphisation and recrystallisation technique generates crystal structures by evolving initially a crystalline seed, which then becomes an attractive substrate onto which (amorphous) ions surrounding the seed condense. We expect that, analogous to experiment, every so often the attaching ion may not necessarily attach exactly at the lattice position required to extend the crystal structure and a 'mistake' is made and can result in the formation of a microstructural feature such as a dislocation, grain-boundary or point defect. For example, a vacancy may form as other ions condense around it or, alternatively, perhaps an ion gets caught in an interstitial position and is 'trapped' once other ions condense around it.

Our simulations require the system to crystallise within the duration of a few nanoseconds (otherwise the computational costs become too high), which results in crystallisation speeds of approximately a metre per second. This is very fast and therefore our atomistic models are therefore likely to have high microstructural contents or defect concentrations. In particular, inspection of the unreduced nanocrystal, CeO_2 (U), reveals about 10 oxygen vacancies. It is important to note that if 10 oxygen vacancies evolve, then 5 Ce^{4+} vacancies must also evolve to

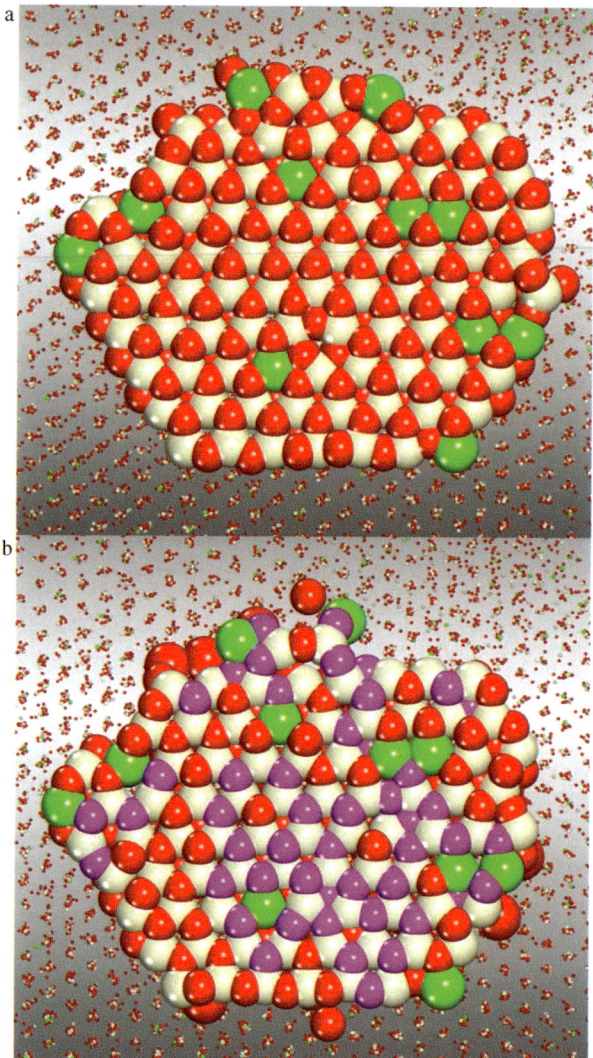

Fig. 11 Sphere model representation of the atom positions comprising a (111) surface associated with the $Ce_{0.1}^{3+}Ce_{0.9}^{4+}O_{1.95}$ (M) nanocrystal (a) at the start of the MD simulation and (b) after 1800 ps of MD simulation (NVT) performed at 2500 K. Ce^{3+} species are coloured green, Ce^{4+} are white and oxygen is red. In 'b' the purple ions are oxygen species that have migrated to the surface from within the bulk. In particular, during the dynamical simulation, most of the surface oxygen ions have migrated into the bulk and have been replaced by other oxygen ions (labelled purple) from deeper inside the nanocrystal.

facilitate charge neutrality. Indeed, inspection of the CeO_2 (U) nanocrystal revealed cation vacancies—the Schottky defects are observed in the atomistic model to exist as both isolated and associated defects.

It is clear from our simulations that the mechanism for ionic transport in ceria nanocrystals is predominantly vacancy driven. In particular, oxygen transport in the unreduced CeO_2 nanocrystal, CeO_2 (U), was so small that our methods could only detect it above 2000 K within the duration of the MD simulation. This nanocrystal only has about 10 vacancies (as determined by inspecting the nanocrystal using molecular graphics) compared with over 266 for the reduced and doped

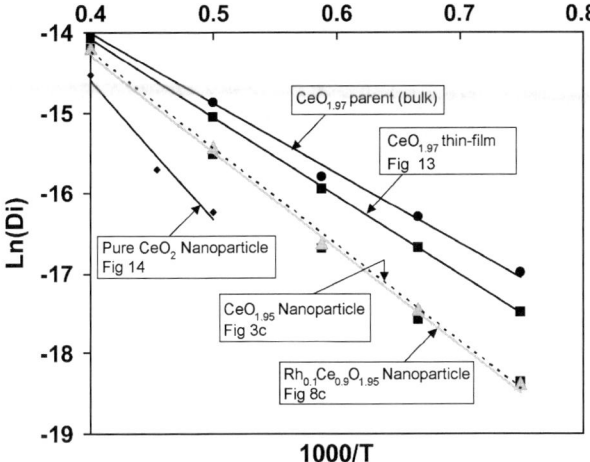

Fig. 12 Oxygen diffusion coefficients, calculated as a function of temperature. D_i has units $cm^2 \, s^{-1}$.

nanocrystals. In addition, animations of the diffusing oxygen ions did not reveal enhanced diffusion at the surface of the nanocrystal (Fig. 9). Neither did the nanocrystals that comprised grain-boundaries exhibit enhanced oxygen transport.

Table 2 Mean square displacements, calculated during an MD simulation performed at 2000 K. Activation energies, E_{act}, calculated for the Rh^{3+}-doped, reduced and unreduced ceria nanocrystals. Experimental data and simulation data for ceria thin films and the unreduced and doped 'bulk' material are given as a comparison

| | MSD/$\mathring{A}^2 \, ps^{-1} \times 10^{-3}$ | | | | |
	Centre	Surface	Mixed	E_{act}/eV	Ref.
Nanocrystal					
$Rh_{0.1}Ce_{0.9}O_{1.95}$	7.5	10.4	11.9	1.04[a]	This work
$CeO_{1.95}$	9.7	8.1	11.0	1.04[a]	This work
$CeO_2(U)$	—	—	—	1.48	This work

[a] These values are calculated for the mixed (M) nanocrystals.

	Technique	E_{act}/eV	Ref.
Thin Films			
$CeO_{1.97}$ thin film (CeO_2/YSZ heterolayers)	Simulation	0.84	27
Bulk material			
CeO_2	Experiment	0.9	34
$Y_{0.1}Ce_{0.9}O_{2-\delta}$	Experiment	0.788	2
$Cu_{0.1}Cu_{0.9}O_{2-\delta}$	Experiment	0.67	28
$Y_{2x}Ce_{1-2x}O_{2-x}$	Simulation	0.65–0.2 (for $x = 0$ to 0.3)	24
$CeO_{1.97}$ parent	Simulation	0.75	27

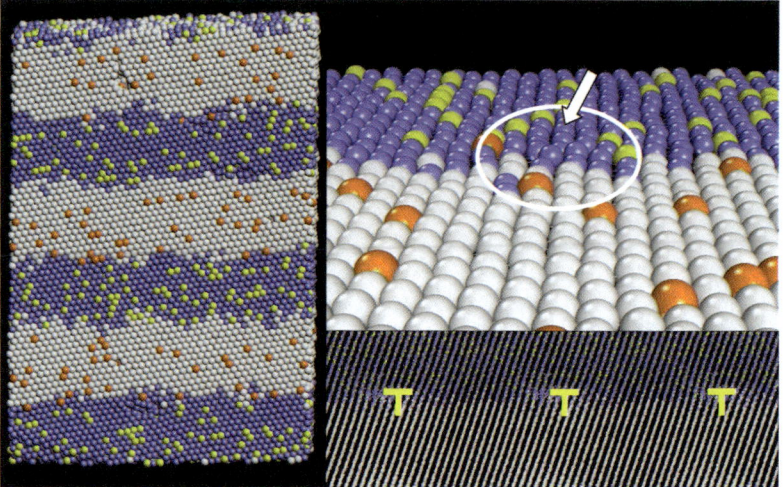

Fig. 13 Molecular graphics representations of the atom positions comprising the thin film $CeO_{1.97}/YSZ$ heterolayered system, taken from ref. 27. Left: slice cut through the system showing the $CeO_{1.97}$ and YSZ thin films; top right: an edge dislocation at the near interfacial region; bottom right: periodic arrays of edge dislocations.

5. Conclusions

Ceria is emerging as a very important and versatile material with applications spanning catalysis to fuel cells. Central to its application in these areas is its oxygen transport properties, which, as is clear from the wealth of literature data, can be influenced by a variety of factors including, for example, intrinsic and extrinsic doping,[34] polycrystallinity,[25] nanocrystalline systems,[35] heterolayered thin films and interfaces.[15] Accordingly, much effort has been directed at rationalising this behaviour. Here we have shown how atomistic computer modelling and simulation can be used to gain tremendous insights to help complement experimental work. Specifically, we have considered ionic transport in bulk, thin film, or nanocrystals and explored the influence of microstructural features, including unreduced and reduced ceria, intrinsic and extrinsic dopants, grain-boundaries and dislocations, epitaxy and morphology.

Our calculations indicate that oxygen diffusion coefficients increase in the order CeO_2 (U) nanocrystal $<$ $CeO_{1.95}$ nanocrystal $\approx$ $Rh_{0.1}Ce_{0.9}O_{1.95}$ nanocrystal $<$ $CeO_{1.97}/YSZ$ thin film $<$ $CeO_{1.97}$ (bulk) parent material and that the mechanism is predominantly vacancy driven. Our findings predict that reducing one- (thin film) or especially three- (nanocrystal) dimensions to the nanoscale may prove deleterious to oxygen transport.

The computational cost of undertaking these simulations is high (with respect to current computational resources) and therefore we were only able to consider a limited number of configurations—albeit ones that comprised a wealth of microstructural features that have been shown to influence the ionic conductivity. Accordingly, we advocate caution in generalising these results without further simulation and experimental data.

Our atomistic models of the ceria nanocrystals were found to be in close structural agreement with experiment, including the morphology, surfaces exposed, edges and corners and microstructural features including line dislocations and grain-boundaries.

In its capacity as a component of an oxidative catalyst, the ceria based system must supply surface oxygens (to oxidise CO) resulting in the formation of surface

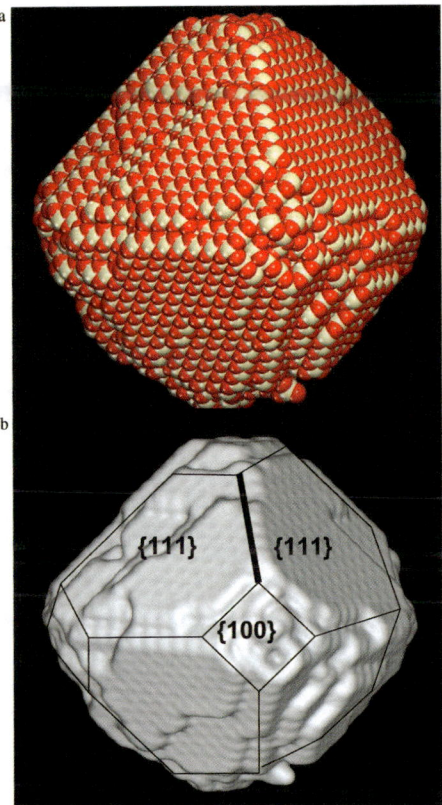

Fig. 14 Final, low temperature structure of the unreduced ceria nanocrystal: CeO$_2$(U). (a) Sphere model representation of the atom positions; (b) surface rendered representation showing more clearly the {111} and {100} surfaces. Images taken from T. X. T. Sayle, S. C. Parker and D. C. Sayle, *Chem. Commun.*, 2004, 2438–2439–Reproduced by permission of The Royal Society of Chemistry.[20].

vacancies. The oxygen 'debt' at the surface must then be repaid by oxygen ions segregating to the surface to fill the vacant positions or the catalyst would be depleted (provided by NO$_2$ reducing to N$_2$). Here, we have shown that vacancy formation and annihilation is dynamic; animations of the simulations reveal oxygen ions segregate from the bulk of the nanocrystal to the surface and *vice versa*. Coupled with the results of a previous simulation study, where we predicted that it was easier to remove an oxygen ion from the surface of a ceria nanocrystal[26] compared with the bulk material[6] and has recently been confirmed experimentally,[36] we predict that ceria nanomaterials will facilitate effective catalysis. However, will oxidative catalysis, using CeO$_2$ nanocrystals, be better than using the (bulk) parent material? Two factors are pivotal:

• Oxygen ions must be easily given up at the surface (oxygen vacancy formation energy)

• The surface oxygen vacancy must be replenished (oxygen transport)

Our calculations predict that oxygen ions are more easily extracted from a nanocrystal compared with the (bulk) parent material,[26] which enhances the catalysis. Conversely, we also find (in this present study) that oxygen transport is reduced in a nanocrystal compared with the parent, which is deleterious to the catalysis.

We describe a framework simulation procedure that can be used to aid experiment explore ionic transport in materials where one (thin films) or more (nanocrystals) dimensions exist at the nanoscale. In particular, it can be used to determine and predict those factors that might enhance or inhibit ionic transport such as size and shape of the nanocrystals, intrinsic or extrinsic dopant ions or other microstructural features such as dislocation and grain-boundary density, interface configuration *etc*.

With the explosive growth in computational power coupled with grid computing, such simulations, which are computationally expensive at present, are likely to become routine in future and help complement and reduce the number of expensive experiments that need to be performed in the search for improved or new materials.

Acknowledgements

EPSRC (GR/S48431/1 and GR/S48448/01) for funding one of us (TXTS), Cambridge-Cranfield HPCF for computational facilities and EPSRC (GR/S84415/01) for computational facilities.

References

1 W. Z. Zhu and S. C. Deevi, A review on the status of anode materials for solid oxide fuel cells, *Mater. Sci. Eng., A*, 2003, **362**(1–2), 228–239.
2 T. S. Zhang, J. Ma, L. B. Kong, S. H. Chan and J. A. Kilner, Aging behavior and ionic conductivity of ceria-based ceramics: a comparative study, *Solid State Ionics*, 2004, **170**(3–4), 209–217.
3 J. A. Lane and J. A. Kilner, Oxygen surface exchange on gadolinia doped ceria, *Solid state Ionics*, 2000, **136**, 927–932.
4 N. Oishi, A. Atkinson, N. P. Brandon, J. A. Kilner and B. C. H. Steele, Fabrication of an anode-supported gadolinium-doped ceria solid oxide fuel cell and its operation at 550 °C, *J. Am. Ceram. Soc.*, 2005, **88**(6), 1394–1396.
5 R. Di Monte and J. Kaspar, On the role of oxygen storage in three-way catalysis, *Top. Catal.*, 2004, **28**(1–4), 47–57.
6 T. X. T. Sayle, S. C. Parker and C. R. A. Catlow, The role of oxygen vacancies on ceria surfaces in the oxidation of carbon monoxide, *Surf. Sci.*, 1994, **316**(3), 329–336.
7 J. Guzman, S. Carrettin and A. Corma, Spectroscopic evidence for the supply of reactive oxygen during CO oxidation catalyzed by gold supported on nanocrystalline CeO_2, *J. Am. Chem. Soc.*, 2005, **127**(10), 3286–3287.
8 S. Gritschneder, Y. Namai, Y. Iwasawa and M. Reichling, Structural features of CeO_2(111) revealed by dynamic SFM, *Nanotechnology*, 2005, **16**(3), S41–S48.
9 Y. Namai, K. Fukui and Y. Iwasawa, The dynamic behaviour of CH_3OH and NO_2 adsorbed on CeO_2(111) studied by noncontact atomic force microscopy, *Nanotechnology*, 2004, **15**(2), S49–S54.
10 N. Sata, K. Eberman, K. Eberl and J. Maier, Mesoscopic fast ion conduction in nanometre-scale planar heterostructures, *Nature*, 2000, **408**(6815), 946–949.
11 D. C. Sayle, J. A. Doig, S. C. Parker and G. W. Watson, Synthesis, structure and ionic conductivity in nanopolycrystalline BaF_2/CaF_2 heterolayers, *Chem. Commun.*, 2003, (15), 1804–1806.
12 D. C. Sayle, J. A. Doig, S. C. Parker, G. W. Watson and T. X. T. Sayle, Computer aided design of nano-structured materials with tailored ionic conductivities, *Phys. Chem. Chem. Phys.*, 2005, **7**(1), 16–18.
13 C. H. Chen, T. Kiguchi, A. Saiki, N. Wakiya, K. Shinozaki and N. Mizutani, Characterization of defect type and dislocation density in double oxide heteroepitaxial CeO_2/YSZ/Si(001) films, *Appl. Phys. A: Mater. Sci. Process.*, 2003, **76**(6), 969–973.
14 C. M. Wang, S. Azad, V. Shutthanandan, D. E. McCready, C. H. F. Peden, L. Saraf and S. Thevuthasan, Microstructure of ZrO_2–CeO_2 hetero-multi-layer films grown on YSZ substrate, *Acta Mater.*, 2005, **53**(7), 1921–1929.
15 S. Azad, O. A. Marina, C. M. Wang, L. Saraf, V. Shutthanandan, D. E. McCready, A. El-Azab, J. E. Jaffe, M. H. Engelhard, C. H. F. Peden and S. Thevuthasan, Nanoscale effects on ion conductance of layer-by-layer structures of gadolinia-doped ceria and zirconia, *Appl. Phys. Lett.*, 2005, **86**(13).
16 A. M. Walker, J. D. Gale, B. Slater and K. Wright, Atomic scale modelling of the cores of dislocations in complex materials part 1: methodology, *Phys. Chem. Chem. Phys.*, 2005, **7**(17), 3227–3234.
17 A. M. Walker, B. Slater, J. D. Gale and K. Wright, Predicting the structure of screw dislocations in nanoporous materials, *Nat. Mater.*, 2004, **3**(10), 715–720.

18   D. M. Duffy and P. W. Tasker, Computer-simulation of (001) tilt grain-boundaries in nickel-oxide, *Philos. Mag. A*, 1983, **47**(6), 817–825.

19   T. X. T. Sayle, C. R. A. Catlow, R. R. Maphanga, P. E. Ngoepe and D. C. Sayle, Generating MnO_2 nanocrystals using simulated amorphization and recrystallization, *J. Am. Chem. Soc.*, 2005, **127**(37), 12828–12837.

20   T. X. T. Sayle, S. C. Parker and D. C. Sayle, Shape of CeO_2 nanocrystals using simulated amorphisation and recrystallisation, *Chem. Commun.*, 2004, (21), 2438–2439.

21   F. Zhang, P. Wang, J. Koberstein, S. Khalid and S. W. Chan, Cerium oxidation state in ceria nanocrystals studied with X-ray photoelectron spectroscopy and absorption near edge spectroscopy, *Surf. Sci.*, 2004, **563**(1–3), 74–82.

22   J. Ma, T. S. Zhang, L. B. Kong, P. Hing, Y. J. Leng and S. H. Chan, Preparation and characterization of dense $Ce_{0.85}Y_{0.15}O_2$-delta ceramics, *J. Eur. Ceram. Soc.*, 2004, **24**(9), 2641–2648.

23   G. Balducci, M. S. Islam, J. Kaspar, P. Fornasiero and M. Graziani, Bulk reduction and oxygen migration in the ceria-based oxides, *Chem. Mater.*, 2000, **12**(3), 677–681.

24   G. Balducci, M. S. Islam, J. Kaspar, P. Fornasiero and M. Graziani, Reduction process in CeO_2-MO and CeO_2-M_2O_3 mixed oxides: A computer simulation study, *Chem. Mater.*, 2003, **15**(20), 3781–3785.

25   A. Tschope, Interface defect chemistry and effective conductivity in polycrystalline cerium oxide, *J. Electroceram.*, 2005, **14**(1), 5–23.

26   T. X. T. Sayle, S. C. Parker and D. C. Sayle, Oxidising CO to CO_2 using ceria nanocrystals, *Phys. Chem. Chem. Phys.*, 2005, **7**(15), 2936–2941.

27   T. X. T. Sayle, S. C. Parker and D. C. Sayle, Ionic conductivity in nano-scale CeO_2/YSZ heterolayers, *J. Mater. Chem.,* DOI: 10.1039/B511547G.

28   T. X. T. Sayle, S. C. Parker and C. R. A. Catlow, Surface segregation of metal-ions in cerium dioxide, *J. Phys. Chem.*, 1994, **98**(51), 13625–13630.

29   W. Smith, T. R. Forester, DL_POLY is a package of Molecular Simulation routines written by Smith and Forester Copyright by the council for the Central Laboratory of the Research Councils, Daresbury Laboratory, Daresbury, Warrington, UK, 1996, http://www.dl.ac.uk/TCSC/Software/DLPOLY.

30   G. V. Lewis and C. R. A. Catlow, Potential models for ionic oxides, *J. Phys. C*, 1985, **18**(6), 1149–1161.

31   Z. L. Wang and X. D. Feng, Polyhedral shapes of CeO_2 nanoparticies, *J. Phys. Chem. B*, 2003, **107**(49), 13563–13566.

32   F. Claeyssens, C. L. Freeman, N. L. Allan, Y. Sun, M. N. R. Ashfold and J. H. Harding, Growth of ZnO thin films—experiment and theory, *J. Mater. Chem.*, 2005, **15**(1), 139–148.

33   G. Balducci, J. Kaspar, P. Fornasiero, M. Graziani, M. S. Islam and J. D. Gale, Computer simulation studies of bulk reduction and oxygen migration in CeO_2–ZrO_2 solid solutions, *J. Phys. Chem. B*, 1997, **101**(10), 1750–1753.

34   A. Gayen, K. R. Priolkar, A. K. Shukla, N. Ravishankar and M. S. Hegde, Oxide-ion conductivity in $Cu_xCe_{1-x}O_2$-delta ($0 \leq x \leq 0.10$), *Mater. Res. Bull.*, 2005, **40**(3), 421–431.

35   S. Kim and J. Maier, On the conductivity mechanism of nanocrystalline ceria, *J. Electrochem. Soc.*, 2002, **149**(10), J73–J83.

36   H. X. Mai, L. D. Sun, Y. W. Zhang, R. Si, W. Feng, H. P. Zhang, H. C. Liu and C. H. Yan, Shape-selective synthesisand oxygen storage behavior of ceria nanopolyhedra, nanorods, and nanocubes, *J. Phys. Chem. B*, 2005, **109**(51), 24380–24385.

General Discussion

Dr Gray-Weale opened the discussion of Dr Nastar's paper: In your paper, you mention the use of an effective Hamiltonian to calculate the probability of non-equilibrium configurations. I understand the non-equilibrium case you're looking at is a gradient of chemical potential, but surely if you use a Hamiltonian to get a probability, then you are looking at a spontaneous fluctuation about equilibrium? Or is the chemical potential gradient small enough that you can use equilibrium fluctuations?

Dr Nastar answered: We are calculating the dissipation of a system submitted to a small gradient of chemical potential close to equilibrium. According to the fluctuation–dissipation theorem it corresponds to equilibrium fluctuations.

Dr Stolwijk asked: Your last example was concerned with extreme differences between the atom–vacancy exchange rates W_1 and W_3. One of these limiting cases seems to imply that the vacancy-binding energy is high whereas in the opposite case it should be low. In metal systems impurity–vacancy binding energies appear to be rather low. Do you know any practical cases which show the proposed (implicit) binding between the type of atom and the vacancy, and concurrently, the vacancy drag effects observed in the numerical calculations?

In oxide systems vacancy–ion interactions can be very strong, so it seems worthwhile to apply your theory on diffusion problems to oxides.

Dr Nastar replied: Indeed an inversion of the atomic flux in relation to the vacancy flux becomes spectacular only if the vacancy-binding energy is high enough. Recent *ab initio* calculations predict this inversion phenomenon in metal dilute systems such as Fe(P)[1] and Fe(Cu).[2] It would be interesting to use our SCMF theory to predict the behaviour of these systems at higher concentrations of P and Cu.

In oxide systems, I agree that this effect must be stronger and should be studied in detail.

1   A. V. Barashev, *Philos. Mag.*, 2005, **85**, 1539.
2   A. V. Barashev and A. C. Arokiam, *Philos. Mag. Lett.*, 2006, **86**, 321.

Dr Zhukovskii asked: Could you clarify the structural stability of the dumbbell model? What does it mean physically—co-existence of two metal atoms in the center of a densely packed unit cell? In such a case drastic lattice reconstruction around this unit cell could happen but you consider a rigid lattice model.

Dr Nastar replied: In general a dumbbell configuration induces much more relaxation of the atomic positions than the vacancy one. In systems we studied, such as Fe–Ni–Cr, Fe(P) and Fe(Cu), it does not seem too bad to keep a rigid lattice representation and to represent the enthalpy associated with a configuration by a sum of pairwise interactions.

Professor Maier asked:
(1) You mentioned in the paper that on-site rotation systematically enhances transport. Can you give a qualitative explanation and could you comment on the magnitude of the effect?
(2) How far can these considerations be generalised to ionic materials?

Dr Nastar responded:
(1) A dumbbell interstitial can rotate on its site or jump. If jump frequencies were all equal to zero, dumbbells would not migrate. In other words, an on-site rotation does not intervene in the non-correlated displacement of the dumbbells. However an on-site rotation affects the correlation part of the diffusion. Indeed, after an on-site rotation

the probability that the defect jumps back to its previous site is smaller because the previous site might become inaccessible by the defect within a new orientation. Therefore a high value of the rotation frequency makes the defect lose its memory of the previous jumps, decreases the correlation effects and therefore enhances the transport *via* the dumbbells. The magnitude of this effect is discussed in the paper.

(2) The arguments used to explain the enhancement of interstitial transport by an on-site rotation are mainly geometrical. If the mechanism of interstitial migration in ionic materials involves an orientation of the defect, a set of sites accessible after a jump which depends on the defect orientation and an on-site rotation mechanism, then the same arguments could be applied to ionic materials.

Professor Islam asked:

(1) Irradiation creates vacancy and interstitial defects. Are they present as "intrinsic" defects in the alloy system at ambient temperatures?

(2) You mentioned the interstitial "dumbbell" configuration—is there any experimental evidence (*e.g.*, from electron microscopy) for such configurations?

Dr Nastar replied:

(1) Vacancy represents the main point defect at ambient temperature. Dumbbell interstitials are usually associated with very high formation enthalpies and are practically absent at ambient temperature.

(2) Diffraction experiments[1] and mechanical or magnetic relaxation techniques[2,3] performed on pure metals give direct information about the position, atomic structure and orientation of these dumbbell configurations.

1   P. Ehrart, *J. Nucl. Mat.*, 1978, **69/70**, 200.
2   P. Moser, *J. Microsc..*, 1973, **16**, 157.
3   A. S. Nowick and R. Feder, *Phys. Rev. B*, 1972, **5**, 1238.

Professor Heitjans said: You mentioned that there is practically no experimental evidence or data for interstitial diffusion in alloys. However, I understood that part of the motivation to calculate the phenomenological coefficients for atomic transport *via* interstitials along with that *via* vacancies is that both mechanisms are supposed to play a role in alloys irradiated with particles. My question is whether your theory also copes with diffusion under irradiation.

Dr Nastar responded: The main motivation of this work is to calculate the atomic fluxes induced by irradiation *via* both vacancies and interstitials and to study the effect of these fluxes on the kinetics and the non-equilibrium steady states reached by the alloys under irradiation. Diffusion under irradiation occurs with the same microscopic mechanisms except that the population of point defects is strongly modified by irradiation with, for example, a large increase in the dumbbell population.

Professor Haile opened the discussion of Dr Nakamura's paper: Can you distinguish between the motion of lithium ions and that of neutral lithium species?

Dr Nakamura answered: We observed narrowed ^{7}Li NMR spectra and ^{7}Li nuclear relaxation rates due to the migration of Li$^+$ ion vacancies. The frequency shift of NMR spectra (Knight shift) of ^{7}Li is very small in LiGa, and this means that the observed Li species are in an ionic state. Thus, it is considered that we are detecting only the motion of Li$^+$ ions.

Professor Heitjans commented: You ascribed the diffusion induced spin–lattice relaxation of ^{7}Li in β-LiGa with 50 at% Li, showing up in a broad asymmetric rate peak at about 175 K, to ^{7}Li quadrupole interaction. In the case of the isostructural compound β-LiAl it was argued that the corresponding ^{7}Li relaxation rate is primarily due to dipolar interaction.[1,2] This also applies to the diffusion induced spin–lattice relaxation rate of the β-emitter ^{8}Li, which has a quadrupole moment similar to that of ^{7}Li and was measured by the β-NMR technique.[3] However, I agree that the relaxation mechanism, whether quadrupolar or dipolar, should have no influence on the value of the activation energy extracted from the data.

1 T. Tokuhiro, S. Susman, T. O. Brun and K. J. Volin, *J. Phys. Soc. Jpn.*, 1989, **58**, 2553.
2 D. Clausen, I. Burmester, P. Heitjans and A. Schirmer, *Solid State Ionics*, 1994, **70/71**, 482.
3 A. Schirmer, P. Heitjans, W. Faber and D. Clausen, *Solid State Ionics*, 1992, **53–56**, 426.

Dr Nakamura replied: As pointed out, the spin–lattice relaxation in LiAl is discussed on the basis of the dipole relaxation. In the present study, we tried to analyse the results based on the quadrupole relaxation mechanism, which was applied to interpret the spin–lattice relaxation of ^{7}Li in Li_3N.[1] Though it is important to discuss whether the relaxation of ^{7}Li is dipole or quadrupole, we confined ourselves in this paper to the estimation of the activation energy and the discussion about the non-BPP like relaxation behavior.

1 D. Brinkmann, M. Mali and J. Roos, *Phys. Rev. B*, 1982, **26**, 4810.

Dr De Souza opened the discussion of Dr Savini's paper: Dislocations are sources and sinks of point defects. Even in pure, stoichiometric samples one may expect segregation of silicon or carbon vacancies to be the core. In real samples there are going to be some impurities that may segregate to the dislocation core. Can you ignore these effects when considering defect electronic levels and dislocation mobility?

Dr Savini answered: Yes, dislocations are sources and sinks for point defects. The amount of decoration of a dislocation by point defects will depend on the history of the sample. Heavily dislocated or networked material will suffer enormously from this problem. However, the devices are generally carefully grown, rather pure and not deformed, so these problems are less likely.

In addition, decorated dislocations are usually pinned, *i.e.*, point defects decrease the dislocation mobility, but freshly produced dislocations can glide and, in gliding, dislocations discard point-defect atmospheres (the so-called Cottrell atmosphere). Therefore, we consider it appropriate to discuss electronic levels of clean stoichiometric glide dislocations.

Professor Haile asked: You've argued that point defects are not important because they would pin dislocations rather than enhance their mobility. Can you consider the opposite possibility—that plasma irradiation frees the dislocations from pinning defects and this is the reason for enhanced dislocation mobility?

Dr Savini responded: This is an age-old argument in the world of dislocations in semiconductors. Nowadays the concept of dislocation glide controlled by the Peierls relief (lattice friction) is accepted for a number of reasons.

Firstly, dislocations tend to follow low index directions when moving under high stresses, and this is a consequence of anisotropic mobilities typical of the Peierls relief. Secondly, first principles calculations of activation energies for glide are very close to experimental measurements in most tetrahedral semiconductors.

However, we do accept the principle that electron–hole plasma can enhance the rate of escape from pinning points. Nevertheless, this argument alone cannot explain why dislocations start to glide at room temperature, nor describe the electrical activity of the mobile dislocations.

Professor Heitjans opened a general discussion, addressing Dr Nastar: The findings that, according to your calculations, an on-site rotation mechanism enhances the transport in alloys reminds me of the 'paddle wheel mechanism' discussed for certain ionic conductors like, *e.g.*, Na_2SO_4 or Na_3PO_4. As probably known by several in the audience, it has been proposed that there is an enhancement of the diffusivity of the alkali ions by dynamical coupling between the translation of the cation and the rotation of the polyatomic anion (see, *e.g.*, ref. 1 and references therein). In my opinion these findings might help to fertilize the exchange of ideas between the 'metal and alloys community' and the 'solid state ionics community', being the majority here.

1 T. Springer and R. E. Lechner, in *Diffusion in Condensed Matter - Methods, Materials, Models*, ed. P. Heitjans and J. Kärger, Springer, Berlin/Heidelberg, 2005, p. 93.

Dr Nastar replied: Our mean-field theory predicts that an on-site rotation of a dumbbell interstitial decreases the correlation effects and since it does not affect the uncorrelated transport coefficients it systematically enhances the macroscopic transport of the dumbbell interstitials.

Professor Islam asked:
(1) For your complex mechanisms of defect migration, have you derived activation energies for the separate steps? Is there a rate-limiting step?
(2) Is there any experimental diffusion data for comparison with the calculated activation energies for these alloy systems?

Dr Nastar answered:
(1) The vacancy and interstitial migration is thermally activated. A saddle point energy has to be overcome by the system in order to make the defect jump or the dumbbell rotate on its site. The activation energy corresponds to the difference between the saddle point energy and the initial energy before the jump.
(2) For the vacancy mechanism, the kinetic parameters of our atomic jump frequency model are adjusted to reproduce activation energies and attempt jump frequencies extracted from high temperature tracer diffusion and inter-diffusion experiments. Due to its high formation energy value, dumbbell type defects occur in metals mainly under irradiation. Therefore, the few available experimental data are electrical resistivity measurements at low temperature performed on irradiated alloys. They provide an effective migration energy averaged over the different compositions of the dumbbell. For both point defects, the tendency now is to use *ab initio* calculations, which are now able to compute atomic jump frequencies, although mainly in dilute alloys.[1–5]

1 C. Fu and F. Willaime, *Phys. Rev. B*, 2005, **72**, 064117.
2 C. Fu, F. Willaime and P. Ordejón, *Phys. Rev. Lett.*, 2004, **92**, 175503.
3 C. Domain and C. S. Becquart, *Phys. Rev. B*, 2005, **71**, 214109.
4 E. Clouet, M. Nastar and C. Sigli, *Phys. Rev. B*, 2004, **69**, 064109.
5 E. Clouet, L. Laé, T. Epicier, W. Lefebvre, M. Nastar and A. Deschamps, *Nat. Mater.*, 2006, **5**, 482.

Dr Stolwijk said: Possibly I misunderstood your method of calculation. Do you need the irradiation and the continuous production of vacancies and intersitials to determine your model parameters in a steady-state non-equilibrium situation? Can your model also be applied to the more common case of vacancy diffusion under thermal equilibrium?

Dr Nastar replied: Our theory starts from a microscopic model of the point defect jump mechanism. This model does not depend on the external conditions and can be applied to any system at equilibrium or not. What is changing under irradiation is the population of point defects, which results from the balance between the point defect production created by irradiation, the diffusion, the elimination of point defects at sinks like surfaces, grain-boundaries, dislocations and the recombination of some of the vacancies and interstitials nearby.

Professor Heitjans asked:

Are there alternative approaches to the calculation of phenomenological coefficients in alloys available in the literature?

Dr Nastar replied: The main alternative approach which takes into account short-range order explicitly is the path probability method (PPM),[1] which is also a mean field theory starting from a master equation.

For the vacancy, the PPM[1] and its modified version[2] are limited to body centred cubic alloys. And for the interstitial, the SCMF theory is the first theory to propose a calculation of the transport coefficients for the dumbbell mechanism in bcc concentrated alloys.

1 H. Sato and R. Kikuchi, *Phys. Rev. B*, 1983, **28**, 648.
2 N. A. Stolwijk, *Phys. Status Solidi B*, 1981, **105**, 223.

Professor Maier addressed Dr Savini:

(1) Kinetically, dislocations usually play an intermediate role in "getting equilibrated": While point defects may be easily brought into equilibrium, interfaces we typically consider as meta-stable long-term. How well can dislocations be annealed off in SiC?

(2) What is the crystallographic role of dislocations in SiC in relation to the stacking fault "heterostructures"?

Dr Savini answered:

(1) Normally there is no 'equilibrium' concentration of dislocations. Dislocations anneal out relatively easily in crystals which only have one dominant slip plane (this is often the case for hexagonal crystals). My feeling is that dislocations can be annealed out of SiC but I fear that heat treatment is necessary, plus the effect of their gliding could cause transformation at least locally from one polytype to another.

(2) The partial dislocations we describe are the bounding dislocations for stacking faults. The various polymorphs of SiC could be interconverted in principle by glide of these partial dislocations on their slip planes, but this would need to occur on a regularly spaced pattern of planes.

Professor Yashima addressed Dr Nastar:

(1) Can you estimate the occupancy factor at the interstitial site? If so, is it consistent with experimental diffraction data?

(2) Your method is based on thermodynamics. Can the parameter sets reproduce the phase diagram and the enthalpy?

Dr Nastar replied:

(1) Interstitials are produced by irradiation. The most stable configuration is usually the dumbbell configuration, *e.g.*, two atoms sharing a substitutional site. It is characterized by a composition and an orientation. Irradiation conditions impose the concentration of the dumbbells but the repartition of each composition depends on the thermodynamic parameters and irradiation conditions and can be calculated. Some diffraction experiments[1] performed on pure metals give direct information about the position, the atomic structure of point defects and its orientation. But in multi-component systems, interpretation of such experiments becomes much more complex and, for example, it is not obvious to determine the repartition of the dumbbell compositions.

(2) Yes, our method is able to treat kinetic properties and thermodynamics in a self-consistent way. Without an external force the kinetic equations converge to a stationary state equivalent to the equilibrium state determined by the thermodynamic parameters. These parameters are fitted to reproduce the main characteristics

of the phase diagram, the formation enthalpy of point defects, the surface tensions, *etc.*

1 P. Ehrart, *J. Nucl. Mat.*, 1978, **69/70**, 200.

Professor Heitjans addressed Dr Nakamura: In your ^{7}Li NMR measurements, performed at 10 MHz, you found a cusp-like peak of the spin–lattice relaxation rate in the β-LiGa sample with, *e.g.*, 47 at% Li, and attributed it to the order–disorder transformation of Li vacancies. It might be useful to study the spin–lattice relaxation rate, interpreted in terms of critical slowing down, at other NMR frequencies, too. By the way, in β-LiAl with 47 at% Li, studied at various ^{7}Li NMR frequencies in the range from 16–100 MHz, we only observed the usual broad asymmetric peak of the diffusion induced spin–lattice relaxation rate.[1] So with respect to vacancy ordering β-LiGa and β-LiAl appear to behave differently.

1 D. Clausen, I. Burmester, P. Heitjans and A. Schirmer, *Solid State Ionics*, 1994, **70/71**, 482.

Dr Nakamura replied: There might exist different behaviors with respect to vacancy-ordering in β-LiGa and β-LiAl. β-LiGa with 47 at% Li shows anomalous behavior at 230 K in heat capacity, while β-LiAl with 47 at% Li shows that at 97 K.[1,2] Additionally, the number of vacancies is larger in LiGa than LiAl. This means that the correlation strength between vacancies in LiGa is much stronger than that in LiAl. So, one might be able to observe anomalous behavior in the spin–lattice relaxation below 100 K in LiAl. As you indicated, we also think that the field dependence of the spin–lattice relaxation is important in elucidating the vacancy-ordering effects.

1 H. Hamanaka, S. Kaidou, K. Kuriyama and M. Yahagi, *Solid State Ionics*, 1998, **113–115**, 69.
2 K. Kuriyama, S. Yanada, T. Nozaki and T. Kamijoh, *Phys. Rev. B*, 1981, **24**, 6158.

Dr Ruiz-Trejo opened the discussion of Professor Yashima's paper: Your experiments at 1400 °C will probably reduce the cerium oxide. Did you see any loss of oxygen? Did you consider this change of oxygen stoichiometry in your calculations?

Professor Yashima answered: Yes, there is some loss of oxygen in the ceria lattice at 1400 °C, but the quantity of the loss is small and is less than 0.1%. In the structural refinement, the error bar of the occupancy is more than 0.1%. Thus, it is not a problem to assume this oxygen stoichiometry in the Rietveld refinements.

Dr De Souza asked: Have you calculated the activation barriers for the two diffusion paths from the nuclear density maps?

Professor Yashima replied: No, I did not calculate the activation energy. But it is possible to estimate it.

Professor Vannier asked:
(1) Are you able to derive the pseudo-potential from the PDF you obtained by MEM?
(2) What are the basics of MEM, are anharmonic thermal parameters introduced in the minimisation?
(3) Could you tell us more about your experimental set up? Did you carry out the experiments on dense pellets under a particular atmosphere?

Professor Yashima answered:
(1) I have not derived it, but it is possible in principle.

(2) In MEM, the information entropy is maximized. Thus, the final nuclear distribution includes anharmonic thermal motions.

(3) I have used both dense pellets and pellets which are not dense. The density does not influence the results.

Professor Aranda asked: Why do you use such a long wavelength, if peak overlapping is not a problem? This material is cubic with only 11 peaks in the measured diffraction pattern.

Professor Yashima replied: Unfortunately, the wavelength available in Japan is this long. As you point out, it might be better to use shorter wavelengths. Fortunately, for studies of the diffusion path, it does not matter so much, because the lowest 2θ 111 peak has the most influence on the results.

Dr Andreev asked: How many parameters did you use to derive the relatively complex density pattern? How does this number compare to the number of Bragg reflections in the diffraction pattern?

Professor Yashima answered: There were only 2 structural parameters during the Rietveld refinement. There were about 11 peaks in the profile. Thus, this was not a problem in the Rietveld refinement.

Dr Wilson said: Fig. 2 (unit cell length against temperature) shows an interesting discontinuity in the gradient at around 600 °C. I imagine that this temperature is many hundreds of degrees below the superionic transition temperature (at which point the oxygen sublattice effectively "melts"). Can the authors point to the physical origin of this discontinuity?

Professor Yashima replied: Fig. 1 shows the temperature dependence of the gradient of the unit cell parameter against temperature. The gradient increases with increasing temperature. However, no discrete points were observed within the error bar. Thus, there is no correlation between the gradient and the superionic transition.

Professor Ishihara asked:
(1) With increasing temperature, the lattice is expanded according to the data. How does this expansion of the unit lattice affect the diffusion path?
(2) From theoretical considerations, an increase in the lattice free volume should affect the diffusion path. No changes in the diffusion path suggests that the resolution of the measurement is not high enough. Is this correct?

Professor Yashima answered: The lattice expansion is about 1% or less, thus the effect on the diffusion path is not significant. Even if the resolution of the diffractometer is improved, it might be difficult to detect the possible difference.

Professor Islam said: As an initial comment, it is good to see that there are experimental attempts to probe the atomistic migration pathways in oxides. You note that you observe a curved path in doped $LaGaO_3$ perovskite in accord with our earlier predictions from atomistic simulation studies of $LaGaO_3$ perovskites[1] and other $LaMO_3$ perovskites[2,3]. Related to this, can you analyse "interstitial-type" oxide-ion migration in materials such as K_2NiF_4-type La_2NiO_4 or apatite–silicates?

1   M. S. Khan, M. S. Islam and D. Bates, *J. Phys. Chem. B*, 1998, **102**, 3099.
2   M. Cherry, M. S. Islam and C. R. A. Catlow, *J. Solid State Chem.*, 1995, **118**, 125.
3   M. S. Islam, *J. Mater. Chem.*, 2000, **10**, 1027.

Professor Yashima responded: Yes, it is possible to find the mobile ions at the interstitial site in "interstitial-type" ionic conductors through the MEM technique.

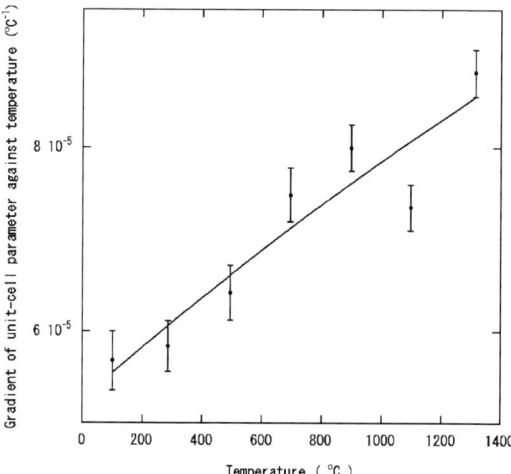

Fig. 1 Temperature dependence of the gradient of the unit cell parameter of the $Ce_{0.93}Y_{0.07}O_{1.96}$ against temperature.

Mr O'Dell opened the discussion of Dr Sayle's paper: Your paper reports that, in going from the bulk to the nanophase in ceria, it becomes easier to remove oxygen from the surface and more difficult for oxygen to come out from the bulk (interior of the nanocrystal). Which of these effects is the overriding factor?

Dr Sayle replied: For ceria-based materials to be exploited in oxidative catalysts and solid oxide fuel cells, oxygen extraction and transport are pivotal. If both of these processes were facile at the nanoscale, compared with the parent (bulk) material, then one would advocate testing nanophase ceria in such applications. However, our calculations reveal that this is not the case and therefore one has to determine whether enhanced oxygen extraction outweighs inhibited oxygen transport. We expect that the balance between the two is complex and, moreover, will be different for different applications and therefore a *general* answer is not possible.

Professor Navrotsky commented: Experimentally we see about 3% of the Ce as Ce^{3+} in nanoparticles, but we do not know if it is uniformly distributed or at the surface. Also at low temperature the particles are strongly hydrated. It would be interesting to simulate the effects of both these factors.

Dr Sayle replied: Water can have a profound influence upon the detailed atomistic structures of crystal surfaces. Indeed, Zhang and co-workers showed recently that a nanoparticle of ZnS changes its crystal structure when immersed in water.[1] If CeO_2 nanocrystals undergo structural changes when immersed in water, these will likely impact upon oxygen transport.

For lower 'concentrations' of water, hydroxyl groups attached to the CeO_2 surfaces may also influence the atomistic relaxations of surface ions. Moreover, for small nanoparticles I can envisage that this could impinge upon the transport of oxygen. In particular, I would expect that the effect will depend upon how deep the ionic relaxations penetrate.

We are in a position to simulate the influence of water on the structure and properties of ceria nanocrystals, and although we have not initiated such simulations, they are planned for future work. We will explore the influence of hydroxyl groups adsorbed on the surface, and the complete immersion of the nanocrystal in water.

1   H. Z. Zhang, B. Gilbert, F. Huang and J. F. Banfield, Water-driven structure transforma-
    tion in nanoparticles at room temperature, *Nature*, 2003, **424**(6952), 1025–1029.

Professor Navrotsky then said: We are measuring the energetics of water adsorption on zirconia, ceria, and YSZ by calorimetry and indeed have evidence for some
very strongly (dissociatively) bound water.

Mr O'Dell said: Your paper reported no enhancement in oxygen diffusion across
the surface of the nanocrystals. However, a poster at this conference reported an
enhanced conductivity at the surface. Would you like to comment on this?

Dr Sayle replied: Our calculations reveal that oxygen diffusion proceeds *via* a
vacancy mechanism in all three systems: the bulk parent material, thin CeO_2 films
sandwiched between YSZ and doped CeO_2 nanocrystals. Indeed, when we looked at
movies showing the oxygen ions diffusing (during the MD simulations) we did not
see any evidence for oxygen diffusing faster at the surface.

Mr O'Toole asked:
  (1) How did you include the correlation factor when calculating the diffusion
coefficient from the mean square displacement?
  (2) How did you test the interaction potentials for your system?t
  (3) Would you prefer to use a core-shell representation?

Dr Sayle answered:
  (1) Molecular dynamics simulation can be used to extract important atomistic
transport properties. In particular, the ionic self diffusion can be determined from
calculated mean square displacements (MSDs) following:

$$\text{MSD} = \langle r_i^2(t) \rangle = \frac{1}{N} \sum_1^N [r_i(t) - r_i(0)]^2$$

The diffusion coefficient, D_i, can then be derived from the MSDs following:

$$\langle r_i^2(t) \rangle = 6D_i + B$$

One problem we encountered, which hindered our ability to extract the MSD from
the ceria nanocrystals was that they rotated during the simulation. To prevent them
rotating we held three cerium atoms in fixed positions.
  A second issue that we considered in calculating MSDs for nanocrystals is that
oxygen ions on the surface of the nanocrystal can only move within the surface or
deeper into the nanocrystal. In particular, for long-duration simulations, the
gradient of the MSD, calculated as a function of time, will shallow and, ultimately,
converge because the oxygen ions can never move further than the diameter of the
nanocrystal. The mobility of the oxygen ions, calculated in this study, was extracted
from *linear* MSD *vs.* time traces. In addition, particles comprising the 'real'
nanomaterial are likely to be aggregated and therefore the oxygen ion mobility
can proceed from one nanocrystal to another (depending upon the interfacial area
between the nanocrystals). We did not estimate the influence of this—the calculated
MSD are for isolated nanocrystals. We are at present generating models for
aggregated CeO_2 nanocrystals with various interfacial areas and this will enable us
to extract such diffusion data.
  (2) The potential parameters have been used extensively previously to explore: the
structure of the surfaces of CeO_2; vacancy formation energies; Ce^{4+}/Ce^{3+} reduction
in CeO_2–ZrO_2 solid solutions and ionic conductivity in CeO_2/YSZ heterolayers. We
therefore suggest they are well suited to this present study.

(3) A shell-model representation would certainly be preferable to the rigid-ion model that we used in this study. In particular, a shell-model would describe point defects better than a rigid ion model—perhaps the absolute values of our diffusion coefficients would change. However, the main aim of this study was to compare oxygen ion mobility in the bulk, thin films and nanocrystals and therefore we suggest that the *relative* ordering, as calculated using a rigid-ion model description, would not change compared to analogous models that include shells.

The cost of shell-model simulations is, for these types of systems, computationally prohibitive. For example a typical simulation, to calculate an MSD at one temperature, required about 2000 cpu hours. To run shell-model MD we would need to reduce the time-step from 0.005 to 0.001 ps or less and double the effective number of species. We estimate that these simulations would be 10–50 times more expensive (about 20–100 000 hours each). Clearly, as computational resources increase, it would be prudent to repeat these simulations using a shell-model description.

Mr Martin asked: Have you been able to use your models to compare oxygen atom extraction at corners, edges and the different faces of the ceria nanoparticles?

Dr Sayle answered: The number of atoms comprising the ceria nanocrystals is about 16 000, many of which are located at the surface. To calculate the energy required to extract each oxygen atom from the surface would require more computational resources than we have available. Accordingly, a 'single' calculation was performed in which oxygen ions were extracted from a variety of positions. In particular, 266 oxygen atoms, chosen at random, were removed from the surface. For each oxygen ion removed, 2 Ce^{4+} ions were reduced to Ce^{3+} to maintain charge neutrality—again the positions were chosen at random. The positions of the oxygen vacancies included: plateau, step and corner from all faces of the nanocrystal. The positions of the Ce^{3+} ions were: 20% {111} terraces, 26% decorate {111} step (17%) or corner (9%) positions and 11% occupy positions on {100}. The remaining 43% of the Ce^{3+} species occupied positions within the bulk of the nanoparticle.

Extraction of oxygen from the surface was calculated to be easier compared with extraction from the extended $CeO_2(111)$ surface of the corresponding bulk material, which is the most stable ceria surface.[1]

This prediction was later confirmed experimentally by Mai and co-workers[2] who found that ceria nanocrystals, nanorods, nanopolyhedra and nanocubes are more reactive (for CO oxidation, *via* oxygen vacancy) compared with the bulk parent material. This enhanced catalytic activity was attributed to the preferential exposure of thermodynamically 'unstable' surfaces, which are inherently more reactive towards CO oxidation, compared with more stable, yet less reactive, surfaces proffered by the parent material.[2]

1 T. X. T. Sayle, S. C. Parker and D. C. Sayle, Oxidising CO to CO_2 using ceria nanoparticles, *Phys. Chem. Chem. Phys.*, 2005, **7**(15), 2936–2941.
2 H. X. Mai, L. D. Sun, Y. W. Zhang, R. Si, W. Feng, H. P. Zhang, H. C. Liu and C. H. Yan, Shape-selective synthesis and oxygen storage behaviour of ceria nanopolyhedra, nanorods, and nanocubes, *J. Phys. Chem. B*, 2005, **109**(51), 24380–24385.

Professor Maier said:

(1) Regarding the "ease of taking out oxygen from the surface", dealing with catalysis implies considering the detailed kinetics. As to changing the surface stoichiometry, this implies the necessity of studying the detailed kinetics of oxygen interaction (adsorption, ionisation, transfer *etc.*), while your argument was an equilibrium argument (see ref. 1).

(2) When particle size decreases, the tendency of $V_O^{\cdot\cdot}$ and e' to associate (*e.g.* forming V_O^x which is non-conductive) should increase because of proximity (see ref. 2).

This journal is © The Royal Society of Chemistry 2006

1   R. Merkle and J. Maier, *Phys. Chem. Chem. Phys.*, 2002, **4**, 4140.
1   J. Maier, *Nat. Mater.*, 2005, **4**, 805.

Dr Sayle replied:

(1) Oxidative catalysis of, for example, CO to CO_2 facilitated by ceria, rests ultimately on the 'ease of taking out oxygen from the surface' of the CeO_2. A complete study of this catalysis would, of course, involve exploring both kinetic and thermodynamic aspects, including the synergy between the two. In our study, we explored the thermodynamics—we calculated the oxygen vacancy formation energy on the surface of the CeO_2 nanocrystal and compared it with the oxygen vacancy formation energy calculated for the $CeO_2(111)$ surface of the (bulk) parent material. Many atomistic simulations involve calculations similar to those we performed in this study, and they have been found to be very useful in exploring the catalysis.[1] However, we went a step further in our calculations and attempted to include the kinetics in the same simulation. In particular, the location of the oxygen vacancies and charge compensating Ce^{3+} species was directed *via* the crystallisation (kinetic) of the nanoparticle; rather than chosen by the 'simulator'. In addition, the molecular dynamics enabled us to trace the positions of the oxygen vacancies, as a function of time, and hence observe how oxygen vacancies that form on the surface on the CeO_2 nanocrystal are replenished by oxygen ions migrating from the bulk—again a kinetic aspect. Our model is not sufficiently sophisticated to include oxygen molecules adsorbing onto the surface, ionising/electron transfer and cleavage of the oxygen–oxygen bond *etc*. Such a model would, clearly, prove very valuable. However, I do not anticipate such a complete model being available in the near future because a quantum mechanical treatment would be required and, together with the size of the CeO_2 nanocrystal, this would likely prove beyond the computational facilities available at present. Major developments in this area are being made with much success—using a combined quantum mechanical/molecular mechanical approach (see the paper by Dr Alexey Sokol and co-workers).

(2) Certainly we see examples of this within our atomistic models. In particular, we have seen a variety of isolated and associated point defects including, for example: a Ce^{4+} vacancy together with two oxygen vacancies: $[V_{Ce}'', 2V_{O}^{\bullet\bullet}]$; Ce^{3+} species at the surface of the nanocrystal bound by two adjacent Ce^{4+}, which are reduced to Ce^{3+} $[2Ce_{Ce}', V_{O}^{\bullet\bullet}]$, and larger (exotic) defect clusters such as $[V_{Ce}'', 3V_{O}^{\bullet\bullet}]^{2-}$.

We stress that further to Professor Maier's first comment, these defect clusters were not constructed 'by the simulator' rather they 'evolved' during the crystallisation. Moreover, they are transient and can move or rearrange—even annihilate—during the molecular dynamics simulations.

1   T. X. T Sayle, S. C. Parker and C. R. A. Catlow, The role of oxygen vacancies on ceria surfaces in the oxidation of carbon monoxide, *Surf. Sci.*, 1994, **316**(3), 329–336.

Professor Haile commented: I think Joachim (Professor Maier) was asking about electronic effects, for example:

$V_{O}^{\bullet\bullet} + e' \rightarrow V_{O}^{\bullet}$, and this could not be accounted for in your model.

Professor Islam commented: In reduced ceria, you have both oxygen vacancies ($V_{O}^{\bullet\bullet}$) and Ce^{3+} (Ce_{Ce}') species. As you reduce the size of your system to nanocrystal size, there might be greater oxygen–vacancy–Ce^{3+} association or binding that could reduce the oxide-ion vacancy mobility (by adding a binding energy term). Such binding (association) energies are worth investigating in the ceria nanocrystal *versus* the bulk crystal.

Dr Sayle replied: The mobility of oxygen vacancies in CeO_2 will be influenced by the presence of electronic defects. In our atomistic models, each oxygen vacancy is compensated by two electronic holes, which reside on Ce. These charge compensating Ce^{3+} ions are dispersed throughout the CeO_2 lattice—their position is not strictly random; rather they are directed to (low energy) positions during the recrystallisation of the CeO_2. However, once the nanoparticle has crystallised, the mobility of the Ce^{3+} ions compared with the oxygen ions is negligible and therefore they cannot move in response to the (mobile) oxygen vacancies. Given the strong binding energy between Ce^{3+} and oxygen vacancies, this is a limitation of the model.

Clearly, introducing a mechanism facilitating mobility of the Ce^{3+} during the small timescales accessible to typical molecular dynamical simulations is an intriguing possibility. Moreover, some kind of Monte Carlo sampling, every few steps of the dynamical simulation, is certainly viable. An alternative is to calculate the Madelung potential at each time-step and thus identify those Ce sites that the electrons holes would 'prefer' to reside on. We will explore the possibility of introducing this feature into future simulations.

Dr De Souza asked:

(1) How large does a particle have to be, or equivalently, how thick does a film have to be, for one to observe bulk behaviour?

(2) How do you explain the variation in diffusion coefficients and migration energies with dimensionality?

Dr Sayle answered:

(1) For the parent CeO_2 material, the bulk structure is observed a few atomic layers from the surface although this does depend heavily upon the surface exposed—particularly dipolar surfaces such as (100). Atomistic simulation is a very useful and accurate tool to explore ionic relaxation at surfaces.

For thin films, this is a difficult question to answer because it depends upon many interrelating factors such as the misfit of the two interfaced materials, their physical properties (*i.e.*, elastic constants), the planes exposed by the thin films at the interface, and epitaxial configurations. The misfit is the major driving force: for low misfit between the two materials, the *critical* thickness (thickness before dislocations evolve to help quench the strain and return the material back to its natural 'bulk' lattice parameter) can be hundreds of atomic layers. Conversely, for high misfit, the critical thickness can be reached in a single monolayer. Certainly, I would expect it to be possible to fine-tune the thickness of the thin films to facilitate desirable (optimum) transport properties for ceria-based materials (as has been demonstrated for BaF_2/CaF_2).[1]

To calculate the thickness of a nanocrystal to ensure bulk behaviour is intriguing. I expect that computer simulation would be an ideal forum to establish this although we have not attempted this as yet. Experimental studies reveal that the lattice parameters of ceria crystals increase as one traverses to the nanoscale.[2]

(2) Our calculations reveal that the oxygen ions diffuse in the nanocrystals and thin films *via* a vacancy mechanism as in the parent bulk material.

For the thin films we think that epitaxial constraints influence the structure of the CeO_2 thin films, making mobility more difficult. This argument suggests that there is an 'optimum' structure, which facilitates maximum mobility. Indeed, this is echoed in the many studies that show doping the CeO_2 can influence, sometimes profoundly, its transport properties—the size of the dopant influences the relaxation of the CeO_2 lattice local to the dopant. We have found that interfaces can have a similar influence.

To extract a definitive answer from our calculations, we would need to generate interfacial structures with a variety of epitaxial configurations and thin film thicknesses and calculate oxygen mobility within the thin films. Certainly, our paper has demonstrated our ability to perform such calculations, we simply await the computational resources to perform many more of these kinds of simulation.

For the nanocrystals, the influence of surface relaxations may have a similar influence upon the structure.

1 N. Sata, K. Eberman. K. Eberl and J. Maier, Mesoscopic fast ion conduction in nanometre-scale planar heterostructures, *Nature*, 2000, **408**(6815), 946–949.
2 V. Perebeinos, S. W. Chan and F. Zhang, 'Madelung model' prediction for dependence of lattice parameter on nanocrystal size, *Solid State Commun.*, 2002, **123**(6–7), 295–297.

Dr Sokol asked: Could you comment on the importance of electron hopping for the oxygen vacancy migration in reduced cerias? Technically you could model this by allowing exchange of Ce^{3+} and Ce^{4+} ions during MD simulations using some kind of Monte Carlo procedure. At present, the positions of all Ce^{3+} ions in your simulations are preset.

Dr Sayle answered: For the nanocrystals, the ceria was reduced from CeO_2 to $CeO_{1.95}$, *i.e.*, 10% of the Ce^{4+} ions were reduced to Ce^{3+}. At these concentrations, assuming a random distribution of Ce^{3+} at cation sites, there is a low probability that an oxygen will move without 'feeling' the influence of a Ce^{3+} in a nearest neighbour or next nearest neighbour position. However, in Fig. 10, you can see an area that is relatively devoid of Ce^{3+} species and if one observes closely the mobility of the oxygen ions in this region under MD simulation, one can judge the mobility to be somewhat reduced. The ability to 'move' the Ce^{3+} ions (electron hopping) would shed some light on this. Certainly, some kind of Monte Carlo procedure for exchanging the Ce^{3+}/Ce^{4+} would prove a valuable addition to the model.

Professor Haile asked: Could you clarify the difference between 'extraction' and 'transport' when you conclude that oxygen extraction is easier from the nanocrystals yet the transport is harder? Is 'extraction' simply from the surface?

Dr Sayle answered: Yes. The 'transport' was the new work reported in this present study, while the 'extraction' was calculated and reported previously.

Mr Martin asked: With the formation of edges, corners and different faces on the ceria particles are there plans to use a QM/MM approach to investigate vacancies at these structurally differing areas of the ceria nanoparticle?

Dr Sayle replied: Yes. Our overall project aim was to 'simulate automobile catalysis'—basically, one can envisage a CO molecule arriving at the surface of CeO_2 and extracting an oxygen to form CO_2. For this we need a quantum mechanical description to probe the reaction mechanism. However, the mechanism will be highly dependent upon the detailed atomistic *microstructure* of the CeO_2, which will include, for example: morphology—surfaces, steps, corners, grain-boundaries, defects, epitaxy, dislocations *etc*. In addition, we also needed to explore transport of oxygen through the lattice to replace oxygen given up from the surface to oxidise CO.

Such a system is too large to be considered quantum mechanically and therefore we span the length-scales by combining atomistic simulation to generate the models (reported here) with a quantum mechanics/molecular mechanics approach to explore the mechanism. The latter is being investigated by Dr Graeme Watson using atomistic models derived in this present study.[1]

1 For example, see: M. Nolan, S. C. Parker and G. W. Watson, CeO_2 catalysed conversion of CO, NO_2 and NO from first principles energetics, *Phys. Chem. Chem. Phys.*, 2006, **8**(2), 216–218.

Professor Haile opened a general discussion, addressing Professor Yashima: You've said that you'd like to evaluate the activation energies for jumps between various sites from your nuclear density maps. Will this require a fundamental theoretical breakthrough or is this a simple evaluation that you just haven't had a chance to get to yet?

Professor Yashima replied: I have no experience of the evaluation, but according to the literature, it is possible to evaluate the activation energy assuming a simple potential model.

Professor Bruce asked a general question: I would like to address the issue of ion transport in the presence of high defect concentrations like we see in structures with a high degree of order/association. In LISICON, $Li_{2+2x}Zu_{1-x}CeO_4$, the interstitial Li^+ ions responsible for conduction exist in defect clusters. Ion transport induces cooperative rearrangement of the defect clusters, resulting in Li^+ ion transport. Does anyone have a comment on the mechanisms of ion transport in other highly associated systems?

Dr Sayle answered: For the CeO_2/YSZ heterolayers and nanocrystals reported in our study, the concentration of oxygen vacancies, and hence charge compensating Ce^{3+} species, is high and therefore the oxygen transport we have calculated involved predominantly associated species.

It would be interesting to perform analogous simulations where any such association is eliminated. This can be achieved at low defect concentration, or, alternatively, using a 'mean-field' strategy where the charge balance associated with forming an oxygen vacancy is spread (artificially) over all the Ce ions in the system.

Professor Heitjans addressed Professor Bruce: Is there experimental evidence for the interstitial Li ion conduction mechanism you showed previously?

Professor Bruce replied: Here is a diagram of the mechanism that you are referring to (Fig. 2). The mechanism of defect formation and the subsequent mechanism of ion transport was presented to illustrate one possibility. The low concentration of defects makes it very difficult to obtain direct evidence for a specific conduction mechanism.

Professor Yashima asked: It is difficult to determine the exact positions of interstitial sites. Did you examine various positions before your conclusions? You would need to examine this very carefully. The Fourier map may be helpful to find the interstital sites.

Professor Bruce replied: As I explained in my answer to Professor Heitjans, the number of defects is too small to be detected directly.

Professor Heitjans commented: As I understood it, the interstitial Li ion conduction mechanism you propose involves Li sites with different ^7Li quadrupole coupling constants. In this case the experimental method of choice to elucidate the diffusion pathway would be ^7Li spin alignment echo (SAE) NMR. We recently used this technique for the study of layered Li_xTiS_2,[1] the well known prototype 2D conductor, and its cubic polymorph,[2] where diffusion was shown to be 3D by ^7Li spin–lattice relaxation NMR.[3] Very recently we evaluated the potential of ^7Li SAE NMR for other Li ion conductors, including $LiNbO_3$.[4,5]

1 M. Wilkening, W. Küchler and P. Heitjans, *Phys. Rev. Lett.*, 2006, **97**, 065901.
2 M. Wilkening and P. Heitjans, *Diffusion Fundamentals*, 2005, **2**, 60.
3 W. Küchler, P. Heitjans, A. Payer and R. Schöllhorn, *Solid State Ionics*, 1994, **70/71**, 434.
4 M. Wilkening and P. Heitjans, *Solid State Ionics*, 2006, DOI: 10.1016/j.ssi.2006.07.037.
5 M. Wilkening and P. Heitjans, *J. Phys.: Condens. Matter*, 2006, **18**, 9849.

Professor Bruce said: The Arrhenius plots for LISICON are curved at high temperature and defect clusters change.

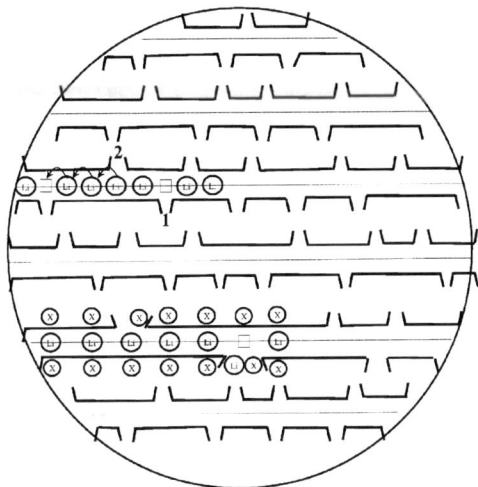

Fig. 2 Possible mechanisms of defect formation and conduction in the crystalline PEO$_6$:LiXF$_6$ (X = P, As, Sb) polymer electrolytes. Black lines represent PEO chains, dotted lines show axis of tunnels; encircled Li and X represent the cations and anions respectively; squares represent Li$^+$ vacancies. The conduction mechanism shown in the upper region of the crystallite involves a Li$^+$ vacancy at a chain end (1) migrating to the next chain end (2) by cooperative displacement of Li$^+$ ions.

Professor Islam addressed Professor Yashima: From your neutron diffraction MEM method, can you obtain information on dopant–defect association or clustering in doped oxides such as Y/CeO$_2$?

Professor Yashima replied: It is difficult to observe directly the defect association or clustering using the neutron diffraction MEM method, which gives the average structure. But the MEM map gives the disorder of mobile/immobile ions and diffusion path of mobile ions, which are influenced by the association and clustering.

Mr O'Toole addressed Dr Sayle: Why did you use a spherical representation that does not allow the use of periodic boundary conditions? Could you not use a series of surface calculations and apply statistics?

Dr Sayle responded: All the simulations we performed involved periodic boundary conditions. For the CeO$_2$/YSZ thin films, the interfacial plane was defined and repeated by simulation vectors x,y and the heterolayers were repeated in the z direction, facilitating full space-filling models.

For the CeO$_2$ nanocrystals, periodic boundary conditions were again applied. However, the size of the simulation cell was increased to ensure no interaction between neighbouring images. This is perhaps a rather strange (unnecessary) construction. However, future work involves aggregated nanocrystals, which is achieved simply by reducing the size of the simulation cell such that the nanocrystal can 'feel' and respond to its neighbours. It was prudent therefore to develop our methods using this approach.

Simulating a series of surfaces and then applying statistics is a valuable simulation strategy in that it allows one to deconvolute each of the many competing factors. To include all surfaces and microstructural features in a single simulation cell, as we have done, is more computationally demanding, but the results will incorporate any synergic effects. Accordingly, the predictions (*i.e.*, oxygen transport) will be directly comparable to experiment—assuming that our model is sufficiently realistic. The power of this approach is that it can be used predictively for screening, thus reducing the need for expensive experiments. Indeed, Professor Kilner requested that we (simulators) develop such a simulation tool.

Professor Bruce commented: In materials with high defect concentrations there will be strong correlations between mobile ions so that when an ion hops, the other ions must reorganise, this results in frequency dependent conductivity and we must use this as a means of probing ion transport in the systems described in this conference, correlating the results with the structural studies to understand ion transport in such correlated systems.

Professor Maier added: As to studying displacement on different time and length scales there is intensive activity around the world, typically reflected by the conference session on "Relaxation in Solids" [K. Ngai, K. Furke, H. Jain, A. Bunte, W. Dietrich, P. Heitjans and others].

Professor Bruce replied: Yes, I agree. Over the last 20 years the significance of cooperative effects on ion transport in solids with a large concentration of charge carriers has been recognised in both simulation and experiment. Klaus Funke's work[1] on the dispersion in conductivity over a wide frequency range, from dc to the far IR, is a case in point. Yet in other areas, especially anion conductors, the significance of such co-operative effects and especially their study has been more limited. This is the first time that the interaction between mobile ions and its effect on ion transport has been mentioned at this meeting (*i.e.*, that the hopping of one ion must be influenced by the ion atmosphere formed by the other mobile ions, analogous to the Debye–Hückel ion atmosphere in liquid electrolytes).

1   K. Funke and R. D. Banhatti, *Solid State Ionics*, 2005, **176**, 1971.

Dr Sayle said: We thank Professor Maier for directing us along this avenue.

Professor Aranda addressed Professor Yashima: In order to decorrelate (at least partially) the thermal vibration and the positional disorder the usual way is to cool the sample down to very low temperature and to check the residual thermal vibration parameters. Have you observed peak broadening at high temperature in the neutron patterns due to the presence of defects?

Professor Yashima replied: No, I have not observed peak broadening at high temperatures in the neutron diffraction patterns. The disorder in the crystalline materials decreases the peak intensity at higher 2θ regions due to the larger atomic displacement parameters.
Also, the disorder can be separated into two parts. One is dynamic disorder and the other is static disorder. You can distinguish one from the other by analyzing the thermal parameters and the disorder down to lower temperatures.

Professor Vannier asked: Did you notice a change (modulations) in your background on reaction diffractograms when increasing the temperature? This is a further proof of the dynamics of your material.

Professor Yashima answered: Yes, we observed a modulated background with a complicated structure, due to diffusion scattering. This is direct evidence for the local ordering/clusters associations. They are ascribed to both dynamic and static disorder.

Concluding Remarks

Alan V. Chadwick

Received 30th August 2006, Accepted 30th August 2006
First published as an Advance Article on the web 16th November 2006
DOI: 10.1039/b612508p

It is difficult to summarize the work presented in the two full and packed days of this Faraday Discussion. The meeting has been an outstanding success and the Faraday Discussion format, although new to many of the attendees, has totally fulfilled the objective of providing a thorough discussion of each contribution and has stimulated wider discussions of atomic transport and defect phenomena in solids. There were many highlights, too many to cover in this brief summary. Those that I will cover are clearly aligned to my own personal research interests and whereas other delegates may have different preferences I am sure they will agree that the meeting provided an excellent overview of contemporary work in this field.

As this is a Faraday Discussion on a topic that originated from the work of Michael Faraday it is appropriate to give a brief review of the history of this field. In the 1830s Faraday was studying the electrical conductivity of a wide range of materials using what would now be regarded as very basic apparatus, *i.e.* Volta piles and a gold leaf electroscope. His diaries[1] show that on 21st February 1833 he noted the 'very extraordinary behaviour of silver sulfide, Ag_2S, which conducted electricity in the solid state and showed electrolytic decomposition'. Almost exactly two years later, on 19th February 1835, he found that solid lead fluoride, PbF_2, at 'dull red heat' also conducted electricity. The genius of Faraday was to link the two observations and record in his diary that other materials 'when sought for' would probably behave in a similar manner, *i.e.* conduct in the solid state. Thus he foresaw the field of *superionic conductors, fast-on conductors* and *solid electrolytes*, which came to prominence some 130 years later in the 1960s.

The next development was due to Kohlrausch[2] who developed the a.c. resistance bridge. As a result electrical resistance became the most accurate physico-chemical parameter that could be measured. Although Kohlrausch is best known for his work on electrolyte solutions, his bridge is the forerunner of the a.c. bridges used in current conductivity studies, the mainstay of much of the transport information. Various studies appeared during the latter half of the 19th century and it is interesting that they appear to have been the preserve of physicists.[3] Wiedemann's[4] standard physics text, "The Science of Electricity" (1893/98) included chapters entitled "Conductivity of Solid Salts", "Determination of the Electromotive Force—Two Metals and Solid Electrolytes" and "Electrolysis of Solid Electrolytes". However, in Ostwald's[5] "Treatise on Electrochemistry" (1885/1917) solid electrolytes are not mentioned. The term '*solid electrolyte*' appears to have been in general use by the end of the 19th century and in 1899 Nernst filed his patent application for the *Nernst glower*, the first commercial application of a solid electrolyte and based on rare earth doped zirconia.

In the first half of the 20th century experimental studies of ionic solids were pursued mainly in Germany by Lorenz, Jost, Tubandt and others, which provided information on conductivity, transport numbers and diffusion. The subject of Jost's Ph.D. thesis, in the 1920s, was diffusion in ionic crystals. The quality of the ionic conductivity measurements that were being made during that period can be seen in the work of Phipps and co-workers.[6] They were using procedures that are in use

Functional Materials Group, School of Physical Sciences, University of Kent, Canterbury, Kent, UK CT2 7NH

today, *e.g.* graphite electrodes, a.c. bridges, *etc.*, and the precision is as good as can be obtained in modern work. The theoretical breakthroughs came with the developments of models of lattice defect in the work of Frenkel,[7] Schottky and Wagner.[8] The pioneering use of radiotracer by von Hevesy in the 1920s added a new experimental tool for the direct determination of diffusion coefficients.

After the Second World War there was very rapid progress. On the experimental front this was aided by the ready availability of synthetic single crystals, radiotracers and good commercial equipment. The bank of data that was produced allowed the testing and development of theoretical models and further progress in the understanding of point defects and atomic transport. By 1957, the time of the article by Alan Lidiard[9] in *Handbuch der Physik*, the understanding and models had reached a very high level of sophistication. This seminal article, which is now a citation classic, not only provided a thorough overview of the field but also introduced concepts that are still extant in the present day. It is pleasing to note that Lidiard is still active and that he has a poster presentation at this Faraday Discussion (Poster P19, "Computation of defect heats of transport", by P. J. Grout and A. B. Lidiard). In 1959 there was sufficient interest to hold a Faraday Discussion (FD28, Crystal imperfections and the chemical reactivity of solids) on the topics. One of the best examples of the utilisation of the theoretical models in interpreting experimental data was the development of computer least squares fitting procedures by Jacobs and Beaumont[10] to analyse the conductivity data for single crystals. This made full use of the basic high accuracy of the raw data, which with the inclusion of known corrections to the simple theory (*e.g.* allowance for crystal expansion, Debye–Hückel–Lidiard activity corrections, Onsager–Pitt drag correction, *etc.*), yielded accurate values of the defect formation and migration enthalpies and entropies. It is worth noting that the experimental accuracy usually obtained from complex impedance spectroscopy of compacted pellets rarely allows the use of such a sophisticated analysis.

The current interest in atomic transport and defect phenomena in ionic solids dates back to the late 1960s. This was the point at which it was realised that there were many solid electrolytes with high ionic conductivities and these had potential technological applications. The need for more efficient energy usage and storage, emphasized by the Oil Crisis of the 1970s, increased the interest on these materials with their applications as electrolyte membranes in fuel cells, batteries and sensors. This interest in solids with real or potential technological applications has been a major theme of this meeting and provides a direct contrast with the topics discussed at the earlier Faraday Discussion, FD28. The opening lecture by John Kilner was a fine example of the best of the current approaches to the study of technologically important materials, in this case oxides for solid oxide fuel cell applications.[11] This involved a very thorough characterisation of materials using a range of techniques to provide a detailed understanding of the roles of structure and composition on the transport properties.

The contributions covered a wide range of materials types and techniques, both experimental and theoretical. The materials discussed, along with potential applications, included:

• Fluorite structured materials; well known for their fast ion transport due to the open nature of the structure.

• Various lithium ion conductors (both crystalline and amorphous); important as potential materials for lithium battery applications.

• Oxygen ion conductors; for applications in fuel cells and oxygen separators.

• Electronic conductors.

• Polymer electrolytes, particularly those based on polyethylene oxide; for applications as battery electrolyte membranes.

• Carbon nanotubes; the effect of filling the tubes with simple ionic salts.

• Metallic alloys.

The techniques discussed included:

• Impedance spectroscopy.

- Tracer diffusion.
- Diffraction, both X-ray and neutron.
- Nuclear magnetic resonance spectroscopy.
- X-Ray absorption spectroscopy.
- Calorimetry
- Computer simulations.

I identified three themes that ran through the meeting, namely nanomaterials, materials for energy production and storage, and computer modelling. The last few years has seen an enormous interest in anything that is nanoscale, including the atomic transport properties. In fact one of the earliest 'unusual' properties of nanomaterials was exceptionally high diffusion coefficients in metals.[12] However, until relatively recently there was some scepticism over the experiments reporting enhanced diffusion in ionic solids.[13] A real breakthrough was the work of Maier and co-workers[14] with an experimental study of films of alternating layers films of CaF_2 and BaF_2. This work showed the conductivity was dramatically increased when the layers were of nanometre dimensions. The enhanced diffusion in nanocrystals is now accepted, as pointed out in the opening lecture by Kilner and several papers at this meeting considered aspects of the phenomenon. For example, further experimental evidence of enhanced diffusion was presented by Heitjans and co-workers in their paper on lithium niobate.[15] This very careful study using a range of techniques emphasised the role of sample preparation on the properties and the microstructure of the nanocrystals, particularly the difference between sol-gel and ball-milling routes. Sample preparation was also a key feature of the work by O'Dell and co-workers who used NMR methods to probe the effect of pinning agents used to stabilise oxide nanocrystals at elevated temperatures.[16] Potential technological applications were also considered in the papers by Maier (mass storage at interfaces) and Nazar and co-workers (in battery applications).[17,18] The discussion of the above papers raised a number of key issues when considering ionic materials. Firstly, the relative roles of the space-charge and texture (surface mismatch) at the interfaces, in the origin of the enhanced diffusion need to be clarified. Secondly, there is still a dearth of reliable information and more studies using a range of experimental techniques are required, with NMR methods showing considerable potential. In addition, it is essential that the same samples are used in the different experiments if comparisons of the results are to be meaningful.

Efficient energy production and storage is a key world wide issue and improvements rely on understanding of defect and diffusion phenomena in the materials. A range of systems and materials were considered at the meeting, including uses in fuel cells and batteries. In addition to the opening lecture by Kilner fuel cell electrolytes were covered in the paper by Irvine and co-workers.[19] Several interesting papers covered battery materials and introduced new materials and concepts. The work of Nazar and co-workers indicated that a material's battery performance could be improved by engineering the structure of the grain boundaries and the introduction of glassy layers.[18] Polymer electrolytes based on polyethylene oxide have been the subject of research for many years since the early demonstration of applications in lithium batteries by Armand,[20] however there is still considerable debate about the mechanisms of ionic transport. Two papers at this meeting reported new data for these systems that will provide progress in their understanding. Firstly the paper by Stolwijk and co-workers presented accurate radiotracer diffusion data which combined with conductivity can be subjected to detailed analysis to reveal the nature of the migrating species.[21] Secondly, the work of Bruce and co-workers showed that high conductivity is not restricted to amorphous polymers and revealed new strategies to obtain good conductivity in crystalline polyethylene oxide–salt complexes.[22] In all the discussions of the materials for energy application a recurring theme was the importance of the relative roles of 'size', 'interactions' and 'order' in controlling the transport.

Computer modelling is now a well established technique in the study of materials[23] and much of the work originated in transport and defect studies. Thus it is not surprising that it was a key theme of this meeting, with a wide range of materials and methods being presented. These included work on ionic solids using molecular mechanics methods (Kendrick and co-workers, de Leeuw and co-workers), density functional theory (de Leeuw and co-workers), and embedded cluster hybrid quantum mechanical/molecular mechanical approaches (Sokol and co-workers).[24,25,26] Simulations of nanomaterials were considered for ionic salts inside carbon nanotubes (Wilson) and for oxide nanocrystals (Sayle and co-workers).[27,28] Diffusion in metallic alloys was treated in the paper by Nastar and co-workers[29] and the structure of dislocations in 4H-SiC was covered in the paper by Savini and co-workers.[30] All the papers involving computer modelling emphasized the benefits of linking the calculations with experimental studies to provide information on defect structures and diffusion mechanisms. At several points in the discussion of the papers a number of key questions from the experimentalists present recurred, namely: How accurate are the simulations? How reliable are the simulations? Can we rely on predictions concerning trends? The answers to the first two questions from the modellers were very much dependent on the systems. However, the prediction of trends from the simulations was generally accepted as being reliable.

My final remarks must be to thank the various people involved in the organisation of the meeting. The RSC staff, led by Morwena Gilbert, did excellent work in the preparation for the meeting and during the meeting were tireless in their efforts to maintain smooth running. In the latter aspect they were helped by the very energetic and enthusiastic local team, namely Peter Slater, Craig Fisher and Julia Percival. Final thanks must go to Saiful Islam for initiating the meeting and carrying it through to such a successful conclusion. A test of the success will be the published issue. I have treasured my copy FD28 as a valuable source of reference. I am confident that FD134 will be similarly prized by future researchers.

References

1 (a) M. Faraday, *Experimental Researches in Electricity*, R. and J. E. Taylor, London, 1940; (b) M. Faraday, *Faraday's Diaries 1820–1862*, G. Bell, London, 1939, vol. II.
2 F. Kohlrausch and W. A. Nippoldt, *Pogg. Ann.*, 1868, **138**, 280–370.
3 H.-H. Möbius, *J. Solid State Electrochem.*, 1997, **1**, 2.
4 G. Wiedemann, *Die Lehre von der Electricität*, Vieweg u Sohn, Braunschweig, 2nd edn, 1893/98, vol I, pp. 553–561, pp. 815–819; vol II, pp. 491–493.
5 W. Ostwald, *Lehrbuch der Allgemeinen Chemie*, Engelmann, Leipzig, 1st edn, 1885/87.
6 (a) T. E. Phipps, W. D. Lansing and T. G. Cooke, *J. Am. Chem. Soc.*, 1926, **48**, 112; (b) T. E. Phipps and E. G. Partridge, *J. Am. Chem. Soc.*, 1929, **51**, 1331.
7 J. Frenkel, *Z. Phys.*, 1926, **35**, 632.
8 W. Schottky and C. Wagner, *Z. Phys. Chem. B*, 1930, **11**, 163.
9 A. B. Lidiard, in *Handbuch der Physik*, ed. S. Flügge, Springer, Berlin, 1957, vol. 20, p. 246.
10 J. H. Beaumont and P. W. M. Jacobs, *J. Phys. Chem.*, 1965, **45**, 1496.
11 J. Kilner, *Faraday Discuss.*, 2007, **134**, DOI: 10.1039/b615306m.
12 H. Gleiter, *Adv. Mater.*, 1992, **4**, 474.
13 H. L. Tuller, *Solid State Ionics*, 2000, **131**, 143.
14 N. Sata, K. Ebermann, K. Eberl and J. Maier, *Nature*, 2000, **408**, 946.
15 P. Heitjans, M. Masoud, A. Feldhoff and M. Wilkening, *Faraday Discuss.*, 2007, **134**, DOI: 10.1039/b602887j.
16 L. A. O'Dell, S. L. P. Savin, A. V. Chadwick and M. E. Smith, *Faraday Discuss.*, 2007, **134**, DOI: 10.1039/b601928e.
17 J. Maier, *Faraday Discuss.*, 2007, **134**, DOI: 10.1039/b603559k.
18 B. Ellis, P. Subramanya Herle, Y.-H. Rho, L. F. Nazar, R. Dunlap, L. K. Perry and D. H. Ryan, *Faraday Discuss.*, 2007, **134**, DOI: 10.1039/b602698b.
19 J. T. S. Irvine, J. W. L. Dobson, T. Politova, S. G. Martín and A. Shenouda, *Faraday Discuss.*, 2007, **134**, DOI: 10.1039/b604441g.
20 M. B. Armand, J. M. Chabango and M. J. Duclot, in *Fast Ion Transport in Solids*, ed. P. Vashishta, J. N. Mundy and G. K. Shenoy, North-Holland, Amsterdam, 1979, p. 131.

21 N. A. Stolwijk, M. Wiencierz and Sh. Obeidi, *Faraday Discuss.*, 2007, **134**, DOI: 10.1039/b602143n.

22 E. Staunton, Yu. G. Andreev and P. G. Bruce, *Faraday Discuss.*, 2007, **134**, DOI: 10.1039/b601945e.

23 *Handbook of Materials Modelling*, ed. S. Yip, Springer, New York, 2005.

24 E. Kendrick, J. E. H. Sansom, J. R. Tolchard, M. S. Islam and P. R. Slater, *Faraday Discuss.*, 2007, **134**, DOI: 10.1039/b602258h.

25 N. H. de Leeuw, J. R. Bowe and J. A. L. Rabone, *Faraday Discuss.*, 2007, **134**, DOI: 10.1039/b602012g.

26 A. A. Sokol, S. A. French, S. T. Bromley, C. R. A. Catlow, H. J. J. van Dam and P. Sherwood, *Faraday Discuss.*, 2007, **134**, DOI: 10.1039/b607406e.

27 M. Wilson, *Faraday Discuss.*, 2007, **134**, DOI: 10.1039/b602488b.

28 T. X. T. Sayle, S. C. Parker and D. C. Sayle, *Faraday Discuss.*, 2007, **134**, DOI: 10.1039/b601521b.

29 M. Nastar and V. Barbe, *Faraday Discuss.*, 2007, **134**, DOI: 10.1039/b605834e.

30 G. Savini, M. I. Heggie and S. Öberg, *Faraday Discuss.*, 2007, **134**, DOI: 10.1039/b603920k.

Poster titles

Effect of processing atmosphere on the dielectric properties of $BaCo_{1/3}Nb_{2/3}O_3$ ceramics **Ming Li, Antonio Feteira** and **Derek C. Sinclair,** *University of Sheffield, UK*

Atomistic study of the $CaTiO_3$-based mixed conductor: defects, cluster formation and oxide-ion migration **Glenn C. Mather, M. Saiful Islam** and **Filipe M. Figueiredo,** *Instituto de Cerámica y Vidrio, Consejo Superior de Investigaciones Científicas, Spain*

Conductivity of mixed-anion crystalline polymer electrolytes **Scott J. Lilley, Yuri G. Andreev** and **Peter G. Bruce,** *University of St Andrews, UK*

Synthesis and characterization of nanoparticles and nanoceramics of $Ce_{0.9}Gd_{0.1}O_{2-d}$ **Enrique Ruiz-Trejo, Francisco Gómez-García, Ruben Vilchis-Morales, Jaime Santoyo-Salazar, Carlos Flores-Morales** and **José A. Chávez-Carvayar,** *Universidad Nacional Autónoma de México, Mexico*

The effect of the preparation route on the properties of proton ceramic nanopowders **Zohreh Khani, Mélanie Jacquin, Gilles Taillade, Yan Jing, Deborah J. Jones, Jacques Roziére** and **Mathieu Marrony,** *Universite Montpellier II, France*

Electronically conducting oxides for solid oxide fuel cells **Hilary Wood, Andrey Berenov** and **Alan Atkinson,** *Imperial College London, UK*

Investigation of oxide ion and proton conduction in gallates **E. Kendrick, M. Saiful Islam** and **Peter Slater,** *University of Surrey, UK*

Mechanism of oxygen incorporation into $SrFe_xTi_{1-x}O_{3-\delta}$: kinetic and electrochemical studies **R. Merkle** and **J. Maier,** *Max-Planck Institute for Solid State Research, Germany*

Crystal structure and positional disorder of the ceria–zirconia solid solution $CeZrO_4$ **Takahiro Wakita, Masatomo Yashima, Yoshiaki Ando, Takayuki Tsuji, Qi Xu, Toshikazu Ueda, Yoichi Kawaike** and **Roushown Ali,** *Daiichi Kigenso Kagaku Kogyo Co. Ltd, Japan*

Dense membranes BIMEVOX catalysts for alkanes oxidation **Hervé Bodet, César Steil, Caroline Pirovano, Rose-Noëlle Vannier, Axel Löfberg** and **Elisabeth Bordes-Richard,** *Unité de Catalyse et de Chimie du Solide, CNRS, France*

Investigations of olivine and nasicon-type phases for lithium battery applications **J. Percival, P. Slater** and **M. S. Islam,** *University of Surrey, UK*

Theoretical modeling of proton diffusion in acceptor doped $BaZrO_3$ based on density-functional calculations **Per G. Sundell, Mårten E. Björketun** and **Göran Wahnström,** *Chalmers University of Technology, Sweden*

A comparative study of cubic $PbZrO_3$ and $SrTiO_3$ perovskites containing single F centers: *ab initio* simulations **Yu. F. Zhukovskii, S. Piskunov, E. A. Kotomin, E. Heifets** and **D. E. Ellis,** *Institute of Solid State Physics, Latvia*

Role of surface F_s centers in formation of ultra-thin Ag and Cu films on the $MgO(001)$ substrate **D. Fuks, Yu. F. Zhukovskii, E. A. Kotomin** and **D. E. Ellis,** *Institute for Solid State Physics, Latvia*

Defects and ion transport in the Li battery material LiFePO$_4$ **Craig A. J. Fisher, Daniel J. Driscoll, Peter R. Slater** and **M. Saiful Islam**, *University of Surrey, UK*

Mixed ionic-electronic conducting strontium ferrites: atomic-scale studies of defects, dopants and oxide-ion migration **Craig A. J. Fisher** and **M. Saiful Islam**, *University of Surrey, UK*

Chemical capacitance of oxide anodes for solid oxide fuel cells **T. Nakamura, K. Yashiro, A. Kaimai, K. Sato, J. Mizusaki** and **T. Kawada**, *Tohoku University, Japan*

Solid solution formation in saline hydrides **M. C. Verbraeken** and **J. T. S. Irvine**, *University of St Andrews, UK*

Computation of defect heats of transport **P. J. Grout** and **A. B. Lidiard**, *University of Oxford, UK*

Defect structure of A site- and B site-doped lanthanum chromites **Keiji Yashiro, Mamoru Hasegawa, Masatsugu Oishi, Kazuhisa Sato, Atsushi Kaimai, Takanori Otake, Tatsuya Kawada** and **Junichiro Mizusaki**, *Tohoku University, Japan*

Solvation structure and transport of acidic protons ionic liquids **Mario G. Del Pópolo, Jorge Kohanoff** and **Ruth M. Lynden-Bell**, *Queen's University Belfast, UK*

The defect structure of nickel doped strontium titanate **Michael Paul, C. Richard A. Catlow, Richard J. Oldman** and **Sam A. French**, *Royal Institution of Great Britain, UK*

Conductivity enhancement in organic ionic plastic crystals induced by nano-sized oxide particles **Youssof Shekibi, Josefina Adebahr, Natalie Ciccosillo, Douglas R. MacFarlane, Anita J. Hill** and **Maria Forsyth**, *Monash University, Australia*

Optimization of ionic conductivity in doped ceria from first principles **David Andersson, S. I. Simak, N. V. Skorumova, I. A. Abrikosov** and **B. Johansson**, *Royal Institute of Technology, Sweden*

The structure and electrochemical properties of (La,Sr)(Ti,Al)O$_{3+\delta}$ perovskites **David N. Miller** and **John T. S. Irvine**, *University of St Andrews, UK*

Atomic transport and defects in Ba$_2$In$_2$O$_5$-based oxide ion conductors **A. Rolle, C. A. J. Fisher, P. Roussel, S. Daviero-Minaud, M. S. Islam** and **R. N. Vannier**, *Ecole Nationale Supérieure de Chimie de Lille/Université des Sciences et Technologies de Lille, France*

High temperature diffusion of oxygen in chromite **Stephen O'Toole, Nicholas Stevens, Andrew Willetts** and **Mark Bankhead**, *University of Manchester, UK*

Oxygen permeation property of doped Pr$_2$NiO$_4$ based oxide for partial oxidation of CH$_4$ **Tatsumi Ishihara, Shogo Miyoshi, Tetsuro Furuno** and **Hiroshige Matsumoto**, *Kyushu University, Japan*

Computational modelling of oxygen mobility at ceria surfaces **Paul Martin** and **Stephen C. Parker**, *University of Bath, UK*

The Skinner prize for best poster was jointly awarded to Ms Julia Percival from the University of Surrey, UK, for her poster on investigations of olivine and nasicon-type phases for lithium battery applications and Mr Per Sundell from Chalmers University of Technology, Sweden, for his poster on the theoretical modelling of proton diffusion in acceptor doped BaZrO$_3$ based on density-functional calculations.

List of Participants

Mr D. Andersson, *Royal Institute of Technology, Sweden*
Dr Yu. Andreev, *University of St Andrews, United Kingdom*
Professor M. Aranda, *University of Malaga, Spain*
Dr S. Batten, *Royal Society of Chemistry, United Kingdom*
Professor Dr H. P. Beck, *Saarland University, Germany*
Mr S. Bhuhi, *Elite Thermal Systems Ltd, United Kingdom*
Mr M. Björketun, *Chalmers University of Technology, Sweden*
Professor P. Bruce, *University of St. Andrews, United Kingdom*
Professor R. Catlow, *University College London, United Kingdom*
Professor A. Chadwick, *University of Kent, United Kingdom*
Professor D. Clary, *University of Oxford, United Kingdom*
Dr N. de Leeuw, *University College London, United Kingdom*
Dr R. De Souza, *RWTH Aachen, Institute of Physical Chemistry, Germany*
Professor R. Dieckmann, *Cornell University, USA*
Dr S. Dixon, *Royal Society of Chemistry, United Kingdom*
Professor J. Drennan, *The University of Queensland, Australia*
Mr C. Fisher, *University of Bath, United Kingdom*
Professor M. Forsyth, *Monash University, Australia*
Miss M. Gilbert, *Royal Society of Chemistry, United Kingdom*
Dr A. Gray-Weale, *University of Sydney, Australia*
Dr P. Grout, *University of Oxford, United Kingdom*
Professor S. Haile, *Caltech, USA*
Miss S. Harrison, *Royal Society of Chemistry, United Kingdom*
Professor P. Heitjans, *University of Hanover, Germany*
Professor J. Irvine, *University of St Andrews, United Kingdom*
Professor T. Ishihara, *Kyushu University, Japan*
Professor S. Islam, *University of Bath, United Kingdom*
Dr A. Jones, *University of Bath, United Kingdom*
Dr E. Kendrick, *University of Surrey, United Kingdom*
Miss Z. Khani, *Universite Montpellier II, France*
Professor J. Kilner, *Imperial College London, United Kingdom*
Dr J. Kohanoff, *Queens University Belfast, United Kingdom*
Mr H. Kusumi, *Toyota Motor Europe, Belgium*
Dr A. Kybett, *Royal Society of Chemistry, United Kingdom*
Mr M. Li, *University of Sheffield, United Kingdom*
Dr J. Lord, *ISIS Facility, CCLRD, United Kingdom*
Professor Dr J. Maier, *Max-Planck Institute Fur Festkorperforschung, Germany*
Mr B. Mangili, *Cranfield University, United Kingdom*
Mr P. Martin, *University of Bath, United Kingdom*
Dr G. Mather, *Ceramics and Glass Institute, CSIC, Spain*
Dr R. Merkle, *MPI for Solid State Research, Germany*
Dr D. Miller, *University of St Andrews, United Kingdom*
Dr K. Nakamura, *University of Tokushima, Japan*
Mr T. Nakamura, *Tohoku University, Japan*
Dr M. Nastar, *CEA Saclay, France*
Professor A. Navrotsky, *University of California, Davis, USA*
Professor L. Nazar, *University of Waterloo, Canada*
Mr L. O'Dell, *University of Warwick, United Kingdom*
Dr P. O'Sullivan, *Royal Society of Chemistry, United Kingdom*
Mr S. O'Toole, *MPC University of Manchester, United Kingdom*
Dr M. Paul, *Royal Institution of Great Britain, United Kingdom*
Miss J. Percival, *University of Surrey, United Kingdom*

Dr A. Rolle, *University of St Andrews, United Kingdom*
Dr E. Ruiz-Trejo, *UNAM Facultad De Quimica, Mexico*
Dr G. Savini, *University of Sussex, United Kingdom*
Dr D. Sayle, *Cranfield University, United Kingdom*
Dr P. Slater, *University of Surrey, United Kingdom*
Dr A. Sokol, *Royal Institution, United Kingdom*
Dr N. Stolwijk, *University of Muenster, Germany*
Mr P. Sundell, *Chalmers University of Technology, Sweden*
Miss J. Thomson, *Royal Society of Chemistry, United Kingdom*
Mr H. Thorpe, *Elite Thermal Systems Ltd, United Kingdom*
Professor R. Vannier, *UCCS, Université des Sciences et Technologies de Lille, France*
Mr M. Verbraeken, *University of St Andrews, United Kingdom*
Mr T. Wakita, *Daiichi Kigenso Kagaku Kogyo Co. Ltd, Japan*
Dr M. Wilson, *University College London, United Kingdom*
Miss H. Wood, *Imperial College London, United Kingdom*
Professor M. Yashima, *Tokyo Institute of Technology, Japan*
Mr K. Yashiro, *IMRAM, Tohoku University, Japan*
Dr Yu. Zhukovskii, *University of Latvia, Latvia*

Index of contributors*

* The page numbers in **bold** type indicate papers submitted for discussions.

Mass and charge transport in the PEO–NaI polymer electrolyte system: effects of temperature and salt concentration

N. A. Stolwijk, M. Wiencierz and Sh. Obeidi

Faraday Discuss., 2007, **134**, (DOI: 10.1039/b602143n). **Amendment published 24th October 2006**

Eqn (5) should read:

$$\frac{r_p}{(1 - r_p)^2} = k_p[1 - z\tilde{c}_s(1 - r_p)]^{-1}$$

The Royal Society of Chemistry apologises for these errors and any consequent inconvenience to authors and readers.

Additions and corrections can be viewed online by accessing the original article to which they apply.
